T0137444

Communications in Computer and Information Science 1414

More information about this series at http://www.springer.com/series/7899

Robert M. Corless · Jürgen Gerhard ·
Ilias S. Kotsireas (Eds.)

Maple in Mathematics Education and Research

4th Maple Conference, MC 2020
Waterloo, Ontario, Canada, November 2–6, 2020
Revised Selected Papers

Springer

Editors
Robert M. Corless
Western University
London, ON, Canada

Jürgen Gerhard
Maplesoft
Waterloo, ON, Canada

Ilias S. Kotsireas
Wilfrid Laurier University
Waterloo, ON, Canada

ISSN 1865-0929 ISSN 1865-0937 (electronic)
Communications in Computer and Information Science
ISBN 978-3-030-81697-1 ISBN 978-3-030-81698-8 (eBook)
https://doi.org/10.1007/978-3-030-81698-8

This Springer imprint is published by the registered company Springer Nature Switzerland AG
The registered company address is: Gewerbestrasse 11, 6330 Cham, Switzerland

Foreword

As a global pandemic has been gripping the world for over a year now, many aspects of our lives have changed: The way we work; the way we play; the way we meet friends and family. And so the way the Maple community gathered on November 2–6, 2020 was very different as well. We met virtually. And so there was no conference dinner, no coffee breaks, no handshakes or hugs with friends that we had not seen in a while.

But we coped. Our social event was a virtual tour of the Tom Thomson Art Gallery. We met at informal breakfast and lunch "tables" to chat with old and new members of the community. We were fortunate to have some brilliant keynote speakers like Dr. Gabor Domokos who took us behind the scenes of the discovery of the Gömböc and Dr. Juana Sendra Pons who introduced us to the magic of Bohemian matrices.

And indeed, despite the difficult circumstances that the world finds itself in, some things were better. With no constraints from travel budget and week-long scheduling conflicts, an unprecedented proportion of the Maple community was able to participate, making the event more engaging and inspiring than ever with a large number of excellent contributions drawn from research to education to interesting and novel applications of Maple. In addition, the new format of combining pre-recorded talks with live Q&A sessions allowed participants to watch presentations at their own pace and to have the time to absorb the material before bringing thoughtful questions and comments to the authors during the live session.

What is driving us at Maplesoft is the belief that Math Matters, stemming from the realization that mathematics drives the world around us. Our mission is to build engaging tools that help gain insight into mathematical concepts and that not just provide solutions but also provide the inspiration to dig deeper and discover not only the usefulness but also the beauty of mathematics. With this in mind, it was great to see another first for a Maple Conference: We hosted a very lively and interesting panel discussion with a group of social media influencers that share our vision of making math accessible and enjoyable. Sneak peeks at our Maple Calculator Mobile App as well as our new online solution for learning and teaching math, Maple Learn, rounded out this part of the program.

The virtual format also allowed a record number of Maplesoft staff from around the world to attend the conference and share their expertise with all attendees. There is always something new to discover in the world of Maple and I would go out on a limb and say that every participant learned something new in the course of this week.

Overall, we had an exceptional event and it was great to see the Maple community come together. I was able to greet many familiar faces and introduce myself to an even larger number of people that attended for the first time.

Finally, I would like to express a huge thank you to the Program Chairs Rob Corless and Jürgen Gerhard and the rest of the Program Committee as well as to the countless people at Maplesoft who made the event a resounding success.

Laurent Bernardin

Preface

The Maple Conference 2020 happened under stressful circumstances, mostly to do with the global COVID-19 pandemic, but important political events (all over the world) also happened during the conference. Political events are not usually explicitly acknowledged in the proceedings of any scientific conference, on the basis that science is above politics, or else on the basis that mathematics and science are apolitical. The truth of a mathematical theorem or the functioning of a piece of software is indeed normally independent of what any given group of humans has decided to do.

As always, the truth is more complex. The science and engineering that gets accomplished by a community depends strongly on how that community is organized. It depends in a long-term way—on a cycle of decades at least—on how the members of the society are educated, and on the intellectual infrastructure and social capital available to thinkers, educators, and doers—who may or may not be the same people.

The Maple Conference 2020 happened at a time of severe crisis. Millions of people were being struck down, and many dying, of an infectious disease. The personal impact of this crisis was of course huge, and continues to be huge: we swim in grief for family, friends, loved ones, even for people we do not know. To protect vulnerable people and medical institutions, many governments imposed Non-Pharmaceutical Interventions (NPIs) which, among other impacts, forbade or reduced travel and in-person interactions.

In spite of the impact of this crisis and of the NPIs taken to mitigate the impact, not all of these impositions seem to have been wholly bad for science in general, or for the Maple conference in particular. Perhaps as a side effect of going virtual, the Maple Conference 2020 had over 700 registrants, the most of any Maple conference to date. On any given day of the conference, which took place during November 2–6, 2020, we might have had 400 people actively participating. By all normal measures, the conference was a resounding success. For this we have to thank the efforts of many working behind the scenes: specifically Kathleen McNichol, Eithne Murray, Jen Iorgulescu, and Rochelle Angyal. Their very hard work and adaptability made the virtual conference a success, both by long preparation ahead of time and by putting out the inevitable "last-minute" fires with the software platform we were using. As a result, the conference ran very smoothly.

One concludes in general that the impact of the crisis on science is as yet unclear, because people have worked so hard to adapt. Because of these efforts, we now know that remote collaboration and conferencing are not only possible but have some advantages, as this conference proved. New tools for remote learning—such as Maple Learn, perhaps—may help even more in the future.

This proceedings provides a tangible archive of that success. In this volume you will find a selection of papers based on work presented at the conference: on mathematical research, applications of Maple, and on mathematics education. There is another archive: all the talks were recorded and they are, at the time of writing, still available on the original website; after November 2021 they will move to a YouTube channel. We encourage you to watch the videos of the talks, not just read the papers in these proceedings.

In particular, the invited talks by Professor Gabor Domokos, Professor Juana Sendra Pons, and Dr. Laurent Bernardin all remain available, and we highly recommend watching them. We thank all of them for their discussions of fascinating work. Professor Sendra Pons also contributed a paper to the proceedings, for which we also thank her.

The "Meet the Developers" panel, consisting of Laurent Bernardin, Paulina Chin, Paul DeMarco, Jürgen Gerhard, Erik Postma, Karishma Punwani, and Andrew Smith, was lively and engaging, and we thank them all for their time and expertise.

The rise of YouTube "influencers", among them Online Kyne, Bobby Seagull, and Tom Crawford (The Naked Mathematician), was possibly predictable but the reality is so amazing—they have not contributed to the proceedings but we urge you to watch the panel discussion in the video on the conference website—that we would actually be shocked had anyone predicted it, or predicted just how popular math and science videos would turn out to be. We thank these influencers for their very entertaining and thought-provoking panel discussion.

The three workshops, presented by Paulina Chin, Erik Postma, and Stephen Forrest, were well-attended and extremely valuable. These workshops are not part of the formal proceedings, and are not attached to the video record of the conference, but we thank the presenters for their contributions to the conference at the time.

Several Maplesoft personnel gave presentations at the conference and these are part of the video record: Dr. Robert Lopez, "Analytic Approximation for the Dirichlet Problem;" Thomas Richard, "Application of the Identify Command to Special Functions;" Valery McKay-Crites, "Generate Captivating Visualizations with Maple;" Karishma Punwani, "Introducing Maple Calculator and Maple Learn; " Dr. Stephen Forrest, "Machine Learning in Maple;" Samir Khan, "Maple Whiteboard - tactile, responsive calculations for science, engineering and technical analysis;" and Samir Khan and Karishma Punwani, "Our Favorite Things: Maple 2020 Gems You May Have Missed." These presentations were extremely useful and enlightening, and are still available; we hope that at least some of them will be reprised at the Maple Conference 2021!

We thank our Program Committee and all the reviewers for all their hard work, especially during this time of the COVID-19 pandemic. Refereeing is one of the most critical, but thankless, jobs of an academic. Everyone is just expected to do it. Our referees put in a very significant amount of work, providing feedback to our authors and presenters, going well above the norm which made a significant difference to the quality of the papers.

Of course, we also thank all our presenters and authors. They, too, worked hard; in preparing their videos (sometimes for the first time ever for a conference), in taking questions, in writing their papers, and in taking the constructive criticism of the referees and using it to improve their papers.

Science takes time, and social stability, and education, and other things. That the Maple 2020 Conference went so well, with participants from 70 countries, from Australia to Zambia, is a mark of hard work and persistence, and of the resilience of the supporting institutions and personal resilience of the participants. We believe that these proceedings show evidence of this high-water mark, and we hope that all the participants feel justified pride in their achievements in the face of the truly difficult circumstances that they faced.

May 2021

Robert M. Corless
Jürgen Gerhard
Ilias S. Kotsireas

Organization

Program Chairs

Rob Corless	Western University, Canada
Jürgen Gerhard	Maplesoft, Canada

Program Committee

Andrew Arnold	Google, USA
David Bailey	University of California, Davis, USA
Michel Beaudin	ETS Montreal, Canada
Murray Bremner	University of Saskatoon, Canada
Curtis Bright	Carleton University, Canada
Neil Calkin	Clemson University, USA
Eunice Chan	Western University, Canada
Bruce Char	Drexel University, USA
Shaoshi Chen	Chinese Academy of Sciences, China
Paulina Chin	Maplesoft, Canada
Jean-Guillaume Dumas	Université Grenoble Alpes, France
Matthew England	University of Coventry, UK
Laureano Gonzalez-Vega	CUNEF, Spain
Jonathan Hauenstein	University of Notre Dame, USA
Silvana Ilie	Ryerson University, Canada
David Jeffrey	Western University, Canada
Jeremy Johnson	Drexel University, USA
Manuel Kauers	Johannes Kepler University Linz, Austria
Ilias Kotsireas	Wilfrid Laurier University, Canada
George Labahn	University of Waterloo, Canada
Gilbert Labelle	Université du Québec à Montréal, Canada
Wen-shin Lee	University of Antwerp, Belgium
David Linder	Maplesoft, Canada
Austin Lobo	Washington College, USA
Robert Martin	University of Manitoba, Canada
John May	Maplesoft, USA
Douglas B. Meade	University of South Carolina, USA
Michael Monagan	Simon Fraser University, Canada
Guillaume Moroz	Inria, France
Judy-anne Osborn	University of Newcastle, Australia
Veronika Pillwein	Johannes Kepler University Linz, Austria
Erik Postma	Maplesoft, Canada
Alban Quadrat	Inria, France
Georg Regensburger	Johannes Kepler University Linz, Austria

Thomas Richard	Maplesoft, Canada
Rafael Sendra	University of Alcala, Spain
Brandilyn Stigler	Southern Methodist University, USA
M. Pilar Velez	Nebrija University, Spain
Thomas Wolf	Brock University, Canada
Lihong Zhi	Chinese Academy of Sciences, China

Local Organizing Committee

Rob Corless	Western University, Canada
Jürgen Gerhard	Maplesoft, Canada
Kathleen McNichol	Maplesoft, Canada
Eithne Murray	Maplesoft, Canada
Jennifer Iorgulescu	Maplesoft, Canada
Rochelle Angyal	Maplesoft, Canada

Proceedings Editors

Rob Corless	Western University, Canada
Jürgen Gerhard	Maplesoft, Canada
Ilias Kotsireas	Wilfrid Laurier University, Canada

Contents

Keynote Presentation

Bohemian Matrices: Past, Present and Future

Juana Sendra(✉)

Dpto. de Matemática Aplicada a las TIC, Universidad Politécnica de Madrid, Madrid, Spain
juana.sendra@upm.es

Abstract. A matrix family is called Bohemian if its entries come from a fixed finite discrete (and hence bounded) set, usually integers, called the "population" P. We look at Bohemian matrices, specifically those with entries from $\{-1, 0, +1\}$. The name is a mnemonic for Bounded Height Matrix of Integers. Such families arise in many applications (e.g. compressed sensing) and the properties of matrices selected "at random" from such families are of practical and mathematical interest. An overview of some of our original interest in Bohemian matrices can be found in [6,7]. In this paper we present a Bohemian Matrices tour, exposing their appearance in the past, their promising present and their hopeful future.

Keywords: Bohemian matrices · Maple · Eigenvalues · Symbolic computation

1 Introduction and Terminology

A family of Bohemian matrices is a set of structured matrices where the entries are from a finite set of integers. These families of matrices are interesting by themselves but they can appear in many applications. For instance, in signal processing, where they use Bernoulli matrices, or error correcting codes working with Hadamard matrices. Other fields where they can be applied are combinatorics or Graph Theory, and in this same frame it is interesting the application in Spectral Graph Theory, among others.

Dealing with this family of matrices, the question arises as: why Bohemian Matrices? The original motivation of the authors of [7] was test problems for various algorithms. Focusing on this type of representation we could give interesting results and analyzing extreme behaviors. The basic idea is to develop algorithms for computing discrete families of Bohemian (or brute force if necessary) to analyse the behavior of certain facts to be used to conjecture properties, and hopefully prove them. In this context, the first drawbacks appear.

The author is partially supported by FEDER/Ministerio de Ciencia, Innovación y Universidades - Agencia Estatal de Investigación/MTM2017-88796-P (Symbolic Computation: new challenges in Algebra and Geometry together with its applications).

R. M. Corless et al. (Eds.): MC 2020, CCIS 1414, pp. 3–16, 2021.
https://doi.org/10.1007/978-3-030-81698-8_1

The number of possible Bohemian matrices of dimension n is typically quadratically exponential ($\exp(cn^2)$ for some c), depending on the matrix structure. For instance, with a population $\{-1, 0, 1\}$ the number of general 5×5 Bohemians is: $847, 288, 609, 443$.

Analyzing such matrices leads to many unanswered questions. For instance, for a given dimension and population, the set of Bohemian matrices is finite. But how many are singular? or how many distinct characteristic polynomials does the family have? or how many distinct eigenvalues does the family contain? or how many distinct Jordan canonical forms are there? In this sense, a lot of challenges arise and, in turn, provide new opportunities.

Through extensive experimental work, the authors have discovered many properties of families of Bohemian matrices and their eigenvalues which lack obvious explanations. These matrices are studied including the distributions of their eigenvalues, and integer sequences arising from properties of the families providing connections to other areas of mathematics and have been archived in the Characteristic Polynomial Database, http://www.bohemianmatrices.com/cpdb/. Currently the database contains $1, 762, 728, 065$ characteristic polynomials from $2, 366, 960, 967, 336$ matrices.

By plotting the distributions of the eigenvalues of all matrices in a Bohemian family over the complex plane, many interesting discrete structures appear. For example, in Fig. 1, distinct "holes" appear in the distribution. Other families exhibit fractal like structures and diffraction patterns, see [11]. By studying these families in greater detail we are able to understand why some structures appear. Many examples of the discrete structures that appear in the distributions of eigenvalues can be found at http://www.bohemianmatrices.com. Pictures in this paper have been taken from http://www.bohemianmatrices.com/gallery/.

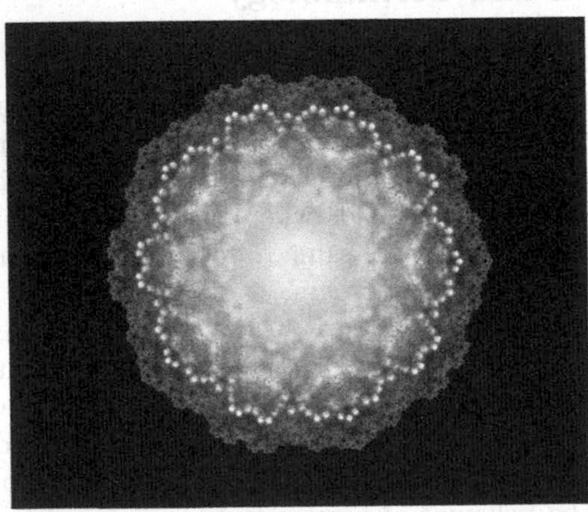

Fig. 1. Density plot in the complex plane of the eigenvalues of a Bohemian matrix. Image of the front cover of Newsletter of the London Math Society Gazette. Issue 491, November 2020

Our experimental work is only possible thanks to advances in the processing power of common personal computers. This has allowed us to explore families containing upwards of 1 trillion matrices on a laptop. Through brute-force computation we are able to answer questions such as how many 7×7 matrices with entries from the set $\{-1, 0, +1\}$ are nilpotent? The answer is $1,138,779,265$. Further, these computations have helped us make connections among properties of matrices that we may not have made otherwise. We have used Maple 2019 for the experiments with small dimension (up to 5 for instance). Maple becomes a good tool to analyze exact properties and it help us to check conjectures and it is a great help in our theoretical analysis. On the other hand, we use Matlab and Python for bigger computations and as a good tool for plotting and visualizing. We finish this section introducing the notation used throughout this paper.

Definition 1 (BOunded HEight Matrix of Integers: BOHEMI). *Matrices where the free entries are from a finite population of small integers (or other population) or even a bounded subset.*

Definition 2. *The* **Population** *of a Bohemian family is the set (usually finite and discrete, hence bounded) of possible entries for each matrix in the family.*

Definition 3. Bohemian Eigenvalues *are the eigenvalues of Bohemian matrices.*

Definition 4. Eigenvalue Exclusion Zone *are distinct region in the complex plane where no eigenvalues fall.*

Definition 5. Rhapsody Matrix *is a Bohemian matrix whose inverse is Bohemian with respect to the same population.*

Definition 6. *The* **Height** *of a matrix A is the infinity norm of vec(A).*

Definition 7. *The* **Characteristic Height** *of a matrix A is the infinity norm of the vector of coefficients of its characteristic polynomial.*

2 The Past

In this paper, we will travel from the past to the future, anchoring the present in nowadays. Let's start this fascinating travel by taking a look at the past. We can go back to Leonhard Euler in the 18th century, passing through Ronald Fisher, John E. Litlewood, Olga Taussky and H. J. Ryser in 19th and 20th century and more recently we can mention C.W Gear, Borwein & Jorgenson or T. Tao, V. Vu, among others. Their work has had repercussions in several areas, for instance statistics, number theory, computer science, matrix theory, random matrices, root finding and eigenvalues problems, and a great etc. We will take as starting point the work of L. Euler who use the *latin square* using latin characters as symbols. More formally, a Latin Square is an $n \times n$ array filled with n different symbols, each occurring exactly once in each row and exactly once in each column.

A	B	C	D
B	C	A	D
C	D	B	A
D	A	C	B

In recreational mathematics, a square array of numbers, usually positive integers, is called a magic square if the sums of the numbers in each row, each column, and both main diagonals are the same, called magic constant. Magic squares have a long history. At various times they have acquired occult or mythical significance, and have appeared as symbols in works of art. Főr instance in the *Sagrada Familia* of Barcelona, while appreciating the Passion facade and the sculpture of Judas Kiss, we find a magic square with magic constant 33. This magic square is inspired in the engraving entitle Melancholia I, of the German artist Albrecht Dürer. We can see the year 1514 in the last row, see Fig. 2.

Fig. 2. Dürer's Melancholia I (1514) includes an order 4 square with magic sum 34. Pictures taken from wikipedia.

Let us mention the work of Andrey Andreyevich Markov (1856–1922) who use the stochastic matrix to describe the transition of a Markov chain. Each of its entries is a nonnegative real number representing a probability and with each row and/or column summing to 1. The stochastic matrix was first developed by A. Markov, and has found use throughout a wide variety of scientific fields, including probability theory, statistics, mathematical finance and linear algebra, as well as computer science and population genetics. Another interesting contribution precursor of the bohemians is due to John E. Littlewood (1885–1977). More precisely, a **Littlewood Polynomial** is a polynomial all of whose coefficients are $+1$ or -1. **Littlewood's problem** asks how large the module of

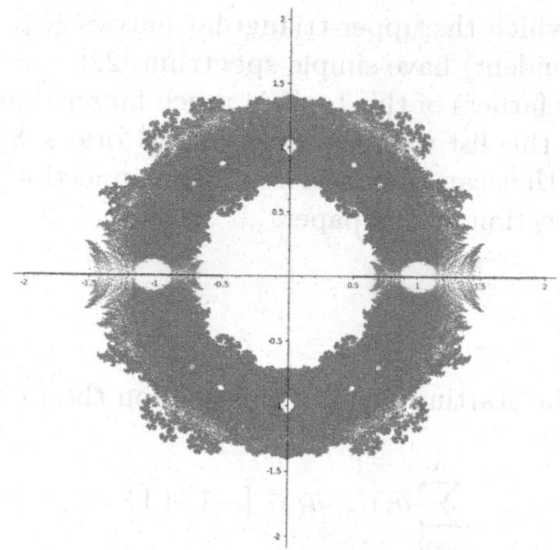

Fig. 3. Roots of all the Littlewood polynomials of degree 15.

the values of such polynomials can be when taking values on the unit circle in the complex plane, see Fig. 3. Bohemian families have been studied for a long time, although not under that name. For instance, Odlyzko and Poonen in their work from 1993 [19], studied the zeros of polynomial with 0 and 1 coefficients (also called Newman polynomials)and they obtained bounds for such zeros. Let me also mention the work of Olga Taussky (1905–1995) [23,24]. She is famous for her research papers in algebraic number theory, and integral matrices, in particular she worked in the computational stability of complex matrices. These papers show her work in computational problems about matrices with integer coefficients. Taussky begins by saying

"*This subject is very vast and very old. It includes all of the arithmetic theory of quadratic forms, as well as many of other classical subjects, such as Latin squares and matrices with elements +1 or −1 which enter into Euler's, Sylvester's or Hadamard's famous conjectures.*"

The contribution by H. J. Ryser (1923 – 1985) is another example. He worked with matrices of 0 and 1 in the frame of combinatorics. C.W. Gear in [16] investigates the eigenvalues of certain kind of matrices with 0 and 1, see [21]. Another interesting contributions are the works of Borwein and Jorgenson [2] related with computer graphics, and James Guyker of 2007 (see [13]) where he uses characteristic polynomials to determine when magic squares have magic inverse. Let us metion Matthew Lettington and his work [17] about fleck's congruence involving magic squares and a zeta identity. On the other hand, the idea of visualizing the eigenvalues of random samples of matrices is not new. L. N. Trefethen in [27] uses this idea to visualize the pseudospectra of several test matrices. Related to the eigenvalues of matrices, many authors have studied the zeros of polynomials whose coefficients belong to discrete sets of integers. Early work by Odlyzko and Poonen [19] studies the zeros of polynomials with coefficients in $\{0, 1\}$. And Tao and Vu have shown that random matrices (more specifically real symmetric random matrices in which the upper-triangular entries $\xi_{i,j}$, $i < j$ and diagonal entries $\xi_{i,j}$ are independent) have simple spectrum [22].

The list of the forefathers of this topic is much larger than the one presented here. But let us close this list with the work of R. Corless [8], about generalized companion matrices, that can be considered as a connection bridge between the first and the second section of this paper.

3 The Present

We can consider as the starting point of this section the well-known Littlewood polynomial defined by

$$\sum_{i=0}^{n} a_i x^i, \quad a_i \in \{-1, +1\}$$

whose roots produce interesting pictures.

More recently, the distributions of the roots of Littlewood polynomials [18] have been studied [1,2,20]. In Fig. 4, the distribution of the roots of all degree 25 Littlewood polynomials are visualized. This is, of course, equivalent to the eigenvalues of a Bohemian family of (Frobenius) companion matrices with entries from the set $\{-1, +1\}$.

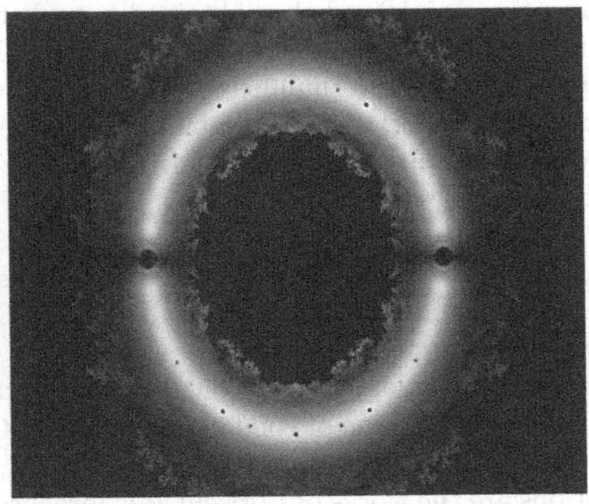

Fig. 4. The roots of all degree 25 polynomials (expressed in the monomial basis) with ± 1 coefficients.

In [12], R.M. Corless and S.E. Thornton focus on the questions raised when visualizing the distributions of Bohemian eigenvalues over the complex plane. In particular, they analyze eigenvalue exclusion zones (i.e. distinct regions within the domain of the eigenvalues where no eigenvalues exist), computational methods for visualizing eigenvalues, and some results on eigenvalue conditioning over distributions of random matrices. By plotting the distributions of the eigenvalues of all matrices in a Bohemian family over the complex plane, many interesting discrete structures appear. For example, in the image in Fig. 5 raises many questions, ranging from whether the set is a fractal or about the holes, and specially their regularity of the boundary whose shape suggests a dragon curve. Some of these questions could have some significance in number theory. Many examples of the discrete structures that appear in the distributions of eigenvalues can be found at http://www.bohemianmatrices.com/gallery/. In this context, in [15] the authors study the geometry of algebraic numbers in the complex plane called Algebraic Numbers Starscapes. More precisely, they describe the geometry of the map from the coefficient space of polynomials to the root space, focussing on the quadratic and cubic cases.

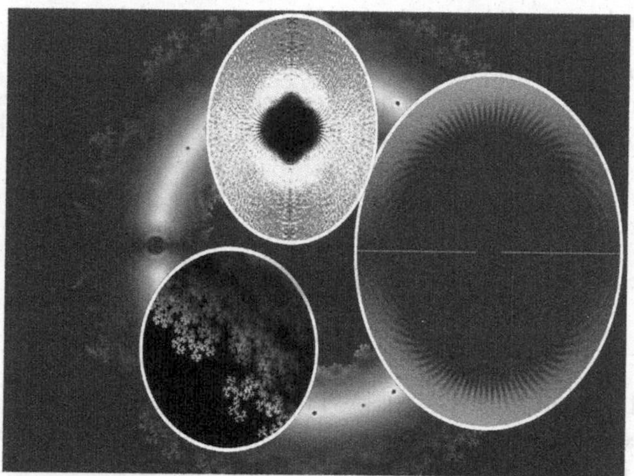

Fig. 5. Details of the eigenvalue exclusion zones and regularity of the boundary.

Corless used a generalization of the Littlewood polynomial (to Lagrange bases). In his paper [8], he gave a new kind of companion matrix for polynomials expressed in a Lagrange basis. He used generalized Littlewood polynomials as test problems for his algorithm.

In Chan's Master's thesis [3], she extended Piers W. Lawrence's construction of the companion matrix for the Mandelbrot polynomials, [9,10], to other families of polynomials, mainly the Fibonacci-Mandelbrot polynomials and the Narayana-Mandelbrot polynomials. What is relevant here about this construction is that these matrices are upper Hessenberg and contain entries from a constrained set of numbers $\{-1, 0\}$, and therefore fall under the category of being Bohemian upper Hessenberg. Both the Fibonacci-Mandelbrot matrices and Narayana-Mandelbrot matrices are also Bohemian upper Hessenberg, but the set that the entries draw from is $\{-1, 0, +1\}$.

These new constructions led Chan and Corless to a new kind of companion matrix for polynomials of the form $c(z) = za(z)b(z) + c_0$. A first step towards this was first proved using the Schur complement in [4]. D. Knuth then suggested that Chan and Corless look at the Euclid polynomials [5], based on the Euclid numbers. It was the success of this construction that led to the realization that this construction is general, and gives a genuinely new kind of companion matrix. Similar to the previous three families of matrices, the Euclid matrices are also upper Hessenberg and Bohemian, as the entries are comprised from the set $\{-1,0,+1\}$. In addition, an interesting property of these companion matrices is that their inverses are also Bohemian with the same population, i.e. Rhapsody matrices. As an extension of this generalization, Chan et al. [6] showed how to construct linearizations of matrix polynomials, particularly of the form

$$za(z)d_0 + c_0, a(z)b(z), a(z) + b(z)$$

$$\textit{when } \deg(b(z)) < \deg(a(z)), \textit{ and } za(z)d_0 b(z) + c_0,$$

using a similar construction.

Different matrix structures produce remarkably different pictures. One structure useful in eigenvalue computation is the Upper Hessenberg matrix,

$$H_n = \begin{pmatrix} h_{11} & h_{12} & h_{13} & \cdots & h_{1n} \\ s_1 & h_{21} & h_{22} & \cdots & h_{2n-1} \\ 0 & s_2 & h_{31} & \cdots & h_{3n-2} \\ \vdots & \ddots & \ddots & \ddots & \vdots \\ 0 & \cdots & 0 & s_{n-1} & h_{n1} \end{pmatrix}$$

with $s_k \in \{-1,1\}$ and $h_{ij} \in \{-1,0,1\}$. The main reason to analyze these kind of matrices is because these arise naturally in eigenvalue computation because the QR iteration is cheaper for matrices in Hessenberg form. For other families of matrices, such as upper Hessenberg Toeplitz matrices, there is no compression at all because each matrix has a distinct characteristic polynomial, see [26].

We present some result proved in [7] by considering Bohemian upper Hessenberg matrices. We begin with two recursive formulae for the characteristic polynomials $Q_n(z) = \det(zI - H_n)$. Later we will specialize the population P to contain only zero and numbers of unit magnitude, usually $\{-1,0,+1\}$.

Theorem 1. *The characteristic polynomial can be express as*

$$Q_n(z) = zQ_{n-1}(z) - \sum_{k=1}^{n} \left(\prod_{j=n-k+1}^{n-1} s_j \right) h_{n-k+1,n} Q_{n-k}(z) \tag{1}$$

with the conventions that an empty product is 1, $Q_0(z) = 1$ *and* $H_0 = [\,]$, *that is the empty matrix.*

Theorem 2. *Expanding* $Q_n(z)$ *as*

$$Q_n(z) = q_{n,n}z^n + q_{n,n-1}z^{n-1} + \cdots + q_{n,0},$$

we can express the coefficients recursively by

$$q_{n,n} = 1,$$

$$q_{n,j} = q_{n-1,j-1} - \sum_{k=1}^{n-j} \left(\prod_{j=n-k+1}^{n-1} s_j \right) h_{n-k+1,n} q_{n-k,j} \quad \text{for} \quad 1 \le j \le n-1,$$

$$q_{n,0} = -\sum_{k=1}^{n} \left(\prod_{j=n-k+1}^{n-1} s_j \right) h_{n-k+1,n} q_{n-k,0} \quad \text{for} \quad n > 0, \quad \text{and}$$

$$q_{0,0} = 1\,.$$

The natural question now is, which matrices reach maximal characteristic height?

Proposition 1. *For populations P with maximal height 1, the maximal characteristic height of H_n occurs when $\left(\prod_{j=n-k+1}^{n-1} s_j\right) h_{i,i+k-1} = -1$ for $1 \le i \le n-k+1$ and $1 \le k \le n$.*

Remark 1. When all $s_j = 1$ and $h_{i,j} = -1$ for all $1 \le i \le j \le n$, and similarly for $s_j = -1$ and $h_{i,j} = 1$ we attain maximal characteristic height. Both of these cases correspond to Upper Hessenberg matrices with a Toeplitz structure.

This motivates the interest in Upper Hessenberg Toeplitz matrices with subdiagonal $s \in \{\pm 1\}$.

$$M_n = \begin{pmatrix} t_1 & t_2 & t_3 & \cdots & t_n \\ s & t_1 & t_2 & \cdots & t_{n-1} \\ 0 & s & t_1 & \cdots & t_{n-2} \\ \vdots & \ddots & \ddots & \ddots & \vdots \\ 0 & \cdots & 0 & s & t_1 \end{pmatrix}$$

Let us denote by $\mathcal{M}^{n\times n}(P) = \{n \times n$ Bohemian Upper Hessenberg Toeplitz matrices with upper triangle population $P = \{-1, 0, 1\}$ and subdiagonal $s = 1\}$.

The characteristic polynomial recurrence from Theorem 1 can be written for upper Hessenberg Toeplitz matrices in $\mathcal{M}^{n\times n}(P)$ as follows.

Theorem 3 (Recurrence relation for the characteristic polynomial). *Let us denote $P_n(z)$ the characteristic polynomial of $\mathcal{M}^{n\times n}(P)$. Then*

$$P_n(z) = zP_{n-1}(z) - \sum_{k=1}^{n} t_k P_{n-k}(z)$$

with the convention that $P_0 = 1$ and $M_0 = [\,]$ the empty matrix.

Corollary 1. *The characteristic polynomial recurrence form can be written for the previous upper Hessenberg Toeplitz matrices as*

$$p_{n,n} = 1,$$
$$p_{n,j} = p_{n-1,j-1} - \sum_{k=1}^{n-j} t_k p_{n-k,j} \quad \textit{for} \quad 1 \leq j \leq n-1,$$
$$p_{n,0} = -\sum_{k=1}^{n} t_k p_{n-k,0}, \textit{ and}$$
$$p_{0,0} = 1$$

Theorem 4. *The set of distinct characteristic polynomials for all matrices $M_n \in \mathcal{M}^{n\times n}$ has cardinality $(\#P)^n$, which is the same as the cardinality of $\mathcal{M}^{n\times n}$. That is, each matrix in $\mathcal{M}^{n\times n}$ has a different characteristic polynomial.*

Next two theorems stablish two kind of matrices of the family in which we reach maximal characteristic height.

Proposition 2. *The characteristic height of $M_n \in \mathcal{M}^{n\times n}$ is maximal when $t_k = -1$ for $1 \leq k \leq n$.*

$$M_n = \begin{pmatrix} -1 & -1 & -1 & \cdots & -1 \\ 1 & -1 & -1 & \cdots & -1 \\ 0 & 1 & -1 & \cdots & -1 \\ \vdots & \ddots & \ddots & \ddots & \vdots \\ 0 & \cdots & 0 & 1 & -1 \end{pmatrix}$$

Proposition 3. *$M_n \in \mathcal{M}^{n\times n}$ also reaches maximal characteristic height when $t_k = (-1)^{k-1}$ for $1 \leq k \leq n$.*

$$M_n = \begin{pmatrix} 1 & -1 & 1 & \cdots & (-1)^{n-1} \\ 1 & 1 & -1 & \cdots & * \\ 0 & 1 & 1 & \cdots & * \\ \vdots & \ddots & \ddots & \ddots & \vdots \\ 0 & \cdots & 0 & 1 & 1 \end{pmatrix}$$

Let us now introduce $\overline{\mathcal{M}}^{n\times n} = \{\overline{M} \in \mathcal{M}^{n\times n}(P)$ of maximal characteristic height$\}$. Let τ_n be the characteristic height of $\overline{M}$ and let μ_n be the largest degree of the term of the characteristic polynomial of $\overline{M}$ whose coefficient gives the height.

Theorem 5. *For fixed n, μ_n is the same for all $\overline{\mathbf{M}}_n \in \overline{\mathcal{M}}^{n\times n}$.*

Theorem 6. *$\overline{\mathcal{M}}^{n\times n}$ contains $2 \cdot 3^{\mu_n}$ matrices.*

Theorem 7. *The maximum characteristic height, τ_n, of any upper Hessenberg Bohemian with population $\{-1, 0, 1\}$ lies between the following bounds:*

$$\frac{F_{2n+1}}{n+1} < \tau_n < F_{2n+1} \, .$$

Here F_k is the kth Fibonacci number, with the conventional numbering given by $F_{n+1} = F_n + F_{n-1}$ with $F_0 = 0$ and $F_1 = 1$.

Conjecture 1. The maximum characteristic height, τ_n, approaches the exponentially growing function $CF_{2n+1}/\sqrt{n+1}$ as $n \to \infty$ for some constant C. Our experiments indicate that $C \doteq 0.7701532$.

Many properties of Bohemian families are included in the Characteristic Polynomial Database in www.bohemianmatrices.com, see [25]. Finally, we end with 21 conjectures related to integer sequences arising from the properties of Bohemian families. Lastly, the work in [14] should be mentioned, which analyses the determinants of normalized Bohemian upper Hessenberg Matrices.

4 The Future

Many questions relating to Bohemian families remain unanswered. Future work is focused on a few main problems. For instance, the exploration of the distributions of eigenvalue condition numbers are of interest. Are the eigenvalues within some families or for certain structures inherently better conditioned than in other families? Inverse eigenvalue problems are also of interest. That is, given a characteristic polynonial and a Bohemian family, identify a matrix in the family with the given characteristic polynonial.

The class of upper Hessenberg Bohemian matrices gives a useful way to study Bohemian matrices in general. Many questions remain unanswered about general Bohemian matrices and new families remain to be explored. The Characteristic Polynomial Database, http://www.bohemianmatrices.com/cpdb/, will continue to expand to include many new families including structured matrices such as symmetric, circulant, etc. As new families are explored new questions will arise, see Fig. 7. The list of properties computed on these families will continue to grow and new algorithms will be required as family sizes grow. New conjectures will appear and the list of conjectures will continue to expand. Authors of [7] are working in several promising work lines, see Fig. 6.

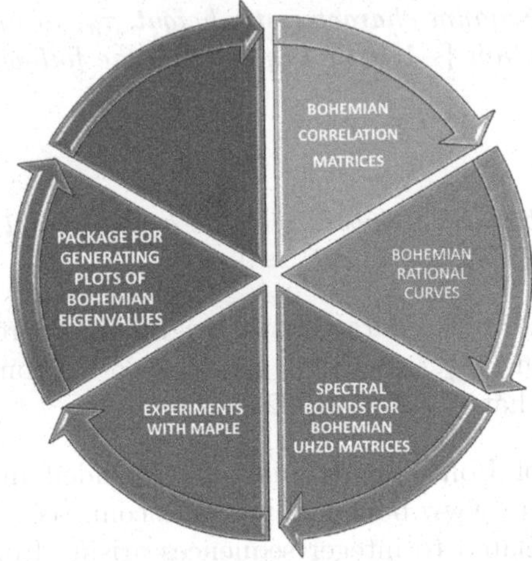

Fig. 6. Work in progress by the author of [7].

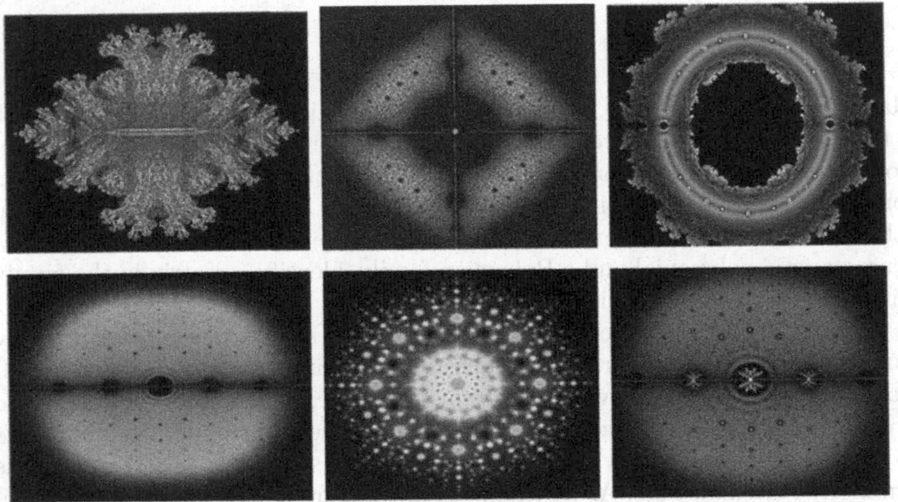

Fig. 7. Density plots in the complex plane of the Bohemian eigenvalues of a sample of matrices with entries in a discrete set. Images implemented by S. Thorton and E. Chan.

References

1. Baez, J.: The beauty of roots (2011). https://johncarlosbaez.wordpress.com/2011/12/11/the-beauty-of-roots/
2. Borwein, P., Jorgenson, L.: Visible structures in number theory. Am. Math. Mon. **108**, 897–910 (2001)

3. Chan, E.Y.S.: A comparison of solution methods for Mandelbrot-like polynomials, Electronic thesis and Dissertation Repository, The University of Western Ontario (2016). https://ir.lib.uwo.ca/etd/4028
4. Chan, E.Y.S., Corless, R.M.: A new kind of companion matrix. Electron. J. Linear Algebra **32**, 335–342 (2017)
5. Chan, E.Y.S., Corless, R.M.: Minimal height companion matrices for Euclid polynomials. Math. Comput. Sci. (2018). https://doi.org/10.1007/s11786-018-0364-2
6. Chan, E.Y.S., Corless, R.M., Gonzalez-Vega, L., Sendra, J.R., Sendra, J.: Algebraic linearizations of matrix polynomials. Linear Algebra Appl. **563**, 373–399 (2019)
7. Chan, E.Y.S., Corless, R.M., Gonzalez-Vega, L., Sendra, J.R., Sendra, J., Thornton, S.E.: Upper Hessenberg and Toeplitz Bohemians. Linear Algebra Appl. **601**, 372–100 (2020)
8. Corless, R.M.: Generalized companion matrices in the Lagrange basis. In: Proceedings EACA, Universidad de Cantabria, Santander, Spain, 2004, pp.317–322 (2004)
9. Corless, R.M., Lawrence, P.W.: Mandelbrot polynomials and matrices, in preparation
10. Corless, R.M., Lawrence, P.W.: The largest roots of the Mandelbrot polynomials. In: Bailey, D. et al. (eds.) Computational and Analytical Mathematics., pp. 305–324. Springer, New York (2013). https://doi.org/10.1007/978-1-4614-7621-4_13
11. Corless, R.M., Thornton, S.: Visualizing eigenvalues of random matrices. ACM Commun. Comput. Algebra **50**, 35–39 (2016). https://doi.org/10.1145/2930964.2930969
12. Corless, R.M., Thornton, S.E.: The Bohemian eigenvalue project. ACM Commun. Comput. Algebra **50**, 158–160 (2016)
13. Guyker, J.: Magic squares with magic inverses. Int. J. Math. Educ. Sci. Technol., 683–688 (2007)
14. Fasi, M., Negri, G.M.: Determinants of normalized bohemian upper Hessenberg matrices. Electron. J. Linear Algebra **36**, 352–366 (2020). ISSN 1081–3810
15. Harris, E., Stange, K.E., Trettel, S.: Algebraic number starscapes (2020). https://arxiv.org/pdf/2008.07655.pdf
16. Gear, C.: A simple set of test matrices for eigenvalue programs. Math. Comput. **23**, 119–125 (1969)
17. Lettington, M.C.: Fleck's congruence, associated magic squares and a zeta identity. Funct. Approx. Comment. Math. **45**, 165–205 (2011)
18. Littlewood, J.E.: On polynomials $\sum^n \pm z^m, \sum^n e^{\alpha_m i} z^m, z = e^{\theta i}$. J. London Math. Soc. **41**, 367–376 (1966)
19. Odlyzko, A.M., Ponnen, B.: Zeros of polynomials with 0,1 coefficients. In: Salvy, B. (ed.) Algorithms Seminar, vol. 2130, pp. 169–172, December 1993
20. Reyna, R., Damelin, S.: On the structure of the Littlewood polynomials and their zero sets. arXiv preprint arXiv:1504.08058 (2015)
21. Ryser, H.J.: Matrices of zeros and ones. Bull. Amer. Math. Soc. **66**(6), 442–464 (1960)
22. Tao, T., Vu, V.: Random matrices have simple spectrum. Combinatorica **37**, 539–553 (2017)
23. Taussky, O.: Matrices of rational integers. Bull. Am. Math. Soc. **66**, 327–345 (1960)
24. Taussky, O.: Some computational problems involving integral matrices. J. Res. Natl. Bur. Stand. B Math. Sci. **65**, 15–17 (1961)
25. Thornton, S.E.: The characteristic polynomial database, 7 September 2018. http://bohemianmatrices.com/cpdb

26. Thornton, S.E.: Algorithms for Bohemian Matrices, Electronic Thesis and Dissertation Repository, The University of Western Ontario (2019). https://ir.lib.uwo.ca/etd/6069/
27. Trefethen, L.N.: Pseudospectra of matrices. Numer. Anal. **91**, 234–266 (1991)

Accepted Papers Alphabetically by Author

The TruncatedSeries Package for Solving Linear Ordinary Differential Equations Having Truncated Series Coefficients

S. A. Abramov, D. E. Khmelnov(✉), and A. A. Ryabenko

Dorodnicyn Computing Centre, Federal Research Center "Computer Science and Control" of the Russian Academy of Sciences, Vavilova, 40, Moscow 119333, Russia

Abstract. We consider linear ordinary differential equations with power series in the role of coefficients. It is assumed that some or all of the series are truncated. A series of the form $\Sigma\, a_i x^i$ can also be given completely using an algorithm that computes a_i from i. The equation may contain both types of coefficients—truncated and represented algorithmically. Algorithms and commands that implement them in Maple as the `TruncatedSeries` package are proposed, which make it possible to find Laurent, regular and exponential-logarithmic solutions. In cases where, due to the presence of truncated coefficients, the information about the equation is incomplete, commands of our package find the maximum possible number of terms of those series that are involved in the solutions. If all the coefficients of the given equation are algorithmically represented series then the commands allow finding any specified number of initial terms of the series involved in the solutions.

Keywords: Differential equations · Truncated power series · Algorithmically represented infinite formal series

1 Introduction

Power and Laurent series are important and convenient tools of representing linear ordinary differential equations with variable coefficients as well as of representing solutions to these equations. This is reflected in theoretical studies (see, e.g., [17–23]) and found numerous application in computer algebra (see, e.g., [1–6,15,16,24]).

Linear ordinary differential equations with coefficients in the form of truncated power series have been considered by us in [7–14]. Concerning the original differential equation we have incomplete information in this case: for a power series, only a finite number of initial terms are known. We are interested in the information on the solutions of the equation given in this form that is invariant under all possible prolongations of all the truncated series that represent the coefficients of the equation (the *prolongation* is a series whose initial terms

Supported by RFBR grant, project 19-01-00032.

R. M. Corless et al. (Eds.): MC 2020, CCIS 1414, pp. 19–33, 2021.
https://doi.org/10.1007/978-3-030-81698-8_2

coincide with the known initial terms of the original truncated series). First, we have investigated what can be learned about the solutions in the field of Laurent formal series (we call them *Laurent solutions*) (see [7,8]). Then a similar question has been discussed for *regular solutions* in [10]. In both cases, the proposed algorithms construct the maximum possible number of invariant initial terms of the series involved in the solutions.

The approach that we use in the algorithms for computing Laurent and regular solutions, has allowed us, in combination with the well-known algorithm of Newton polygons, to construct *formal exponential-logarithmic solutions* of linear ordinary differential equations having coefficients in the form of truncated power series (see [12,13]). The series which appear in the solutions have also only a finite number of known initial terms.

Linear ordinary differential equations with the coefficients that are either algorithmically represented power series, or truncated power series have been considered as well in [11]. For such a mixed case, the problem of the construction of the maximum possible number of terms of the involved in the solutions series is algorithmically undecidable (for some such equations, the information is sufficient for computing any number of terms of the series). This undecidability is, so to speak, not too burdensome. If we are interested in all solutions with a truncation degree not exceeding a given integer d then the proposed algorithm allows to construct all of them.

All the developed algorithms are implemented by us as the `TruncatedSeries` package in Maple. Some examples of the use of the package procedures have been already presented in the preceding works [7–14], the corresponding algorithms being presented and justified there as well. In this work, we outline the current state of the package and present more examples to demonstrate its up to date key capabilities. We do not repeat the descriptions and justifications of the implemented algorithms. In the future we plan to extend the package possibilities, in particularly, to the case of the systems of linear ordinary differential equations having truncated series coefficients.

The Maple library with the `TruncatedSeries` package and Maple worksheets with examples of using its commands are available from

http://www.ccas.ru/ca/truncatedseries.

2 Short Specification of the Package

The `TruncatedSeries` package provides three commands `LaurentSolution`, `RegularSolution` and `FormalSolution`:

```
> with(TruncatedSeries);
```

$$[\,FormalSolution, LaurentSolution, RegularSolution\,]$$

Calling sequence of these three commands:

```
LaurentSolution(ode, var, opts),
RegularSolution(ode, var, opts),
FormalSolution(ode, var, opts)
```

with parameters

`ode` – a homogeneous linear ordinary differential equation;
`var` – a dependent variable, for example $y(x)$;
`opts` – a sequence of optional arguments of the form `keyword=value`.

The equation `ode` for $y(x)$ may be given in the *diff-form*:

$$a_r(x)\,\frac{\mathrm{d}^r}{\mathrm{d}x^r}\,y(x) + \cdots + a_1(x)\,\frac{\mathrm{d}}{\mathrm{d}x}\,y(x) + a_0(x)\,y(x) = 0$$

or in the *theta-form*:

$$a_r(x)\,\theta^r y(x) + \cdots + a_1(x)\,\theta\,y(x) + a_0(x)\,y(x) = 0.$$

where r is a positive integer and $\theta\,y(x) = x\,\frac{\mathrm{d}}{\mathrm{d}x}\,y(x)$. The derivative $\theta^k y(x)$ is specified as `theta(y(x), x, k)` for $k \geq 1$. The derivative $\frac{\mathrm{d}^k}{\mathrm{d}x^k}\,y(x)$ is specified using the ordinary Maple `diff` command.

Coefficients $a_r(x), \ldots, a_1(x), a_0(x)$ of the equation may be of two types. The first type is an algorithmically represented power series in one of the following forms:

- A polynomial in x over the algebraic number field.
- A finite power sum $\sum\limits_{k=k_0}^{N} f(k)\,x^k$ with a summation index k, a non-negative integer low limit of summation k_0 and a non-negative integer upper limit of summation $N \geq k_0$. It has to be specified by means of the Maple inert `Sum` command. The coefficient $f(k)$ of x^k may be given by an arbitrary expression of the index k which gives an algebraic number for all $k \geq k_0$.
- An infinite power sum $\sum\limits_{k=k_0}^{\infty} f(k)\,x^k$ with k_0, $f(k)$ as described above.
- A sum of a polynomial and power sums described above.

The other type of coefficients is a truncated power series in one of the following forms:

- $O(x^{t+1})$, where t is an integer, $t \geq -1$.
- $a(x) + O(x^{t+1})$, where $a(x)$ is a polynomial in x over the algebraic number field and t is an integer greater than or equal to the degree of $a(x)$.

The integer t is called the *truncation degree.* In the presented package, all algebraic numbers have to be represented as `RootOf(expr, x, 'index'=i)` where `expr` is an irreducible polynomial in x with rational number coefficients.

The following optional arguments can be used:

- `'top'=d`, where `d` is an integer;
- `'threshold'='h'`, where `h` is a name of a variable.

Below we present the use of the commands with optional arguments `'top'` and `'threshold'`.

3 LaurentSolution

For an equation whose all coefficients are algorithmically represented power series, the `LaurentSolution` command determines a finite set of all integers i_0 such that the equation has Laurent series solutions with the *valuation* i_0, i.e. the equation has solutions in the form $\sum_{i=i_0}^{\infty} v(i)\, x^i$ where $v(i_0) \neq 0$.

If the option `'top' = d` is given, the `LaurentSolution` command computes the initial terms of Laurent series solutions to the degree `d` or greater for each valuation i_0. The `LaurentSolution` command returns a list $[s_1, s_2, \ldots]$ of truncated Laurent series solutions for all found valuations. The elements of the list involve parameters of the form $\_c_1, \_c_2, \ldots$ For each element s_j these parameters can take any such values that the valuation of s_j does not change.

Below is an equations whose all coefficients are algorithmically represented power series:

```
> f := proc(i)
           if i::'integer' then 0 else 'procname'(i) end if;
       end proc:
  Ex1 := x^9*diff(y(x), x$5)+
          (x^7+Sum(k^2*x^k/2, k = 9 .. infinity))*diff(y(x), x$4)+
          (2*x^5+x^2)*diff(y(x), x$2)+
          (2*x^10+x^4+3*x)*diff(y(x), x)+
          Sum(f(k)*x^k, k = 0 .. infinity)*y(x) = 0;
```

$$
\begin{aligned}
Ex1 := x^9 \left(\frac{d^5}{dx^5}\, y(x)\right) + \left(x^7 + \left(\sum_{k=9}^{\infty} \frac{k^2 x^k}{2}\right)\right)\left(\frac{d^4}{dx^4}\, y(x)\right) + \\
(2\,x^5 + x^2)\left(\frac{d^2}{dx^2}\, y(x)\right) + (2\,x^{10} + x^4 + 3\,x)\left(\frac{d}{dx}\, y(x)\right) + \\
\left(\sum_{k=0}^{\infty} f(k)\, x^k\right) y(x) = 0
\end{aligned}
$$

This equation has polynomial coefficients x^9, $2x^5 + x^2$. The coefficient for the third derivative is zero. There are also two infinite sums: one with the explicitly defined coefficients $\frac{k^2}{2}$ and the other with the coefficients defined by the Maple-procedure `f`.

For the equation `Ex1` we obtain the list of two elements of solutions with the truncation degree 3 which is set by the option `'top'`:

```
> LaurentSolution(Ex1, y(x), 'top' = 3);
```

$$
\left[\frac{\_c_1}{x^2} + \_c_2 - \frac{130\,x\_c_1}{3} + 90\,\_c_1 x^2 - 324\,x^3\_c_1 + O(x^4),\ \_c_2 + O(x^4)\right]
$$

The truncation degree of the result may be d1 which is greater than d if it is needed to compute initial terms up to the degree d1 to determine if Laurent solutions with valuation i_0 exist. Below we obtain the list of two elements of Laurent solutions with the truncation degree 0. This is needed to determine if there are Laurent solutions with the valuation -2 and with the valuation 0 :

```
> LaurentSolution(Ex1, y(x), 'top' = -1);
```

$$\left[\frac{\_c_1}{x^2} + \_c_2 + O(x),\ \_c_2 + O(x)\right]$$

The same result will be obtained if the option 'top' = d is not given:

```
> LaurentSolution(Ex1, y(x));
```

$$\left[\frac{\_c_1}{x^2} + \_c_2 + O(x),\ \_c_2 + O(x)\right]$$

For an equation whose all coefficients are truncated series, the LaurentSolution command investigates what can be learned from the equation about its Laurent solutions. The command constructs the maximum possible number of initial terms of solutions which are defined uniquely by known terms of coefficients of the given equation. The maximum truncation degree may be different in the solutions with different valuations, that is why the LaurentSolution command forms the solution for each valuation separately. The greatest truncation degree of Laurent solutions is called the *threshold* of the given equation.

For example, the equation whose all coefficients are truncated power series with various truncation degrees:

```
> Ex2 := O(x^9)*diff(y(x), x$5)+
        (x^7+81/2*x^9+50*x^10+O(x^11))*diff(y(x), x$4)+
        O(x^7)*diff(y(x), x$3)+
        (2*x^5+x^2+O(x^7))*diff(y(x), x$2)+
        (x^4+3*x+O(x^5))*diff(y(x), x)+
        O(x^6)*y(x) = 0;
```

$$Ex2 := O(x^9)\left(\frac{d^5}{dx^5}y(x)\right) + \left(x^7 + \frac{81\,x^9}{2} + 50\,x^{10} + O(x^{11})\right)\left(\frac{d^4}{dx^4}y(x)\right) + O(x^7)\left(\frac{d^3}{dx^3}y(x)\right) + (2\,x^5 + x^2 + O(x^7))\left(\frac{d^2}{dx^2}y(x)\right) + (x^4 + 3\,x + O(x^5))\left(\frac{d}{dx}y(x)\right) + O(x^6)\,y(x) = 0$$

We know only several initial terms of all coefficients. The coefficients of $\frac{d^5}{dx^5}y(x)$, $\frac{d^3}{dx^3}y(x)$ and $y(x)$ are $O(x^9)$, $O(x^7)$ and $O(x^6)$, and we don't know if they are zero or not. For Ex2 we obtain:

```
> LaurentSolution(Ex2, y(x));
```

$$\left[\frac{\_c_1}{x^2} + \_c_2 - \frac{130\,x\_c_1}{3} + O(x^2),\ \_c_2 + O(x^6)\right]$$

The first element of the returned list has the valuation -2. The maximum possible number of initial terms for it is equal to 4 , the truncation degree is equal to 1. The second one has the valuation 0 and the maximum possible number of the initial terms is equal to 6, the truncation degree is equal to 5. So, the threshold for Ex2 is equal to 5.

If the option `'top' = d` is given, then the `LaurentSolution` command handles `d` in the same way as described for equations whose all coefficients are algorithmically represented power series.

```
> LaurentSolution(Ex2, y(x), 'top' = 3, 'threshold' = 'h');
```

$$\left[\frac{\_c_1}{x^2} + \_c_2 - \frac{130\,x\_c_1}{3} + O(x^2),\ \_c_2 + O(x^4)\right]$$

Using the option `'threshold'='h'` we can obtain the information whether the given `d` is greater than the threshold of the `ode`. If it isn't then `h` is set equal to FAIL:

```
> h;
```

FAIL

Otherwise, `h` is set equal to the threshold. Below `h` is set equal to 5:

```
> LaurentSolution(Ex2, y(x), 'top' = 8, 'threshold' = 'h');
  'h' = h;
```

$$\left[\frac{\_c_1}{x^2} + \_c_2 - \frac{130\,x\_c_1}{3} + O(x^2),\ \_c_2 + O(x^6)\right]$$

$$h = 5$$

For an equation whose coefficients are of both types, in general, it's impossible to determine the greatest degree of truncated Laurent solutions. The threshold may be a finite number or infinity. Then if the option `'top'` is not given, the `LaurentSolution` command computes exactly as many initial terms of the solutions as needed to find a set of all valuations of the Laurent solutions of the given equation. For the equation

```
> Ex3 := O(x^9)*diff(y(x), x$5)+
        (x^7+Sum((1/2)*k^2*x^k, k = 9 .. infinity))*diff(y(x), x$4)+
        (2*x^5+x^2)*diff(y(x), x$2)+(x^4+3*x+O(x^5))*diff(y(x), x)+
        Sum(f(k)*x^k, k = 0 .. infinity)*y(x) = 0;
```

$$Ex3 := O(x^9)\left(\frac{d^5}{dx^5}y(x)\right) + \left(x^7 + \left(\sum_{k=9}^{\infty}\frac{k^2x^k}{2}\right)\right)\left(\frac{d^4}{dx^4}y(x)\right) + (2x^5 + x^2)\left(\frac{d^2}{dx^2}y(x)\right) + (x^4 + 3x + O(x^5))\left(\frac{d}{dx}y(x)\right) + \left(\sum_{k=0}^{\infty} f(k)\,x^k\right)y(x) = 0$$

we obtain

```
> LaurentSolution(Ex3, y(x));
```

$$\left[\frac{\_c_1}{x^2} + \_c_2 + O(x),\ \_c_2 + O(x)\right]$$

If the option `'top' = d` is given, then the `LaurentSolution` command tries to compute all Laurent solutions to the truncation degree `d`. For `Ex3` it's only possible for the solutions having valuation 0 (see the second element of the returned list):

```
> LaurentSolution(Ex3, y(x), 'top' = 4);
```

$$\left[\frac{\_c_1}{x^2} + \_c_2 - \frac{130\,x\_c_1}{3} + O(x^2),\ \_c_2 + O(x^5)\right]$$

If the threshold of the equation is greater than or equal to `d` and the option `'threshold'='h'` is given, then `h` is set equal to FAIL:

```
> LaurentSolution(Ex3, y(x), 'top' = 4, 'threshold' = 'h'):
  h;
```

$$FAIL$$

In fact, the threshold of `Ex3` is equal to ∞. This equation has the solution $y(x) = \_c_2$, where $\_c_2$ is an arbitrary constant. The trailing coefficient of `Ex3` is $\sum_{k=0}^{\infty} f(k)\,x^k$ and the `LaurentSolution` command can check any finite number of values of $f(k)$:

```
> {seq(f(k), k = 0 .. 100)};
```

$$\{0\}$$

but there is no algorithm to check that $f(k) = 0$ for all integer $k \geq 0$.

If the given equation has no nonzero Laurent solution (the set of valuations of Laurent solutions is empty), then the `LaurentSolution` command returns the empty list:

```
> LaurentSolution(x^2*diff(y(x), x)+y(x), y(x));
```

$$[\,]$$

And it returns FAIL if the known terms of the coefficients of the given equation are not sufficient to find a set of valuations of Laurent solutions:

```
> LaurentSolution(O(x)*diff(y(x), x)+y(x), y(x));
```

FAIL

4 RegularSolution

For an equation with power series coefficients a formal regular solution is a finite sum of expressions in this form:

$$x^{\lambda}\left(\sum_{k=0}^{m}\left(\sum_{i=i_{k,0}}^{\infty} v_k(i)\,x^i\right)\ln^k x\right)$$

where λ is an algebraic number, m is a non-negative integer, $i_{0,0}, \ldots, i_{m,0}$ are integers and $v_k(i_{k,0}) \neq 0$ for $k = 0, 1, \ldots, m$.

Same as for the case of Laurent solutions, the definition of the threshold of the equation is introduced. For the `'top'` and `'threshold'` options, the `RegularSolution` command works in the same way as the `LaurentSolution` command.

Below, we obtain the truncated regular solutions with $\lambda = 0$ (the truncation degree is 4) and $\lambda = \frac{1}{3}$ (the truncation degree is 1). The threshold is computed, it is equal to 4:

```
> Sol := RegularSolution((-3+x+O(x^2))*theta(y(x), x, 2)+
                         (1+x+O(x^2))*theta(y(x), x, 1)+
                         (x^4+O(x^5))*y(x), y(x), 'threshold' = 'h');
'h' = h;
```

$$Sol := \left[\_c_1 + \frac{x^4\_c_1}{44} + O(x^5) + x^{1/3}\left(\_c_2 + \frac{x\_c_2}{9} + O(x^2)\right)\right]$$

$$h = 4$$

Note that if the result `Sol` is combined in one series, for example using the `series` command, then the number of the initial terms is less then the maximum possible one (the term $\frac{x^4\_c_1}{44}$ is lost):

```
> series(Sol[1], x, infinity);
```

$$\_c_1 + \_c_2\, x^{1/3} + \frac{\_c_2\, x^{4/3}}{9} + O\left(x^{7/3}\right)$$

Below is an equation which has regular solutions with $\ln x$:

```
> RegularSolution((4+x+O(x^2))*theta(y(x), x, 2)+
                  O(x^2)*theta(y(x), x, 1)+
                  (x^3+O(x^4))*y(x), y(x), 'threshold' = 'h');
'h' = h;
```

$$\left[\_c_2 + O(x^2) + \ln(x)\left(\_c_1 - \frac{x^3\_c_1}{36} + O(x^4)\right),\ \_c_2 - \frac{x^3\_c_2}{36} + O(x^4),\right.$$
$$\left. O(x^2) + \ln(x)\left(\_c_1 - \frac{x^3\_c_1}{36} + O(x^4)\right)\right]$$

$$h = 3$$

The first element of the returned list is the truncated regular solutions with $\lambda = 0$ and $m = 1$, having two series with the valuation 0 and the truncation degrees 1 and 3. The second one is the truncated regular solutions with $\lambda = 0$ and $m = 0$, having one series with the valuation 0 and the truncation degree 3. The third one has the logarithm-free part with the valuation which is greater than 1. The threshold is computed, it is equal to 3.

If the equation has at least one completely given coefficient (below it is the coefficient of $\theta(y(x), x, 1)$ which is equal to 0) then we can use the command with different values of `d` in the option `'top' = d` to obtain the threshold.

Below we obtain that `h` is equal to FAIL if `'top' = 2`:

```
> Ex4 := (4+x+O(x^2))*theta(y(x), x, 2)+(x^3+O(x^4))*y(x):
  RegularSolution(Ex4, y(x), 'top' = 2, 'threshold' = 'h');
  'h' = h;
```

$$[\_c_2 + O(x^3) + \ln(x)\,(\_c_1 + O(x^3)),$$
$$\_c_2 + O(x^3),\ O(x^3) + \ln(x)\,(\_c_1 + O(x^3))]$$

$$h = \mathit{FAIL}$$

and we obtain that the threshold is equal to 3 if `'top' = 4`:

```
> RegularSolution(Ex4, y(x), 'top' = 4, 'threshold' = 'h');
  'h' = h;
```

$$\left[\_c_2 + x^3\left(-\frac{\_c_2}{36} + \frac{\_c_1}{54}\right) + O(x^4) + \ln(x)\left(\_c_1 - \frac{x^3\_c_1}{36} + O(x^4)\right),\right.$$
$$\left.\_c_2 - \frac{x^3\_c_2}{36} + O(x^4),\ \frac{x^3\_c_1}{54} + O(x^4) + \ln(x)\left(\_c_1 - \frac{x^3\_c_1}{36} + O(x^4)\right)\right]$$

$$h = 3$$

The following equation has no nonzero formal regular solution (the set of possible λ of regular solutions is empty), the result is the empty list:

```
> RegularSolution(x*theta(y(x), x, 2)+y(x), y(x));
```

$$[\,]$$

For the following one, the known terms of the coefficients are not sufficient to determine the set of λ, the result is FAIL:

```
> RegularSolution(x*theta(y(x), x, 2)+
                  O(1)*theta(y(x), x, 1)+y(x), y(x));
```

$$FAIL$$

5 FormalSolution

A formal exponential-logarithmic solution has the form

$$e^{Q(x)}\, x^{\lambda} \left(\sum_{k=0}^{m} \left(\sum_{i=i_{k,0}}^{\infty} v_k(i)\, x^{i/q} \right) \ln^k x \right)$$

where q is a positive integer, $Q(x)$ is a polynomial in $x^{-1/q}$, λ is an algebraic number, m is a non-negative integer, $i_{0,0}, \ldots, i_{m,0}$ are integers and $v_k(i_{k,0}) \neq 0$ for $k = 0, 1, \ldots, m$. Laurent and regular solutions are special cases of formal exponential-logarithmic solutions.

To construct all formal solutions for an equation with completely given coefficients, the `DEtools[formal_sol]` command can be used. For an equation whose coefficients may be truncated series the `FormalSolution` command of the presented `TruncatedSeries` package computes the maximum possible terms of the exponent $Q(x)$. If $Q(x)$ is obtained completely, the `FormalSolution` command then computes λ and initial terms of series which are components of solutions (if they are invariant to all possible prolongations of the given equation).

For the following equation, the given initial terms are only sufficient to obtain the one term of the exponent $Q(x)$. The unknown part of solutions is denoted by $y_1(x)$:

```
> Ex5 := (x^3 + O(x^4))*diff(y(x), x)+(2 + O(x))*y(x):
  FormalSolution(%, y(x));
```

$$\left[e^{\frac{1}{x^2}}\, y_1(x) \right] \qquad (1)$$

The following equation is a prolongation of Ex5 (extra new known terms are added to the series coefficients). We obtain the second term of the exponent $Q(x)$. The notation $y_{reg}(x)$ in the result means that the exponential part of formal solutions is obtained completely:

```
> (x^4+x^3+O(x^5))*diff(y(x), x)+(2+x+O(x^2))*y(x):
  FormalSolution(%, y(x));
```

$$\left[e^{\frac{1}{x^2} - \frac{1}{x}} y_{reg}(x) \right] \quad (2)$$

Another prolongation of Ex5 leads to another result:

```
> (x^3+(1/2)*x^4+O(x^5))*diff(y(x), x)+(2+x+O(x^2))*y(x):
  FormalSolution(%, y(x));
```

$$\left[e^{\frac{1}{x^2}} y_{reg}(x) \right] \quad (3)$$

Both (2) and (3) are prolongations of (1). It shows that (1) presents the maximum possible information about the solution which is invariant to all possible prolongations of Ex5.

Again and again, increasing the number of known terms in Ex5 we obtain more information about solutions:

```
> (x^5+x^4+x^3+O(x^6))*diff(y(x), x)+(-x^2+x+2+O(x^3))*y(x):
  FormalSolution(%, y(x));
```

$$\left[e^{\frac{1}{x^2} - \frac{1}{x}} x^2 (\_c_1 + O(x)) \right]$$

```
> (x^5+x^4+x^3+(3/2)*x^6+O(x^7))*diff(y(x), x)+
      (-x^2+x+2+O(x^4))*y(x):
  FormalSolution(%, y(x));
```

$$\left[e^{\frac{1}{x^2} - \frac{1}{x}} x^2 (\_c_1 + O(x^2)) \right]$$

```
> (x^5+x^4+x^3+(3/2)*x^6+(1/4)*x^7+O(x^8))*diff(y(x), x)+
      (-x^2+x+2+O(x^5))*y(x):
  FormalSolution(%, y(x));
```

$$\left[e^{\frac{1}{x^2} - \frac{1}{x}} x^2 \left(\_c_1 - \frac{3\_c_1 x^2}{2} + O(x^3) \right) \right]$$

The result of the **FormalSolution** command may contain the following expressions: $y_{reg}(x^{1/q})$, $y_{irr(p)}(x)$, $y_{irr}(x)$, $y_i(x)$, where y, x are given via the second parameter of the command, q and i are positive integers, p is a rational number.

As mentioned above, the notation $y_{reg}(x^{1/q})$ in the result means that the exponent $Q(x)$ (together with the number q) is obtained completely but the algebraic number λ is not invariant to all possible prolongations of the given equation (see (2) and (3) where q is equal to 1).

If the result has a term in the form

$$e^{Q_1(x)}\, y_{irr(p)}(x)$$

then it means that all prolongations of the given equation have formal solutions with the exponential

$$Q(x) = Q_1(x) + \frac{b}{x^p} + Q_2(x) \tag{4}$$

where $Q_1(x)$ and p are invariant to all possible prolongations (and the command computes them) but $b \neq 0$ is not invariant.

If the result has a term in the form

$$e^{Q_1(x)}\, y_{irr}(x)$$

then it means that all prolongations of the given equation have formal solutions with the exponential (4) but p is not invariant as well as $b \neq 0$.

If the result has a term in the form

$$e^{Q_1(x)}\, y_i(x)$$

where i is an integer it means that there are prolongation of the given equation having solutions with the exponent (4) and $b \neq 0$, and there are other ones having solutions with the exponent (4) and $b = 0$ and $Q_2(x) = 0$.

If different terms with the same expressions $y_{reg}(x^{1/q})$, or $y_{irr(p)}(x)$, or $y_{irr}(x)$ appear in the result, then such expressions are additionally indexed as follows: $y_{reg,1}(x^{1/q})$, $y_{reg,2}(x^{1/q})$, etc.

For example,

```
> Ex6 := (x^5 + O(x^6))*diff(y(x), x$3) +
         (-3*x^3 + O(x^4))*diff(y(x), x$2) +
         O(x)*diff(y(x), x) + (2 + O(x))*y(x) = 0:
  FormalSolution(Ex6, y(x));
```

$$\left[y_1(x) + y_{irr}(x) + y_{irr(1)}(x)\right]$$

The given initial terms are only sufficient to obtain the following information:

- all prolongations of **Ex6** have a three-dimensional linear space of formal exponential-logarithmic solutions;

– the first term $y_1(x)$ of the result means that there are prolongations of Ex6 that have a one-dimensional space of regular solutions, and there are prolongations that do not have regular solutions;
– the second term $y_{irr}(x)$ means that all prolongations of Ex6 have such irregular solutions that the exponent $Q(x)$ has no invariant terms;
– the last term $y_{irr(1)}(x)$ means that all prolongations of Ex6 have at least a one-dimensional space of irregular solutions with an exponent (4), where $Q_1(x) = 0$ and $p = 1$ but b is not invariant.

Below are two prolongations of Ex6 confirming the above:

```
> (x^5 + O(x^6))*diff(y(x), x$3) +
    (-3*x^3 + O(x^4))*diff(y(x), x$2) +
      (2*x + O(x^2))*diff(y(x), x) + (1 + O(x))*y(x) = 0:
  FormalSolution(%, y(x));
```

$$\left[\frac{\_c_1 + O(x)}{\sqrt{x}} + \mathrm{e}^{-\frac{2}{x}} y_{reg,1}(x) + \mathrm{e}^{-\frac{1}{x}} y_{reg,2}(x)\right]$$

```
> (x^5 + O(x^6))*diff(y(x), x$3) +
    (-3*x^3 + O(x^4))*diff(y(x), x$2) +
      O(x^2)*diff(y(x), x) + (1 + O(x))*y(x) = 0:
  FormalSolution(%, y(x));
```

$$\left[\mathrm{e}^{-\frac{2 RootOf(3\_Z^2-1, index=1)}{\sqrt{x}}} y_{reg,1}\left(\sqrt{x}\right) + \mathrm{e}^{-\frac{2 RootOf(3\_Z^2-1, index=2)}{\sqrt{x}}} y_{reg,2}\left(\sqrt{x}\right) + \mathrm{e}^{-\frac{3}{x}} y_{reg,3}(x)\right]$$

The following equation is also a prolongation of Ex6. It contains enough information to construct the exponential parts of the solutions completely. The solution involves series in fractional powers of x:

```
> (x^5 + x^6 + O(x^7))*diff(y(x), x$3) +
    (-3*x^3 - x^4 + O(x^5))*diff(y(x), x$2) +
      (1 + x + O(x^2))*y(x) = 0:
  FormalSolution(%, y(x));
```

$$\left[\mathrm{e}^{-\frac{2 RootOf(3\_Z^2-1, index=1)}{\sqrt{x}}} x^{29/36}\left(\_c_1 + \frac{191\, RootOf\left(3\_Z^2-1, index=1\right) \_c_1 \sqrt{x}}{432} - \frac{82679\, \_c_1 x}{1119744} + O\left(x^{3/2}\right)\right)\right.$$

$$+ e^{-\frac{2RootOf(3\_Z^2-1,index=2)}{\sqrt{x}}} x^{29/36}\Bigg(\_c_2 + \frac{191\,RootOf\left(3\_Z^2-1, index=2\right)\_c_2\sqrt{x}}{432} - \frac{82679\,\_c_2 x}{1119744} + O\left(x^{3/2}\right)\Bigg) + e^{-\frac{3}{x}} x^{17/9}\left(\_c_3 + O\left(x\right)\right)\Bigg]$$

References

1. Abramov, S.A., Barkatou, M.A.: Computable infinite power series in the role of coefficients of linear differential systems. In: Gerdt, V.P., Koepf, W., Seiler, W.M., Vorozhtsov, E.V. (eds.) CASC 2014. LNCS, vol. 8660, pp. 1–12. Springer, Cham (2014). https://doi.org/10.1007/978-3-319-10515-4_1
2. Abramov, S., Barkatou, M., Khmelnov, D.: On full rank differential systems with power series coefficients. J. Symbolic Comput. **68**(1), 120–137 (2015). https://doi.org/10.1016/j.jsc.2014.08.010
3. Abramov, S.A., Barkatou, M.A., Pflügel, E.: Higher-order linear differential systems with truncated coefficients. In: Gerdt, V.P., Koepf, W., Mayr, E.W., Vorozhtsov, E.V. (eds.) CASC 2011. LNCS, vol. 6885, pp. 10–24. Springer, Heidelberg (2011). https://doi.org/10.1007/978-3-642-23568-9_2
4. Abramov, S., Bronstein, M., Petkovšek, M.: On polynomial solutions of linear operator equations. In: Proceedings ISSAC 1995, pp. 290–296 (1995). https://doi.org/10.1145/220346.220384
5. Abramov, S.A., Khmelnov, D.E.: Regular solutions of linear differential systems with power series coefficients. Program. Comput. Softw. **40**(2), 98–106 (2014). https://doi.org/10.1134/S0361768814020029
6. Abramov, S.A., Ryabenko, A.A., Khmelnov, D.E.: Procedures for searching local solutions of linear differential systems with infinite power series in the role of coefficients. Program. Comput. Softw. **42**(2), 55–64 (2016). https://doi.org/10.1134/S036176881602002X
7. Abramov, S., Khmelnov, D., Ryabenko, A.: Laurent solutions of linear ordinary differential equations with coefficients in the form of truncated power series. In: Computer Algebra, Moscow, 17–21 June 2019, International Conference Materials, pp. 75–82 (2019)
8. Abramov, S.A., Ryabenko, A.A., Khmelnov, D.E.: Linear ordinary differential equations and truncated series. Comput. Math. Math. Phys. **59**(10), 1649–1659 (2019). https://doi.org/10.1134/S0965542519100026
9. Abramov, S.A., Ryabenko, A.A., Khmelnov, D.E.: Procedures for searching Laurent and regular solutions of linear differential equations with the coefficients in the form of truncated power series. Program. Comput. Softw. **46**(2), 67–75 (2020). https://doi.org/10.1134/S0361768820020024
10. Abramov, S.A., Ryabenko, A.A., Khmelnov, D.E.: Regular solutions of linear ordinary differential equations and truncated series. Comput. Math. Math. Phys. **60**(1), 1–14 (2020). https://doi.org/10.1134/S0965542520010029

11. Abramov, S.A., Khmelnov, D.E., Ryabenko, A.A.: Truncated and infinite power series in the role of coefficients of linear ordinary differential equations. In: Boulier, F., England, M., Sadykov, T.M., Vorozhtsov, E.V. (eds.) CASC 2020. LNCS, vol. 12291, pp. 63–76. Springer, Cham (2020). https://doi.org/10.1007/978-3-030-60026-6_4
12. Abramov, S.A., Ryabenko, A.A., Khmelnov, D.E.: Truncated series and formal exponential-logarithmic solutions of linear ordinary differential equations. Comput. Math. Math. Phys. **60**(10), 1609–1620 (2020). https://doi.org/10.1134/S0965542520100024
13. Abramov, S., Khmelnov, D., Ryabenko, A.: Truncated series (in Russian). In: Differential Equations and Related Questions of Mathematics, Works of the XII Prioksky Scientific Conference, 19–20 June 2020, pp. 8–19 (2020). http://kolomnamath.ru/download/Kolomna_Sbornik_2020.pdf
14. Abramov, S.A., Ryabenko, A.A., Khmelnov, D.E.: Procedures for constructing truncated solutions of linear differential equations with infinite and truncated power series in the role of coefficients. Program. Comput. Software **47**(2), 144–152 (2021). https://doi.org/10.1134/S036176882102002X
15. Abramov, S., Petkovšek, M.: Special power series solutions of linear differential equations. In: Proceedings FPSAC 1996, pp. 1–8 (1996)
16. Barkatou, M.A.: Rational Newton algorithm for computing formal solutions of linear differential equations. In: Gianni, P. (ed.) ISSAC 1988. LNCS, vol. 358, pp. 183–195. Springer, Heidelberg (1989). https://doi.org/10.1007/3-540-51084-2_17
17. Coddington, E., Levinson, N.: Theory of Ordinary Differential Equations. Krieger (1984)
18. Frobenius, G.: Über die Integration der linearen Differentialgleichungen durch Reihen. Journal für die reine und angewandte Mathematik **76**, 214–235 (1873). https://doi.org/10.1515/crll.1873.76.214
19. Heffter, L.: Einleitung in die Theorie der linearen Differentialgleichungen mit einer unabhängigen Variablen. BG Teubner, Leipzig (1894)
20. Ince, E.: Ordinary Differential Equations. Longmans, London, New York, Bombay (1926)
21. Kamke, E.: Differentialgleichungen. Lösungsmethoden und Lösungen I. Gewöhnliche Differentialgleichungen, Leipzig (1942)
22. Malgrange, B.: Sur la réduction formelle des équations différentielles à singularités irrégulières. Université Scientifique et Médicale de Grenoble (1979)
23. Schlesinger, L.: Handbuch der Theorie der linearen Differentialgleichungen, vol. 1. Teubner, Leipzig (1895)
24. Singer, M.F.: Formal solutions of differential equations. J. Symbolic Comput. **10**(1), 59–94 (1990). https://doi.org/10.1016/S0747-7171(08)80037-5

Computation of the Observed Spectral Sequence Spectrum for Nucleotide Sequence Alignments

Ernesto Álvarez González[1(✉)] and Ricardo Balam-Narváez[2]

[1] Departamento de Álgebra, Geometría y Topología, Facultad de Ciencias Matemáticas, Universidad Complutense de Madrid, 28040 Madrid, Spain
eralva01@ucm.es

[2] Laboratorio de biodiversidad, Escuela de Ciencias, Universidad Autónoma 'Benito Juárez' de Oaxaca, Oaxaca de Juárez, 68120 Oaxaca, Mexico
https://www.ucm.es, http://www.uabjo.mx

Abstract. Phylogenetics deals with the task of reconstructing all the ancestral relations among a set of lineages. For any given phylogenetic tree that explains their evolution, subject to the Kimura Three Parameters Model, Hadamard Conjugation is an equation that relates its tree weights (within a matrix Q) to its substitution patterns distribution (within a matrix P) on leaves. In practice, the latter matrix is approximated via the DNA sequence alignment. Some challenges have to be faced in the process such as filling the gaps when they exist. Throughout this manuscript, it is provided fully explanation of how the matrix P can be approximated. The authors contribute to the scientific community with one library running on Maple to lead with these tasks. We conclude this work with a fully developed example focused on three spieces of orchids of the genus Lophiarella distributed in southwestern Mexico and northwestern Mesoamerica for which we obtained the matrix P.

Keywords: Phylogenetic reconstruction · Hadamard Conjugation · Maximum likelihood estimation

1 Phylogenetic Reconstruction

1.1 Introduction

Phylogenetics deals with the task of reconstructing all the ancestral relations among a set of lineages [12]. Those relations can be illustrated by a phylogenetic tree. DNA nucleotide sequences are currently used to hypothesize relationships between members of the study group [6]. Several methods are known to answer evolutionary questions at the macroevolutionary level [6,7]. This often requires the use of a nucleotide substitution rate model. In a phylogenetic tree, branch lengths stand for the expected number of substitutions (tree weights). Restricted

R. M. Corless et al. (Eds.): MC 2020, CCIS 1414, pp. 34–47, 2021.
https://doi.org/10.1007/978-3-030-81698-8_3

to the Kimura Three Parameters Model [13], there are three kind of substitutions: transitions, type I transversions, and type II transversions. Hadamard Conjugation is a tool that relates the set of tree weights to the substitution patterns distribution on leaves [11]. The latter set of data is stored within a spectral matrix P. Applications of Hadamard Conjugation are inferring trees from sequence data, analyzing primate DNA sequences and inferring tree weights, among others [1,9,10]. In real applications, upon the knowledge of a multiple sequence alignment of current lineages, this matrix can be approximated. The main contribution to this article is one library, written on the computer system Maple 2015, that computes the observed substitution patterns distribution on leaves in different contexts (within the corresponding spectral matrix P as in Theorem 1). It is available in zenodo.org [3] as a Maple 2015 worksheet. It can be accessed to it since the Maple version 2015.

Section 1.3 provides some definitions to understand Hadamard Conjugation. It also provides some simple examples of what is pursued. Section 2 provides concrete examples of how the library has to be implemented for obtaining results in Sects. 1.3 to 1.5. Section 3 provides a fully developed example of how the matrix P can be obtained by using the library.

1.2 Problem Statement

For any given set of lineages, we assume a phylogenetic tree that best explains all their ancestral relations. We take its tree weights (edge lengths) as parameters. We also assume the Kimura Three Parameters Model as the molecular evolution model governing the tree.

Hadamard Conjugation is an equation that relates the set of tree weights (within a matrix Q called Edge Length Spectrum) to the substitution patterns distribution (within a matrix P called Spectral Sequence Spectrum) on leaves. In real applications, the approximation to the latter distribution is done after the process of sequence alignment. If the sequence alignment comprises equally length sequences and if it has no gaps, the computation of the observed substitution patterns distribution can be obtained directly. Throughout this article, it is said that a sequence alignment is complete if it comprises equally length sequences and if it has no gaps.

Chor, Hendy and Snir [1] make use of Hadamard Conjugation, applied to a rooted phylogenetic tree, taking its tree weights as the parameters, to infere the parameter values maximizing the likelihood function of substitution patterns distribution. In spite that Hendy and Snir [11] gave a proof of Hadamard Conjugation restricted to the Kimura Three Parameters Model as the molecular evolution model, McBee and Pentilla [2] expanded the number of nucleotide substitution models that can be used with Hadamard Conjugation. In the present work, we restrict to the Kimura Three Parameters Model.

Library Description. The contribution by the authors is one library whose main procedures are *SSS* and *FillingGAPS*. The former one can be implemented on complete sequence alignments to compute their Spectral Sequence Spectrum as in [11] (illustrated in the Example 1.31). The latter one must be implemented on sequence alignments that are not complete to make them be complete by inserting characters under the condition of reducing the variance among character patterns. In either case, for the implementation of these procedures, there is no restriction to the number of sequences nor to their length, except the executing time and memory demand.

Section 3 is interesting as it includes an example, where the procedures *FillingGAPS* and *SSS* were implemented in that order to the taxa *Lophiarella flavovirens*, *Lophiarella microchila* and *Lophiarella splendida*.

The procedure *SSS* is linked to other four procedures, whose dependence order is as follows: $Patterns \rightarrow SPL \rightarrow SPLM \rightarrow SPF \rightarrow SSS$.

Patterns computes all substitution patterns present in the alignment; they are included within the list N_I. For any given substitution pattern, *SPL* provides its location within the matrix S (to be computed later) as an ordered pair. *SPLM* includes into the list PAIRS the location of every substitution pattern in N_I within the matrix S as ordered pairs. *SPF* associates to each ordered pair in PAIRS the absolute frecuency of the corresponding substitution pattern as an element of another matrix, M. *SSS* standarizes each entry in M to get the matrix S: all entries in S sum to 1. This matrix is the observed spectral sequence spectrum.

The procedure *FillingGAPS* is linked to other thirteen procedures, whose dependence order is as follows:

$$\begin{array}{ccccccc}
GAPSL & \longrightarrow & separation & \longrightarrow & GAPSA & \longrightarrow & GAPSITE \\
 & & & & & & \downarrow \\
SLI & \longleftarrow & SD_to_SCHPS & \longleftarrow & DCHPS & \longleftarrow & CPCL \\
\downarrow & & & & & & \\
CCHPS & \longrightarrow & CHPR & \longrightarrow & SHR & \longrightarrow & RCHP \\
 & & & & & & \downarrow \\
 & & & & FillingGAPS & \longleftarrow & RecSeq
\end{array}$$

Gaps are identified by empty characters along an n-sequence. Let L be an n-sequence. *GAPSL* provides positions of its gaps within a list. For any given m-sequence alignment of n-sequences, a, {*separation*, *GAPSA*} call *GAPSL* for each of its n-sequences. *GAPSITE* provides within the list Siteset all sites showing gaps in the m-sequence alignment. *CPCL* classifies all character patterns that are present in a as those character patterns with gaps (within the matrix GM) and as those character patterns with no gaps (within the matrix NGM). Every character pattern has an index as a column position either in GM or in

NGM. For any suitable pair (i, j), *DCHPS* compares nonempty entries GM(k,i) to entries NGM(k,j): if corresponding entries equal, *DCHPS* returns 0; otherwise, it returns 1. *DCHPS* does this for each integer k with $1 \leq k \leq m$ and adds over ones. *DCHPS* measures distances between character patterns. For any suitable integer s, SD_to_SCHP compares the s-th character pattern in GM to every character pattern in NGM by calling *DCHPS* each time. *SLI* stores within the list LOWER the nearest character patterns in NGM from the s-th character pattern in GM. For any suitable pair of integers (s_1, s_2), *CCHPS* returns 1 if the s_1-th and s_2-th character patterns in NGM equal. Otherwise, it returns 0. *CHPR* stores within the list REPETITION the absolute frequency (minus one) of each character pattern in LOWER. *SHR* provides the indexes of those character patterns in LOWER with highest absolute frequency according to the list REPETITION. For the s-th character patter in GM, *RCHP* substitutes its gaps by the corresponding row characters from the previously selected character pattern in LOWER, according to the highest index in *SHR*. {*RecSeq*, *FillingGAPS*} fill all gaps in a by calling the last procedures in certain order.

1.3 Definitions

Hadamard Conjugation assumes an apriori phylogenetic tree as a taxa evolution model [11].

As an illustration of a phylogenetic tree, the Fig. 1 shows a graph with six vertices. The outest vertices stand for the current taxa σ_1, σ_2, σ_3 and σ_4, whose nucleotide sequences appear in Table 1. The inner vertex represented by a dot is the most recent ancestral taxa for σ_3 and σ_4. The inner vertex R, as being the most recent ancestral taxa for σ_1, σ_2, σ_3 and σ_4, is the root. In the literature, the outest vertices are called leaves. The arrows in Fig. 1 express the taxa evolving direction. More over, the arrow lengths stand for the expected number of substitutions from certain ancestral taxa to their nearest descendants. Sometimes, these arrow lengths are called tree weights.

Another restriction for holding Hadamard Conjugation is the molecular evolution model: the Kimura Three Parameters Model [11]. As the Kimura Two Parameters Model and the Jukes-Cantor Model are special cases of the Kimura Three Parameters Model [13], they can also be used as molecular evolution models to hold Hamadard Conjugation.

Hadamard Conjugation involves two spectral matrices: the Edge Lenght Spectrum, Q, and the Spectral Sequence Spectrum, P. The former one includes all the tree weights in the tree. The latter one includes all probabilities of observing the different substitution patterns on the alignment. In real applications, tree weights are parameters for the model [1]. The latter probabilities are approximated via relative frequencies. Throughout this section, we will construct the spectral matrices for an artificial example.

Character Patterns. Let $\sigma_1, \sigma_2, \sigma_3$ and σ_4 be the sequences of nucleotides of Table 1, whose phylogenetic tree is the following:

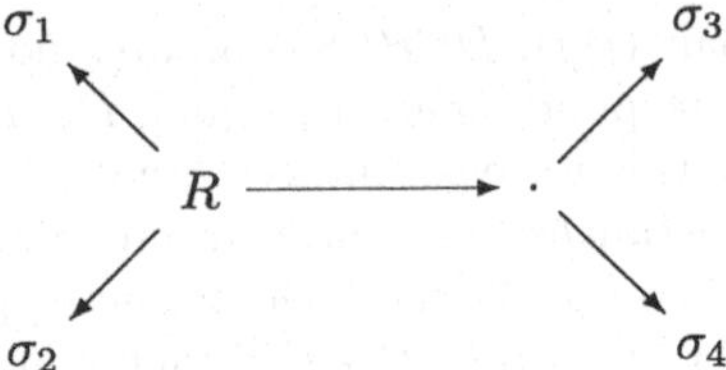

Fig. 1. Phylogenetic tree for the aligned sequences in Table 1.

In Fig. 1, R stands for the root; the internal dot stands for the common ancestor of lineages 3 and 4. The root R is the common ancestor of lineages 1, 2, 3 and 4.

Table 1. Example of four aligned sequences with sixteen sites.

Site	1	2	3	4	5	6	7	8	9	10	11	12	13	14	15	16
$\sigma_1 =$	C	C	A	T	C	A	A	A	C	G	T	G	T	G	A	C
$\sigma_2 =$	A	C	A	G	C	A	A	T	G	T	T	A	T	C	T	C
$\sigma_3 =$	C	C	A	T	T	G	A	A	G	A	T	G	C	G	T	T
$\sigma_4 =$	A	C	A	G	T	A	G	T	G	T	T	A	C	C	A	G

Table 1 shows a sequence alignment with 16 sites. Each site corresponds to a column that is called a character pattern. Character patterns' nucleotides are ordered from top to bottom according to the observed lineage on the tree's leaves of Fig. 1.

For the alignment of Table 1 in the Example 1.32, its observed spectral sequence spectrum P will be computed. It will also be given details on its construction, then it also will be explained how it is obtained by using the procedure *SSS* in Sect. 2.1.

Substitution Patterns. We call a substitution the transformation from one nucleotide to another. According to the Kimura Three Parameters Model, we distinguish three kind of substitutions and refer to them as the following figure shows (Fig. 2):

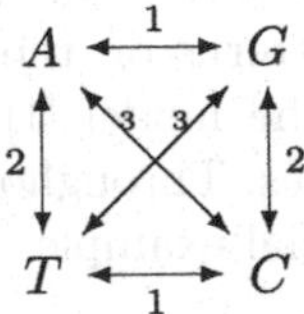

Fig. 2. Classes of substitution patterns: 1 stands for transitions; 2 stands for type I transversions and 3 stands for type II transversions.

If we select a row on Table 1, say row 2, we can produce from this one the substitutions to the rest rows as the Table 2 shows:

Table 2. Substitution patterns.

Site	1 2 3 4 5 6 7	8	9 10 11 12 13 14 15	16
$\sigma_2 \to \sigma_1$	3 0 0 3 0 0 0	$T \to A = 2$	2 3 0 1 0 2 2	$C \to C = 0$
$\sigma_2 \to \sigma_3$	3 0 0 3 1 1 0	$T \to A = 2$	0 2 0 1 1 2 0	$C \to T = 1$
$\sigma_2 \to \sigma_4$	0 0 0 0 1 0 1	$T \to T = 0$	0 0 0 0 1 0 2	$C \to G = 2$

In Table 2, 0 stands for no change, 1 stands for transitions, 2 stands for type I transversions and 3 stands for type II transversions.

Chor, Hendy and Snir [1] make use of another notation for substitution patterns:

Select one reference lineage, say $i \in [n] = \{1, 2, 3, \ldots n\}$. Take $A, B \subset [n] \setminus \{i\}$. The ordered pair (A, B) is the substitution pattern such that

$A \backslash B$: set of lineages obtained by transitions.
$B \backslash A$: set of lineages obtained by type I transversions.
$A \cap B$: set of lineages obtained by type II transversions.
$[n] \backslash (A \cup B)$: set of lineages sharing the same character as row i.

As examples of this notation, we have: for sites 8 and 15 from Table 2, $(\emptyset, \{1, 3\})$; for site 16 from Table 2, $(\{3\}, \{4\})$. Their relative frequencies are reported in the matrix P of the Example 1.31.

Chor, Hendy and Snir [1] take advantage of this last notation to include within a 2^{n-1} square matrix, whose rows and columns are indexed by subsets of the family of leaves (minus the reference lineage), all frequencies for different substitution patterns.

Spectral Sequence Spectrum. We summarize the relative frequencies of all substitution patterns in Table 2 within the matrix P:

Example 1.

$$P = \begin{array}{c} \\ \emptyset \\ \{1\} \\ \{3\} \\ \{1,3\} \\ \{4\} \\ \{1,4\} \\ \{3,4\} \\ \{1,3,4\} \end{array} \begin{array}{c} \begin{array}{cccccccc} \emptyset & \{1\} & \{3\} & \{1,3\} & \{4\} & \{1,4\} & \{3,4\} & \{1,3,4\} \end{array} \\ \begin{pmatrix} \frac{3}{16} & \frac{1}{16} & 0 & \frac{2}{16} & 0 & \frac{1}{16} & 0 & 0 \\ 0 & 0 & 0 & \frac{1}{16} & 0 & 0 & 0 & 0 \\ \frac{1}{16} & 0 & 0 & 0 & \frac{1}{16} & 0 & 0 & 0 \\ \frac{1}{16} & 0 & 0 & \frac{2}{16} & 0 & 0 & 0 & 0 \\ \frac{1}{16} & 0 & 0 & 0 & 0 & 0 & 0 & 0 \\ 0 & 0 & 0 & 0 & 0 & 0 & 0 & 0 \\ \frac{2}{16} & 0 & 0 & 0 & 0 & 0 & 0 & 0 \\ 0 & 0 & 0 & 0 & 0 & 0 & 0 & 0 \end{pmatrix} \end{array}$$

Observe that lineage σ_3 takes σ_2's place in correspondence to the lexicographical order for the subsets of $\{1, 2, 3, 4\}$*:*

$$\begin{array}{cccccccc} \emptyset & \{1\} & \{3\} & \{1,3\} & \{4\} & \{1,4\} & \{3,4\} & \{1,3,4\} \\ \updownarrow & \updownarrow & \updownarrow & \updownarrow & \updownarrow & \updownarrow & \updownarrow & \updownarrow \\ 0 & 1 & 10 & 11 & 100 & 101 & 110 & 111 \end{array}$$

For example, to locate the substitution pattern $(\{3\}, \{4\})$*, we change* $3 \to 2$ *and* $4 \to 3$*:* $(\{3\}, \{4\}) \to (2^{2-1} + 1, 2^{3-1} + 1) = (3, 5)$.

Edge Length Spectrum. The 2^{n-1} square matrix Q in theorem 1 is called the Edge Length Spectrum. It reserves a unique entry for each expected number of substitutions on the branches for the selected phylogenetic tree. It is typically not known by prior. That's why Chor, Hendy and Snir [1] take its entries as parameters. In spite that its construction is not our goal in this article, we include the corresponding Edge Length Spectrum for the phylogenetic tree in Fig. 1. Its structure is evident once the phylogenetic tree's branches are in correspondence to subsets of its leaves: Take any leaf as a reference. Any branch cut produces two subsets of leaves. The subset that omits the reference leaf (called bipartition) corresponds to the given branch.

Example 2. With respect to the third lineage in Fig. 1, the sequence of subsets in $\{1, 2, 4\}$ *(set of bipartitions), ordered lexicografically, are:*

$$\begin{aligned} 0 &\to \emptyset \\ 1 &\to \{1\} \\ 10 &\to \{2\} \\ 11 &\to \{1,2\} \\ 100 &\to \{4\} \\ 101 &\to \{1,4\} \\ 110 &\to \{2,4\} \\ 111 &\to \{1,2,4\} \end{aligned}$$

The bipartitions $\{1, 4\}$ *and* $\{2, 4\}$ *do not correspond to any leaves in Fig. 1. That is why the corresponding rows and columns in the matrix* Q *are fulfilled with zeroes.*

$$Q = \begin{array}{c} \\ \emptyset \\ \{1\} \\ \{2\} \\ \{1,2\} \\ \{4\} \\ \{1,4\} \\ \{2,4\} \\ \{1,2,4\} \end{array} \begin{array}{c} \begin{array}{cccccccc} \emptyset & \{1\} & \{2\} & \{1,2\} & \{4\} & \{1,4\} & \{2,4\} & \{1,2,4\} \end{array} \\ \begin{pmatrix} -K & q_1(\beta) & q_2(\beta) & q_{12}(\beta) & q_4(\beta) & 0 & 0 & q_{124}(\beta) \\ q_1(\alpha) & q_1(\gamma) & 0 & 0 & 0 & 0 & 0 & 0 \\ q_2(\alpha) & 0 & q_2(\gamma) & 0 & 0 & 0 & 0 & 0 \\ q_{12}(\alpha) & 0 & 0 & q_{12}(\gamma) & 0 & 0 & 0 & 0 \\ q_4(\alpha) & 0 & 0 & 0 & q_4(\gamma) & 0 & 0 & 0 \\ 0 & 0 & 0 & 0 & 0 & 0 & 0 & 0 \\ 0 & 0 & 0 & 0 & 0 & 0 & 0 & 0 \\ q_{124}(\alpha) & 0 & 0 & 0 & 0 & 0 & 0 & q_{124}(\gamma) \end{pmatrix} \end{array}$$

Hadamard Conjugation

Theorem 1 *[11]. Let Q be the Edge Length Spectrum for any given phylogenetic tree with n leaves. Let P be the corresponding Spectral Sequence Spectrum. Under the Kimura Three Paramaters Model, it holds that*

$$H_n P H_n = \exp(H_n Q H_n), \tag{1}$$

where H_n is the $2^n \times 2^n$ Hadamard matrix obtained inductively as follows:

1.

$$H_1 = \begin{pmatrix} 1 & 1 \\ 1 & -1 \end{pmatrix}$$

2. $H_n = H_1 \otimes H_{n-1}$, where the symbol $\otimes$ stands for the Kronecker product of matrices.

The exponential function in Theorem 1 acts termwise on $H_n Q H_n$.

1.4 Existence of Gaps

In practice, lineage sequences do not splice well: sites do not match. This is due to the evolving process itself that forced the ancestral lineages diverge in different ways. This divergence produces gaps:

Table 3. Four aligned sequences with gaps (indicated with an asterisk)

Site	1	2	3	4	5	6	7	8	9	10	11	12	13	14	15	16
$\sigma_1 =$	C	C	A	T	C	A	A	A	C	G	T	G	T	G	*	C
$\sigma_2 =$	A	C	*	G	C	A	*	T	G	T	T	A	T	C	T	C
$\sigma_3 =$	C	C	*	T	T	G	A	A	G	A	T	G	C	G	T	T
$\sigma_4 =$	A	C	A	G	T	A	G	T	G	T	T	A	C	C	A	G

In the presence of gaps, the assumption for the sequences to obtain the spectral sequence spectrum does not hold. Some work has to be done on the sequences to overcome that limit: fill the gaps.

The criteria used by the authors to fill gaps on character patterns is that of reducing the variance among the present character patterns on the alignment.

2 Procedures

Throughout the following subsections, it will be shown how the procedures *SSS* and *FillingGAPS* need to be implemented on the computer system Maple 2015. As it also will be noted, there are a few Maple standard libraries that have to be run before executing the procedures *SSS* and *FillingGAPS* as the latter ones work with lists, sets, arrays and matrices. A short note on the variables usage deserves attention: they appear according to the flow schemes in the Sect. 1.2. More over, as it can be seen in the following two subsections, some of them are named the same way to keep track the information.

2.1 Computation of the Observed Spectral Sequence Spectrum

The following lines show how to compute the matrix P in Example 1.31 by calling the procedure *SSS*.

```
> with(LinearAlgebra):
> with(RandomTools):
> with(StringTools):
> with(ArrayTools):
> with(ListTools):
> PATTERNS(m, n, L):
> SPL(r, R):
> COLUMNS(m,n):
> SPLM(N_I):
> SPF(N_I):
> SSS(M,n):
> L[1] := "CCATCAAACGTGTGAC":
> L[4] := "ACAGCAATGTTATCTC":
> L[2] := "CCATTGAAGATGCGTT":
> L[3] := "ACAGTAGTGTTACCAG":
> L := Join([L[1],L[2],L[3],L[4]]):
> for i from 1 to 3 by 1 do
        L := Delete(L,17*(4 - i)..17*(4 - i)):
  end do:
> m := 4: n := 16:
> PATTERNS(m,n,L):
> SPLM(N_I):
> SPF(N_I):
> SSS(M,16);
```

Observe the trick of changing the sequence ordering: the procedure *SSS* always computes all the observed substitution patterns with respect to the last sequence. As it was the intention to compute substitution patterns with respect to the second sequence, we re ordered the sequences: σ_1 kept the same position; σ_3, σ_4 and σ_2 took positions second, third and fourth, respectively.

2.2 Filling the Gaps of an Alignment

Table 3 shows an alignment with gaps in sites three, seven and fifteen. The following lines on Maple show how to fill those gaps by using the procedure *FillingGAPS*:

```
> with(LinearAlgebra):
> with(RandomTools):
> with(StringTools):
> with(ArrayTools):
> with(Logic):
> GAPSL(n,L):
> separation(m,n,L):
> GAPSA(m, n, chain2):
> GAPSITE(m, n, The_gaps):
> CPCL(m,n,L,siteset):
> DCHPS(m,n,i,j,site_diference_set,GM,NGM):
> SD_to_SCHP(m,n,s,site_diference_set,GM,NGM):
> MINOR_INTEGER(n1,n2):
> SLI(m,n,s,site_diference_set,GM,NGM):
> GRATEST_INTEGER(n1,n2):
> CCHPS(m,s1,s2,NGM):
> CHPR(m, MINOR_LIST, NGM):
> SHR(Repetition_list):
> RCHP(m, n, is_s,the_integer,GM,NGM):
> RecSeq(m, n, is_GAPchps, siteset, L, The_gaps):
> FillinGAPS(m,n,L):
> l[1] := "CCATCAAACGTGTG C":
> l[2] := "AC GCA TGTTATCTC":
> l[3] := "CC TTGAAGATGCGTT":
> l[4] := "ACAGTAGTGTTACCAG":
> L := Join([l[1],l[2],l[3],l[4]]):
> for i from 1 to 3 by 1 do
      L := Delete(L,17*(4 - i)..17*(4 - i)):
  end do:
> FillinGAPS(4,16,L);
```

Table 4. Re alignment after implementing the library Spectral Sequence Spectrum on Table 3.

Site	1	2	3	4	5	6	7	8	9	10	11	12	13	14	15	16
$\sigma_1 =$	C	C	A	T	C	A	A	A	C	G	T	G	T	G	T	C
$\sigma_2 =$	A	C	A	G	C	A	T	T	G	T	T	A	T	C	T	C
$\sigma_3 =$	C	C	G	T	T	G	A	A	G	A	T	G	C	G	T	T
$\sigma_4 =$	A	C	A	G	T	A	G	T	G	T	T	A	C	C	A	G

FillingGAPS produces (Table 4):

Observe the following: Nucleotides in Table 3 were recovered with no change. Gaps were filled with nucleotides that reduce distances from the new full-filled character patterns to the rest. This criteria did not recover the deleted nucleotides in Table 2.

3 Real Application

The genus *Lophiarella*(Orchidaceae: Oncidiinae) is a monophyletic clade integrated by taxa *Lophiarella microchila*, *L. flavovirens* and *L. splendida* [4]. This section develops a real case involving those taxa, whose nucleotide sequences appear in the Appendix.

3.1 Molecular Data

Nucleotide sequences and coded gaps for the nrDNA ITS region of 3 taxa of *Lophiarella* were used, whose vouchers and GenBank accession number are listed in the Appendix. Gaps of ITS sequences were aligned manually with Winclada program [5] by the simple indel coding method paradigm of Simmons and Ochoterena [8] in the context of phylogenetic tree proposed by Carnevali et al. [4]: This method represents gaps by short dashes. In fact, they are inserted between nucleotides on a sequence by separating some of these as part of the alignment procedure itself.

The implementation of the procedures *FillingGAPS* and *SSS* (in that order) to the taxa *Lophiarella flavovirens*, *L. microchila* and *L. splendida* (reported in the Appendix) produces the following 4-dimensional Spectral Sequence Spectrum:

$$P = \begin{pmatrix} \frac{717}{740} & \frac{1}{740} & \frac{1}{740} & \frac{1}{740} \\ 0 & \frac{3}{740} & 0 & 0 \\ \frac{2}{185} & 0 & \frac{1}{740} & 0 \\ \frac{1}{148} & 0 & 0 & \frac{3}{740} \end{pmatrix}.$$

4 Conclusions

The contribution of this article is a library for obtaining the observed Spectral Sequence Spectrum for a nucleotide sequence alignment with or without gaps, no matter the number of sequences nor the length sequences (in spite of the time execution and memory demand). The construction algorithm for the spectral matrix is given in [11]. The utility of the Spectral Sequence Spectrum is evident after the application of the Theorem 1: either it imposes conditions to the tree weights for an apriori phylogenetic tree as a model of taxa evolution or it helps to find the best phylogenetic tree that explains all the ancestral relations of the current data [1,9,10]. In this sense, the matrix P of the Sect. 3.1 can be the departure for future research.

Appendix

Voucher information and GenBank accesion numbers on nrITS DNA for taxa used in Sect. 3:

Lophiarella flavovirens, Mexico, Colima, Carnevali 7270 (CICY), JQ319734;
Lophiarella microchila, Mexico, Chiapas, Carnevali 7643 (CICY), JQ319735;
Lophiarella splendida, Guatemala, Carnevali 7232 (CICY), JQ319736.

Nucleotide sequences, as they were used on Maple, are:

```
LophiarellaFlavovirens :=

"GGTGAACCTGCGGAAGGATCATTGTCGAGA-CCGAAAAATAT--ACCGAGCG-ATTCGGACAACC

CGTGAAAATGAGCGTTTTGTA-CTGCTAT-CCTGG-TCGTCGCCCT--CGCTTTCCTTCAGGGGGG

--AGGGGGCACGGCGGAGGTGGATGAAC----CACAAACCGGCGCAGCATCGCGCCAAGGGAAGAT

TGGAATGCACGAGCCCCGCGTCGGGCTCGGAGGCGTGGAGTGCTGTTGCACGCCATGCGGTTGGAC

ACGACTCTCGGCAATGGATATCTCGGCTCTCGCATCGATGAAGAGCGCAGCGAAATGCGATACGTG

GTGTGAATTGCAGAATCCCGTGAACCATCGAGTCTTTGAACGCAAGTTGCGCCCGAGGCCAGCCGG

CCAAGGGCACGCCCGCCTGGGCGTCAAACATTGCGTCGCTCCGTGCCAC-CGCCGGCCCTCCAATG

GTCGTGCCGGTTGCGGCTTGGATGTGCAGAGTGGCCCGTCGCGCCTGCCGG-CGCGGCGGGTTGAA

GAGTTGGTTTCGTCTCGCTGGCCGCGAACAACAA-GGGGTGGGTGAAAGCTATGAGCGCAA-CCTG

CGTTGTCTCCGCGCCGGCCCGAAAGACGGCTTGTGCCTTTTATGTGATCCC-GGACCAA-GCCCCG

ATCGAC----ATGCGGCGGC-TTGGAATGCGACCCC----AGG-ATGGGCGAGGCC-ACCCGCTGA
```

```
GTTTAAGCATATCAA-":

LophiarellaMicrochila :=
"GGTGAACCTGCGGAAGGATCATTGTCGAGA-CCGAAAAATAT--ACCGAGCG-ATTCGGACAACC
CGTGAAA-TTAGCGTTTTGTA-CTGCTAT-CCTGG-TCGTCGCCCT--CGCTTTCCTTAAGGGGGG
--AGGGGGCACGGCGGAGGTGGATGAAC----CACAAACCGGCGCAGCATCGCGCCAAGGGAAGAT
TGGAATGCACGAGCCCCGCGTCGGGCTCGGTGGCGTGGAGTGCTTTTGCACTCCATGCGGTTGGAC
ACGACTCTCGGCAATGGATATCTCGGCTCTCGCATCGATGAAGAGCGCAGCGAAATGCGATACGTG
GTGCGAATTGCAGAATCCCGTGAACCATCGAGTCTTTGAACGCAAGTTGCGCCCGAGGCCAGCCGG
CCAAGGGCACGCCCGCCTGGGCGTCAATCGTTGCGTCGCTCCGTGCCAC-CGCCGGCCCTCCAATG
GTCGTGCCGGTTGCGGCTTGGATGTGCAGAGTGGCCCGTCGTGCCTGTCGG-CGCGGCGGGTTGAA
GAGCTGGTTTCGTCTCGCTGGCCGCGAACAACAA-GGGGTGGGTGAAAGCTATGAGCGCAA-CCTG
CGTTGTCTCCGCGCCGGCCCGAGAGACGGTTTGTGCCTTTTATGTGATCCC-GGACCAA-GCCCCG
ATCGAC----ATGCGGTGGC-TTGGAATGCGACCCC----AGG-ATGGGCGAGGCC-ACCCGCTGA
GTTTAAGCATATCAA-":

LophiarellaSplendida :=
"GGGGGGTnT-CGGAAGGATCATTGTCGAGA-CCGAAAAATATATACCGAGCGGATTCGGACAACC
CGTGAAA-TGAGCGTTTTGTA-CTGCTAT-CCTGG-TCGTCGCCCT--CGCTTTCCTTCAGGGGGG
--AGGGGGCACGGCGGAGGTGGATGAAC----CACAAACCGGCGCAGCATCGCGCCAAGGGAAGAT
TGGAATGCACGAGCCCCGCGTCGGGCTCGGTGGCGTGGAGTGCTGTTGCACTCCATGCGGTTGGAC
ACGACTCTCGGCAATGGATATCTCGGCTCTCGCATCGATGAAGAGCGCAGCGAAATGCGATACGTG
GTGCGAATTGCAGAATCCCGTGAACCATCGAGTCTTTGAACGCAAGTTGCGCCCGAGGCCAGCCGG
CCAAGGGCACGCCCGCCTGGGCGTCAAACGTTGCGTCGCTCCGTGCCAC-CGCCGGCCCTCCAATG
```

GTCGTGCCGGTTGCGGCTTGGATGTGCAGAGTGGCCCGTCGTGCCTGTCGG-CGCGGCGGGTTGAA

GAGCTGGTTTCGTCTCGCTGGCCGCGAACAACAA-GGGGTGGGTGAAAGCTATGAGCGCAA-CCTG

CGTTGTCTCCGCGCCGGCCCGAGAGGCGGTTTGTGCCTTTTATGTGATCCC-GGACCAA-GCCCCG

ATCGAC----ATGCGGTGGC-TTGGAATGCGACCCCCCCAAGGGCGAGGCC------ACCCGCTGA

GTTTAAGCATATC---":

References

1. Chor, B., Hendy, M.D., Snir, S.: Maximum likelihood jukes-cantor triplets: analytic solutions. Mol. Biol. Evol. **23**(3), 626–632 (2006). https://doi.org/10.1093/molbev/msj069
2. McBee, C.D., Penttila, T.: Remarks on Hadamard conjugation and combinatorial phylogenetics. Australas. J. Comb. **66**(2), 177–191 (2016)
3. Álvarez, E.: Spectral Sequence Spectrum for an m nucleotide sequence alignment with or without GAPS (2021). https://doi.org/10.5281/zenodo.4587007
4. Carnevali, G., Cetzal-Ix, W., Balam, R.N., Leopardi, C., Romero-González, G.A.: A combined evidence phylogenetic re-circumscription and a taxonomic revision of lophiarella (Orchidaceae: Oncidiinae). Syst. Botany **38**(1), 46–63 (2013). https://doi.org/10.1600/036364413X661926
5. Nixon, K.C., WinClada ver. 1.00. 08. Published by the author, Ithaca, NY (2002). http://www.cladistics.com/aboutWinc.htm
6. Casanellas, M.: El modelo evolutivo de Kimura: un enlace entre el álgebra, la estadística y la biología. La Gaceta de la RSME. **2**, 241–257 (2018)
7. Holder, M., Lewis, P.O.: Phylogeny estimation: traditional and Bayesian approaches. Nat. Rev. **4**, 275–284 (2003). https://doi.org/10.1038/nrg1044
8. Simmons, M.P., Ochoterena, H.: Gaps as characters in sequence-based phylogenetic analyses. Syst. Biol. **49**, 369–381 (2000). https://doi.org/10.1093/sysbio/49.2.369
9. Hendy, M.D., Charleston, M.A.: Hadamard conjugation: a versatile tool for modelling nucleotide sequence evolution. New Zealand J. Botany. **31**, 231–237 (1993). https://doi.org/10.1080/0028825X.1993.10419500
10. Hendy, M.D., Penny, D., Steel, M.A.: A discrete Fourier analysis for evolutionary trees. Proc. Natl. Acad. Sci. USA **91**, 3339–3343 (1994). https://doi.org/10.1073%2Fpnas.91.8.3339
11. Hendy, M.D., Snir, S.: Hadamard conjugation for the kimura 3ST model: combinatorial proof using path sets. IEEE/ACM Trans. Comput. Biol. Bioinf. **5**(3), 461–471 (2008). https://doi.org/10.1109/TCBB.2007.70227
12. Steel, M.A., Hendy, M.D., Székely, L.A., Erdős, P.L.: Spectral analysis and a closest tree method for genetic sequences. Appl. Math. Lett. **5**(6), 63–67 (1992). https://doi.org/10.1016/0893-9659(92)90016-3
13. Kimura, M.: Estimation of evolutionary sequences between homologous nucleotide sequences. Proc. Nat. Acad. Sci., USA **78**, 454–458 (1981). https://doi.org/10.1073/pnas.78.1.454

Multivariate Power Series in Maple

Mohammadali Asadi[1,2(✉)], Alexander Brandt[1], Mahsa Kazemi[1], Marc Moreno Maza[1], and Erik J. Postma[2]

[1] The University of Western Ontario, 1151 Richmond Street, London, ON, Canada
{masadi4,abrandt5,mkazemin}@uwo.ca, moreno@csd.uwo.ca

[2] Maplesoft, 615 Kumpf Drive, Waterloo, ON, Canada
epostma@maplesoft.com

Abstract. We present `MultivariatePowerSeries`, a MAPLE library introduced in MAPLE 2021, providing a variety of methods to study formal multivariate power series and univariate polynomials over such series. This library offers a simple and easy-to-use user interface. Its implementation relies on lazy evaluation techniques and takes advantage of MAPLE's features for object-oriented programming. The exposed methods include Weierstrass Preparation Theorem and factorization via Hensel's lemma. The computational performance is demonstrated by means of an experimental comparison with software counterparts.

Keywords: Multivariate power series · Weierstrass preparation theorem · Hensel's lemma · Factorization · Lazy evaluation

1 Introduction

In elementary courses on univariate calculus, power series are often introduced as limits of sequences of the form "the first n terms of a given sequence". This leads students to the study of analytic functions and the use of power series in computing function limits. While the extension of those notions to the multivariate case is a standard topic in advanced calculus courses, the availability of multivariate power series and multivariate analytic functions in computer algebra systems is somehow limited.

In MAPLE [11], SAGEMATH [16], and MATHEMATICA [19], power series are restricted to being either only univariate or truncated, that is, reduced modulo a fixed power of the ideal $\langle X_1, \ldots, X_n \rangle$ generated by the variables of those power series. A truncated implementation, while simple, may be insufficient for, or computationally more expensive in, some particular circumstances. For instance, modern algorithms for polynomial system solving require the intensive use of modular methods based on Hensel lifting. In those lifting procedures, degrees of truncation may not be known a priori, thus leading to truncated power series being ineffective.

Considering that a power series has potentially an infinite number of terms naturally suggests to represent it as a procedure which, given a particular (total)

R. M. Corless et al. (Eds.): MC 2020, CCIS 1414, pp. 48–66, 2021.
https://doi.org/10.1007/978-3-030-81698-8_4

degree, produces the terms of that degree. This leads to a so-called lazy evaluation scheme, where the terms of any power series are produced only as needed, via such a *generator* function.

The usefulness of lazy evaluation in computer algebra has been studied for a few decades. In particular, see the work of Karczmarczuk [10], discussing different mathematical objects with an infinite length; Burge and Watt [7], and van der Hoeven [17], discussing lazy univariate power series; and Monagan and Vrbik [12], discussing lazy arithmetic for polynomials.

In this paper, we present `MultivariatePowerSeries`, which is among the new features released in MAPLE 2021 and publicly available in [1]. This library, written in the MAPLE language, provides the ability to create and manipulate multivariate power series with rational or algebraic number coefficients, as well as univariate polynomials whose coefficients are multivariate power series. Through lazy evaluation techniques and a careful implementation, our library achieves very high performance. These power series and univariate polynomials over power series (UPoPS) are employed in optimized implementations of Weierstrass Preparation Theorem and factorization of UPoPS via Hensel's lemma.

Our implementation follows the lazy evaluation scheme of multivariate power series in the BPAS library [3]. The multivariate power series of BPAS, written in the C language, is discussed in [6] and extends upon the work of the `PowerSeries` subpackage of the `RegularChains` MAPLE library [2,13]. The `PowerSeries` package is the only preexisting implementation of multivariate power series integrated in MAPLE. In [6], it is shown that the BPAS implementation provides exceptional performance, surpassing that of the `PowerSeries` package, the basic MAPLE function `mtaylor`, and the multivariate power series available in SageMath [16] by multiple orders of magnitude.

A key design element of our library, in addition to lazy evaluation techniques, is the use of MAPLE *objects* and object-oriented programming. An object in MAPLE is a special kind of module which encapsulates together data and procedures manipulating that data, just like objects in any other object-oriented language; see [5, Chapters 8, 9]. To the best of our knowledge, few MAPLE libraries make use of those objects, which, as our report suggests, are worth considering for improving performance. In particular, objects allow for the overloading of existing builtin MAPLE functions in order to integrate these new custom objects with existing MAPLE library code. Our results show that `MultivariatePowerSeries` is comparable in performance to the implementation of BPAS, is thus similarly several orders of magnitude faster than other existing implementations. These experimental results are discussed in Sect. 6.

The remainder of this paper is organized as follows. We begin in Sect. 2 with reviewing definitions of formal power series, and univariate polynomial over power series, followed by a brief discussion about the basic arithmetic, Weierstrass preparation theorem and factorization via Hensel's lemma. Section 3 presents an overview of the `MultivariatePowerSeries` package, while Sect. 4 explores its underlying design principles. Implementation details are discussed in Sect. 5, followed by our experimentation in Sect. 6. Finally, we conclude and present future works in Sect. 7.

2 Background

In this section we review the basic properties of formal power series and univariate polynomials over those series, following G. Fischer in [8]. While various proofs of Theorems 1 of 2 can be found in the literature, the proofs given in [6] are constructive and support our implementation. Throughout this paper, $\mathbb{N}$ denotes the semi-ring of non-negative integers and $\mathbb{K}$ an algebraic number field.

2.1 Power Series

Given a positive integer n, we denote by $\mathbb{K}[[X_1, \ldots, X_n]]$ the set of multivariate formal power series with coefficients in $\mathbb{K}$ and variables $X_1, \ldots, X_n$. Let $f = \sum_{e \in \mathbb{N}^n} a_e X^e \in \mathbb{K}[[X_1, \ldots, X_n]]$ and $d \in \mathbb{N}$ where $X^e = X_1^{e_1} \cdots X_n^{e_n}$ and $e = (e_1, \ldots, e_n) \in \mathbb{N}^n$. The *homogeneous part* and *polynomial part* of f in degree d are respectively defined by $f_{(d)} := \sum_{|e|=d} a_e X^e$ and $f^{(d)} := \sum_{k \leq d} f_{(k)}$, where $|e| = e_1 + \cdots + e_n$. The sum (resp. difference) of two formal power series $f, g \in \mathbb{K}[[X_1, \ldots, X_n]]$ is defined by the sum (and resp. difference) of their homogeneous parts of the same degree; thus we have: $f \pm g = \sum_{d \in \mathbb{N}} f_{(d)} \pm g_{(d)}$. The product $h = f \cdot g$ can be defined as $h = \sum_{d \in \mathbb{N}} h_{(d)}$ with $h_{(d)} = \sum_{k+l=d} f_{(k)} \, g_{(l)}$. With the above addition and multiplication, the set $\mathbb{K}[[X_1, \ldots, X_n]]$ is a local ring with $\mathcal{M} := \langle X_1, \ldots, X_n \rangle$ as maximal ideal; $\mathbb{K}[[X_1, \ldots, X_n]]$ is also a unique factorization domain (UFD). The order of the power series f, denoted by $\mathrm{ord}(f)$, is defined as $min\{d \in \mathbb{N} \mid f_{(d)} \neq 0\}$ if $f \neq 0$, and as ∞ otherwise. We observe that $\mathcal{M}^k = \{f \in \mathbb{K}[[X_1, \ldots, X_n]] \mid \mathrm{ord}(f) \geq k\}$ holds for every $k \geq 1$. If f is a unit, that is, if $f \notin \mathcal{M}$ (or equivalently, if $\mathrm{ord}(f) = 0$) then the sequence $(h_m)_{m \in \mathbb{N}}$, where $h_m = c^{-1}(1 + g + \cdots + g^m)$, $c = f_{(0)}$, and $g = 1 - c^{-1} f$, converges to the *inverse* of f. This convergence is the sense of Krull topology, see [8] for details.

2.2 Univariate Polynomials over Power Series

We denote by $\mathbb{A}$ and $\mathcal{M}$ the power series ring $\mathbb{K}[[X_1, \ldots, X_n]]$ and its maximal ideal. We allow $n = 0$, in which case we have $\mathcal{M} = \langle 0 \rangle$. Let $f \in \mathbb{A}[[X_{n+1}]]$, written as $f = \sum_{i=0}^{\infty} a_i X_{n+1}^i$ with $a_i \in \mathbb{A}$ for all $i \in \mathbb{N}$. Then, Weierstrass Preparation Theorem (WPT) states the following.

Theorem 1. *Assume $f \not\equiv 0 \mod \mathcal{M}[[X_{n+1}]]$. Let $d \geq 0$ be the smallest integer such that $a_d \notin \mathcal{M}$. Then, there exists a unique pair (α, p) satisfying the following:*

i α is an invertible power series of $\mathbb{A}[[X_{n+1}]]$,
ii $p \in \mathbb{A}[X_{n+1}]$ is a monic polynomial of degree d,
iii writing $p = X_{n+1}^d + b_{d-1} X_{n+1}^{d-1} + \cdots + b_1 X_{n+1} + b_0$, we have $b_{d-1}, \ldots, b_0 \in \mathcal{M}$,
iv $f = \alpha p$ holds.

Moreover, if f is a polynomial of $\mathbb{A}[X_{n+1}]$ of degree $d + m$, for some m, then α is a polynomial of $\mathbb{A}[X_{n+1}]$ of degree m.

Since $\mathbb{A}$ is a UFD, then Gauss' lemma implies that the polynomial ring $\mathbb{A}[X_{n+1}]$ is also a UFD. Hensel's lemma shows how factorizing a polynomial in $\mathbb{A}[X_{n+1}]$ can be reduced to factorizing a polynomial in $\mathbb{K}[X_{n+1}]$.

Theorem 2 (Hensel's Lemma). *Assume that f is a polynomial of degree k in $\mathbb{A}[X_{n+1}]$. We define $\overline{f} = f(0, \ldots, 0, X_{n+1}) \in \mathbb{K}[X_{n+1}]$. We assume that f is monic in X_{n+1}, that is, $a_k = 1$. We further assume that $\mathbb{K}$ is algebraically closed. Thus, there exists positive integers $k_1, \ldots, k_r$ and pairwise distinct elements $c_1, \ldots, c_r \in \mathbb{K}$ such that we have $\overline{f} = (X_{n+1}-c_1)^{k_1}(X_{n+1}-c_2)^{k_2} \cdots (X_{n+1}-c_r)^{k_r}$. Then, there exists $f_1, \ldots, f_r \in \mathbb{A}[X_{n+1}]$, all monic in X_{n+1}, such that we have:*

i $f = f_1 \cdots f_r$,
ii the degree of f_j *is* k_j, *for all* $j = 1, \ldots, r$,
iii $\overline{f_j} = (X_{n+1} - c_j)^{k_j}$, *for all* $j = 1, \ldots, r$.

3 An Overview of the User-Interface

From the point of view of the end-user, the `MultivariatePowerSeries` package is a collection of commands for manipulating multivariate power series and univariate polynomials over multivariate power series. The field of coefficients of all power series created by the command `PowerSeries` consists of all complex numbers that are constructible in MAPLE, thus including rational numbers and algebraic numbers. The main algebraic functionalities of this package deal with arithmetic operations (addition, multiplication, inversion, evaluation), for both multivariate power series and univariate polynomials over multivariate power series (UPoPS), as well as factorization of such polynomials. The list of the exposed commands is given in Fig. 1.

```
> with(MultivariatePowerSeries);
[Add, ApproximatelyEqual, ApproximatelyZero, Copy, Degree, Display, Divide,
    EvaluateAtOrigin, Exponentiate, GeometricSeries, GetAnalyticExpression,
    GetCoefficient, HenselFactorize, HomogeneousPart, Inverse, IsUnit,
    MainVariable, Multiply, Negate, PowerSeries, Precision,
    SetDefaultDisplayStyle, SetDisplayStyle, Subtract, SumOfAllMonomials,
    TaylorShift, Truncate, UnivariatePolynomialOverPowerSeries,
    UpdatePrecision, Variables, WeierstrassPreparation]
```

Fig. 1. List of the commands of `MultivariatePowerSeries`.

The commands `PowerSeries` and `UnivariatePolynomialOverPowerSeries` create power series and univariate polynomials over multivariate power series, respectively, from objects like polynomials, sequences, and functions which produce homogeneous parts of a power series, as illustrated in Figs. 2 and 3. The commands `GeometricSeries` and `SumOfAllMonomials` respectively create the geometric series and sum of all monomials for an input list of variables.

```
> a := PowerSeries(1 + x + x·y + x^2);
```
$$a := \left[\text{PowerSeries: } 1+x+x^2+xy\right] \quad (1)$$
```
> b := 1/a;
```
$$b := \left[\text{PowerSeries of } \frac{1}{x^2+xy+x+1} : 1 + \ldots\right] \quad (2)$$
```
> Truncate(b, 5);
```
$$-2x^4y - 3x^3y^2 - x^4 - x^3y + x^2y^2 + x^3 + 2x^2y - xy - x + 1 \quad (3)$$
```
> c := PowerSeries(d -> (x^d/d!), analytic=exp(x));
```
$$c := \left[\text{PowerSeries of } e^x : 1 + \ldots\right] \quad (4)$$
```
> Truncate(c, 5);
```
$$1 + x + \frac{1}{2}x^2 + \frac{1}{6}x^3 + \frac{1}{24}x^4 + \frac{1}{120}x^5 \quad (5)$$
```
> Truncate(c, 10);
```
$$1 + x + \frac{1}{2}x^2 + \frac{1}{6}x^3 + \frac{1}{24}x^4 + \frac{1}{120}x^5 + \frac{1}{720}x^6 + \frac{1}{5040}x^7 + \frac{1}{40320}x^8 + \frac{1}{362880}x^9 + \frac{1}{3628800}x^{10} \quad (6)$$

Fig. 2. Creating power series from a polynomial or an anonymous function.

```
> a := GeometricSeries([x, y]) :
> GetAnalyticExpression(a);
```
$$\frac{1}{1-x-y} \quad (7)$$
```
> b := 1/PowerSeries(3 + 2·x + y);
```
$$b := \left[\text{PowerSeries of } \frac{1}{3+2x+y} : \frac{1}{3} + \ldots\right] \quad (8)$$
```
> e := PowerSeries(d -> (x^d/d!), analytic=exp(x));
```
$$e := \left[\text{PowerSeries of } e^x : 1 + \ldots\right] \quad (9)$$
```
> f := UnivariatePolynomialOverPowerSeries([a, b, e], z) :
> Truncate(f, 3);
```
$$\left(1 + x + \frac{1}{2}x^2 + \frac{1}{6}x^3\right)z^2 + \left(\frac{1}{3} - \frac{1}{9}y - \frac{2}{9}x + \frac{1}{27}y^2 + \frac{4}{27}xy + \frac{4}{27}x^2 - \frac{1}{81}y^3 - \frac{2}{27}xy^2 - \frac{4}{27}x^2y - \frac{8}{81}x^3\right)z + x^3 + 3x^2y + 3xy^2 + y^3 + x^2 + 2xy + y^2 + x + y + 1 \quad (10)$$
```
> GetAnalyticExpression(f);
```
$$\frac{1}{1-x-y} + \frac{z}{3+2x+y} + e^x z^2 \quad (11)$$

Fig. 3. Creating a univariate polynomial over power series

```
> a := GeometricSeries([x, y]) + SumOfAllMonomials([x, y]);
```
$$a := \left[\text{PowerSeries of } \frac{1}{1-x-y} + \frac{1}{(1-x)(1-y)} : 2 + 2x + 2y + \ldots\right] \quad (12)$$
```
> Display(a);
```
$$\left[\text{PowerSeries of } \frac{1}{1-x-y} + \frac{1}{(1-x)(1-y)} : 2 + 2x + 2y + \ldots\right] \quad (13)$$
```
> Truncate(a, 10) :
> Display(a, [maxterms=20, precision=5]);
```
$$\left[\text{PowerSeries of } \frac{1}{1-x-y} + \frac{1}{(1-x)(1-y)} : 2 + 2x + 2y + 2x^2 + 3xy + 2y^2 + 2x^3 + 4x^2y + 4xy^2 + 2y^3 + 2x^4 + 5x^3y + 7x^2y^2 + 5xy^3 + 2y^4 + \left(2x^5 + 6x^4y + 11x^3y^2 + 11x^2y^3 + 6xy^4 + \ldots\right) + \ldots\right] \quad (14)$$
```
> Display(a, [precision=5]);
```
$$\left[\text{PowerSeries of } \frac{1}{1-x-y} + \frac{1}{(1-x)(1-y)} : 2 + 2x + 2y + 2x^2 + 3xy + 2y^2 + 2x^3 + 4x^2y + 4xy^2 + 2y^3 + 2x^4 + 5x^3y + 7x^2y^2 + 5xy^3 + 2y^4 + 2x^5 + 6x^4y + 11x^3y^2 + 11x^2y^3 + 6xy^4 + 2y^5 + \ldots\right] \quad (15)$$

Fig. 4. Controlling the output format of a multivariate power series.

Whenever possible, the package associates every power series with its so-called *analytic expression*. For each power series `s`, created by the command `PowerSeries` as the image of a polynomial `p` (under the natural embedding from $\mathbb{C}[X_1, \ldots, X_n]$ to $\mathbb{C}[[X_1, \ldots, X_n]]$) the polynomial `p` is the analytic expression of `s`. If a power series is defined by the sequence of its homogeneous parts, as illustrated on Fig. 3, the user can optionally specify the *sum* of that series which is then set to its analytic expression. Power series that have an analytic expression are closed under addition, multiplication and inversion. Propagating that information provides the opportunity to speed up some computations and make decisions that could not be made otherwise. For instance, the command `HenselFactorize` needs to decide whether its input polynomial has an invertible leading coefficient; to do it starts by checking whether the analytic expression of that leading coefficient is known and equal to one.

The commands `Display`, `SetDefaultDisplayStyle` and `SetDisplayStyle` control the output format of multivariate power series and UPoPS. Meanwhile, the commands `HomogeneousPart`, `Truncate`, `GetCoefficient`, `Precision`, `Degree`, `MainVariable` access data from a power series or a univariate polynomial over power series, as illustrated by Fig. 4.

The commands `Add`, `Negate`, `Multiply`, `Exponentiate`, `Inverse`, `Divide`, `EvaluateAtOrigin`, and `TaylorShift` perform arithmetic operations on multivariate power series and univariate polynomials over multivariate power series. The functionality of the first six commands can also be accessed using the standard arithmetic operators. As will be discussed in Sects. 4 and 5, the implementation of every arithmetic operation, such as addition, multiplication, inversion builds the resulting power series (sum, product or inverse) "lazily", by creating its generator from the generators of the operands, which are called *ancestors* of the resulting power series.

```
> f := UnivariatePolynomialOverPowerSeries([PowerSeries(x), GeometricSeries(y), PowerSeries(1),
      1/PowerSeries(1 + x + y)], z);
```

$$f := [\text{UnivariatePolynomialOverPowerSeries: } (x) + (1 + y + \ldots) z + (1) z^2 + (1 + \ldots) z^3] \quad (16)$$

```
> p, a := WeierstrassPreparation(f);
```

$$p, a := [\text{UnivariatePolynomialOverPowerSeries: } (1)], [\text{UnivariatePolynomialOverPowerSeries: } (-6) + (11 + x) z + (-6 + x) z^2 + (1) z^3] \quad (17)$$

```
> UpdatePrecision(p, 5);
```

$$[\text{UnivariatePolynomialOverPowerSeries: } (x + x^2 - x y + x^3 - 3 x^2 y + x^4 - 5 x^3 y + 3 x^2 y^2 - 6 x^4 y + 9 x^3 y^2 - x^2 y^3 + \ldots) + (1) z] \quad (18)$$

```
> a;
```

$$[\text{UnivariatePolynomialOverPowerSeries: } (1 + y - x + x y + y^2 + y^3 + x^4 - x^3 y + x^2 y^2 + y^4 + \ldots) + (1 - x + 2 x y - x^3 + x^2 y - 2 x y^2 + \ldots) z + (1 - x - y + x^2 + 2 x y + y^2 + \ldots) z^2] \quad (19)$$

```
> h := p·a;
```

$$h := [\text{UnivariatePolynomialOverPowerSeries: } (x + \ldots) + (1 + y + y^2 + y^3 + \ldots) z + (1 + \ldots) z^2 + (1 - x - y + x^2 + 2 x y + y^2 + \ldots) z^3] \quad (20)$$

```
> ApproximatelyEqual(f, h, 20);
```

$$true \quad (21)$$

Fig. 5. Factoring univariate polynomials using `WeierstrassPreparation`.

```
> f := UnivariatePolynomialOverPowerSeries((z − 1)·(z − 2)·(z − 3) + x·(z^2 + z), z);
```

$$f := [\text{UnivariatePolynomialOverPowerSeries:}\ (-6) + (11 + x)\, z + (-6 + x)\, z^2 + (1)\, z^3] \tag{22}$$

```
> F := HenselFactorize(f);
```

$$F := [[\text{UnivariatePolynomialOverPowerSeries:}\ (-1 + \ldots) + (1)\, z], [\text{UnivariatePolynomialOverPowerSeries:}\ (-2 + \ldots) + (1)\, z], [\text{UnivariatePolynomialOverPowerSeries:}\ (-3 + \ldots) + (1)\, z]] \tag{23}$$

```
> map(UpdatePrecision, F, 5);
```

$$\Big[\Big[\text{UnivariatePolynomialOverPowerSeries:}\ \Big(-1 + x - 3x^2 + \frac{27x^3}{2} - \frac{291x^4}{4} + \frac{3465x^5}{8} + \ldots\Big) + (1)\, z\Big], [\text{UnivariatePolynomialOverPowerSeries:}\ (-2 - 6x - 30x^2 - 402x^3 - 5610x^4 - 93390x^5 + \ldots) + (1)\, z], \Big[\text{UnivariatePolynomialOverPowerSeries:}\ \Big(-3 + 6x + 33x^2 + \frac{777x^3}{2} + \frac{22731x^4}{4} + \frac{743655x^5}{8} + \ldots\Big) + (1)\, z\Big]\Big] \tag{24}$$

```
> h := F[1]·F[2]·F[3] − f;
```

$$h := [\text{UnivariatePolynomialOverPowerSeries:}\ (0 + \ldots) + (0 + \ldots)\, z + (0 + \ldots)\, z^2 + (0)\, z^3] \tag{25}$$

```
> ApproximatelyZero(h, 100);
```

$$\mathit{true} \tag{26}$$

```
> g := UnivariatePolynomialOverPowerSeries(y^2 + x^2 + (y + 1)·z^2 + z^3, z);
```

$$g := [\text{UnivariatePolynomialOverPowerSeries:}\ (x^2 + y^2) + (0)\, z + (1 + y)\, z^2 + (1)\, z^3] \tag{27}$$

```
> G := HenselFactorize(g);
```

$$G := [[\text{UnivariatePolynomialOverPowerSeries:}\ (0 + \ldots) + (0 + \ldots)\, z + (1)\, z^2], [\text{UnivariatePolynomialOverPowerSeries:}\ (1 + \ldots) + (1)\, z]] \tag{28}$$

```
> map(UpdatePrecision, G, 8);
```

$$[[\text{UnivariatePolynomialOverPowerSeries:}\ (x^2 + y^2 - x^2 y - y^3 - x^4 - x^2 y^2 + 4x^4 y + 7x^2 y^3 + 3y^5 + 3x^6 - x^4 y^2 - 10x^2 y^4 - 6y^6 - 21x^6 y - 43x^4 y^3 - 24x^2 y^5 - 2y^7 - 12x^8 + 36x^6 y^2 + 145x^4 y^4 + 135x^2 y^6 + 38y^8 + \ldots) + (-x^2 - y^2 + 2x^2 y + 2y^3 + 2x^4 + x^2 y^2 - y^4 - 10x^4 y - 16x^2 y^3 - 6y^5 - 7x^6 + 9x^4 y^2 + 34x^2 y^4 + 18y^6 + 56x^6 y + 98x^4 y^3 + 34x^2 y^5 - 8y^7 + 30x^8 - 132x^6 y^2 - 436x^4 y^4 - 363x^2 y^6 - 89y^8 + \ldots)\, z + (1)\, z^2], [\text{UnivariatePolynomialOverPowerSeries:}\ (1 + y + x^2 + y^2 - 2x^2 y - 2y^3 - 2x^4 - x^2 y^2 + y^4 + 10x^4 y + 16x^2 y^3 + 6y^5 + 7x^6 - 9x^4 y^2 - 34x^2 y^4 - 18y^6 - 56x^6 y - 98x^4 y^3 - 34x^2 y^5 + 8y^7 - 30x^8 + 132x^6 y^2 + 436x^4 y^4 + 363x^2 y^6 + 89y^8 + \ldots) + (1)\, z]] \tag{29}$$

```
> h := G[1]·G[2];
```

$$h := [\text{UnivariatePolynomialOverPowerSeries:}\ (x^2 + y^2 + \ldots) + (0 + \ldots)\, z + (1 + y + \ldots)\, z^2 + (1)\, z^3] \tag{30}$$

```
> ApproximatelyEqual(g, h, 20);
```

$$\mathit{true} \tag{31}$$

Fig. 6. Factoring univariate polynomials using `HenselFactorize`.

The commands `WeierstrassPreparation` and `HenselFactorize` factorize univariate polynomials over multivariate power series. Thanks to their implementation based on lazy evaluation, each of these factorization commands returns the factors as soon as enough information is discovered for initializing the data structures of the factors; see Figs. 5 and 6.

The precision of each returned factor, that is, the common precision of its coefficients (which are power series) is zero. However the generator (see Sect. 4 for this term) of each coefficient is known and, thus, the computation of more coefficients can be resumed when a higher precision is requested. Such a request can be explicit by calling `UpdatePrecision`, or implicit, when requesting data of a higher precision than has been previously requested through, e.g., `Truncate` or `HomogeneousPart`.

4 Design Principles

In this section we examine several design principles underpinning the implementation of the `MultivariatePowerSeries` library. Foremost is lazy evaluation: an algorithmic technique where the computation of data is postponed until explicitly required (Sect. 4.1). The eventual implementations of these lazy-evaluation algorithms make deliberate efforts to use appropriate MAPLE data structures and built-in functions to optimize performance (Sect. 4.2). Lastly, in support of software quality and integration with existing MAPLE library code, we employ MAPLE's object-oriented mechanisms (Sect. 4.3).

4.1 Lazy Evaluation

Lazy evaluation is an optimization technique most commonly appearing in the study of functional programming languages [9]. The lazy evaluation or "call-by-need" refers to delaying the call to a function until its result is genuinely needed. This is often complemented by storing the result for later look-up.

In the case of power series, consider a bivariate geometric series $f = \sum_{d=0}^{\infty} f_{(d)}$ where $f_{(0)} = 1$, $f_{(1)} = x + y$, $f_{(2)} = x^2 + 2xy + y^2$, ..., $f_{(d)} = (x + y)^d$. One can prove that f converges to $\frac{1}{1-x-y}$. Of course, in practice, it is impossible to store an infinite number of terms on a computer with finite memory. A naïve implementation then suggests storing $f^{(d)}$ for some large and predetermined d. Thus, one can approximate power series as multivariate polynomials. Such an implementation could be called *truncated power series*.

While this representation of power series is easy to implement, it leads to notable restrictions for the study of formal power series. First, one must a priori determine the *precision*, i.e. the particular value of d. Second, in a most naïve implementation, previously-computed homogeneous parts must be recomputed whenever a new, greater precision is required. For example, the polynomial $f^{(d+1)}$ is likely to be constructed "from scratch" despite the polynomial $f^{(d)}$ possibly being already computed. Third, storing and manipulating the polynomial part of a power series up to a degree d needs a large portion of memory. This latter problem is exacerbated when the predetermined precision is not a tight upper bound on the required precision.

To combat the challenges of a truncated power series implementation, we take advantage of lazy evaluation. Every power series is represented by a unique procedure to compute a homogeneous part for a given degree. For example, Listing 1.1 shows such a procedure for the bivariate geometric series which converges to $\frac{1}{1-x-y}$. As we will see, this lazy evaluation design can be paired with an array of polynomials storing the previously computed homogeneous parts.

```
generator := proc(d :: nonnegint)
    return expand((x+y)^d);
end proc;
```

Listing 1.1. A MAPLE implementation of $f_{(d)}$ in $\frac{1}{1-x-y} = \sum_{d=0}^{\infty} f_{(d)}$.

4.2 Maple Data Structures and Built-in Functions

Using an appropriate data structure for encoding and manipulating data is critical for performance, particularly in high-level and interpreted programming languages like Maple. In Maple, modifying an existing list or set—such as by appending, replacing, or deleting an element—leads to the creation of a new list or set, rather than modifying the original one in-place. In contrast, an `Array` is a low-level and mutable data-structure which allows for in-place modification of its elements. These functionalities provide much better performance than lists or sets when the collection is frequently changed or when the elements being modified are themselves large in size. This fact is clear from the overwhelming improvement in performance of our library compared against the existing `PowerSeries` library which uses lists to encode homogeneous parts; see Sect. 6.

Looking more closely at the `Array` data structure, an n-dimensional `Array` is stored as a n-dimensional rectangular block named `RTABLE`. The length of the associated `RTABLE` is $2n+d$ where d is maximum number of elements that may be stored, i.e., the allocation size of the `Array`; see [5, Appendix 1]. For the storage of homogeneous parts of a power series, and the power series coefficients of a UPoPS, we utilize 1-dimensional `Arrays`. Listing 1.2 in the next section shows this as the variables `hpoly` and `upoly`, respectively.

To further improve performance, we make use of low-level built-in functions. Such functions are provided as compiled code within the Maple kernel, and therefore not written in the Maple language. Most notably, instead of using Maple for-loops and the typical `+` and `*` syntaxes for addition and multiplication, respectively, we reduce the cost of summations and multiplications remarkably by taking advantage of built-in Maple functions, `add` and `mul`. These built-in functions, respectively, add or multiply the terms of an entire sequence of expressions together to return a single sum or product. These functions avoid a large number of high-level function calls and reduce memory usage by avoiding copying and re-allocation of data.

4.3 Maple Objects

An often overlooked aspect of Maple is its object-oriented capability. An object allows for variables and procedures operating on that data to be encapsulated together in a single entity. In Maple, a class—the definition of a particular type of object—can be declared by including the option `object` in a module declaration. Evaluating this declaration returns an object of that class. This new object is often a so-called "prototype" object which, when passed to the `Object` routine, returns a new object of the same class. See [5, Chapter 9] for further details on object-oriented programming in Maple.

Our power series and UPoPS types are implemented using these object-oriented features of Maple. The classes for each are named, respectively, `PowerSeriesObject` and `UnivariatePolynomialOverPowerSeriesObject`.

The use of object-oriented programming in MAPLE has two key benefits: (*i*) the organization object-oriented code provides better software quality through modularity and maintainability; and (*ii*) allows for the overloading of built-in functions, thus allowing objects to be integrated with, and used natively by, existing MAPLE library functions.

```
MultivariatePowerSeries := module()
option package;
    local PowerSeriesObject,
            UnivariatePolynomialOverPowerSeriesObject;
    # create a power series:
    export PowerSeries := proc(...)
    # create a UPoPS:
    export UnivariatePolynomialOverPowerSeries := proc(...)
    #Additional procedures to interface these two classes

    module PowerSeriesObject()
    option object;
        local hpoly :: Array,
                precision :: nonnegint,
                generator :: procedure;
        # other members and methods
    end module;

    module UnivariatePolynomialOverPowerSeriesObject()
    option object;
        local upoly :: Array, vname :: name;
        # other members and methods
    end module;
end module;
```

Listing 1.2. An overview of the `MultivariatePowerSeries` package.

The `MultivariatePowerSeries` library contains a package of the same name which groups together those two aforementioned classes along with additional procedures to construct and manipulate objects of those classes. These additional procedures are used to "hide" the object-oriented nature of the library behind simple procedure calls. This keeps the package syntactically and semantically consistent with the general paradigm of MAPLE which does not use object-oriented programming. As an example of such a procedure, `PowerSeries`, as seen in Fig. 2 (Sect. 3), handles various different types of input parameters to correctly construct a `PowerSeriesObject` object through delegation to the correct class method.

Listing 1.2 shows the declaration of our two classes and the `MultivariatePowerSeries` package. The latter is created by using `option package` in a `module` declaration; see [5, Chapter 8]. The implementation of these two classes is further discussed in Sect. 5.

5 Implementation of MultivariatePowerSeries

The `MultivariatePowerSeries` package provides a collection of procedures which form simple wrappers for the methods of the aforementioned classes, `PowerSeriesObject` and `UnivariatePolynomialOverPowerSeriesObject`.

These classes, respectively, define the data structures and algebraic functionalities for creating and manipulating multivariate power series and univariate polynomials over power series. This section discusses those data structures as well as the implementation of basic arithmetic, Weierstrass Preparation Theorem, and factorization via Hensel's lemma, all following a lazy evaluation scheme.

5.1 PowerSeriesObject

The `PowerSeriesObject` class provides basic arithmetic operations, like addition, multiplication, inversion, and evaluation, for multivariate power series, all utilizing lazy evaluation techniques. Let $f \in \mathbb{K}[[X_1, \ldots, X_n]]$ be a non-zero multivariate power series defined as $f = \sum_{d=0}^{\infty} f_{(d)}$. f is encoded as an object of type `PowerSeriesObject`, containing the following attributes.

First, the power series *generator* is the procedure to compute $f_{(d)}$, the d-th homogeneous part of f, for $d \in \mathbb{N}$. Second, the *precision* is a non-negative integer encoding the maximum degree of the homogeneous parts which have so far been computed. Third, the 1-dimensional array storing the previously computed homogeneous parts of f, denoted as `hpoly` in Listing 1.2.

To create a power series object this class provides a variety of constructors. Power series objects may be created from polynomials, algebraic numbers, UPoPS objects, or procedures defining the generator of the power series.

Every arithmetic operation returns a lazily-constructed power series object by creating its generator from the generators of the operands, but without explicitly computing any homogeneous parts of the result. Thus, this is a lazy power series, so that, the homogeneous parts of the result are computed when truly needed. Once homogeneous parts are eventually computed, they are stored in the array `hpoly`. An important aspect of this organization is that the generator of the resulting power series becomes implicitly connected to the generators of the operands; the latter are thus called the *ancestors* of the former. Note that the ancestors are merely stored as references, not copies, thus saving time and memory resources.

Moreover, the addition and multiplication operations are not only binary operations (operations taking two parameters), but are m-ary operations. For multiplication, a sequence of power series $f_1, \ldots, f_m \in \mathbb{K}[[X_1, \ldots, X_n]]$ may be passed to the multiplication algorithm to produce the product $f_1 \cdot f_2 \cdots f_m$ via lazy evaluation. Similarly, addition may take the sequence $f_1, \ldots, f_m$ to return the sum $f_1 + f_2 + \cdots + f_m$. Further, addition may also take as a parameter an optional sequence of polynomial coefficients $c_1, \ldots, c_m \in \mathbb{K}[X_1, \ldots, X_n]$ to return the sum $c_1 f_1 + \cdots + c_m f_m$ constructed lazily.

A key part to the efficiency of lazy evaluation is to not re-compute any data. We have already seen that the `hpoly` array stores previously computed homogeneous parts for a `PowerSeriesObject` object. What is missing is to ensure that the array is accessed where possible rather than calling the generator function. Moreover, one must avoid directly accessing that array for homogeneous parts which are not yet computed. We thus provide the function `HomogeneousPart`(f, d), demonstrated in Listing 1.3, to handle both of these cases. This function returns the d-th homogeneous part of the power series f; if d is greater than the *precision* (`f:-precision`), then this method iteratively calls the *generator* to update `hpoly` and `precision`, otherwise it simply returns the previously computed homogeneous part. From here on we use `hpart` as shorthand for the `HomogeneousPart` function.

```
export HomogeneousPart ::static := proc(f, d :: nonnegint)
    if d > f:-precision then
        f:-hpoly(d+1) := 0; # resize the hpoly array
        for local i from f:-precision + 1 to d do
            f:-hpoly[i] := f:-generator[i];
        end do;
        f:-precision := d;
    end if;
    return f:-hpoly[d];
end proc;
```

Listing 1.3. A simplified version of the `HomogeneousPart` function in `PowerSeriesObject`.

Listing 1.4 shows a simplified implementation of `Divide` that computes the quotient of two power series objects $f, g \in \mathbb{K}[\![X_1, \ldots, X_n]\!]$. In particular, notice the creation of the local procedure `gen` for the generator of the quotient. Note that `EXPAND` is a local macro defined in `MultivariatePowerSeries` to efficiently perform expansion and normalization supporting *algebraic* inputs.

```
export Divide ::static := proc(f, g)
    if hpart(g,0)=0 then
        error "invalid input: not invertible";
    end if;
    local h := Array(0..0,EXPAND(hpart(f,0)/hpart(g,0)));
    local gen := proc(d :: nonnegint)
        local s := hpart(f,d);
        s -= add(EXPAND(hpart(g,i)*hpart(f,d-i)),i=1..d);
        return EXPAND(s/hpart(g,0));
    end proc;
    return Object(PowerSeriesObject,h,0,gen);
end proc;
```

Listing 1.4. A simplified version of the division method in `PowerSeriesObject`.

5.2 `UnivariatePolynomialOverPowerSeriesObject`

The `UnivariatePolynomialOverPowerSeriesObject` class is implemented as a simple dense univariate polynomial with the simple and obvious implementations of associated arithmetic (see, e.g., [18, Chapter 2]). The arithmetic operations are achieved directly from coefficient arithmetic, that is, `PowerSeriesObject` arithmetic. Since the latter is implemented using lazy evaluation techniques, UPoPS arithmetic is inherently and automatically lazy.

For example, the addition of two UPoPS objects $f = \sum_{i=0}^{k} a_i X_{n+1}^i$ and $g = \sum_{i=0}^{k} b_i X_{n+1}^i$ in $\mathbb{K}[[X_1, \ldots, X_n]][X_{n+1}]$ is the summation $(a_i + b_i)X_{n+1}^i$ for all $0 \le i \le k$, where a_i, b_i are `PowerSeriesObject` objects. Other basic arithmetic operations behave similarly. However, there are important operations on UPoPS which are not as straightforward. In the following we explain our implementation of Weierstrass Preparation Theorem, Taylor shift, and factorization via Hensel's lemma for UPoPS, all of which follow lazy evaluation techniques.

Weierstrass Preparation. Let $f, p, \alpha \in \mathbb{K}[[X_1, \ldots, X_n]][X_{n+1}]$ be such that they satisfy the conditions of Theorem 1 and such that $f = \sum_{i=0}^{d+m} a_i X_{n+1}^i$, $p = {X_{n+1}}^d + \sum_{i=0}^{d-1} b_i X_{n+1}^i$, and $\alpha = \sum_{i=0}^{m} c_i X_{n+1}^i$. Equating coefficients in $f = p\alpha$ we derive the two following systems of equations:

$$\begin{cases} a_0 &= b_0 c_0 \\ a_1 &= b_0 c_1 + b_1 c_0 \\ \vdots & \\ a_{d-1} &= b_0 c_{d-1} + b_1 c_{d-2} + \cdots + b_{d-2} c_1 + b_{d-1} c_0 \end{cases} \tag{1}$$

$$\begin{cases} a_d &= b_0 c_d + b_1 c_{d-1} + \cdots + b_{d-1} c_1 + c_0 \\ \vdots & \\ a_{d+m-1} &= b_{d-1} c_m + c_{m-1} \\ a_{d+m} &= c_m \end{cases} \tag{2}$$

To solve these systems we proceed by solving them modulo successive powers of $\mathcal{M}$, following the proof of Theorem 1 in [6]. Notice that solving modulo successive powers of $\mathcal{M}$ is precisely the same as computing homogeneous parts of increasing degree. Thus, this follows our lazy evaluation scheme perfectly. The power series $b_0, \ldots, b_{d-1}$ are generated by Eqs. (1) and $c_0, \ldots, c_m$ by Eqs. (2).

Consider that $b_0, \ldots, b_{d-1}, c_0, \ldots, c_m$ are known modulo $\mathcal{M}^r$ while $a_0, \ldots, a_{d-1}$ are known modulo $\mathcal{M}^{r+1}$; this latter fact is simple since f is the input to Weierstrass Preparation and is fully known. From the first equation in (1), b_0 can be computed modulo $\mathcal{M}^{r+1}$ since $b_0 \in \mathcal{M}$, c_0 is known modulo $\mathcal{M}^r$, and a_0 is known $\mathcal{M}^{r+1}$. Then, the equation $a_1 = b_0 c_1 + b_1 c_0$, that is, $a_1 - b_0 c_1 = b_1 c_0$ can be solved for b_1 modulo $\mathcal{M}^{r+1}$ since, again, $b_1 \in \mathcal{M}$ and the other terms are sufficiently known. We compute all $b_2, \ldots, b_{d-1}$ modulo $\mathcal{M}^{r+1}$ with the same argument. After determining $b_0, \ldots, b_{d-1}$ modulo $\mathcal{M}^{r+1}$, we can compute $c_m, c_{m-1}, \ldots, c_0$ modulo $\mathcal{M}^{r+1}$ from Eqs. (2) with simple power series multiplication and subtraction, working iteratively, in a bottom up fashion. For example, $c_{m-1} = a_{d+m-1} - b_{d-1} c_m$.

As yet, we have not explicitly seen how the coefficients of p and α will be updated. The key idea is that to update a single power series coefficient of p or α requires simultaneously updating all coefficients of p and α. Thus, all the generators of $b_0, \ldots, b_{d-1}, c_0, \ldots, c_m$ simply call a single "Weierstrass update" function to update all power series simultaneously using Equations (1) and (2). Algorithm 1 shows this Weierstrass update function.

Algorithm 1. WEIERSTRASSUPDATE$(p, \alpha, \mathcal{F}, r)$
Given $p = X_{n+1}^d + \sum_{i=0}^{d-1} b_i X_{n+1}^i$, $\alpha = \sum_{i=0}^{m} c_i X_{n+1}^i$, $r \in \mathbb{N}$, and $\mathcal{F} = \{F_i \mid F_i = a_i - \sum_{j=0}^{i-1} b_j c_{i-j}, 0 \le i < d\}$ are all known modulo $\mathcal{M}^r$, returns $b_0, \ldots, b_{d-1}, c_0, \ldots, c_m$ modulo $\mathcal{M}^{r+1}$.

\# update $b_0, \ldots, b_{d-1}$ modulo $\mathcal{M}^{r+1}$
for i **from** 0 **to** $d-1$ **do**
  $s :=$ **add(seq(hpart**$(b_i, r-k)$ $\cdot$ **hpart**(c_0, k), $k = 1 \,..\, r-1$**))**;
  hpart$(b_i, r) :=$ **(hpart**$(F_i, r) - s$**)/hpart**$(c_0, 0)$;
\# ensure $c_0, \ldots, c_m$ are updated modulo $\mathcal{M}^{r+1}$
for i **from** 0 **to** m **do**
  hpart(c_i, r);

In order to update the coefficients of p, we frequently need to compute $a_i - \sum_{j=0}^{i-1} b_j c_{i-j}$ for $0 \le i < d$. To optimize this operation, we a priori create helper power series as the set $\mathcal{F} = \{F_i \mid F_i = a_i - \sum_{j=0}^{i-1} b_j c_{i-j}, i = 0, \ldots, d-1\}$. The power series F_i, following power series arithmetic with lazy evaluation, allows for the efficient computation of homogeneous parts of increasing degree of $a_i - \sum_{j=0}^{i-1} b_j c_{i-j}$. This set $\mathcal{F}$ is passed to the Weierstrass update function to optimize the overall computation.

Finally, the Weierstrass preparation must be initialized before continuing with Weierstrass updates. Namely, the degree of p and the initial values of p and α modulo $\mathcal{M}$ must first be computed. The degree of p, namely d, is set to be the smallest integer i such that a_i is a unit. If $d = 0$, then $p = 1$ and $\alpha = f$, otherwise, m equals the difference between the degree of f and d, and we initialize $b_i = 0$ for $0 \le i < d$. Then, $c_m, \ldots, c_0$ are initialized using power series arithmetic following Eqs. (2). Lastly, the set $\mathcal{F}$ is initialized.

Taylor Shift. This operation takes a UPoPS object $f \in \mathbb{K}[\![X_1, \ldots, X_n]\!][X_{n+1}]$ and performs the translation $X_{n+1} \to X_{n+1} + c$, i.e. $f(X_{n+1} + c)$, for some $c \in \mathbb{K}$. In our implementation, c can be a `numeric` or `algebraic` MAPLE type with the purpose of being used efficiently in factorization via Hensel's Lemma.

Assume $f = \sum_{i=0}^{k} a_i X_{n+1}^i$ is a UPoPS in $\mathbb{K}[\![X_1, \ldots, X_n]\!][X_{n+1}]$ and $c \in \mathbb{K}$. As the `PowerSeriesObject` objects $a_0, \ldots, a_k$ are lazily evaluated power series, we want to also make Taylor shift a lazy operation. Thus, we need to create a generator for the power series coefficients of $f(X_{n+1} + c)$. Let $\mathfrak{T} = (t_{i,j})$ be the lower triangular matrix of the coefficients of ${X_{n+1}}^j$ in the binomial expansion $(X_{n+1} + c)^i$, for $0 \le i \le k$, and $0 \le j \le i$. Let $(b_0, \ldots, b_k)$ be the list of coefficients of $f(X_{n+1} + c)$ in $\mathbb{K}[\![X_1, \ldots, X_n]\!]$. Then, it is easy to prove that for every $0 \le i \le k$, b_i is the inner product of the i-th sub-diagonal of $\mathfrak{T}$ with the lower $k+1-i$ elements of the vector $(a_0, \ldots, a_k)$. This inner product can be computed efficiently by taking advantage of the m-ary addition operation described for the `PowerSeriesObject` (see Sect. 5.1). Since this operation returns a lazily-constructed power series, this precisely defines the lazy construction of the power series $b_0, \ldots, b_k$, thus making Taylor shift a lazy operation.

Factorization via Hensel's Lemma. Hensel's lemma for factorizing univariate polynomials over power series was reviewed in Theorem 2, where $\mathbb{K}$ is algebraically closed and $f \in \mathbb{K}[\![X_1, \ldots, X_n]\!][X_{n+1}]$ is a UPoPS object. Following the ideas of [6], we compute the factors of f in a lazy fashion. Algorithm 2 proceeds through iterative applications of Taylor shift and Weierstrass Preparation Theorem in order to create one factor of f at a time. Those factors are actually computed through lazy evaluation thanks to the lazy behavior of the procedures WEIERSTRASSPREPARATION and TAYLORSHIFT. This Algorithms thus computes and updates the factors modulo the successive powers $\mathcal{M}, \mathcal{M}^2, \mathcal{M}^3, \ldots$ of the maximal ideal $\mathcal{M}$.

Algorithm 2. HENSELFACTORIZE(f)
Given $f = \sum_{i=0}^{k} a_i {X_{n+1}}^i \in \mathbb{K}[\![X_1, \ldots, X_n]\!][X_{n+1}]$, returns a list of factors $\{f_1, \ldots, f_r\}$ so that $f = a_k \cdot f_1 \cdots f_r$, and satisfies Theorem 2.

---

if $a_k \notin \mathcal{M}$ **then**
    $f^* := \frac{1}{a_k} \cdot f$;
else
    error "a_k must be a unit."
$\bar{f} :=$ EVALUATEATORIGIN(f^*);
$c_1, \ldots, c_r :=$ ROOTS($\bar{f}, X_{n+1}$);
for i **from** 1 **to** r **do**
    $g :=$ TAYLORSHIFT(f^*, c_i);
    $p, \alpha :=$ WEIERSTRASSPREPARATION(g);
    $f_i :=$ TAYLORSHIFT($p, -c_i$);
    $f^* :=$ TAYLORSHIFT($\alpha, -c_i$);
return $\{f_1, \ldots, f_r\}$;

Note that the generation of the factors $f_1, \ldots, f_r$ takes place after factorizing $\bar{f} \in \mathbb{K}[X_{n+1}]$. Recall that $\bar{f}$ is obtained by evaluating each X_i to 0 for $1 \leq i \leq n$. This is called EVALUATEATORIGIN in our implementation. To efficiently factor $\bar{f}$, we take advantage of the package `SolveTools` [15], which allows us to compute the splitting field of $\bar{f}$ (which, in practice, is a polynomial with coefficients in some algebraic extension of $\mathbb{Q}$) and factorize $\bar{f}$ into linear factors.

Let $c_1, \ldots, c_r$ be the distinct roots of $\bar{f}$ and $k_1, \ldots, k_r$ their respective multiplicities. To describe one iteration of Algorithm 2, let f^* be the current polynomial to factorize. For a root c_i of $\bar{f}$, and thus f^*, we perform a Taylor shift to obtain $g = f^*(X_{n+1} + c_i)$. Then, we apply Weierstrass preparation on g to obtain p and α where p is monic and of degree k_i. Again, by using Taylor Shift, we apply the reverse shift to p to obtain $f_i = p(X_{n+1} - c_i)$, a factor of f, and

$f^* = \alpha(X_{n+1} - c_i)$, for the next iteration. As mentioned above, since both Taylor shift and Weierstrass preparation are implemented using lazy evaluation, our factorization via Hensel's lemma is inherently lazy.

6 Experimentation

We compare the performance of the `MultivariatePowerSeries` package, denoted `MPS`, with the previous MAPLE implementation of multivariate power series, the `PowerSeries` package, denoted `RCPS`, and the recent implementation of power series via lazy evaluation in the BPAS library. This latter implementation is written in the C language on top of efficient sparse multivariate arithmetic; see [4,6]. It has already been shown in [6] that the implementation in BPAS is orders of magnitude faster than the `PowerSeries` package, MAPLE's `mtaylor` command, and the multivariate power series available in SAGEMATH. As we will see, our implementation performs comparably to that of BPAS.

Throughout this section, we collect our benchmarks on a machine running Ubuntu 18.04.4, MAPLE 2020, and BPAS (ver. 1.652), with an Intel Xeon X5650 processor running at 2.67 GHz, with 12 × 4 GB DDR3 memory at 1.33 GHz.

Figures 7, 8, and 9, respectively, show the performance of division and multiplication algorithms to compute $\frac{1}{f}$ and $\frac{1}{f} \cdot f$ for power series $f_1 = 1 + X_1 + X_2$, $f_2 = 1 + X_1 + X_2 + X_3$, and $f_3 = 2 + \frac{1}{3}(X_1 + X_2)$. It can be seen that `MPS` power series division is 9×, 2100×, and 3× faster than the previous MAPLE implementation for f_1, f_2, and f_3 respectively. The speed-ups for multiplication are significantly higher. Moreover, `MPS` results are comparable with the C implementation of similar algorithms in BPAS. Figure 10 then highlights the efficiency of m-ary addition (see Sect. 5.1), compared to iterative applications of binary addition. Recall that m-ary addition is exploited in the Weierstrass preparation algorithm.

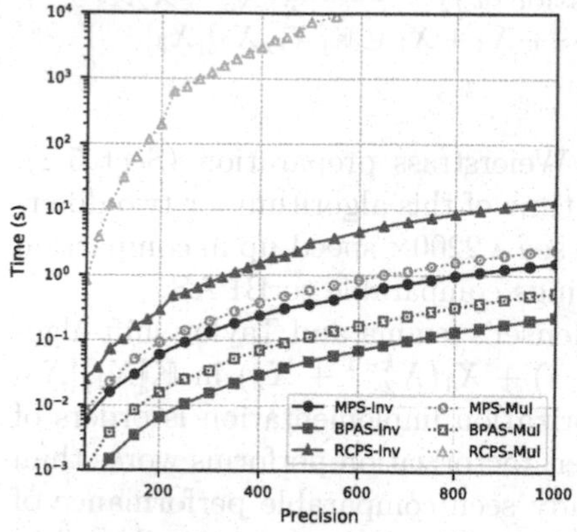

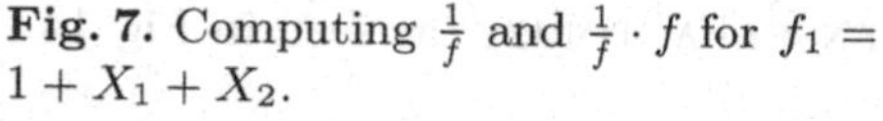
Fig. 7. Computing $\frac{1}{f}$ and $\frac{1}{f} \cdot f$ for $f_1 = 1 + X_1 + X_2$.

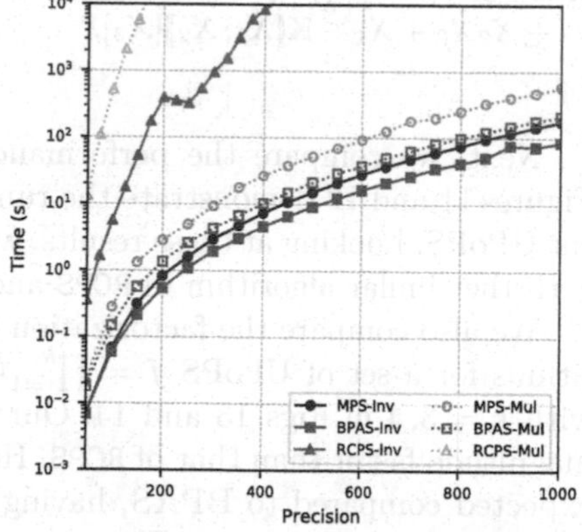

Fig. 8. Computing $\frac{1}{f}$ and $\frac{1}{f} \cdot f$ for $f_2 = 1 + X_1 + X_2 + X_3$.

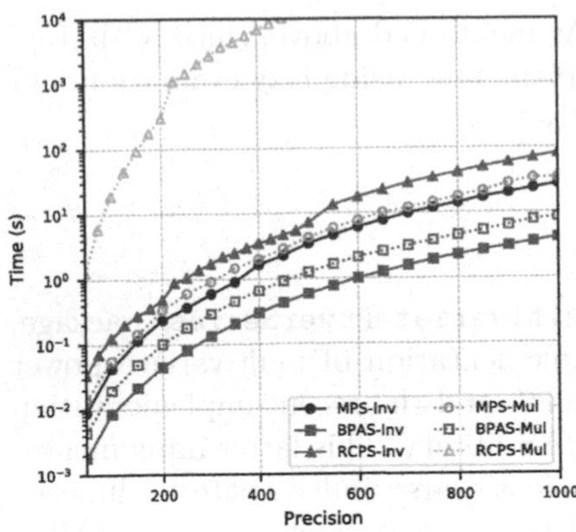

Fig. 9. Computing $\frac{1}{f}$ and $\frac{1}{f} \cdot f$ for $f_3 = 2 + \frac{1}{3}(X_1 + X_2)$.

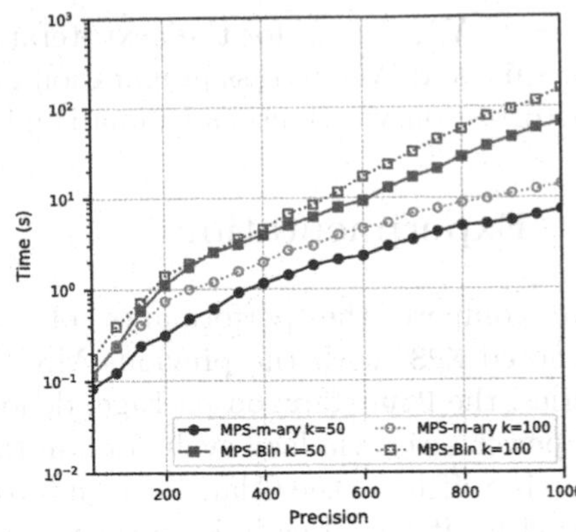

Fig. 10. Computing $f = \sum_{i=1}^{k} \frac{1}{1-x-y}$ using m-ary and binary addition.

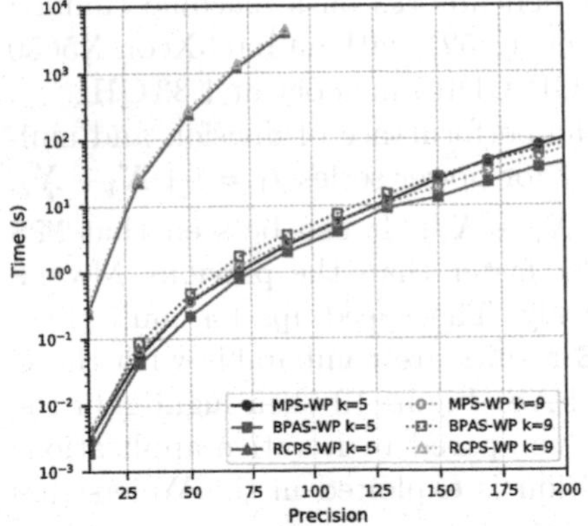

Fig. 11. Computing Weierstrass preparation of $f_1 = \frac{1}{1+X_1+X_2} {X_3}^k + {X_3}^{k-1} + \cdots + X_2X_3 + X_1 \in \mathbb{K}[\![X_1, X_2]\!][X_3]$.

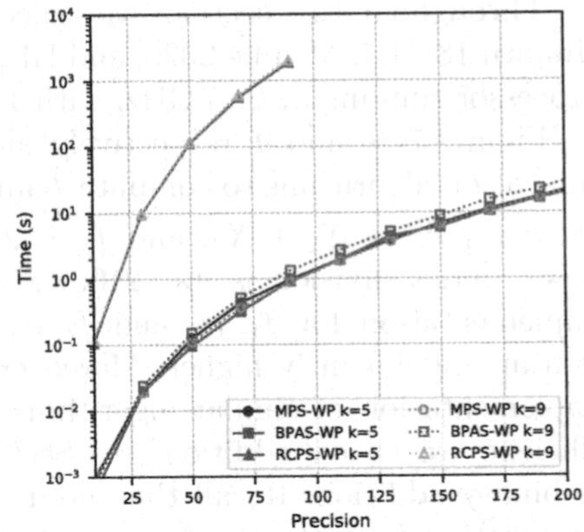

Fig. 12. Computing Weierstrass preparation of $f_2 = \frac{1}{1+X_1+X_2} {X_3}^k + X_2{X_3}^{k-1} + \cdots + X_3 + X_1 \in \mathbb{K}[\![X_1, X_2]\!][X_3]$.

Next, we compare the performance of Weierstrass preparation (Sect. 5.2). Figures 11 and 12 demonstrate the running time of this algorithm for two different UPoPS. Looking at these results, we can see a 2200× speed-up in comparison with the similar algorithm in RCPS and timings comparable to BPAS.

We also compare the factorization via Hensel's lemma and Taylor shift algorithms for a set of UPoPS $f = \prod_{i=1}^{k}(X_2 - i) + X_1(X_2^{k-1} + X_2)$ in $\mathbb{K}[\![X_1]\!][X_2]$ with $k = 3, 4$ in Figs. 13 and 14. Our factorization implementation is orders of magnitude faster than that of RCPS. However, factorization performs worse than expected compared to BPAS, having already seen comparable performance of Weierstrass preparation in Figs. 11 and 12. This difference can be attributed to Taylor shift, the other core operation of HENSELFACTORIZE, as seen in Fig. 14.

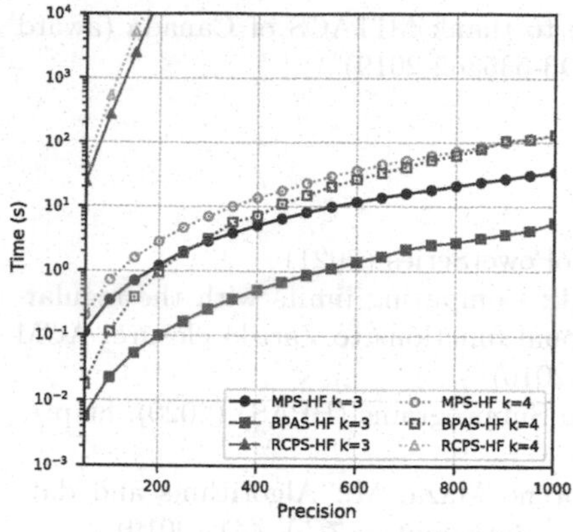

Fig. 13. Computing HENSELFACTORIZE(f) for $f = \prod_{i=1}^{k}(X_2 - i) + X_1({X_2}^{k-1} + X_2)$.

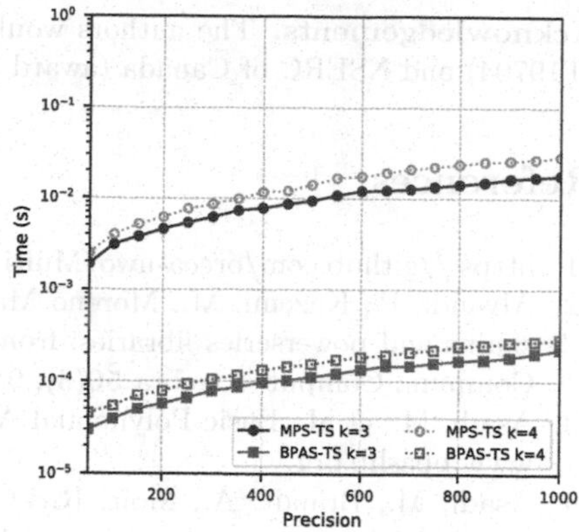

Fig. 14. Computing TAYLORSHIFT$(f, 1)$ for $f = \prod_{i=1}^{k}(X_2 - i) + X_1({X_2}^{k-1} + X_2)$.

The implementation in `MPS` is slower than the same procedure in BPAS by several order of magnitude. This, in turn, can be attributed to using MAPLE matrix arithmetic, rather than the direct manipulation of C-arrays as in BPAS, within the Taylor shift algorithm.

7 Conclusions and Future Work

Throughout this work we have discussed the object-oriented design and implementation of power series and univariate polynomials over power series following lazy evaluation techniques. Basic arithmetic operations for both are examined as well as Weierstrass Preparation Theorem, Taylor shift, and factorization via Hensel's lemma for univariate polynomials over power series. Our implementation in MAPLE is orders of magnitude faster than the existing multivariate power series implementation in the `PowerSeries` package of the `RegularChains` library. Moreover, our implementation is comparable with the C implementation of power series and univariate polynomials over power series in BPAS.

Further work is needed to extend lazy evaluation techniques to more sophisticated algorithms. For example, a general Extended Hensel Construction (EHC) [13], and the Abhyankar-Jung Theorem [14]. As a consequence, it is possible to re-implement the EHC algorithm found in `RegularChains` using this library. Further, as MAPLE supports multithreading, it is possible to apply parallel processing to our algorithms. In particular, the computation of UPoPS coefficients in Weierstrass preparation is embarrassingly parallel. Meanwhile, the successive application of Weierstrass preparation and Taylor shift in HENSELFACTORIZE present an opportunity for *pipelining*. Both should be exploited in to achieve even further performance improvements.

Acknowledgements. The authors would like to thank MITACS of Canada (award IT19704) and NSERC of Canada (award CGSD3-535362-2019).

References

1. https://github.com/orcca-uwo/MultivariatePowerSeries (2021)
2. Alvandi, P., Kazemi, M., Moreno Maza, M.: Computing limits with the regularchains and powerseries libraries: from rational functions to Zariski closure. ACM Commun. Comput. Algebra **50**(3), 93–96 (2016)
3. Asadi, M., et al.: Basic Polynomial Algebra Subprograms (BPAS) (2020). http://www.bpaslib.org
4. Asadi, M., Brandt, A., Moir, R.H.C., Moreno Maza, M.: Algorithms and data structures for sparse polynomial arithmetic. Mathematics **7**(5), 441 (2019)
5. Bernardin, L., et al.: Maple Programming Guide. Maplesoft, a division of Waterloo Maple Inc. (1996–2020)
6. Brandt, A., Kazemi, M., Moreno-Maza, M.: Power series arithmetic with the BPAS library. In: Boulier, F., England, M., Sadykov, T.M., Vorozhtsov, E.V. (eds.) CASC 2020. LNCS, vol. 12291, pp. 108–128. Springer, Cham (2020). https://doi.org/10.1007/978-3-030-60026-6_7
7. Burge, W.H., Watt, S.M.: Infinite structures in scratchpad II. In: Davenport, J.H. (ed.) EUROCAL 1987. LNCS, vol. 378, pp. 138–148. Springer, Heidelberg (1989). https://doi.org/10.1007/3-540-51517-8_103
8. Fischer, G.: Plane algebraic curves. Am. Mathe. Soc. (2001)
9. Harper, R.: Practical foundations for programming languages: Lazy Evaluation, 2nd edn., pp. 323–332. Cambridge University Press (2016)
10. Karczmarczuk, J.: Generating power of lazy semantics. Theor. Comput. Sci. **187**(1–2), 203–219 (1997)
11. Maplesoft, a division of Waterloo Maple Inc.: Maple (2020). www.maplesoft.com/
12. Monagan, M., Vrbik, P.: Lazy and forgetful polynomial arithmetic and applications. In: Gerdt, V.P., Mayr, E.W., Vorozhtsov, E.V. (eds.) CASC 2009. LNCS, vol. 5743, pp. 226–239. Springer, Heidelberg (2009). https://doi.org/10.1007/978-3-642-04103-7_20
13. Moreno Maza, M.: Polynomials over power series and their applications to limit computations (Tutorial at Computer Algebra in Scientific Computing (CASC) (2018). www.csd.uwo.ca/~mmorenom/Publications/Polynomials_over_power_series_and_their_applications_lecture.PDF
14. Parusiński, A., Rond, G.: The Abhyankar-Jung theorem. J. Algebra **365**, 29–41 (2012)
15. The Maple Developers: SolveTools package in Maple 2020, maplesoft, a division of Waterloo Maple Inc. www.maplesoft.com/support/help/Maple/view.aspx?path=SolveTools
16. The Sage Developers: SageMath, the Sage Mathematics Software System (2020). www.sagemath.org
17. van der Hoeven, J.: Relax, but don't be too lazy. J. Symbolic Comput. **34**(6), 479–542 (2002)
18. von zur Gathen, J., Gerhard, J.: Modern Computer Algebra, 3 edn. Cambridge University Press (2013)
19. Wolfram Research Inc.: Mathematica (2020). www.wolfram.com/mathematica

Bernoulli's Problem $x^y = y^x$ and Maple

T. J. Ayoub, K. Basu, and D. J. Jeffrey(✉)

Ontario Research Centre for Computer Algebra, The University of Western Ontario, London, ON N6A 5B7, Canada
djeffrey@uwo.ca

Abstract. We study the problem of $x^y = y^x$, first proposed by Daniel Bernoulli in 1728. We present MAPLE's parametric solution and a solution using the Lambert W function. This leads us to consider an implementation in Maple of new simplifications of the Lambert W function. The method uses a mixture of exact and floating-point computation.

Keywords: Lambert W function · Simplification · Algorithms

1 Bernoulli's Problem

On 29 June 1728, Daniel Bernoulli wrote to Christian Goldbach. Bernoulli was working at the new St Petersburg Academy of Sciences[1], and Goldbach had recently left the same Academy for Moscow. In his letter, Bernoulli considered the equation

$$x^y = y^x \ . \tag{1}$$

Obviously, it has the trivial solution $x = y$, but Bernoulli wrote that he had found the non-trivial solution $x = 2, y = 4$ (and, of course, $y = 2, x = 4$), and further that he had proved that there are no other integer solutions [1].

Goldbach wrote to Bernoulli on 31 January 1729 giving a solution to (1) in parametric form. Goldbach's expressions can be obtained from MAPLE's `solve` command, using the syntax (see MAPLE help for `solve/parametrized`)

```
> solve(x^y=y^x,[x(t),y(t)])
```

$$\left[\left[x = e^{-\frac{\ln\left(\frac{1}{t}\right)}{t-1}}, y = t\,e^{-\frac{\ln\left(\frac{1}{t}\right)}{t-1}}\right]\right] \tag{2}$$

```
> simplify(%)
```

$$\left[\left[x = \left(\frac{1}{t}\right)^{-\frac{1}{t-1}}, y = t\left(\frac{1}{t}\right)^{-\frac{1}{t-1}}\right]\right] \tag{3}$$

[1] The Saint Petersburg Academy of Sciences opened its doors in 1725, shortly after the death of its founder, Peter the Great.

R. M. Corless et al. (Eds.): MC 2020, CCIS 1414, pp. 67–76, 2021.
https://doi.org/10.1007/978-3-030-81698-8_5

Solving using a parameter is a useful MAPLE command that deserves to be more widely known. Goldbach would certainly have been assuming $t > 0$, and so the solution becomes

```
> simplify(%) assuming t > 0
```

$$\left[\left[x = t^{\frac{1}{t-1}}, y = t^{\frac{t}{t-1}}\right]\right] \tag{4}$$

Leonard Euler was in St Petersburg at that time, also working at the Academy, and 20 years later in 1748 (having moved to Berlin) he included this problem and its solution in his famous textbook on analysis [3].

A more conventional use of solve is to ask for y as a function of x. MAPLE returns the expression (using the `alias` command to abbreviate `LambertW` to W)

```
> alias(W = LambertW):
> solve(x^y = y^x,y)
```

$$y = -\frac{x}{\ln x} W\left(-\frac{\ln x}{x}\right) . \tag{5}$$

We recall that the Lambert W function obeys

$$W_k(z)\exp\left(W_k(z)\right) = z , \tag{6}$$

where $k \in \mathbb{Z}$ is the branch label [2]. Its real values are plotted in Fig. 1.

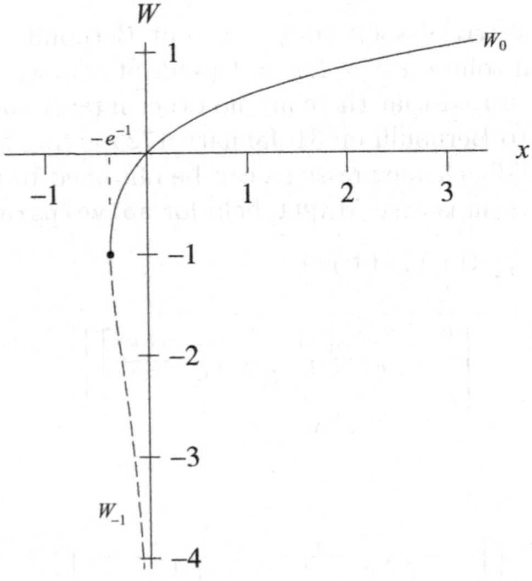

Fig. 1. The real values of W. The solid line is the principal branch $k = 0$ and the dashed line is the $k = -1$ branch.

1.1 Comparing Solutions

Having the two solutions (4) and (5) to Bernoulli's problem invites a comparison between them. Substituting various values for t into (4) produces some typical solutions as (x, y) pairs; see Table 1.

Table 1. Goldbach's solution evaluated for some particular values of t. Bernoulli's own solution is the first entry.

t	x	y
2	2	4
3	$\sqrt{3}$	$\sqrt{27}$
7	$7^{\frac{1}{6}}$	$7^{\frac{7}{6}}$
3/2	9/4	27/8
4/3	64/27	256/81
5/4	625/256	3125/1024

Substituting one member of an (x, y) pair into (5) should produce the other. We can start with Bernoulli's solution $(2, 4)$. Substituting $x = 2$ in (5), we obtain

$$y = -\frac{2}{\ln 2} W\left(-\frac{\ln 2}{2}\right) .$$

MAPLE can simplify this, and returns 2, the trivial solution, because MAPLE assumes, by default, branch $k = 0$. To obtain the non-trivial solution, we must ask for $k = -1$.

$$y = -\frac{2}{\ln 2} W_{-1}\left(-\frac{\ln 2}{2}\right) = 4 , \tag{7}$$

which simplification MAPLE can also do. What happens if we start with $x = 4$?

$$y = -\frac{4}{\ln 4} W\left(-\frac{\ln 4}{4}\right) = -\frac{2}{\ln 2} W\left(-\frac{\ln 2}{2}\right) = 2 .$$

Notice that whether one starts with $x = 2$ or $x = 4$, one arrives at the same expression in terms of W. This underlines the importance of the branch index in the analysis: *different data* can lead to the *same argument* for W, meaning that *only* the branch index can differentiate cases.

Now we try another entry from Table 1 and substitute $x = \sqrt{3}$ in (5) to obtain

$$y = -\frac{\sqrt{3}}{\ln \sqrt{3}} W_k\left(-\frac{\ln \sqrt{3}}{\sqrt{3}}\right) = -\frac{2\sqrt{3}}{\ln 3} W_k\left(-\frac{1}{6}\sqrt{3}\ln 3\right) .$$

MAPLE cannot simplify this expression, for either branch, but we can approximate it using `evalf`.

$$y = -\frac{2\sqrt{3}}{\ln 3} W_0 \left(-\frac{1}{6}\sqrt{3} \ln 3 \right) \approx 1.732050812 \ .$$

The MAPLE `identify` command[2] gives this as $\sqrt{3}$, and it also identifies the $k = -1$ branch as giving $\sqrt{27}$. We are thus led to conjecture the simplifications

$$W_0 \left(-\frac{1}{6}\sqrt{3}\ \ln 3 \right) = -\tfrac{1}{2} \ln 3 \ , \tag{8}$$

$$W_{-1} \left(-\frac{1}{6}\sqrt{3}\ \ln 3 \right) = -\tfrac{3}{2} \ln 3 \ , \tag{9}$$

and we would like MAPLE to incorporate these and similar simplifications into its library.

2 A Class of Simplifications

Guided by the above observations, we develop a class of simplifications for MAPLE's implementation of Lambert W. We first generalize Bernoulli's problem to a problem which has appeared, in various disguises, in a number of mathematical contests. Solve for x

$$x^a b^x = c \ , \tag{10}$$

with the parameters $\{a, b, c\}$ chosen to give x a suitably simple form. The equation has the real solution

$$x = \frac{a}{\ln b} W_k \left(\frac{1}{a} c^{1/a} \ln b \right) \ , \tag{11}$$

with Bernoulli's problem corresponding to $a = y$, $b = 1/y$, and $c = 1$. Our ambition, then, is to detect cases in which (11) can be simplified. Having observed (7), (8) and (9), we aim to decide whether a rational number r exists such that for some branch k

$$W_k \left(\frac{1}{a} c^{1/a} \ln b \right) = r \ln b \ . \tag{12}$$

We limit the domains of the parameters so that the problem is tractable. We restrict a to be an integer n, and assume $c, b, r \in \mathbb{Q}$. This is sufficient to cover solutions to (1) obtained by substituting rational values of t into Goldbach's expressions.

We focus, therefore, on deciding whether r exists, and on calculating it. A first strategy could be to return to the `identify` command we used successfully above. By rewriting (12) as

$$r = \frac{W_k \left(\frac{1}{n} c^{1/n} \ln b \right)}{\ln b} \ ,$$

[2] Another interesting, under-appreciated, feature.

we can evaluate the right side to a floating-point number, and then use `identify` to find the rational number. This conceptually simple approach has, however, practical difficulties. First, the correct identification of a rational number is sensitive to the precision of the floating-point evaluation. With multi-digit fractions, `identify` can struggle, and MAPLE users can be relied upon to pose, sooner or later[3], problems containing large numbers. For example, consider

```
> evalf(1234567/678912)
```

$$1.818449225$$

```
> identify(%)
```

$$\frac{\sqrt{3}}{5} + \frac{5\,\zeta(5)}{8} + \frac{3\ln(3)}{4}$$

```
> evalf[15](1234567/678912)
```

$$1.81844922464178$$

```
> identify(%)
```

$$1.81844922464178$$

```
> evalf[20](1234567/678912)
```

$$1.8184492246417797888$$

```
> identify(%)
```

$$\frac{1234567}{678912}$$

A second problem concerns rounding errors during evaluation [4], which can mean that the number given to `identify` is not accurate.

```
> evalf( W( -1,-189/256*ln(4/3)*sqrt(3))/ln(4/3) );
```

$$-3.500000258$$

```
> identify(%)
```

$$-3.500000258$$

```
> evalf[15]( W( -1,-189/256*ln(4/3)*sqrt(3))/ln(4/3) );
```

$$-3.50000000000461$$

[3] and usually sooner!

```
> identify(%)
```

$$-3.50000000000461$$

The human will exclaim "But it's obvious", and ignore the erroneous digits.

$$W_{-1}\left(-\frac{189}{256}\sqrt{3}\,\ln\frac{4}{3}\right) = -\frac{7}{2}\ln\frac{4}{3}\,. \tag{13}$$

Again, as the numbers grow larger identification will become more difficult. It is possible to modify the approach so that we need only identify an integer. As a step to an improved algorithm, we consider a numerical example: solve for x

$$x^3\left(\frac{9}{4}\right)^x = \frac{3}{16}\,. \tag{14}$$

The solution is $x = \frac{1}{2}$, and this works because $(9/4)^x = (9/4)^{1/2} = 3/2$. Note that we chose $9/4$ because it appears in Table 1, and rational numbers containing pure powers must be expected in simplification problems.

Now we can see that since (12) is equivalent to solving the problem

$$x^n b^x = c\,, \tag{15}$$

and we want all quantities in this equation to be rational, then b^x will have to be rational. Thus we must be able to extract, if necessary, a root from b. Suppose b can be written $b = B^p$, with $B \in \mathbb{Q}$ and $p \in \mathbb{Z}$, with p maximal. Then we can write

$$x^n b^x = x^n B^{px} = c\,,$$

and all terms will be rational, provided $px \in \mathbb{Z}$. This means we can change our search from finding a rational x to finding an integer px.

The first step in our algorithm, therefore, must be to find p. This can be done using MAPLE's `iperfpow` function, which takes a positive integer m, and computes integers q and j, such that $m = q^j$. We extend its functionality in two ways. First, `iperfpow` does not find the maximal power. Thus, `iperfpow(256)` returns 16^2, whereas we need $256 = 2^8$. Secondly, `iperfpow` accepts only integers, whereas we need rational arguments. The calculation is then

$$X = px = \frac{pn}{\ln b} W_k\left(\frac{1}{n}\, c^{1/n} \ln b\right) = \frac{n}{\ln B} W_k\left(\frac{1}{n}\, c^{1/n}\, p \ln B\right)\,. \tag{16}$$

The right-hand side is evaluated in floating point to give a *candidate* integer for X. We return to the numerical example in (13). We convert it to the standard form just given:

$$\begin{aligned} X &= \frac{2}{\ln(4/3)} W_{-1}\left(\frac{1}{2}\left(\frac{107163}{16384}\right)^{1/2} \ln\frac{4}{3}\right) \\ &\approx -7.00000000000922\,. \end{aligned} \tag{17}$$

From this we test the candidate $X = -7$ using the exact computation

$$(-7)^2\left(\frac{4}{3}\right)^{-7} = \frac{107163}{16384}.$$

With this exact verification, we can accept the simplification proposed in (13).

2.1 Limitations to Simplification

A careful study of the concept and the definition of simplification has been made in [5]. Pragmatically, we are following one particular idea from that paper, which is an ordered list of sub-expressions. For the present paper, simplification (12) consists of expressing an instance of W in terms of simpler functions appearing lower in the ordered list.

In the previous section, we restricted ourselves to branches $k = 0, -1$ and real values. The question naturally arises whether complex simplifications are also possible. We briefly consider the option here, but are not tempted to generalize our algorithm. Substituting negative values in (4) gives the pairs $(i, -i)$ and $(2^{-1/3}(-1)^{1/3}, -2^{2/3}(-1)^{1/3})$. For the first pair, we substitute $x = i$ into (5), and get

$$y = -\frac{i}{\ln i} W_k\left(-\frac{\ln i}{i}\right) = -\frac{2W_k(-\pi/2)}{\pi}.$$

The branch behaviour is reversed from previous examples. Now it is the $k = 0$ branch that gives the non-trivial solution,

$$y = -\frac{2W_0(-\pi/2)}{\pi} = -i,$$

which current MAPLE gets, and it is the $k = -1$ branch that gives the trivial solution, which MAPLE also gets.

For the second pair, many people will immediately simplify $(-1)^{1/3} = -1$. In MAPLE, this is obtained using the `surd` command: `surd(-1,3)=-1`. In a MAPLE worksheet, `surd(x,3)` is printed as $\sqrt[3]{x}$, as opposed to $x^{1/3}$ for the principal value. Thus, they would arrive at the pair

$$(x, y) = \left(\frac{-2^{2/3}}{2}, 2^{2/3}\right).$$

Sadly, this tidy solution does not work.

$$x^y = \left(-2^{2/3}/2\right)^{2^{2/3}} ; y^x = \left(2^{2/3}\right)^{-2^{2/3}/2}.$$

Among the infinite complex values of x^y, we compute

$$-0.69297,\ 0.5911 \pm 0.3617i,\ -0.3154 \pm 0.6170i, \ldots$$

and for y^x, we compute

$$1.44306,\ 0.3913 \pm 1.389i,\ -1.231 \pm 0.7532i, \ldots$$

Not even Carathéodory could discover an equality there.

In contrast, the principal value of $(-1)^{1/3} = \frac{1}{2}(1+i\sqrt{3})$ works a treat. Thus

$$\left(\frac{2^{2/3}}{4}(1+i\sqrt{3})\right)^{2^{2/3}(-1-i\sqrt{3})/2} = 4.4145302 - 2.48981i$$

$$\left(\frac{2^{2/3}}{2}(-1-i\sqrt{3})\right)^{2^{2/3}(1+i\sqrt{3})/4} = 4.4145302 - 2.48981i$$

An interesting new feature with the W solution, is that now $k = +1$ gives the non-trivial solution:

$$-\frac{(-1)^{1/3} W_1\left(\ln\left((-1)^{1/3}2^{-1/3}\right)(-1)^{\frac{2}{3}}\sqrt[3]{2}\right)}{2^{1/3}\ln\left((-1)^{1/3}2^{-1/3}\right)} = -(-1)^{1/3}2^{\frac{2}{3}} = -0.7937 - 1.3747i$$

and it is the principal branch that gives the trivial solution.

Simplifications such as these can be expanded indefinitely, and a software system has to draw the line somewhere, and we draw it before complex values.

3 Implementation

The implementation of the simplification proceeds in 3 stages. First, we process the argument to W to write it in the standard form seen in (16). Then there is the computation of the simplification, and finally a new output form that guides users in their requests.

3.1 The Argument

MAPLE's simplification routines use standard forms which are different from the form needed for simplification. For example, suppose the standard form of the argument for simplification is $\frac{1}{3}(3/2)^{1/3}\ln(9/8)$. The following MAPLE session shows how this might be altered.

```
> simplify( (1/3)*(3/2)^(1/3)*ln(9/8) )
```

$$-\frac{\sqrt[3]{3}\,2^{\frac{2}{3}}\left(-2\ln(3)+3\ln(2)\right)}{6}$$

The analysis of the argument assumes a form

$$Arg = a^b c^d \ldots (r\ln s + u\ln v\ldots)\,.$$

This is converted to a list using `convert(Arg, 'list', ''*'')`, and then we use `selectremove(hastype, ArgList, specfunc(rational, ln))` to separate the powers and the logarithms. After that, it is straightforward MAPLE programming to restore the form $(1/n)C^{1/n}\ln B$. If the argument cannot be converted, then simplification fails.

3.2 Testing Simplification

Once the argument is established, X is computed using (16), and the candidate integer obtained using `intX:=round(X)`. If the magnitude of `intX` is large, then the working precision of `evalf` is increased (to mitigate rounding errors) and the evaluation repeated. We check the fractional part of X to ensure the candidate is plausible (`abs(X-intX)` $< 10^{-5}$) and then conduct the test (15). An alternative is to check the definition (6) directly.

3.3 Option Parametric

A long-standing challenge for computer algebra systems has been the question of how to help users obtain useful results. A user who is not familiar with Lambert W, or not used to thinking in terms of branches, might miss a simplification. For example, the principal branch $k = 0$ of $W\left(-\frac{189}{256}\ln(4/3)\sqrt{3}\right)$ does not simplify, but $k = -1$ does. A simple call to simplify, therefore, might not produce a result useful to the user.

Our implementation includes a parametric option, in which both $k = 0$ and $k = -1$ are checked, and the user is given results for all branches. This option is activated by asking for simplification leaving the branch k unassigned. For the current example, this mode produces

$$W_k\left(-\tfrac{189}{256}\ln(4/3)\sqrt{3}\right) = \begin{cases} W_0\left(-\frac{189}{256}\ln(4/3)\sqrt{3}\right)\,, & \text{if } k = 0\,, \\ -\frac{7}{2}\ln\frac{4}{3}\,, & \text{if } k = -1\,. \end{cases}$$

This is to be understood to mean that for $k = 0$ there is no simpler expression for the number being represented than the value of W_0, which can be obtained as a floating-point approximation using `evalf`. For $k = -1$, on the other hand, there is a simpler expression. The user is then in possession of useful information with which to proceed.

4 Concluding Remarks

To repeat a comment made above, the simplifications explored here do not exhaust the list of all possible simplifications. We have concentrated here on covering those that are likely to arise in practice. Any person who wishes to expose gaps in the coverage of simplifications can take inspiration and encouragement from Richardson's famous theorem that zero-recognition is undecidable [6]. For the Bernoulli-Goldbach problem, we considered only real, rational values of the parameter. We have shown in Sect. 2.1 problems that escape the present approach.

In addition to practicality, there are also æsthetic considerations. The equations here have already become complicated. To cover more detailed cases would require even greater length and complexity in the function arguments. We feel we have chosen a good density of coverage without undue complexity.

The basic principal of the simplifications is the fact that W is the inverse of the function $V = ze^z$. A common source of confusion in this regard is the apparent equation $W(xe^x) = x$. This is the result of the standard mathematical education, which teaches that a function f and its inverse f^{-1} obey $f^{-1}(f(x)) = f\left(f^{-1}(x)\right) = x$. For W, however, this is wrong. The correct statement is

$$W_k(xe^x) = y \text{ , \textbf{where} } ye^y = xe^x \ .$$

The choice of y is determined by the branch k as much as by the value of x.

We anticipate that the simplifications described here will be available for all users in MAPLE 2021.

References

1. Sved, M.: On the rational solutions of $x^y = y^x$. Math. Mag. **63**(1), 30–31 (1990)
2. Corless, R.M., Gonnet, G.H., Hare, D.E.G., Jeffrey, D.J., Knuth, D.E.: On the Lambert W Function. Adv. Comput. Math. **5**, 329–359 (1996)
3. Euler, L.: Introductio in Analysin Infinitorum, vol. 2, chap. 21 (English translation by Ian Bruce). http://www.17centurymaths.com/contents/introductiontoanalysisvol1.htm
4. Corless, R.M., Fillion, N.: A Graduate Introduction to Numerical Methods. Springer, New York (2013). https://doi.org/10.1007/978-1-4614-8453-0
5. Carette, J.: Understanding expression simplification. In: Gutierrez, J. (ed.) Proceedings of ISSAC 2004, pp. 72–79. ACM Press, New York (2004)
6. Richardson, D.: Some unsolvable problems involving elementary functions of a real variable. J. Symb. Logic **33**, 514–520 (1968)

Student Satisfaction Determinants in Hybrid Learning Environments Based on MAPLE

Lisa Binkowski(✉), Marcel Dux, and Tilo Wendler(✉)

HTW Berlin (University of Applied Sciences Berlin), Berlin, Germany
{lisa.fischer,marcel.dux,vp.lehre}@htw-berlin.de

Abstract. In Germany mathematics has unfortunately become a subject of fear and remains one of the main reasons for dropping out of university. Therefore, universities need to close the knowledge gap between the mathematical knowledge learned at school and the mathematical knowledge needed in university studies.

To meet the requirements of students at the beginning of their studies, a hybrid course had been implemented. The combination of online lectures to motivate and address more difficult topics, online MAPLE exercises and tests to allow flexible learning attracted many of them. This hybrid type of implementation also meets the current expectations of students by avoiding old-fashioned teaching methods in a highly flexible world, especially from the pandemic perspective.

MAPLE incorporated in the Möbius courseware was used as part of courses to educate new students in mathematics. Lectures, online tutorials, and MAPLE-based online exercises are preceded by a digital placement test, so students are enabled to reflect on their level of knowledge beforehand.

We aim to investigate how intuitive the MAPLE-based learning environment is and to determine the success factors and best practices of a digital bridge course. Please be aware that we present our experiences using Möbius and MAPLE.

Keywords: MAPLE in education · Pre-courses in mathematics · Digital teaching

1 Introduction

The HTW is a university of applied science (HTW Berlin) in Berlin, Germany. Our university has 14,000 Students and any year or more specifically any winter term approximately 2,000 students join us in our over 70 bachelor programmes.

In 2020 we tackled the challenges of the COVID-19 crisis by using Möbius in a hybrid environment [1]. Hybrid typically means on-campus plus online. But our setting was different since we could not welcome hundreds of new students in math courses on campus. So, we decided to combine online instructor-led training with self-paced learning in Möbius.

The next subsections explain the necessities of such pre-courses, why there exists a knowledge gap in mathematics when started studying at university and the challenges regarding pre-courses.

R. M. Corless et al. (Eds.): MC 2020, CCIS 1414, pp. 77–88, 2021.
https://doi.org/10.1007/978-3-030-81698-8_6

1.1 Necessities of Pre-courses in Mathematics

Mathematics is required in almost all faculties of the HTW Berlin. Be it the economists who learn to understand cost, profit and revenue functions, or the engineers who are starting to develop models based on sine and cosine functions, for example. In all cases, mathematics represents the underlying language. The correct use of this language enables a better understanding of complex issues in various fields.

In Germany mathematics has unfortunately become a subject of fear and remains one of the main reasons for dropping out of university [2]. Therefore, universities need to close the knowledge gap between the mathematical knowledge learned at school and the mathematical knowledge required in university studies [3].

By offering a specialized course for approximately 1,000 first-year students, the university wants to help students to be better prepared to tackle the upcoming challenges. To meet the requirements of students at the beginning of their studies, a hybrid course had been implemented. The combination of on-campus lectures to motivate and address more difficult topics, online MAPLE exercises and tests to allow flexible learning attracted many of them. This hybrid type of implementation also meets the current expectations of students by avoiding old-fashioned teaching methods in a highly flexible world, especially from the pandemic perspective.

1.2 Knowledge Gap and Challenges in Mathematics

As described before, there is a gap between the mathematical knowledge learned at school and the knowledge required to successfully complete a study at university. The "cooperation Schule und Hochschule" (cosh) identified the mathematical topics students should be able to understand before starting a study including mathematics [3]. This work was also used by us to structure and create content for our bridge courses.

The knowledge gap differs with respect to the university entrance qualification. In Germany, there are several ways to obtain this entrance qualification, e.g., with high school graduation, on the second education path or through work experience. These different possibilities to start a Bachelor with comes along with different student's level of expertise in mathematics which is an additional challenge when organizing such a learning program.

Furthermore, we observe that students are trained to reproduce basic concepts in mathematics, but we sense a lack of deeper understanding, which we think should be practiced by case studies. That leads to the challenge for new team members in our university that need help to extend and apply their math expertise in such learning materials.

2 Digital Bridge Course Mathematics

2.1 General Course Structure

Since the mathematical requirements are different regarding the study programs, two courses have been offered. One is aimed at students of Economics and the other at students of Engineering & Computer Sciences. The courses only differ in content and not in the structure.

Figure 1 Bridge course structure illustrates the course structure. The dashed frames correspond to the self-study materials offered in the cloud-based learning platform Möbius and the dashed-dotted frames to the digital lectures and tutorials.

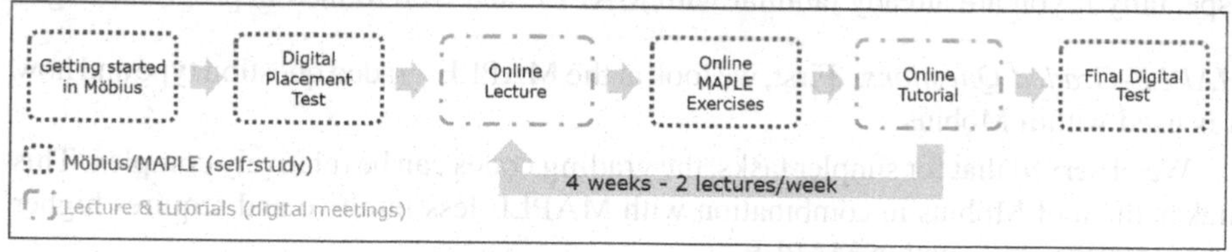

Fig. 1. Bridge course structure

In the beginning of the course students are asked to go through the tool's introduction "Getting started in Möbius", so they know how to enter mathematical expressions within that particular tool. Then the "Digital Placement Test" should be taken to give the students a first impression on their abilities. After the submission of the "Digital Placement Test" students get direct feedback on the solutions of the tasks and can decide which lectures or topics of the pre-course they want to attend. The "Online Lecture", "Online MAPLE Exercises" and "Online Tutorial" will take place over four weeks including two online lectures and two online tutorials per week. To enable students to assess their learning progress, a final test is offered at the end of the course. The self-study material will be available to all registered students for one semester.

The course structure including the course schedule is prepared in the Learning Management System (LMS) Moodle [4]. All further communication is also handled via Moodle.

2.2 Self-study Materials Using Möbius Courseware and MAPLE

This section introduces Möbius Courseware, explains Möbius and MAPLE from an instructor's point of view, including the different question types and related challenges in programming the assignments, and provides information about the possibility of combining Möbius with an LMS.

Möbius Courseware. Möbius is a learning platform with the focus on use in and creating STEM courses [1]. This learning platform allows you to create mathematical content based on the Computer Algebra System MAPLE [5], gives direct feedback on the question solution, makes it possible to randomize question parameters and provides different types of questions, i.e., matching, clickable image, graph sketching, multiple choice, gradeable Math Apps, MAPLE-graded and others.

As presented, Möbius does not only consist of the computer algebra system MAPLE but has much more possibilities to prepare learning materials. However, we focus exclusively on those aspects of Möbius that use MAPLE directly.

Möbius and MAPLE from Teaching Perspective. Since we are interested in how MAPLE is used in education, we concentrate on the question types such as MAPLE-graded, gradeable Math Apps and randomized questions.

In our opinion, it is also important to point out the difficulties we have observed in using Möbius and MAPLE, therefore it will be part of the following paragraphs. We would like to emphasize that this tool generally has great potential and good features, especially if you are already familiar with MAPLE and well trained in programming.

MAPLE-Graded Questions. First, we look at the MAPLE-graded question type and how it is used within Möbius.

We observed that for simpler tasks, the grading codes can be relatively complex. This makes the tool Möbius in combination with MAPLE less intuitive and requires higher programming skills within MAPLE.

To illustrate this problem, consider the following mathematical question:

Apply the distributive law, i.e., exclude the greatest common multiple and calculate without using a calculator:

$$5 \cdot \mathrm{a} + 5 \cdot \mathrm{b} = \text{solution}(5 \cdot (a + b)).$$

Grading Code:

```
resp:= "$RESPONSE":
if StringTools[Search]("5*(", resp) > 0 or
StringTools[Search](")*5",resp) >0
then
evalb(($ANSWER)-(parse(resp))=0)
else 0 end if;
```

The task is short and simple, and the grading code is more complex compared to the task complexity. If we use the standard grading code,

```
evalb(($ANSWER)-($RESPONSE)=0);
```

and type in the response area the question itself, i.e. $5 \cdot \mathrm{a} + 5 \cdot \mathrm{b}$, then this will be scored as correct. Therefore, the upper grading code, searching the string for certain symbols, needs to be used.

This issue was also observed for simplifying fractions. If a student is asked to simplify a fraction the standard grading code will simplify the solution given by the student automatically and scores the response as correct. Here again, a more complex grading code is required.

Gradeable Math Apps. Gradeable Math Apps allow you to include interactive MAPLE documents. That feature makes questions much more interactive and allows the students to find the answer by themselves (Fig. 2). As Fig. 2 shows, we ask which parameters of a parabola influence the function and how. With this question type, students can figure out the solution on their own by varying the parameters using the sliders. This makes the online exercises more interesting and varied.

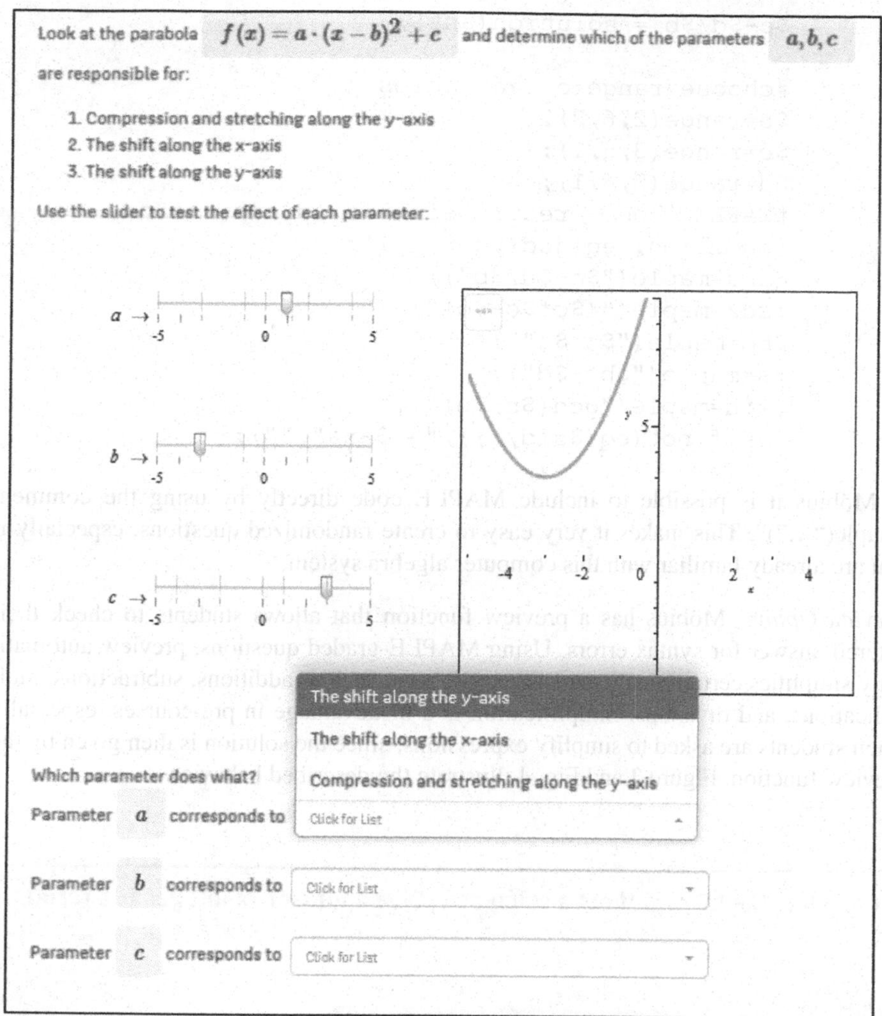

Fig. 2. Example Math Apps in Möbius

Randomized Questions. Actually, randomized questions can be used in all the above-mentioned question types, so it is more a feature of the above-mentioned question types than a question type itself. It allows to create several questions at once and gives students the possibility to repeat the question with different parameters.

Given is the following example, e.g., subtraction of fractions:

```
$c-$d/$b = solution(ans);

#choose range of variables
$b=range(2,6,2);
$c=range(3,5,1);
$d=range(3,6,1);
#MAPLE code - return simplified expression
condition: eq(gcd($b,$d),1);
$ans=maple("$c-$d/$b");
$sd2=maple("($c*$b-$d)";
$bc=maple("$b*$c");
$s=maple("$bc-$d");
$sfd=maple("gcd($s,$b)");
$r=if(not(eq($sfd,1)),"= $ans","");
```

In Möbius it is possible to include MAPLE code directly by using the comment "maple("…")". This makes it very easy to create randomized questions, especially if you are already familiar with this computer algebra system.

Preview Option. Möbius has a preview function that allows students to check their entered answer for syntax errors. Using MAPLE-graded questions, preview automatically simplifies certain mathematical expressions, such as additions, subtractions, multiplications, and divisions. Simplification is a disadvantage in pre-courses, especially when students are asked to simplify expressions, since the solution is then given by the preview function. Figure 3 and Fig. 4 illustrate the described behavior.

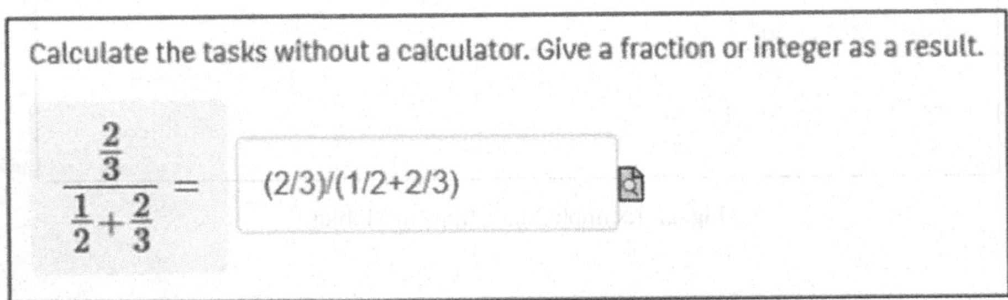

Fig. 3. Example on preview – entering response.

Fig. 4. Example on preview – automatic simplification of the response.

To circumvent this observation, the Cascading Style Sheets (CSS) can be adjusted to hide the preview function. Then, students do not receive any direct help for their entered answer and thus none for the syntax used. Depending on the complexity of the response this can be a disadvantage (Fig. 5).

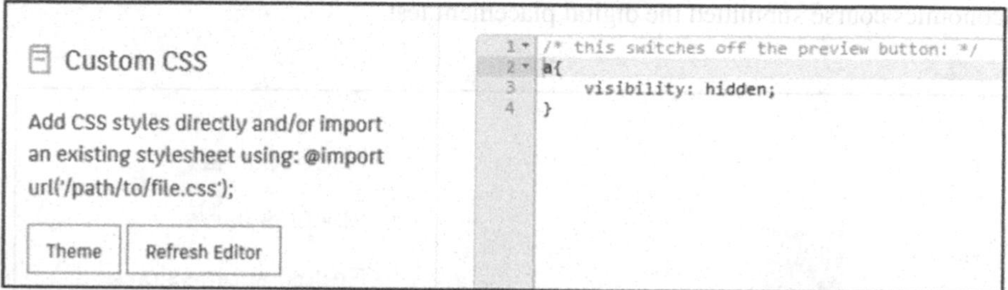

Fig. 5. Hide preview – Custom CSS.

Overall Features and Policies. Once all content has been created, you can design tasks or lessons by arranging the created questions in a specific order, e.g., by difficulty level (easy to difficult). What we call "Online MAPLE exercises" (see Fig. 1) is called assignments in Möbius. In assignments you can adjust the maximum allowed attempts, the time limit and more. Our online exercises can be repeated an unlimited number of times and have no time limit, as they are intended to be used to practice mathematics. It is also possible to design exams, e.g., by inserting time limits, just to name a few possibilities of the tool.

Using Möbius in a Learning Management System (LMS). In general, it is possible to combine different LMS with Möbius **by using the Learning Tools Interoperability (LTI) specification.** This allows students to use the material written in Möbius directly from the university's LMS without the need for additional accounts and registrations. In Germany, the most common LMS is Moodle [4], where an integration of Möbius is theoretically possible.

However, due to the high level of data protection guidelines in Germany, it is not allowed to use Möbius directly from the LMS Moodle, which leads to a participation hurdle and can reduce the number of participants.

By the way, if you combine Möbius with your LMS, the scores are automatically written to the LMS, so all data can be retrieved directly from the LMS.

2.3 Student's Experience

By means of the placement test we want to show how many students have worked at Möbius and explain what the reasons for the differences in the course registrations and the processing of the learning material could be. In addition, we look at the score distribution.

Figure 6 shows how many students are enrolled in Möbius, how many students use Möbius and how many of them have submitted the digital placement test (in short: pretest). You can see that almost 46.15% (252 out of 546) of the students in the Engineering and Computer Science course and 36.30% (110 out of 303) of the students enrolled in the Economics course used Möbius. 75.4% (190 out of 252) of the Möbius users of the Engineering & Computer Sciences and 70.91% (78 out of 110) of the users of the Economics course submitted the digital placement test.

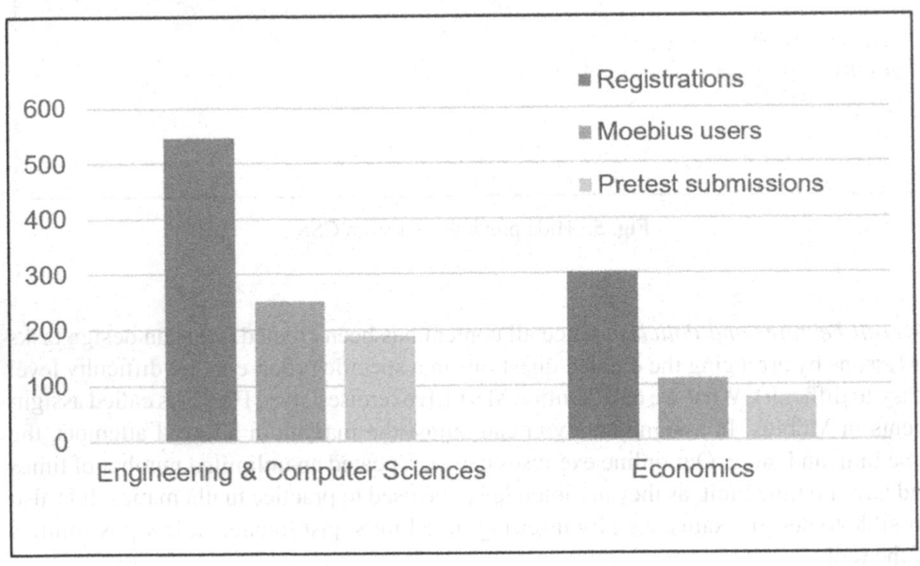

Fig. 6. Number of students enrolled in Möbius, number of students who have used Möbius and number of students who have submitted the digital placemen test.

As mentioned in previous paragraph, the first decrease in participants can be explained by the participation hurdle due to German data protection guidelines. The students were not allowed to use Möbius directly from the LMS and thus had to register separately in Möbius. The second dropout, i.e., the reduction in the submission of the pretest, could be due to the feedback we received regarding usability, i.e., the syntax for entering mathematical expressions. Another reason may be also the normal drop-out rate that university courses have to deal with.

Looking at the syntax of Möbius, it is especially confusing for the German students that they have to enter the English writing of decimal numbers (commas and periods). To help students enter answers, we wrote a note on how to enter the response within the task description, which was helpful on reducing syntax problems.

Figure 7 shows the score distribution from incorrectly to correctly answered questions of the digital placement test for each bridge course. The y-axis gives the percentage of the students reaching a certain percentage of correct answers in the placement test. Due to the good results in the test, i.e., a right-sided distribution, we are considering increasing the difficulty of the test for the next semester. On the other hand, it is important to define your target group, to adjust the level of difficulty accordingly. Is it of interest to encourage prospective students to study, or is it of interest to make it clear that the course of study chosen by new students includes higher level mathematics?

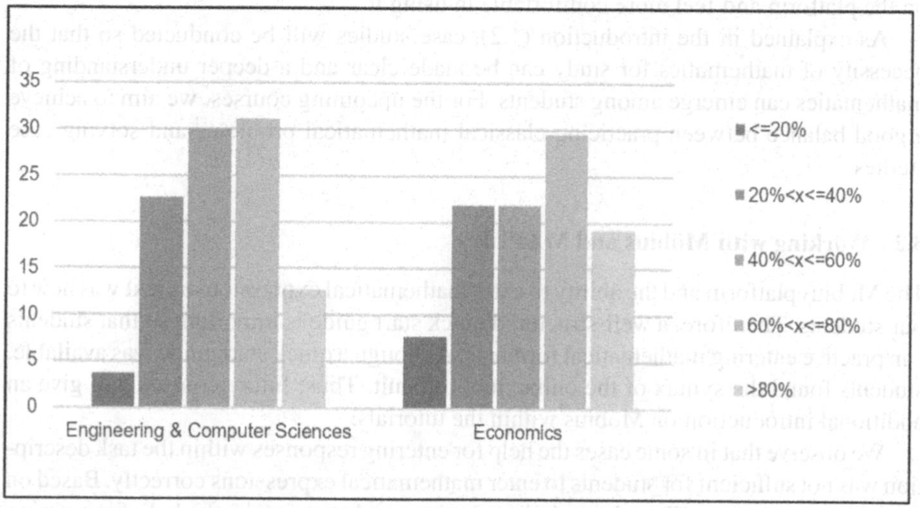

Fig. 7. Score distribution of the digital placement test.

Student's feedback was positive in terms of self-directed learning, the opportunity for immediate feedback to assess their skills, and the opportunity to ask questions about the online tests and exercises in the tutorial.

There are students who are already parents, students who need to work while studying, and other challenges that prevent students from attending their classes. Considering the heterogeneity of the student background at our university, it is of high interest to offer digital learning material to all students so that they can educate themselves in a flexible time.

Our experience shows, when digital learning material is offered, it is beneficial to give direct feedback that goes far beyond right or wrong. A solution path to the set task helps to quickly acquire missing knowledge, as one's own gaps can be better identified.

The combination of the opportunity for direct feedback on digital assignments with the personal attention of tutors provides the student with full support in achieving their learning progress.

3 Acquired Experiences and Knowledge

3.1 Course Organization

Online learning software is not only new to first-year students, but also to many lecturers and tutors at our university. Onboarding of the responsible persons running a course is necessary to increase the willingness to use such software and to enable lecturers and tutors to help students with tool-related questions. An internal support was offered by us organizers to help when there were problems with Möbius, which was very useful.

Previously, our focus was more on empowering students to use the tool rather than empowering the teachers involved. This needs to be changed. Hence, for future courses, we will extend the onboarding such that persons involved in teaching get more insights on the platform and feel more comfortable in using it.

As explained in the introduction (1.2), case studies will be conducted so that the necessity of mathematics for study can be made clear and a deeper understanding of mathematics can emerge among students. For the upcoming courses, we aim to achieve a good balance between practicing classical mathematical problems and solving case studies.

3.2 Working with Möbius and MAPLE

The Möbius platform and the ability to enter mathematical expressions as text was new to our students. Therefore, a well-structured quick start guide is important so that students can practice entering mathematical formulas. Although a quick start guide was available, students found the syntax of the online tool difficult. Thus, future courses will give an additional introduction on Möbius within the tutorials.

We observe that in some cases the help for entering responses within the task description was not sufficient for students to enter mathematical expressions correctly. Based on this observation, we will go through the questions and try to make the help for entering answers clearer and extend this help to questions that previously had none.

Structuring the repository is another challenge. The more tasks there are in the repository, the more difficult it becomes to find the question you are looking for. To avoid this problem, the repository was structured according to the different areas of mathematics, which works relatively well.

The registration hurdle due to data protection guidelines in Germany is a disadvantage that can reduce participation in online course offerings. Therefore, before using a particular online tool, it is highly recommended to be aware of what data protection guidelines the considered country has. However, since participation in the course is voluntary and there is no grading, it is therefore not a disadvantage that the course grades are not linked to the LMS. Möbius itself has a record function about the participation and the grading of the course participants. This allows lecturers to see the progress of the students and makes it possible to repeat certain topics due to test results.

The platform is in many ways intuitive to use. Nevertheless, it is important to have very good programming skills and experience in MAPLE to master this tool. Especially simple school math tasks require special care and make programming unnecessarily cumbersome. The documentation for the software is in progress and is currently being improved, which we consider positive.

There is an idea portal where users can report ideas to the system's developers, who provide direct feedback on the ideas and whether they will be considered for future implementation.

4 Summary

This article discusses a lot of different aspects in digital teaching. A good course structure balancing the self-study experiences and supervised learning in the form of online lectures and online tutorials is required to attract students. The pandemic makes the attraction and motivation of especially new students even more challenging. Therefore, the positive feedback on the course coming from the students allows to implement the same course structure in upcoming semesters. How to use Möbius from student's and lecturer's and tutor's perspective requires more attention. An additional introduction to tutors and lecturers will be offered, so they are enabled to explain how to use the tool to the students during their online sessions.

The data protection guidelines of Germany do not allow us to overcome the participation hurdle. Therefore, it is of interest if the cloud-based software Möbius can be used on a university's server or if the university requires other tools to overcome this problem. Furthermore, the sometimes long training period in Möbius or MAPLE represents a hurdle to use on the part of the teachers.

Nonetheless, the features of Möbius, especially when combined with MAPLE, allow for the creation of many different digital learning materials, providing flexibility for the individual learning scenarios that different universities require.

Acknowledgment. We would like to thank Birgitta Kinscher and Jessica Töpfer for their good cooperation, without whose help in organizing the course would not have been possible.

For the mathematical contents we would like to thank especially Thoralf Chrobok and Akiko Kato. Thanks to the help of the student assistants Annalena Schubert, Maike Günther and Shantanu Ladhwe we were able to quickly create many tasks in our new digital tool and build the structure of the courses as desired.

The exchange with the lecturers Thoralf Chrobok and Christian Germer helped to prepare the course contents accordingly in online exercises and placement and final tests.

Special thanks also go to Sebastian Homer, who was responsible for the preparation and technical support for the LMS Moodle.

Many thanks for the good cooperation.

References

1. Cooperation Schule Hochschule (Cosh). https://lehrerfortbildung-bw.de/u_matnatech/mathematik/bs/bk/cosh/katalog/makv2.pdf. Accessed 7 Dec 2020

2. Digital Ed, https://www.digitaled.com/mobius. Accessed 16 Dec 2020
3. Heublein, U., Richter, J., Schmelzer, R.: https://www.dzhw.eu/pdf/pub_brief/dzhw_brief_03_2020.pdf. Accessed 16 Dec 2020
4. Moodle, https://moodle.de/, last accessed 2020/12/16.
5. MAPLE: www.maplesoft.com/. Accessed 25 Mar 2021

Puiseux Series and Algebraic Solutions of First Order Autonomous AODEs – A MAPLE Package

François Boulier[1], José Cano[2], Sebastian Falkensteiner[3(✉)], and J. Rafael Sendra[4]

[1] Univ. Lille, CNRS, Centrale Lille, Inria, UMR 9189 - CRIStAL, Lille, France
francois.boulier@univ-lille.fr

[2] Dpto. Algebra, análisis matemático, geometría y topología, Universidad de Valladolid, Valladolid, Spain
jcano@agt.uva.es

[3] Research Institute for Symbolic Computation (RISC), Johannes Kepler University Linz, Linz, Austria
falkensteiner@risc.jku.at

[4] Universidad de Alcalá, Dpto. Física y Matemáticas, Alcalá de Henares, Madrid, Spain
rafael.sendra@uah.es

Abstract. There exist several methods for computing exact solutions of algebraic differential equations. Most of the methods, however, do not ensure existence and uniqueness of the solutions and might fail after several steps, or are restricted to linear equations. The authors have presented in previous works a method to overcome this problem for autonomous first order algebraic ordinary differential equations and formal Puiseux series solutions and algebraic solutions. In the first case, all solutions can uniquely be represented by a sufficiently large truncation and in the latter case by its minimal polynomial.

The main contribution of this paper is the implementation, in a MAPLE package named FirstOrderSolve, of the algorithmic ideas presented therein. More precisely, all formal Puiseux series and algebraic solutions, including the generic and singular solutions, are computed and described uniquely. The computation strategy is to reduce the given differential equation to a simpler one by using local parametrizations and the already known degree bounds.

Keywords: Maple · Symbolic computation · Algebraic differential equation · Formal Puiseux series solution · Algebraic solution

1 Introduction

The problem of finding power series solutions of ordinary differential equations has been extensively studied in the literature. A method to compute generalized formal power series solutions, i.e. power series with real exponents, and

R. M. Corless et al. (Eds.): MC 2020, CCIS 1414, pp. 89–103, 2021.
https://doi.org/10.1007/978-3-030-81698-8_7

to describe their properties is the Newton polygon method. A description of this method is given in [11,12] and more recently in [1,6,13]. In [4], the second author, using the Newton polygon method, gives a theoretical description of all generalized formal power series solutions of a non-autonomous first order ordinary differential equation as a finite set of one parameter families of generalized formal power series. This description of the solutions is in general not algorithmic by several reasons. One of them is that there is no bound on the number of terms which have to be computed in order to guarantee the existence of a generalized formal power series solution when extending a given truncation of a determined potential solution. Also the uniqueness of the extension can not be ensured a-priori.

In [5] this problem has been overcome by the authors for autonomous first order differential equations by using a local version of the algebro-geometric approach introduced in [9].

In [15] they derive an associated differential system to find rational general solutions of non-autonomous first order differential equations by considering rational parametrizations of the implicitly defined curve. We instead consider its places and obtain an associated differential equation of first order and first degree which can be transformed into an equation of a very specific type [3]. Using the known bounds for computing places of algebraic curves (see e.g. [7]), existence and uniqueness of the solutions and the termination of our computations can be ensured.

In [2] the results of [9,10] are generalized to algebraic solutions. It is well known that algebraic solutions can be represented as Puiseux series. The advantage is that they can be fully described by its minimal polynomial. In this package we mainly follow [2], but we use an adapted version of the algorithm there for deciding the existence of algebraic solutions and computing all of them in the affirmative case.

2 Theoretical and Algorithmic Framework

In this section we recall the main notions and results that are used in our implementations. For further details we refer to [5] in the case of formal Puiseux series and to [2] in the case of algebraic solutions.

Let $\mathbb{K}$ be a computable field of characteristic zero such as the rational numbers $\mathbb{Q}$ and let us denote by $\overline{\mathbb{K}}$ its algebraic closure. Let us consider the differential equation

$$F(y, y') = 0, \tag{1}$$

where $F \in \mathbb{K}[y, p]$ is square-free and non-constant in the variables y and p. We are looking for formal Puiseux series and algebraic solutions of (1). In the case of formal Puiseux series solutions we will represent the full series by a sufficiently large truncation such that existence and uniqueness are guaranteed. In the case of algebraic solutions we look for its minimal polynomial.

We associate to (1) the affine algebraic curve $C(F) \subset \overline{\mathbb{K}}^2$ defined by the zero set of $F(y, p)$ in $\overline{\mathbb{K}}^2$. We denote by $\mathscr{C}(F)$ the Zariski closure of $C(F)$ in $\overline{\mathbb{K}}_\infty^2$, where $\overline{\mathbb{K}}_\infty = \overline{\mathbb{K}} \cup \{\infty\}$ denotes the one-point compactification of $\overline{\mathbb{K}}$. In the case of formal Puiseux series solutions we will look for local parametrizations of $\mathscr{C}(F)$ and in the case of algebraic solutions for algebraic parametrizations, respectively.

2.1 Formal Puiseux Series Solutions

Formal Puiseux series can either be expanded around a finite point or at infinity. In the first case, since Eq. (1) is invariant under translation of the independent variable, without loss of generality we can assume that the formal Puiseux series is expanded around zero and it is of the form $\varphi(x) = \sum_{j \geq j_0} a_j\, x^{j/n}$, where $a_j \in \overline{\mathbb{K}}$, $n \in \mathbb{Z}_{>0}$ and $j_0 \in \mathbb{Z}$. In the case of infinity we can use the transformation $x = 1/z$ obtaining the (non-autonomous) differential equation $F(y(z), -z^2 y'(z)) = 0$. In order to deal with both cases in a unified way, we will study equations of the type

$$F(y(x), (1-h)x^h y'(x)) = 0, \tag{2}$$

with $h \in \{0, 2\}$ and its formal Puiseux series solutions expanded around zero. We note that for $h = 0$ Eq. (2) is equal to (1) and for $h = 2$ the case of formal Puiseux series solutions expanded at infinity is treated.

We use the notations $\mathbb{L}[[x]]$ for the ring of formal power series, $\mathbb{L}((x))$ for its fraction field and $\mathbb{L}((x))^* = \bigcup_{n \geq 1} \mathbb{L}((x^{1/n}))$ for the field of formal Puiseux series expanded at zero with coefficients in some field $\mathbb{L}$. We call the minimal natural number n such that $\varphi(x)$ belongs to $\mathbb{L}((x^{1/n}))$ the *ramification order* of $\varphi(x)$. Moreover, for $\varphi(x) = \sum_{j \geq j_0} a_j\, x^{j/n}$ with $a_{j_0} \neq 0$ we call $j_0/n \in \mathbb{Q}$ the order of φ, denoted by $\operatorname{ord}_x(\varphi(x))$, and set $\operatorname{ord}_x(\varphi(x)) = \infty$ for $\varphi = 0$.

Additionally to (2) we may require that a formal Puiseux series solution $y(x)$ of (2) fulfills the initial conditions $y(0) = y_0, ((1-h)x^h y'(x))(0) = p_0$ for some fixed $\mathbf{p}_0 = (y_0, p_0) \in \overline{\mathbb{K}}_\infty^2$. In the case where $y(0) = \infty$, $\tilde{y}(x) = 1/y(x)$ is a Puiseux series solution of a new first order differential equation of the same type, namely the equation given by the numerator of the rational function $F(1/y, -(1-h)x^h p/y^2)$, and $\tilde{y}(0) \in \overline{\mathbb{K}}$. Therefore, in the sequel, we may assume that $\mathbf{p}_0 \in \overline{\mathbb{K}} \times \overline{\mathbb{K}}_\infty$.

Formal Parametrizations. Let us recall some classical terminology on local parametrizations of algebraic curves and its algorithmic aspects, for further details see e.g. [7,16].

A *formal parametrization* centered at $\mathbf{p}_0 \in \mathscr{C}(F)$ is a pair of formal Puiseux series $A(t) \in \overline{\mathbb{K}}((t))^2 \backslash \overline{\mathbb{K}}^2$ such that $A(0) = \mathbf{p}_0$ and $F(A(t)) = 0$. In the set of all formal parametrizations of $\mathscr{C}(F)$ we introduce the equivalence relation $\sim$ by defining $A(t) \sim B(t)$ if and only if there exists a formal power series $s(t) \in \overline{\mathbb{K}}[[t]]$ of order one such that $A(s(t)) = B(t)$. A formal parametrization is said to be *irreducible* if it is not equivalent to another one in $\overline{\mathbb{K}}((t^m))^2$ for some $m > 1$. An equivalence class of an irreducible formal parametrization $(a(t), b(t))$ is called a *place* of $\mathscr{C}(F)$ centered at the common center point $\mathbf{p}_0$ and is denoted by $[(a(t), b(t))]$.

Let $\mathrm{IFP}(\mathbf{p}_0)$ denote the set of all irreducible formal parametrizations of $\mathscr{C}(F)$ at $\mathbf{p}_0$ and $\mathrm{Places}(\mathbf{p}_0)$ containing the places of $\mathscr{C}(F)$ centered at $\mathbf{p}_0$. Computationally we have to truncate the formal parametrizations. There are bounds presented in [7,14] such that

1. the truncations of the formal parametrizations $(a(t), b(t))$ at $\mathbf{p}_0$ are in one-to-one correspondence to $\mathrm{Places}(\mathbf{p}_0)$;
2. the orders $\mathrm{ord}_t(a(t) - y_0), \mathrm{ord}_t(b(t))$ are determined;
3. no further extension of the ground field for computing the following coefficients have to be done.

More precisely, in [7], $N = 2(\deg_p(F) - 1)\deg_y(F) + 1$ is proved to be a bound satisfying the above requirements, under the hypothesis that the leading coefficient of $F(y,p)$, seen as polynomial in p and denoted by $\mathrm{lc}_p(F)(y)$, is constant. Her proof can be generalized straightforward to the case where $\mathrm{lc}_p(F)(y)$ is of order zero. The general case is reduced to the previous one by a change of variable $q(y) = y^\nu p(y)$, where ν is the order of $\mathrm{lc}_p(F)(y)$. In this way, the bound above generalizes as

$$\begin{aligned} N &= (2\deg_p F - 1)(\deg_y F + \nu(\deg_p F - 1)) + 1 \\ &\leq (2\deg_p F - 1)\deg_p F \deg_y F + 1. \end{aligned} \tag{3}$$

In the literature other possible bounds exist such as in [14] given in terms of the Milnor number.

Let us note that the solutions of (2) will be independent of the chosen representative of a place. Hence, regarding uniqueness of the prolongation, number of field extensions, etc. it does not matter which local parametrization we chose (for example classical Puiseux parametrizations or rational Puiseux parametrization [7]). For representing the solution parametrizations, which is not the goal of the current paper, however, it would be relevant.

Puiseux Solution Place. Let $\mathrm{Sol}_{\overline{\mathbb{K}}((x))^*}(\mathbf{p}_0)$ be the set containing the non-constant formal Puiseux series solutions of Eq. (2), expanded at zero, with coefficients in $\overline{\mathbb{K}}$ and with $\mathbf{p}_0$ as initial values. Then the mapping $\Delta : \mathrm{Sol}_{\overline{\mathbb{K}}((x))^*}(\mathbf{p}_0) \longrightarrow \mathrm{IFP}(\mathbf{p}_0)$ defined as

$$\Delta(y(x)) = \left(y(t^n), (1-h)t^{hn}\frac{dy}{dx}(t^n)\right),$$

where n is the ramification order of $y(x)$, is well-defined and injective. Moreover, we denote by $\delta : \mathrm{Sol}_{\overline{\mathbb{K}}((x))^*}(\mathbf{p}_0) \longrightarrow \mathrm{Places}(\mathbf{p}_0)$ the map $\delta(y(x)) = [\Delta(y(x))]$.

An irreducible formal parametrization $A(t) \in \mathrm{IFP}(\mathbf{p}_0)$ is called a *solution parametrization* of (1) if A is in the image $\mathrm{Im}(\Delta)$. Similarly, a place in $\mathrm{Im}(\delta)$ is called a *(Puiseux) solution place.*

It can be shown that for solution parametrizations $(a(t), b(t)) \in \mathrm{IFP}(\mathbf{p}_0)$, corresponding to a solution with ramification index n, it holds that

$$n(1-h) = \mathrm{ord}_t(a(t) - y_0) - \mathrm{ord}_t(b(t)). \tag{4}$$

This condition is invariant for the representative of a place. In particular, all Puiseux series solutions in the same solution place have the same ramification order. It turns out that condition (4) is already sufficient for solution places at $\mathbf{p}_0$ with $y_0 \in \overline{\mathbb{K}}$. Let us highlight this statement (see Theorem 10 in [5]):

Theorem 1. *Let* $\mathcal{P} = [(a(t), b(t))] \in \mathrm{Places}(\mathbf{p}_0)$ *and* $h = 0$. *Then* $\mathcal{P}$ *is a solution place if and only if Eq.* (4) *holds for an* $n \in \mathbb{Z}_{>0}$. *In the affirmative case the ramification order of* $\mathcal{P}$ *is equal to* n.

Also the solutions with $h = 2$ can be computed algorithmically. For this purpose let us give in the following some insight into to proof of Theorem 1.

Let $\mathbb{L}$ be a subfield of $\overline{\mathbb{K}}$. For a given parametrization $(a(t), b(t))) \in \mathbb{L}((t))^2$ satisfying (4), our strategy is to find $s(t) \in \overline{\mathbb{K}}[[t]]$ with $\mathrm{ord}_t(s(t)) = 1$ such that $(a(s(t)), b(s(t)))$ satisfies the *associated differential equation*

$$a'(s(t)) \cdot s'(t) = n(1-h)\, t^{n(1-h)-1}\, b(s(t)). \tag{5}$$

Let $k = \mathrm{ord}_t(a(t) - y_0), r = \mathrm{ord}_t(b(t))$ and $n(1-h) = k - r > 0$. By transforming (5) into an equation of Briot-Bouquet type [3], the solutions $s(t) = \sum_{i=1}^{\infty} \sigma_i\, t^i$ fulfill the following items.

1. If $h = 0$, there are exactly n solutions where $\sigma_1^n \in \mathbb{L}$ and $\sigma_i \in \mathbb{L}(\sigma_1)$ are uniquely determined for $i > 1$.
2. If $h = 2$, there is no solution or up to n one-parameter families of solutions with $\sigma_1^n \in \mathbb{L}$, $\sigma_{r-k} \in \mathbb{L}$ is a free parameter; $\sigma_2, \ldots, \sigma_{r-k-1} \in \mathbb{L}(\sigma_1)$ and for $i > r - k$ the coefficients $\sigma_i \in \mathbb{L}(\sigma_1, \sigma_{r-k})$ are uniquely determined.

After computing the solutions $s(t)$ of the associated differential equation, we obtain the solutions of the original differential equation by $a(s(x^{1/n}))$.

Solution Truncations. Since we cannot compute all coefficients of the Puiseux series solution, we have to truncate at some point. A *determined solution truncation* of (2) is an element of $\mathbb{L}[x^{1/n}][x^{-1}]$, for some $n \in \mathbb{Z}_{>0}$, that can be extended uniquely to a formal Puiseux series solution in $\mathbb{L}((x^{1/n}))$ or $\mathbb{L}((x^{-1/n}))$, respectively.

A point $\mathbf{p}_0 = (y_0, p_0) \in \mathscr{C}(F)$ is called a *critical curve point* if $p_0 \in \{0, \infty\}$ or $\frac{\partial F}{\partial p}(\mathbf{p}_0) = 0$ or $y_0 = \infty$. Under our assumptions, the set of critical curve points is finite. The only formal Puiseux series solution with non-critical $\mathbf{p}_0$ as initial tuple is a formal power series and its determined solution truncation is given by $y_0 + p_0 x$.

Assume that $\mathbf{p}_0 \in \mathscr{C}(F)$ is a critical curve point. Then, by the properties of the solutions of the associated differential equations, the bound N as in (3) also holds for the computation of the determined solution truncations. In particular, Eq. (4) can be checked, no further extensions of the ground field for computing the coefficients are necessary and the ramification index is determined.

Algorithm 1. PuiseuxSolve

Input: A first-order AODE $F(y, y') = 0$, where $F \in \mathbb{K}[y, p]$ is square-free with no factor in $\mathbb{K}[y]$ or $\mathbb{K}[p]$.

Output: A set consisting of all determined solution truncations of $F(y, y') = 0$ (expanded around zero and around infinity).

1: If $(\infty, \infty) \in \mathscr{C}(F)$, then perform the transformation $\tilde{y} = 1/y$ and apply the following steps additionally to the numerator of $F(1/y, -p/y^2)$ and $\mathbf{p}_0 = (0, 0)$ in order to obtain the solutions of negative order.

2: Compute the set of critical curve points $\mathcal{B}(F)$ (for $y_0 \in \overline{\mathbb{K}}$) and $\mathbb{V}(F(y, 0))$ (for $y_0 = \infty$).

3: For every point $(y_0, p_0) \in \mathscr{C}(F) \setminus \mathcal{B}(F), y_0 \neq \infty$, a determined solution truncation is $y_0 + p_0 x$.

4: Add to the output the constant solutions $y(x) = y_0$ corresponding to $(y_0, 0) \in \mathscr{C}(F), y_0 \neq \infty$.

5: For every place centered at a critical curve point $\mathbf{p}_0 = (y_0, p_0) \in \mathcal{B}(F)$, compute the first N terms of a formal parametrization $(a(t), b(t))$.

6: For solutions expanded around zero (resp. around infinity), check equation (4) with $h = 0$ (resp. $h = 2$).

7: In the negative case, $[(a(t), b(t)]$ is not a solution place. In the affirmative case, compute the first N terms of the solutions $s(t)$ of (5).

8: If $h = 0$ and $n > 0$, there exist exactly n solutions. If $h = 2$ and $n > 0$, the associated differential equation is either unsolvable or contains a free parameter.

9: The first N terms of $a(s(x^{1/n}))$ are the solution truncations with $\mathbf{p}_0$ as initial tuple.

For solutions expanded around zero we are able to ensure uniqueness of the extension of the truncated Puiseux series solutions (see also [5] [Theorem 14]). In the case where the expansion point is infinity, some truncations may coincide for some specific values of the free parameter coming from the solution of the associated differential equation.

2.2 Algebraic Solutions

In this section we consider a subclass of formal Puiseux series, namely algebraic series. These are $y(x) \in \overline{\mathbb{K}}((x))^*$ such that there exists a non-zero $G \in \overline{\mathbb{K}}[x, y]$ with $G(x, y(x)) = 0$. Since the field of formal Puiseux series is algebraically closed, all algebraic solutions can be represented as (formal) Puiseux series.

In [2] a bound on the degree of algebraic general solutions is given. There the authors indicate how to use these results in order to compute all algebraic solutions of such a given differential equation. A more detailed proof of this fact can be found in [8].

The first important observation is that if there exists one non-constant algebraic solution of (1), then all of them can be found easily by a shift in the minimal polynomial (see [8] [Theorem 4.1.22]).

Theorem 2. *Let $F \in \mathbb{K}[y, y']$ be irreducible and let $y(x)$ be a non-constant solution of $F = 0$ algebraic over $\overline{\mathbb{K}}(x)$ with minimal polynomial $G \in \overline{\mathbb{K}}[x, y]$.*

Then all formal Puiseux series solutions $\mathrm{Sol}_{\overline{\mathbb{K}}((x))^*}(F)$ *are algebraic and given by* $G(x+c, y)$*, where* $c \in \overline{\mathbb{K}}$.

The second important computational aspect is the degree bound on the solutions [2] [Theorem 3.4, Theorem 3.8]:

Theorem 3. *Let* $F \in \mathbb{K}[y, y']$ *be irreducible and let* $y(x)$ *be a non-constant solution of* $F = 0$ *algebraic over* $\overline{\mathbb{K}}(x)$ *with minimal polynomial* $G \in \overline{\mathbb{K}}[x, y]$*. Then*

$$\deg_x(G) = \deg_p(F), \quad \deg_y(G) \leq \deg_y(F) + \deg_p(F).$$

The third result is used to construct candidates of algebraic solutions:

Lemma 1. *Let* $G(x, y) \in \mathbb{K}[x, y]$ *be an irreducible polynomial with* $d_x = \deg_x G, d_y = \deg_y G$*. Let* $y(x)$ *be a Puiseux series solution of* $G(x, y) = 0$ *expanded at* $x = 0$ *with* $\mathrm{ord}_x(y(x)) = \nu$*. Let* $\nu' = \min\{\nu, 0\}$ *and write* $y(x) = \bar{y}(x) + \varphi(x)$ *with* $\mathrm{ord}_x(\varphi(x)) > N > 0$ *where*

$$N \geq 2\, d_x\, d_y - 2\, \nu'\, (d_y - 1). \tag{6}$$

Assume that $A(x, y) \in \mathbb{K}[x, y], \deg_x A \leq d_x, \deg_y A \leq d_y$ *is minimal with respect to the lexicographical order* $y > x$ *such that*

$$\mathrm{ord}_x(A(x, \bar{y}(x)) > 2\, d_x\, d_y - \nu'(d_y - 1), \tag{7}$$

holds. Then $A(x, y)$ *is, up to a constant factor, equal to* $G(x, y)$.

Proof. Let $R(x)$ be the resultant of $G(x, y)$ and $A(x, y)$ with respect to y. It is well known that there exist polynomials $B(x, y)$, $C(x, y)$ with $\deg_y B < d_y$, $\deg_y C < d_y$ such that

$$G(x, y)\, B(x, y) + A(x, y)\, C(x, y) = R(x).$$

Evaluating at $\bar{y}(x)$ we obtain

$$G(x, \bar{y}(x))\, B(x, \bar{y}(x)) + A(x, \bar{y}(x))\, C(x, \bar{y}(x)) = R(x). \tag{8}$$

Since $\nu' \leq 0$, it follows that $\mathrm{ord}_x\, C(x, \bar{y}(x)) \geq \nu' \deg_y C \geq \nu'(d_y - 1)$ and similarly for $B(x, \bar{y}(x))$. Hence, by (7), we have that

$$\mathrm{ord}_x(A(x, \bar{y}(x))\, C(x, \bar{y}(x))) > 2\, d_x\, d_y.$$

Let us proof that $\mathrm{ord}_x(G(x, \bar{y}(x)) > \nu'(d_y - 1) + N$. Taking the Taylor series of $G(x, \bar{y}(x) + \varphi(x))$ and because $G(x, \bar{y}(x) + \varphi(x)) = 0$, we have:

$$G(x, \bar{y}(x)) = -\sum_{j=1}^{d_y} \frac{1}{j!} \frac{\partial^j G}{\partial y^j}(x, \bar{y}(x))\, \varphi(x)^j.$$

The order in x of each term on the right hand side of above equation is greater than $\nu'(d_y - 1) + N$, so it is for the left hand side. Now, because of (6), we have that

$$\operatorname{ord}_x(G(x, \bar{y}(x))\, B(x, \bar{y}(x))) > \nu'(d_y - 1) + N + \nu'(d_y - 1) \geq 2\, d_x\, d_y.$$

Hence, the left hand side of (8) has order greater than $2\, d_x\, d_y$ and the right hand side is a polynomial of degree less than or equal to $2\, d_x\, d_y$. Hence, $R(x) = 0$, and therefore, $G(x, y)$ and $A(x, y)$ have a common factor. Since $G(x, y)$ is an irreducible polynomial, it is a factor of $A(x, y)$. Then, by the degree conditions on $A(x, y)$, the statement follows. □

In [2] the method of detecting candidates $G(x, y)$ for algebraic solutions of the differential equations $F(y, y') = 0$ consists by computing $\bar{y}(x)$, the first N terms of a power series solution $y(x)$ of the differential equations $F(y, y') = 0$, with a regular curve point of $\mathscr{C}(F)$ as initial tuple. Hence, in this case the solution $y(x)$ is of order 0 and $\nu' = 0$. Choose $N > d_x\, d_y$ and construct, by solving a linear system of equations, a polynomial A fulfilling the properties (6) and (7). This approach reduced the number of formal power series solutions that we can use to construct a candidate. Lemma 1 allows to choose any Puiseux series solutions of the differential equations and reduce the computational cost.

Once a candidate $A(x, y)$ is detected, we can check whether it is an actual algebraic solution of the differential equation $F(y, y') = 0$ by checking whether the differential pseudo remainder of $F(y, y')$ with respect $A(x, y)$ is zero. These results lead to the following algorithm.

Algorithm 2. AlgebraicSolve

Input: A first-order AODE $F(y, y') = 0$, where $F \in \mathbb{K}[y, p]$ is irreducible over $\overline{\mathbb{K}}(y)$.
Output: The minimal polynomial of an algebraic solution of $F(y, y') = 0$, describing all solutions, if it exists.

1: Compute the minimal number of terms of all Puiseux solutions of $F(y, y') = 0$ using PuiseuxSolve and choose one of them, denote it by $\hat{y}(x)$. Let ν be its order, $\nu' = \min(\nu, 0)$ and n its ramification index.
2: Let $d_x = \deg_p F$ and $d_y = \deg_y F + \deg_p F$.
3: Compute the prolongation $\bar{y}(x)$ of $\hat{y}(x)$ up to order $N = 2\, d_x\, d_y - 2\, \nu'(d_y - 1) + 1/n$.
4: Compute $A(x, y) \in \overline{\mathbb{K}}[x, y]$ fulfilling the required conditions from Lemma 1 by an ansatz of unknown coefficients and solving the resulting linear system.
5: Check whether $\operatorname{prem}(F, A) = 0$. If so, then $A(x, y)$ is an actual solution. Otherwise there exists no algebraic solution.

---

3 The Package FirstOrderSolve

In this section, we present the structure and content of the `MAPLE` package `FirstOrderSolve`. It consists several procedures that implement in particular the algorithms PuiseuxSolve and AlgebraicSolve described above. This package computes the Puiseux series solutions and algebraic solutions of first order autonomous AODEs with coefficients in an algebraic extension field of $\mathbb{Q}$.

3.1 Overview of the Software Structure

The created `MAPLE` package is initialized by the command

```
> with(FirstOrderSolve):
```

The main procedures are

- `SolutionTruncations`: for computing all formal Puiseux series solutions (Algorithm PuiseuxSolve);
- `AlgebraicSolution`: for computing the minimal polynomial of the algebraic solutions (Algorithm AlgebraicSolve);
- `GenericSolutionTruncation`: for computing a truncation of the solutions with non-critical initial tuple;
- `ProlongSolutionTruncation`: for prolonging the solution truncations up to a higher degree.

These four commands are public to the user. The package is divided into several sub-packages `BriotBouquetSolve, LocalSolve, AlgebraicSolve`, which are not accessible for the user, and uses the hierarchy sketched below (Fig. 1).

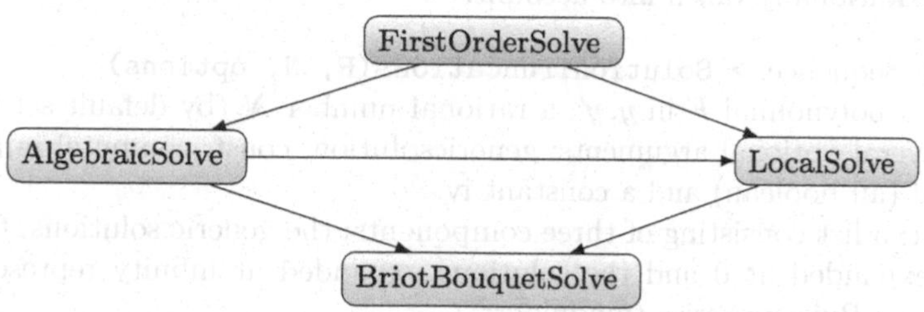

Fig. 1. The hierarchy of the package.

The main commands in the sub-packages are the following.

- `ParametrizationSetAlgCurve`: for computing truncations of the formal parametrizations of an implicitly defined algebraic curve by using the command `algcurves:-puiseux`;
- `ReparametrizationSet`: for computing the solutions of the associated differential equation by using `BriotBouquetSolve`;
- `BriotBouquetSolve`: for computing the unique solution of a first-order differential equation in quasi-solved form (which is called an equation of Briot-Bouquet type [3] [Section 80,86]); this procedure is using a Newton type algorithm for solving the resulting linear system in several variables;

In the following, we give a description of the procedures in the package `FirstOrderSolve`, available at https://risc.jku.at/sw/firstordersolve/. There can be found more detailed information on the commands in the included help.

3.2 Description of the Software Components

`> SolutionTruncations`
Computes all Puiseux series solutions of a given first order autonomous ordinary differential equation.

Since the equation is autonomous, the translation of the independent variable by any constant in a solution is again a solution. Hence, the only relevant expansion points are 0 and infinity.

The solutions expanded at 0 can be split into two sets: a generic solution and a set of particular solutions. The generic solution consists of all solutions starting with a non-critical curve point and is addressed in `GenericSolutionTruncation`. Each critical curve point corresponds to a set (that could be empty) of particular Puiseux series solutions.

The command computes the generic solution, all particular solutions expanded at 0 and all solutions expanded at infinity. The solutions are represented as truncations such that existence and uniqueness is ensured. In other words, the truncations are in one-to-one correspondence to the solutions. By setting the optional arguments genericsolution, const, computeFinite, computeInf to false, the corresponding subsets of the solution set can be suppressed. The remaining option iv $= y_0$ represents an initial condition of the format $y(0) = y_0$, where y_0 is an element of the ground field or an algebraic extension field of it, which is additionally taken into account.

- ◇ Calling Sequence: `> SolutionTruncations(F, N, options)`
- ◇ Input: a polynomial F in y, y', a rational number N (by default set to zero) and several optional arguments: genericsolution, const, computeFinite, computeInf (all boolean) and a constant iv.
- ◇ Output: a list consisting of three components: the generic solutions, the solutions expanded at 0 and the solutions expanded at infinity represented as truncated Puiseux series (modulo x^N).

`> GenericSolutionTruncation`
The first order differential equation has a generic local solution $y(x) = y_0 + y_1 x + \mathcal{O}(x^2)$, where $F(y_0, y_1) = 0$. If F is irreducible as polynomial and (y_0, y_1) is a regular affine point of the curve implicitly defined by F, the extension of $y_0 + y_1 x$ to a solution $y(x)$ is guaranteed and unique. The command `GenericSolutionTruncation` computes the first terms of the generic (formal) power series solutions, expanded around 0, of the given differential equation.

Note that for every irreducible component one generic solution is computed. Thus, all generic solutions of $F = 0$ are given by the union of the generic solutions of the components. If the given differential equation is known to be irreducible, the optional argument irreducible = true (see below) can be used in order to speed up computations.

The output of the command is a set of lists with two entries: a polynomial in x representing the solution computed modulo x^N involving an unspecified parameter $\_CC$ and a set of exceptional values for $\_CC$. For these values the generic solution would in general not lead to a solution of the given differential equation or might involve fractional exponents. Finally, let us mention that, if the precision of the output is not high enough, it is possible to use the command `ProlongSolutionTruncation`; see below.

- ◇ Calling Sequence: `> GenericSolutionTruncation(F, N, options)`
- ◇ Input: a polynomial F in y, y', a rational number N (by default set to zero), and optionally `irreducible` as boolean.
- ◇ Output: is a set of lists with two entries: a polynomial in x representing the solution computed modulo x^N involving an unspecified parameter $\_CC$ and a set of exceptional values for $\_CC$.

`> ProlongSolutionTruncation`

For the given first order differential equation, if an appropriate change of variables $z(x) = y(x) + s(x)$ is performed, the resulting equation

$$G(x, z(x), z'(x)) = F(y(x) + s(x), y'(x) + s'(x)) = 0$$

might be of Briot-Bouquet type. In case that $s(x)$ is such a solution truncation of $y(x)$, existence and uniqueness of the solution $z(x)$ of G are ensured and the following coefficients can be found by a Newton type algorithm. In particular, this is the case when $s(x)$ is an output element of `GenericSolutionTruncation` or `SolutionsTruncations`.

In this situation, the command `ProlongSolutionTruncation` prolongs the first terms of a truncated Puiseux series solution $s(x)$ of $F(y(x), y'(x)) = 0$.

- ◇ Calling Sequence: `> ProlongSolutionTruncation(F, s, N, x0)`
- ◇ Input: a polynomial F in y, y', a polynomial s, a rational number N and $x0$ equals 0 or infinity (by default set to zero)
- ◇ Output: it is again a truncated Puiseux series computed until the order x^N (or $1/x^N$)

`> AlgebraicSolution`

Algebraic solutions of the first order autonomous differential equation are represented by its minimal polynomisl, say $G(x, y)$. In this case, all the functions $y(x)$ with $G(x, y(x)) = 0$ are solutions of this differential equation and can be represented as Puiseux series.

Assuming that F is an irreducible polynomial, the existence of algebraic solutions can be decided and, in the affirmative case, all solutions are algebraic and are given as shift of the independent variable, namely by $G(x+c, y)$. Therefore, by factorizing the given differential equation, all algebraic solutions can be found using this procedure for every component.

The command `AlgebraicSolution` decides the existence of algebraic solutions of the given differential equation. Furthermore, if a solution exists the output is the minimal polynomial of the solution. The other solutions then can be easily found by shifting x. The solutions are found by checking whether a particular solution is algebraic. Efficiency of the algorithm highly depends on the chosen initial value. The procedure is using a formal power series solution, which means non-negative integer exponents for the solution, with a relatively small number of algebraic extensions of the ground field. Similarly to the command `GenericSolutionTruncation`, if the given differential equations is known to be irreducible, this can be specified by the optional argument irreducible $=$ true.

- ◇ Calling Sequence: `> AlgebraicSolution(F, options)`
- ◇ Input: F is a first-order differential polynomial, and `irreducible` is a boolean option.
- ◇ Output: the decision on the existence of algebraic solutions of the differential equation. If a solution exists the output is the minimal polynomial of the solution.

3.3 Usage of the Package

In order to use the package, download the file FirstOrderSolve.m from https://risc.jku.at/sw/firstordersolve/ and save it as your local folder. After starting Maple you redefine the variable libname as

```
> libname:=libname, 'path of user local folder';
```

Then, after executing the command

```
> with(FirstOrderSolve);
```

The package can be used. In the appendix we provide a Maple Worksheet illustrating the usage of our package for the computation of all Puiseux series solutions of first-order autonomous ordinary differential equations. We provide at https://risc.jku.at/sw/firstordersolve/ an extended version of this file.

```
> with(FirstOrderSolve);
```

$$[AlgebraicSolution, GenericSolutionTruncation, ProlongSolutionTruncation, SolutionTruncations] \quad (1)$$

A list of possible examples is the following.

```
> Examples := [diff(y(x), x) * y(x)^2 + y(x) - 1,
      diff(y(x), x) - y(x)^4 - y(x)^2,
      diff(y(x), x) - y(x)^3 - y(x)^2,
      diff(y(x), x)^3 + y(x)^2 - diff(y(x), x),
      diff(y(x), x)^2-4*y(x)^3,
      diff(y(x), x)^3 * (y(x)^6 + 2*y(x) + 1) - (12*y(x)^5 + 9*y(x)^4 - 1) * diff(y(x),
    x)^2 + 27*y(x)^8 + 54*y(x)^7 + 27*y(x)^6 + 4*y(x)^3
    ]:
```

We start with the first entry of the list and compute all solutions and use the option of minimal output length by not specifying the truncation order.

```
> F1 := Examples[1];
```

$$F1 := \left(\frac{d}{dx}\, y(x)\right) y(x)^2 + y(x) - 1 \quad (2)$$

```
> sol1 := SolutionTruncations(F1);
```

$$sol1 := \left[\left\{\left[\_CC - \frac{(-1 + \_CC)\, x}{\_CC^2}, \{0\}\right]\right\}, \{1, RootOf(\_Z^3 - 3)\, x^{1/3}\}, \varnothing\right] \quad (3)$$

Let us prolong the non-constant solution.

```
> ProlongSolutionTruncation(F1, sol1[2, 2], 7/3);
```

$$RootOf(\_Z^3 - 3)\, x^{1/3} - \frac{RootOf(\_Z^3 - 3)^2\, x^{2/3}}{4} - \frac{3x}{80} + \frac{RootOf(\_Z^3 - 3)\, x^{4/3}}{320} + \frac{67\, RootOf(\_Z^3 - 3)^2\, x^{5/3}}{22400} + \frac{603\, x^2}{179200} + \frac{163\, RootOf(\_Z^3 - 3)\, x^{7/3}}{179200} \quad (4)$$

The generic solution can be either prolonged by the same command or by GenericSolutionTruncation itself. The exceptional value is _CC=0.

```
> GenericSolutionTruncation(F1, 3);
```

$$\left\{\left[\_CC - \frac{(-1 + \_CC)\, x}{\_CC^2} - \frac{(\_CC^2 - 3\,\_CC + 2)\, x^2}{2\,\_CC^5}, \{0\}\right]\right\} \quad (5)$$

For the second example we prolong the solution expanded around infinity.

```
> F2 := Examples[2];
```

$$F2 := \frac{d}{dx}\, y(x) - y(x)^4 - y(x)^2 \quad (6)$$

```
> sol2 := SolutionTruncations(F2);
```

$$sol2 := \left[\{[\_CC + (\_CC^4 + \_CC^2)\, x, \varnothing]\}, \left\{0, -I, I, -\frac{RootOf(\_Z^3 + 3)^2}{3\, x^{1/3}}\right\}, \left\{-\frac{1}{x} + \frac{\_CC}{x^2}\right\}\right] \quad (7)$$

```
> ProlongSolutionTruncation(F2, sol2[3, 1], 4, infinity);
```

$$\frac{\_CC}{x^2} - \frac{1}{x} + \frac{-\_CC^2 - 1}{x^3} + \frac{\_CC^3 + 3\_CC}{x^4} \tag{8}$$

The standard procedure of Maple cannot find all of the solutions, since for example the series expansion of the Puiseux series is not covered.

```
> dsolve( {F2 = 0}, y(x) );
```

$$\{y(x) = \tan(RootOf(\_C1 \tan(\_Z) + \_Z \tan(\_Z) + x \tan(\_Z) + 1))\} \tag{9}$$

Solutions at infinity with fractional exponents are found in the next example. The generic solution, constant solutions and the solutions expanded around infinity are suppressed and the precision is set to N=2.

```
> SolutionTruncations(Examples[3], 2, computeInf=false, genericsolution=false, const=false);
```

$$\left[\varnothing, \left\{-\frac{1}{3} - \frac{RootOf(\_Z^2+2)}{2\sqrt{x}} + \frac{RootOf(\_Z^2+2)\sqrt{x}}{12} + \frac{4x}{135} - \frac{RootOf(\_Z^2+2)\, x^{3/2}}{432}\right\}, \varnothing\right] \tag{10}$$

Alternatively one may directly specify the initial value by iv=infinity.

```
> SolutionTruncations(Examples[3], 2, iv = infinity);
```

$$\left\{-\frac{1}{3} - \frac{RootOf(\_Z^2+2)}{2\sqrt{x}} + \frac{RootOf(\_Z^2+2)\sqrt{x}}{12} + \frac{4x}{135} - \frac{RootOf(\_Z^2+2)\, x^{3/2}}{432}\right\} \tag{11}$$

In the next example we compute the solutions with the initial value y(0)=0. The non-constant solutions are not detected by 'dsolve'.

```
> SolutionTruncations(Examples[4], iv = 0);
```

$$\{0, x, -x\} \tag{12}$$

```
> dsolve( {Examples[4] = 0, y(0) = 0}, y(x) );
```

$$y(x) = 0 \tag{13}$$

For obtaining algebraic solutions we run the corresponding command over all examples. We see that not all of them have algebraic solutions.

```
> map(a → AlgebraicSolution(a), Examples);
```

$$[\varnothing, \varnothing, \varnothing, \varnothing, \{x^2 y - 2xy + y - 1\}, \{y x^3 + y^2 + x + y\}] \tag{14}$$

For the last two differential equations we obtain other results by using 'dsolve'. In the latter example the output is very complicated and lengthy whereas the algebraic solution obtained by our code is very compact.

```
> dsolve(Examples[5] = 0);
```

$$y(x) = \mathrm{WeierstrassP}(x + \_C1, 0, 0) \tag{15}$$

```
> dsolve(Examples[6] = 0) :
```

Acknowledgements. The first author was supported by the bilateral project ANR-17-CE40-0036 and DFG-391322026 SYMBIONT. The second author was partially supported by Agencia Estatal de Investigación PID2019-105621GB-I00. The third and fourth authors were partially supported by FEDER/Ministerio de Ciencia, Innovación y Universidades Agencia Estatal de Investigación/MTM2017-88796-P (Symbolic Computation: new challenges in Algebra and Geometry together with its applications). The third author was also supported by the Austrian Science Fund (FWF): P 31327-N32. The fourth author is member of the Research Group ASYNACS (Ref. CT-CE2019/683).

References

1. Aroca, J.: Puiseux Solutions of Singular Differential Equations, pp. 129–145. Birkhäuser Basel, Basel (2000)
2. Aroca, J., Cano, J., Feng, R., Gao, X.S.: Algebraic general solutions of algebraic ordinary differential equations. In: Proceedings of the 2005 International Symposium on Symbolic and Algebraic Computation, pp. 29–36. ACM (2005)
3. Briot, C., Bouquet, J.: Recherches sur les proprietés des équations différentielles. J. de l'Ecole Polytechnique **21**(36), 133–198 (1856)
4. Cano, J.: The newton polygon method for differential equations. In: Li, H., Olver, P.J., Sommer, G. (eds.) GIAE/IWMM-2004. LNCS, vol. 3519, pp. 18–30. Springer, Heidelberg (2005). https://doi.org/10.1007/11499251_3
5. Cano, J., Falkensteiner, S., Sendra, J.R.: Existence and convergence of Puiseux series solutions for first order autonomous differential equations. J. Symb. Comput. (2020). https://doi.org/10.1016/j.jsc.2020.06.010
6. Della Dora, J., Richard-Jung, F.: About the newton algorithm for non-linear ordinary differential equations. In: Proceedings of the 1997 International Symposium on Symbolic and Algebraic Computation, ISSAC 1997, pp. 298–304. ACM, New York (1997). https://doi.org/10.1145/258726.258817
7. Duval, D.: Rational Puiseux expansion. Compositio Mathematica **70**(2), 119–154 (1989)
8. Falkensteiner, S.: Power series solutions of AODEs - existence, uniqueness, convergence and computation. Ph.D. thesis, RISC Hagenberg, Johannes Kepler University Linz (2020)
9. Feng, R., Gao, X.S.: Rational general solutions of algebraic ordinary differential equations. In: Proceedings of the 2004 International Symposium on Symbolic and Algebraic Computation, pp. 155–162. ACM (2004)
10. Feng, R., Gao, X.S.: A polynomial time algorithm for finding rational general solutions of first order autonomous ODEs. J. Symb. Comput. **41**(7), 739–762 (2006)
11. Fine, H.: On the functions defined by differential equations, with an extension of the Puiseux polygon construction to these equations. Am. J. Math. **11**, 317–328 (1889). https://doi.org/10.2307/2369347
12. Fine, H.: Singular solutions of ordinary differential equations. Am. J. Math. **12**, 295–322 (1890). https://doi.org/10.2307/2369621
13. Grigoriev, D., Singer, M.: Solving ordinary differential equations in terms of series with real exponents. Trans. A.M.S. **327**, 329–351 (1991). https://doi.org/10.2307/2001845
14. Stadelmeyer, P.: On the computational complexity of resolving curve singularities and related problems. Ph.D. thesis, RISC, Johannes Kepler University Linz (2000)
15. Vo, N., Grasegger, G., Winkler, F.: Deciding the existence of rational general solutions for first-order algebraic ODEs. J. Symb. Comput. **87**, 127–139 (2018)
16. Walker, R.: Algebraic Curves. Princeton University Press, Princeton (1950)

Algebraic Aspects of a Rank Factorization Problem Arising in Vibration Analysis

Yacine Bouzidi[1], Roudy Dagher[1], Elisa Hubert[2], and Alban Quadrat[3(✉)]

[1] Inria Lille – Nord Europe, Villeneuve-d'Ascq, France
yacine.bouzidi@gmail.com, roudy.dagher@inria.fr

[2] University of Lyon, UJM-St-Etienne, LASPI, 42334 Saint-Etienne, France
elisa.hubert@wanadoo.fr

[3] Inria Paris, Ouragan project-team, IMJ – PRG, Sorbonne University, Paris, France
alban.quadrat@inria.fr

Abstract. This paper continues the study of a rank factorization problem arising in gear fault surveillance [10–13]. The structure of a class of solutions – important in practice – of the rank factorization problem is studied. We show that these solutions can be parametrized. Using module theory and computer algebra methods, the parameter space $\mathcal{P}$ is explicitly characterized and is shown to be the complementary of an algebraic set. Finally, a finite open cover of $\mathcal{P}$ is obtained and for each basic open subset of the cover of $\mathcal{P}$, a closed-form solution is characterized.

Keywords: Polynomial systems · Effective module theory · Demodulation problems · Gearbox vibration signals

1 Introduction

Before stating the mathematical problem studied in this paper, we first introduce a few notations. Let $\mathbb{k}$ denote a field (e.g., $\mathbb{k} = \mathbb{Q}$, $\mathbb{R}$, $\mathbb{C}$), R a commutative ring, $R^{n\times m}$ the R-module (the $\mathbb{k}$-vector space if $R = \mathbb{k}$) formed by all the $n \times m$ matrices with entries in R, $U(R) := \{r \in R \mid \exists\, s \in R:\ r\, s = 1\}$ the group of units of R, $\mathrm{GL}_n(R) := \{U \in R^{n\times n} \mid \det(U) \in \mathrm{U}(R)\}$ the *general linear group* of invertible $n \times n$ matrices with entries in R, and I_n the $n \times n$ identity matrix of $\mathrm{GL}_n(R)$. If $A \in R^{r\times s}$, then we can consider the following R-*homomorphisms*

$$\begin{array}{rcl} .A : R^{1\times r} & \longrightarrow & R^{1\times s} \\ \lambda & \longmapsto & \lambda\, A, \end{array} \qquad \begin{array}{rcl} A. : R^{s\times 1} & \longrightarrow & R^{r\times 1} \\ \eta & \longmapsto & A\, \eta, \end{array}$$

and the following R-*modules* (the $\mathbb{k}$-vector spaces if $R = \mathbb{k}$):

$$\left\{\begin{array}{l} \mathrm{im}_R(.A) := R^{1\times r}\, A, \\ \ker_R(.A) := \{\lambda \in R^{1\times r} \mid \lambda\, A = 0\}, \\ \mathrm{coker}_R(.A) := R^{1\times s}/\mathrm{im}_R(.A), \end{array}\right. \qquad \left\{\begin{array}{l} \mathrm{im}_R(A.) := A\, R^{s\times 1}, \\ \ker_R(A.) := \{\eta \in R^{s\times 1} \mid A\, \eta = 0\}, \\ \mathrm{coker}_R(A.) := R^{r\times 1}/\mathrm{im}_R(A.). \end{array}\right.$$

R. M. Corless et al. (Eds.): MC 2020, CCIS 1414, pp. 104–118, 2021.
https://doi.org/10.1007/978-3-030-81698-8_8

Recall that A is said to have *full column* (resp., *full row*) *rank* if $\ker_R(A.) = 0$ (resp., $\ker_R(.A) = 0$). $A \in \Bbbk^{r\times s}$ has full row rank (resp., full column rank) iff it admits a *right* (resp., *left*) *inverse* $B \in \Bbbk^{s\times r}$, i.e., $A\,B = I_r$ (resp., $B\,A = I_s$).

Motivated by the application of *vibration analysis* to *gearbox fault surveillance* [2,3], a new *demodulation* approach of *gearbox vibration signals* was developed in [10,11]. It yielded the study of the following mathematical problem.

Rank Factorization Problem:
Let $D_1, \ldots, D_r \in \Bbbk^{n\times n}\backslash\{0\}$ and $M \in \Bbbk^{n\times m}\backslash\{0\}$ be such that $\operatorname{rank}_{\Bbbk}(M) \leq r$. Determine – if they exist – $u \in \Bbbk^{n\times 1}$ and $v_1, \ldots, v_r \in \Bbbk^{1\times m}$ satisfying:

$$M = \sum_{i=1}^{r} D_i\,u\,v_i. \tag{1}$$

Note that (1) is a system formed by $m\,n$ polynomial equations in the $n + m\,r$ entries of u and of the v_i's. Thus, (1) belongs to the realm of algebraic geometry.

The rank factorization problem was first solved for $r = 1$ and $D_1 = I_n$ in [11], and then for $r = 2$ and $D_1 = I_n$ in [12]. In [13], the general problem was studied with the assumption that the row vectors v_i's are $\Bbbk$-linearly independent, i.e., that the matrix $v := (v_1^T \;\ldots\; v_r^T)^T$ has full row rank. This assumption, which is motivated by the application, made the characterization of this class of solutions possible using linear algebra methods. These results are reviewed in Sect. 2.

Based on *module theory* and *computer algebra* methods [7,14,17], the first goal of the paper is to develop the algorithmic aspects of the results presented in [13]. We then study the set formed by all the solutions (u, v) of (1) with full row rank matrices v. An important problem in practice is to know how the solutions can vary within the solution space. Hence, we develop the local study of the solution space by proving the existence of local closed-form solutions that can be computed by computer algebra methods. Finally, the existence of global solutions is investigated and we show that this problem is related to well-known difficult problems in module theory (e.g., computing the least number of generator sets of an ideal, recognizing when a stably free module over certain localizations of a polynomial ring is free and if so, computing a basis of the free module) [7,17].

2 The Rank Factorization Problem

In this section, we state again results on the problem obtained in [13]. If we note

$$A(u) := (D_1\,u \;\ldots\; D_r\,u) \in \Bbbk^{n\times r}, \quad v := (v_1^T \;\ldots\; v_r^T)^T \in \Bbbk^{r\times m},$$

then (1) can be rewritten as the following factorization of M (*bilinear system*):

$$M = A(u)\,v. \tag{2}$$

Note that if $(u,\, v)$ is a solution of (2), then so is $(\lambda\,u,\, \lambda^{-1}\,v)$ for all $\lambda \in \Bbbk\backslash\{0\}$.

We also note that Problem (2) is solvable iff there exists $u \in \Bbbk^{n\times 1}$ such that:

$$\operatorname{im}_{\Bbbk}(M.) \subseteq \operatorname{im}_{\Bbbk}(A(u).). \tag{3}$$

Indeed, if (2) holds, then $\zeta \in \mathrm{im}_{\Bbbk}(M.)$ is of the form $\zeta = M\,\eta = A(u)\,(v\,\eta)$ for a certain $\eta \in \Bbbk^{m\times 1}$, which shows that (3) holds. Conversely, if there exists a vector $u \in \Bbbk^{n\times 1}$ such that (3) holds, then for $i = 1, \ldots, m$, the ith column $M_{\bullet i}$ of M belongs to $\mathrm{im}_{\Bbbk}(A(u).)$, and thus, there exists $w_i \in \Bbbk^{r\times 1}$ such that $M_{\bullet i} = A(u)\,w_i$, which yields (2) with $v := (w_1 \ \ldots \ w_m)$.

Using (3), a necessary condition for the solvability of (2) is then:

$$\exists\, u \in \Bbbk^{n\times 1}, \quad l := \mathrm{rank}_{\Bbbk}(M) \leq \mathrm{rank}_{\Bbbk}(A(u)) \leq \min\{r, n\}. \tag{4}$$

Suppose that (2) is solvable with a full row rank matrix v. Then, v admits a right inverse $t \in \Bbbk^{m\times r}$, i.e., $v\,t = I_r$. Hence, (2) yields $A(u) = M\,t$, which yields

$$\mathrm{im}_{\Bbbk}(A(u).) \subseteq \mathrm{im}_{\Bbbk}(M.), \tag{5}$$

and thus, we have:

$$\mathrm{im}_{\Bbbk}(A(u).) = \mathrm{im}_{\Bbbk}(M.). \tag{6}$$

The existence of $u \in \Bbbk^{n\times 1}$ satisfying (6) is then equivalent to:

1. $D_i\,u \in \mathrm{im}_{\Bbbk}(M.)$ for $i = 1, \ldots, r$, i.e., (5).
2. $\mathrm{rank}_{\Bbbk}(A(u)) = l := \mathrm{rank}_{\Bbbk}(M)$, i.e., $\dim_{\Bbbk}(\mathrm{span}\{D_i\,u\}_{i=1,\ldots,r}) = l$, i.e.:

$$\dim_{\Bbbk}(\ker_{\Bbbk}(A(u).) = r - l.$$

Remark 1. If $r = l$, then the last condition becomes $\ker_{\Bbbk}(A(u).) = 0$, i.e., the $D_i\,u$'s are $\Bbbk$-linearly independent, which yields the uniqueness of the matrix v.

Remark 2. If $\mathrm{rank}_{\Bbbk}(M) = \mathrm{rank}_{\Bbbk}(A(u))$, then (3) is equivalent to (6). Using (4), it holds if $l = \mathrm{rank}_{\Bbbk}(M) = r$ or $l = n$.

In this paper, we shall focus on the study of (6), i.e., on the above Conditions 1 and 2. In particular, we shall get the solutions $(u, v_1, \ldots, v_r)$ of (2) which are such that the v_i's are $\Bbbk$-linearly independent. In the demodulation problems for gearbox vibration signals [10], each row vector v_i contains Fourier coefficients of a signal to be estimated. The hypothesis that v has full row rank amounts to saying that the time signals are $\Bbbk$-linearly independent, which is a fair hypothesis in practice. The general rank factorization problem, i.e., (5), is studied in [6].

Let us now state again the approach developed in [13] for studying (2). We first suppose that $\ker_{\Bbbk}(.M) \neq 0$ (if $\ker_{\Bbbk}(.M) = 0$, see Remark 3 below). Let $L \in \Bbbk^{p\times n}$ be a full row rank matrix whose rows define a basis of $\ker_{\Bbbk}(.M)$, i.e.:

$$\ker_{\Bbbk}(.M) = \mathrm{im}_{\Bbbk}(.L), \quad p := \dim_{\Bbbk}(\ker_{\Bbbk}(.M)) = n - \mathrm{rank}_{\Bbbk}(M) = n - l.$$

Hence, we get $L\,M = 0$, which yields $\mathrm{im}_{\Bbbk}(M.) \subseteq \ker_{\Bbbk}(L.)$. Using $\dim_{\Bbbk}(\ker_{\Bbbk}(L.)) = n - p = \mathrm{rank}_{\Bbbk}(M)$, we obtain $\ker_{\Bbbk}(L.) = \mathrm{im}_{\Bbbk}(M.)$. Hence, Condition 1 above is equivalent to $D_i\,u \in \ker_{\Bbbk}(L.)$ for $i = 1, \ldots, r$, i.e., to the following linear system:

$$N\,u = 0, \quad N := ((L\,D_1)^T \ \ldots \ (L\,D_r)^T)^T \in \Bbbk^{p\,r\times n}.$$

If $\ker_{\Bbbk}(N.) = 0$, then $u = 0$, $A(u) = 0$ and (6) is not satisfied since $M \neq 0$.

Let us now suppose that $\ker_{\Bbbk}(N.) \neq 0$ and let $Z \in \Bbbk^{n \times d}$ be a full column matrix whose columns define a basis of $\ker_{\Bbbk}(N.)$, where $d := \dim_{\Bbbk}(\ker_{\Bbbk}(N.))$. The vectors $u \in \Bbbk^{n \times 1}$ satisfying Condition 1 are then defined by:

$$\forall\, \psi \in \Bbbk^{d \times 1}, \quad u = Z\, \psi. \tag{7}$$

Remark 3. If $\ker_{\Bbbk}(.M) = 0$, i.e., $\operatorname{rank}_{\Bbbk}(M) = n$, then $\operatorname{im}_{\Bbbk}(M.) = \Bbbk^{n \times 1}$. Condition 1 is $D_i\, u \in \Bbbk^{n \times 1}$ for $i = 1, \ldots, r$, which is satisfied for all $u \in \Bbbk^{n \times 1}$ and yields $Z = I_n$. Equivalently, if we set $L := 0$, then $N = 0$, and thus, $Z = I_n$.

Using (7), Condition 2, i.e., $\operatorname{rank}_{\Bbbk}(A(u)) = l$, is then equivalent to characterizing the set of all the $\psi \in \Bbbk^{d \times 1}$ which are such that:

$$\operatorname{rank}_{\Bbbk}(A(Z\, \psi)) = l \Leftrightarrow \dim_{\Bbbk}(\ker_{\Bbbk}(A(Z\, \psi).) = r - l. \tag{8}$$

Example 1. Let us consider the following matrices:

$$M = \begin{pmatrix} 3 & 5 \\ 4 & 7 \end{pmatrix}, \quad D_1 = I_2, \quad D_2 = \begin{pmatrix} 1 & 0 \\ 0 & 2 \end{pmatrix}.$$

Then, $l := \operatorname{rank}_{\Bbbk}(M) = r := 2$, which by Remark 3 shows that $Z = I_2$. Hence, (6) holds for all $u = \psi = (\psi_1 \quad \psi_2)^T$ satisfying $\det(A(\psi)) = \psi_1\, \psi_2 \neq 0$.

Let $X \in \Bbbk^{n \times l}$ be a full column rank whose columns define a basis of $\operatorname{im}_{\Bbbk}(M.)$. Since $\operatorname{im}_{\Bbbk}(M.) = \operatorname{im}_{\Bbbk}(X.)$, there exist $T \in \Bbbk^{m \times l}$ and a unique matrix $Y \in \Bbbk^{l \times m}$ such that $X = M\, T$ and $M = X\, Y$. Hence, we get $X\, (I_l - Y\, T) = 0$, which yields $Y\, T = I_l$ because X has full column rank. In particular, Y has full row rank.

By construction, $D_i\, Z\, \psi \in \ker_{\Bbbk}(L.) = \operatorname{im}_{\Bbbk}(M.) = \operatorname{im}_{\Bbbk}(X.)$ for all $\psi \in \Bbbk^{d \times 1}$, which shows that there exists a unique matrix $W_i \in \Bbbk^{l \times d}$ such that $D_i\, Z = X\, W_i$ for $i = 1, \ldots, r$. If we set $B(\psi) := (W_1\, \psi \; \ldots \; W_r\, \psi) \in \Bbbk^{l \times r}$, then we obtain:

$$\forall\, \psi \in \Bbbk^{d \times 1}, \quad A(Z\, \psi) = X\, B(\psi). \tag{9}$$

Using the fact that X has full column rank, we get $\ker_{\Bbbk}(A(Z\, \psi).) = \ker_{\Bbbk}(B(\psi).)$. Hence, using (8), (6) holds iff there exists $\psi \in \Bbbk^{d \times 1}$ such that:

$$\dim_{\Bbbk}(\ker_{\Bbbk}(B(\psi).)) = r - l \Leftrightarrow \operatorname{rank}_{\Bbbk}(B(\psi)) = l.$$

Hence, (6) holds iff the following set

$$\mathcal{P} := \{\psi \in \Bbbk^{d \times 1} \mid \operatorname{rank}_{\Bbbk}(B(\psi)) = l\} \tag{10}$$

is not empty. In particular, if $r = l$, then $\mathcal{P} = \{\psi \in \Bbbk^{d \times 1} \mid \det(B(\psi)) \neq 0\}$.

Let us suppose that $\mathcal{P} \neq \emptyset$ and let us show how to characterize the solutions (u, v) of (2). By construction, $u = Z\, \psi$ for $\psi \in \mathcal{P}$ and using (9), we get $A(Z\, \psi)\, v = X\, B(\psi)\, v = X\, Y$. Now, since X has full column rank, we obtain:

$$B(\psi)\, v = Y. \tag{11}$$

Since $\psi \in \mathcal{P}$, $B(\psi)$ admits a right inverse $E_\psi \in \Bbbk^{r\times l}$, i.e., $B(\psi)\, E_\psi = I_l$. Hence, if the matrix $C_\psi \in \Bbbk^{r\times(r-l)}$ is such that its columns define a basis of $\ker_{\Bbbk}(B(\psi).)$, i.e., $\ker_{\Bbbk}(B(\psi).) = \mathrm{im}_{\Bbbk}(C_\psi.)$, then all the solutions of (11) are given by:

$$\forall\, Y' \in \Bbbk^{(r-l)\times m}, \quad v = E_\psi\, Y + C_\psi\, Y'.$$

Note that $\det((E_\psi \quad C_\psi)) \neq 0$. Hence, v has full row rank iff $Y' \in \Bbbk^{(r-l)\times m}$ is chosen such that the matrix $(Y^T\; Y'^T)^T \in \Bbbk^{r\times m}$ has full row rank. If $r = l$, then we note that $C_\psi = 0$, which shows again that v is unique (see Remark 1).

Theorem 1 ([13]). *With the above notations, (6) holds iff the set $\mathcal{P}$ defined by (10) is not empty. If so, then*

$$\forall\, \psi \in \mathcal{P}, \quad \forall\, Y' \in \Bbbk^{(r-l)\times m}, \quad \begin{cases} u = Z\, \psi, \\ v = (E_\psi \quad C_\psi) \begin{pmatrix} Y \\ Y' \end{pmatrix}, \end{cases} \tag{12}$$

are solutions of (2). Moreover, v has full row rank iff the matrix $Y' \in \Bbbk^{(r-l)\times m}$ is chosen such that $(Y^T\; Y'^T)^T \in \Bbbk^{r\times m}$ has full row rank. Finally, $\mathcal{P}$ does not depend on choices of the bases while defining the matrices L, Z and X.

Remark 4. Note that $0 \notin \mathcal{P}$ since $B(0) = 0$. If $\psi \in \mathcal{P}$ and $\lambda \in \Bbbk\setminus\{0\}$, then $B(\lambda\, \psi) = \lambda\, B(\psi)$, i.e., $\lambda\, \psi \in \mathcal{P}$. Remark 6 of [13] shows that the solutions (12) are stable under the transformations $(u,\, v) \longmapsto (\lambda\, u,\, \lambda^{-1}\, v)$ for all $\lambda \in \Bbbk\setminus\{0\}$.

Note that the matrices X, Y, Z, $W_1, \ldots, W_r$, B of Theorem 1 can be obtained by linear algebra methods as well as the matrices E_ψ and C_ψ for a fixed $\psi \in \mathcal{P}$.

Example 2. We consider again Example 1. Taking $X = M$ and $Y = I_2$, we get:

$$W_1 = \begin{pmatrix} 7 & -5 \\ -4 & 3 \end{pmatrix}, \quad W_2 = \begin{pmatrix} 7 & -10 \\ -4 & 6 \end{pmatrix}, \quad B(\psi) = \begin{pmatrix} 7\,\psi_1 - 5\,\psi_2 & 7\,\psi_1 - 10\,\psi_2 \\ -4\,\psi_1 + 3\,\psi_2 & -4\,\psi_1 + 6\,\psi_2 \end{pmatrix},$$

$$\mathcal{P} = \{\psi \in \Bbbk^{2\times 1} \mid \det(B(\psi)) = \psi_1\, \psi_2 \neq 0\}, \quad C_\psi = 0,$$

$$E_\psi = \frac{1}{\psi_1\, \psi_2} \begin{pmatrix} -4\,\psi_1 + 6\,\psi_2 & -7\,\psi_1 + 10\,\psi_2 \\ 4\,\psi_1 - 3\,\psi_2 & 7\,\psi_1 - 5\,\psi_2 \end{pmatrix}.$$

Hence, the solutions of (2) are then defined by $u = \psi \in \mathcal{P}$ and $v = E_\psi$.

For more explicit examples, see [12,13].

3 Characterization of $\mathcal{P}$

In this section, we characterize the set $\mathcal{P}$ defined by (10). An element $\psi \in \mathcal{P}$ is such that at least one of the $C_r^l := r!/(l!\,(r-l)!)$ $l \times l$-minors $\mathbb{m}_k(\psi)$ of the matrix $B(\psi) := (W_1\, \psi \ \ldots \ W_r\, \psi) \in \Bbbk^{l\times r}$ does not vanish, i.e., we have:

$$\mathcal{P} = \Bbbk^{d\times 1} \setminus \{\psi \in \Bbbk^{d\times 1} \mid \mathbb{m}_k(\psi) = 0,\, k = 1, \ldots, C_r^l\}. \tag{13}$$

Note that $\mathfrak{m}_k$ is either 0 or a *homogeneous polynomial of degree l*, i.e., it satisfies $\mathfrak{m}_k(\lambda\,\psi) = \lambda^l\,\mathfrak{m}_k(\psi)$ for all $\lambda \in \Bbbk\backslash\{0\}$. Note also that C_r^l can be very large. Hence, we have to find a more tractable way to characterize $\mathcal{P}$.

If ψ is considered as an arbitrary vector of $\Bbbk^{d\times 1}$, then $B(\psi)$ can be interpreted as a matrix with polynomial entries in the ψ_i's. A natural framework for the study of $\mathcal{P}$ is thus *module theory* over a polynomial ring [7,14]. Based on module theory and computer algebra methods (*Gröbner bases*) [7,9,17], in this section, we give a characterization of $\mathcal{P}$ which is more tractable in practice. The corresponding algorithm is implemented in the OreModules package [5] but the `homalg` library (GAP) [1] or the `Singular` system [9] can also be used.

Let $R := \Bbbk[x_1, \ldots, x_d]$ be the commutative polynomial ring in $x_1, \ldots, x_d$ with coefficients in the field $\Bbbk$. Moreover, let us consider:

$$x := (x_1 \ldots x_d)^T, \quad B := (W_1\,x \ldots W_r\,x) \in R^{l\times r}.$$

Then, we can define the following *finitely presented* R-module [7,17]:

$$\mathcal{N} := \operatorname{coker}_R(B.) = R^{l\times 1}/\operatorname{im}_R(B.) = R^{l\times 1}/\left(B\,R^{r\times 1}\right).$$

The R-module $\mathcal{N}$ defines the obstruction of the surjectivity of the R-homomorphism $B. : R^{r\times 1} \longrightarrow R^{l\times 1}$, i.e., the obstruction for $B\,R^{r\times 1}$ to be equal to $R^{l\times 1}$.

Remark 5. In Remark 5 of [13], it is shown that, up to invertible matrices, B does not depend on arbitrary choices for the matrices L, X and Z (whose rows or columns define bases of certain $\Bbbk$-vector spaces). Hence, up to isomorphism, the R-module $\mathcal{N}$ is associated with the solvability of Problem (2).

We have the following *finite presentation* of the R-module $\mathcal{N}$ [7,14,17], i.e., the following *exact sequence* of R-modules:

$$0 \longleftarrow \mathcal{N} \xleftarrow{\kappa} R^{l\times 1} \xleftarrow{B.} R^{r\times 1}. \tag{14}$$

For each $\psi \in \Bbbk^{d\times 1}$, we can define the following *maximal ideal* of R

$$\mathfrak{m}_\psi := \langle x_1 - \psi_1, \ldots, x_d - \psi_d\rangle = \left\{\sum_{i=1}^{d} a_i\,(x_i - \psi_i) \mid a_i \in R,\ i = 1, \ldots, d\right\}, \tag{15}$$

i.e., $R/\mathfrak{m}_\psi$ is isomorphic to the field $\Bbbk$, which is denoted by $R/\mathfrak{m}_\psi \cong \Bbbk$ [7,14,17].

Applying the *covariant right exact functor* $(R/\mathfrak{m}_\psi) \otimes_R \cdot$ to (14), we obtain the following exact sequence of $\Bbbk$-vector spaces [7,17]:

$$0 \longleftarrow (R/\mathfrak{m}_\psi) \otimes_R \mathcal{N} \xleftarrow{\mathrm{id}\otimes\kappa} \Bbbk^{l\times 1} \xleftarrow{B(\psi).} \Bbbk^{r\times 1}. \tag{16}$$

Using properties of tensor products [17], $B(\psi). : \Bbbk^{r\times 1} \longrightarrow \Bbbk^{l\times 1}$ is surjective iff

$$\mathcal{N}/(\mathfrak{m}_\psi\,\mathcal{N}) \cong (R/\mathfrak{m}_\psi) \otimes_R \mathcal{N} \cong \Bbbk^{l\times 1}/\left(B(\psi)\,\Bbbk^{r\times 1}\right) = 0,$$

where $\mathfrak{m}_\psi \mathcal{N} := \left\{\sum_{i\in I} a_i\, n_i \mid a_i \in \mathfrak{m}_\psi,\, n_i \in \mathcal{N},\, \sharp I < \infty\right\}$, i.e., iff we have:

$$\mathcal{N} = \mathfrak{m}_\psi \mathcal{N}. \tag{17}$$

Note that $\mathfrak{m}_\psi \subset R$ yields $\mathfrak{m}_\psi \mathcal{N} \subset \mathcal{N}$, i.e., (17) is equivalent to $\mathcal{N} \subset \mathfrak{m}_\psi \mathcal{N}$, i.e.:

$$\mathcal{P} = \left\{\psi \in \Bbbk^{d\times 1} \mid \mathcal{N} \subset \mathfrak{m}_\psi \mathcal{N}\right\}.$$

Nakayama's lemma [7,14,17] gives a necessary condition for (17). Before stating again this well-known result, we rewrite (17) in terms of equations. Let $\kappa : R^{l\times 1} \longrightarrow \mathcal{N}$ be the R-homomorphism which sends $\eta \in R^{l\times 1}$ onto its *residue class* in $\mathcal{N}$, i.e., $\kappa(\eta') = \kappa(\eta)$ if there exists $\zeta \in R^{r\times 1}$ such that $\eta' = \eta + B\,\zeta$ [17]. Let f_j be the jth vector of the standard basis of $R^{l\times 1}$, i.e., the vector defined by 1 at the jth position and 0 elsewhere, and $y_j := \kappa(f_j)$ the residue class of f_j in $\mathcal{N}$. It can be easily show that $\{y_j\}_{j=1,\dots,l}$ is a set of generators of $\mathcal{N}$ [4,16]. Then, (17) is equivalent to the existence of $r_{jk} \in \mathfrak{m}_\psi$ such that $y_j = \sum_{k=1}^{l} r_{jk}\, y_k$ for $j = 1, \dots, l$. Noting $y := (y_1, \dots, y_l)^T$, (17) is equivalent to the existence of $G := (r_{jk}) \in {\mathfrak{m}_\psi}^{l\times l}$ such that $(I_l - G)\, y = 0$, which is then equivalent to the existence of $E \in R^{r\times l}$ such that $I_l = G + B\,E$, and thus:

$$\mathcal{P} = \left\{\psi \in \Bbbk^{d\times 1} \mid \exists\, G \in {\mathfrak{m}_\psi}^{l\times l},\ \exists\, E \in R^{r\times l} :\ I_l = G + B\,E\right\}.$$

Setting $x := \psi$, $I_l = G + B\,E$ yields $B(\psi)\,E(\psi) = I_l$ and $\operatorname{rank}_{\Bbbk}(B(\psi)) = l$.

Now, if $(I_l - G)^{\text{adj}}$ denotes *adjugate matrix* of $I_l - G$, using the standard identity $(I_l - G)^{\text{adj}}\,(I_l - G) = \det(I_l - G)$ [17], then we get $\det(I_l - G)\, y = 0$. Let $p(\lambda) := \det(\lambda\, I_l - G) = \lambda^l + p_1\, \lambda^{l-1} + \ldots + p_l$ be the characteristic polynomial of G. We can check that $p_i \in \mathfrak{m}_\psi$ for $i = 1, \dots, l$, and thus, $\det(I_l - G) = p(1) = 1 + a$ for a certain $a \in \mathfrak{m}_\psi$. Since $1 \notin \mathfrak{m}_\psi$, $\det(I_l - G) \neq 0$ and each generator y_j of $\mathcal{N}$ satisfies the non-trivial equation $(1 + a)\, y_j = 0$ for $j = 1, \dots, l$. Hence, we get

$$0 \neq 1 + a \in \operatorname{ann}_R(\mathcal{N}) := \{b \in R \mid b\,\mathcal{N} = 0\}, \tag{18}$$

where $\operatorname{ann}_R(\mathcal{N})$ is an ideal of R called the *annihilator* of $\mathcal{N}$. Nakayama's lemma asserts (17) implies (18) [7,14,17]. In particular, (18) implies that the R-module $\mathcal{N}$ is *torsion*, namely, $t(\mathcal{N}) := \{n \in \mathcal{N} \mid \exists\, 0 \neq b \in R :\ b\,n = 0\} = \mathcal{N}$ [7,17].

Let us consider a family of generators $\{g_i\}_{i=1,\dots,t}$ of $\operatorname{ann}_R(\mathcal{N})$, i.e.:

$$\operatorname{ann}_R(\mathcal{N}) = \langle g_1, \dots, g_t\rangle := \left\{\sum_{i=1}^{t} a_i\, g_i \mid a_1, \dots, a_t \in R\right\}. \tag{19}$$

A set of generators $\{g_i\}_{i=1,\dots,t}$ of $\operatorname{ann}_R(\mathcal{N})$ can be computed by the command PiPolynomial of OreModules [5]. See also `Homalg` [1] and `Singular` [9]. Note that t is usually much smaller than C_r^l. Now, (18) shows that there exist $q_i \in R$ for $i = 1, \dots, t$ satisfying $1 + a = \sum_{i=1}^{t} q_i\, g_i$. Evaluating this identity at the point $x = \psi$, we obtain the following Bézout identity:

$$\sum_{i=1}^{t} q_i(\psi)\, g_i(\psi) = 1. \tag{20}$$

Hence, $\psi \in \Bbbk^{d\times 1}$ must to be chosen such that the generators $g_1, \ldots, g_t$ of $\operatorname{ann}_R(\mathcal{N})$ do not simultaneously vanish at ψ.

Remark 6. For two finitely generated R-modules $\mathcal{M}$ and $\mathcal{N}$, it can be proved that $\mathcal{M} \otimes_R \mathcal{N} = 0$ implies $\operatorname{ann}_R(\mathcal{M}) + \operatorname{ann}_R(\mathcal{N}) = R$. See, e.g., Corollary 4.9 of [7]. Setting $\mathcal{M} := R/\mathfrak{m}_\psi$ and using $\operatorname{ann}_R(\mathcal{M}) = \mathfrak{m}_\psi$, a necessary condition for $\psi \in \mathcal{P}$ is then $\mathfrak{m}_\psi + \operatorname{ann}_R(\mathcal{N}) = \langle x_1 - \psi_1, \ldots, x_d - \psi_d, g_1, \ldots, g_t\rangle = R$, i.e., $\sum_{i=1}^t q_i\, g_i + \sum_{j=1}^d r_j\,(x_j - \psi_j) = 1$ for certain $q_i, r_j \in R$, $i = 1, \ldots, t$, $j = 1, \ldots, d$, which, by evaluation at $x = \psi$, yields again (20).

If I is an ideal of R, we can define the *algebraic set* of the *affine space* $\Bbbk^{d\times 1}$:

$$V_\Bbbk(I) := \{\psi \in \Bbbk^{d\times 1} \mid \forall\, g \in I:\; g(\psi) = 0\}.$$

If $I = \langle g_1, \ldots, g_t\rangle$, i.e., I is generated by the g_i's, then $V_\Bbbk(I)$ is the common zeros $\psi \in \Bbbk^{d\times 1}$ of all the g_i's, i.e., $V_\Bbbk(I) = \{\psi \in \Bbbk^{d\times 1} \mid g_i(\psi) = 0,\ i = 1, \ldots, t\}$. Hence:

$$V_\Bbbk(\operatorname{ann}_R(\mathcal{N})) = V_\Bbbk(\langle g_1, \ldots, g_t\rangle) = \bigcap_{i=1}^t V_\Bbbk(\langle g_i\rangle). \tag{21}$$

Hence, a necessary condition for (17) to hold is $\psi \in \Bbbk^{d\times 1} \backslash V_\Bbbk(\operatorname{ann}_R(\mathcal{N}))$. This condition is also sufficient as explained in the following remark.

Remark 7. Let $\operatorname{Fitt}_0(\mathcal{N})$ be the 0th *Fitting ideal* of $\mathcal{N}$, namely, the ideal of R defined by all the $l \times l$-minors of B [7]. Proposition 20.7 of [7] then yields:

$$\operatorname{ann}_R(\mathcal{N})^l \subseteq \operatorname{Fitt}_0(\mathcal{N}) \subseteq \operatorname{ann}_R(\mathcal{N}).$$

If $\sqrt{I} := \{a \in R \mid \exists\, n \in \mathbb{Z}_{\geq 0}:\; a^n \in I\}$ denotes the *radical* of I [7,14], then

$$\sqrt{\operatorname{ann}_R(\mathcal{N})} = \sqrt{\operatorname{Fitt}_0(\mathcal{N})} \;\Rightarrow\; V_\Bbbk(\operatorname{ann}_R(\mathcal{N})) = V_\Bbbk(\operatorname{Fitt}_0(\mathcal{N}))),$$

which also shows again (13), i.e., $\mathcal{P} = \Bbbk^{d\times 1} \backslash V_\Bbbk(\operatorname{Fitt}_0(\mathcal{N})))$.

In Sect. 4, we shall give a more useful proof of $\mathcal{P} = \Bbbk^{d\times 1} \backslash V_\Bbbk(\operatorname{ann}_R(\mathcal{N}))$.

Example 3. We consider the following matrices:

$$M = \begin{pmatrix} 0 & 0 \\ -147360 & -96804 \\ 0 & 0 \end{pmatrix}, \quad D_1 = \begin{pmatrix} 0 & 0 & 0 \\ 0 & 54 & -31 \\ 0 & 0 & 0 \end{pmatrix},$$

$$D_2 = \begin{pmatrix} 0 & 0 & 0 \\ 0 & -58 & -77 \\ 0 & 0 & 0 \end{pmatrix}, \quad D_3 = \begin{pmatrix} 0 & 0 & 0 \\ 79 & 0 & 0 \\ 0 & 0 & 0 \end{pmatrix}.$$

We can check that $l := \operatorname{rank}_\Bbbk(M) = 1 < r = 3$,

$$X = \begin{pmatrix} 0 \\ -147360 \\ 0 \end{pmatrix}, \; Y = \begin{pmatrix} 1 & \frac{8067}{12280} \end{pmatrix}, \; L = \begin{pmatrix} 1 & 0 & 0 \\ 0 & 0 & 1 \end{pmatrix}, \; N = 0, \; Z = I_3,$$

and $\psi = (\psi_1 \quad \psi_2 \quad \psi_3)^T$. If $c := 1/(147360)$, then we have:

$$W_1 = c\,(0 \quad -54 \quad 31)\,, \quad W_2 = c\,(0 \quad 58 \quad 77)\,, \quad W_3 = c\,(-79 \quad 0 \quad 0)\,,$$
$$B(\psi) = c\,(-54\,\psi_2 + 31\,\psi_3 \quad 58\,\psi_2 + 77\,\psi_3 \quad -79\,\psi_1)\,.$$

Let $R = \Bbbk[x_1, x_2, x_3]$, $x := (x_1 \quad x_2 \quad x_3)^T$, $B := (W_1\,x \quad W_2\,x \quad W_3\,x) \in R^{1\times 3}$. The R-module $\mathcal{N} = R/\left(B\,R^{3\times 1}\right) = R/I$, where $I = \langle B_1, B_2, B_3\rangle$ is the ideal generated by the three entries B_i's (i.e., 1×1-minors $\mathfrak{m}_k$) of B, is clearly a torsion R-module. The R-module $\mathcal{N}$ is generated by the residue class y of 1 in $\mathcal{N}$ and we can check that $\operatorname{ann}_R(\mathcal{N}) = I = \mathfrak{m}_0 := \langle x_1, x_2, x_3\rangle$. Such a computation can directly be obtained by the PiPolynomial command of the OreModules package [5]. Hence, we get:

$$V_{\Bbbk}(\operatorname{ann}_R(\mathcal{N})) = \{(0\;0\;0)^T\} \;\Rightarrow\; \mathcal{P} = \Bbbk^{3\times 1} \setminus \{0\}.$$

Remark 8. Since the generators g_i's of $\operatorname{ann}_R(\mathcal{N})$ can be chosen to be homogeneous polynomials, $0 \in V_{\Bbbk}(\operatorname{ann}_R(\mathcal{N}))$, which shows that $0 \notin \mathcal{P}$ (see Remark 4).

4 Local and Global Studies of the Solution Space

4.1 Existence of a Local/Global Right Inverse E of B

Let us first study the problem of computing a right inverse E_ψ of $B(\psi)$ for $\psi \in \mathcal{P}$. With the notation (19), let us consider the following integral domain

$$S_{g_i}^{-1}R := \left\{ \frac{a}{g_i^n} \mid a \in R,\ n \in \mathbb{Z}_{\geq 0} \right\},$$

i.e., the *localization* of R at the *multiplicatively closed set* $S_{g_i} := \{g_i^n \mid n \in \mathbb{Z}_{\geq 0}\}$ [7,14,17]. We can then consider the *localization* of $\mathcal{N}$ with respect of the powers of g_i, namely, the $S_{g_i}^{-1}R$-module defined by $S_{g_i}^{-1}\mathcal{N} := \{s^{-1}\,n \mid s \in S_{g_i},\ n \in \mathcal{N}\}$. It is well-known $S_{g_i}^{-1}R$ is a *flat* R-module [7,14,17], which yields the isomorphism

$$S_{g_i}^{-1}\mathcal{N} \cong (S_{g_i}^{-1}R)^{l\times 1}/\left(B\,(S_{g_i}^{-1}R)^{r\times 1}\right)$$

of $S_{g_i}^{-1}R$-modules. Hence, $S_{g_i}^{-1}\mathcal{N}$ can be seen as the $S_{g_i}^{-1}R$-module obtained from $\mathcal{N}$ by extending the scalars from R to $S_{g_i}^{-1}R$. See, e.g., [7,14,17]. By definition (see (19)), we have $g_i\,\mathcal{N} = 0$ and $g_i^{-1} \in S_{g_i}^{-1}R$, which yields $S_{g_i}^{-1}\mathcal{N} = 0$, i.e.:

$$B\,(S_{g_i}^{-1}R)^{r\times 1} = (S_{g_i}^{-1}R)^{l\times 1}, \quad i = 1, \ldots, t.$$

Hence, there exists $E_{g_i} \in (S_{g_i}^{-1}R)^{r\times l}$ such that $B\,E_{g_i} = I_l$, i.e., E_{g_i} is a right inverse of B defined over the *Zariski distinguished/basic open subset* of $\Bbbk^{d\times 1}$ [7]

$$D(g_i) := \Bbbk^{d\times 1} \setminus V_{\Bbbk}(\langle g_i\rangle), \quad i = 1, \ldots, t,$$

i.e., $E_{g_i}(\psi)$ is a right inverse of $B(\psi)$ for all $\psi \in D(g_i)$, where $E_{g_i}(\psi)$ denotes the value of the matrix E_{g_i} evaluated at $x := \psi$. The matrix E_{g_i} can be computed by the LocalLeftInverse command of the OreModules package.

Remark 9. Using (14), we get the *split exact sequence* of $S_{g_i}^{-1}R$-modules [17]:

$$0 = S_{g_i}^{-1}\mathcal{N} \xleftarrow{S_{g_i}^{-1}\kappa} (S_{g_i}^{-1}R)^{l\times 1} \underset{B.}{\overset{E_{g_i}}{\rightleftarrows}} (S_{g_i}^{-1}R)^{r\times 1}.$$

Thus, we have $S_{g_i}^{-1}\,\mathrm{im}_R(B.) = \mathrm{im}_{S_{g_i}^{-1}R}(B.) \cong (S_{g_i}^{-1}R)^{l\times 1}$ for $i = 1,\ldots,t$, i.e., $S_{g_i}^{-1}R$-module $S_{g_i}^{-1}\,\mathrm{im}_R(B.)$ is free of rank l.

From the above results, $\mathrm{rank}_{\Bbbk}(B(\psi)) = l$ for all $\psi \in \Bbbk^{d\times 1} \setminus \bigcap_{i=1}^t V_{\Bbbk}(\langle g_i\rangle)$. Using (21), (2) has solutions in the complementary $\mathcal{P}$ of the Zariski closed subset $V_{\Bbbk}(\mathrm{ann}_R(\mathcal{N}))$ in $\Bbbk^{d\times 1}$. Hence, if $\mathcal{P} \neq \emptyset$ (e.g., $\mathrm{ann}_R(\mathcal{N}) \neq \langle 0\rangle$ and $\Bbbk$ is algebraically closed), then (2) *generically* has solutions in the sense of algebraic geometry, i.e., outside the Zariski closed subset $V_{\Bbbk}(\mathrm{ann}_R(\mathcal{N}))$ of $\Bbbk^{d\times 1}$ [7,14]. Moreover, we have:

$$\begin{aligned}\mathcal{P} &= \Bbbk^{d\times 1} \setminus \bigcap_{i=1}^t V_{\Bbbk}(\langle g_i\rangle) = \bigcup_{i=1}^t (\Bbbk^{d\times 1} \setminus V_{\Bbbk}(\langle g_i\rangle)) = \bigcup_{i=1}^t D(g_i)\\ &= \{\psi \in \Bbbk^{d\times 1} \mid \exists\, i \in [\![1,\ldots,t]\!] : \psi \notin V_{\Bbbk}(\langle g_i\rangle)\}.\end{aligned}$$

Since $\mathcal{P} \cap D(g_i) = D(g_i)$, $D(g_i)$ is also an open subset of $\mathcal{P}$ for the induced *Zariski topology* [7,14]. Finally, $\mathcal{P}$ is an open subset of the *irreducible* affine set $\Bbbk^{d\times 1} = V_{\Bbbk}(\langle 0\rangle)$, i.e., which shows that $\mathcal{P}$ is a *quasi-affine variety* [9].

Theorem 2. *Let $R = \Bbbk[x_1,\ldots,x_d]$, $x = (x_1\ \ldots\ x_d)^T$, $W_i \in \Bbbk^{l\times d}$, $i = 1,\ldots,r$, be the matrices defined in Sect. 2, $B = (W_1\,x\ \ldots\ W_r\,x) \in R^{l\times r}$, the R-module $\mathcal{N} = R^{l\times 1}/\big(B\,R^{r\times 1}\big)$ and its annihilator $\mathrm{ann}_R(\mathcal{N}) = \langle g_1,\cdots,g_t\rangle$. Then, we get:*

$$\mathcal{P} = D(\mathrm{ann}_R(\mathcal{N})) := \Bbbk^{d\times 1} \setminus V_{\Bbbk}(\mathrm{ann}_R(\mathcal{N})). \tag{22}$$

Hence, Problem (2) has solutions in the complementary $\mathcal{P}$ of the closed algebraic set $V_{\Bbbk}(\mathrm{ann}_R(\mathcal{N}))$ in $\Bbbk^{d\times 1}$. Moreover, $\mathrm{ann}_R(\mathcal{N}) = \langle 0\rangle$ yields $\mathcal{P} = \emptyset$ and the converse holds if $\Bbbk$ is algebraically closed.

The quasi-affine variety $\mathcal{P}$ has a finite open cover defined by $\mathcal{P} = \bigcup_{i=1}^t D(g_i)$, where $D(g_i) := \Bbbk^{d\times 1}\setminus V_{\Bbbk}(\langle g_i\rangle)$ is a basic open subset of $\Bbbk^{d\times 1}$ (of $\mathcal{P}$). Finally, there exist $E_{g_i} \in (S_{g_i}^{-1}R)^{r\times l}$ such that $B\,E_{g_i} = I_l$ for $i = 1,\ldots,t$, i.e., for each $D(g_i)$, there exists a smooth right inverse E_{g_i} of B, i.e., $\psi \in D(g_i) \longmapsto E_{g_i}(\psi)$.

Using Theorem 2, $B(\psi)$ admits a global right inverse $E(\psi)$ over $\mathcal{P}$, i.e., $B(\psi)\,E(\psi) = I_l$ for all $\psi \in \mathcal{P}$, iff the ideal $\mathrm{ann}_R(\mathcal{N})$ can be generated by a single element $g \in R$, i.e., $\mathrm{ann}_R(\mathcal{N}) = \langle g\rangle$, in which case $\mathrm{ann}_R(\mathcal{N})$ is principal [7,17]. For instance, it is the case if we have $l = r$ and $g := \det(B) \neq 0$ (see Example 2), or if $d = 1$, i.e., $R = \Bbbk[x_1]$ is a *principal ideal domain*, namely, every ideal of R (e.g., $\mathrm{ann}_R(\mathcal{N})$) can be generated by a single element g of R which can be obtained by Euclidean division [7,17]. Let us now study the general case. Let $\mathrm{ann}_R(\mathcal{N}) = \langle g_1,\ldots,g_t\rangle$, g be a greatest common divisor of all the g_i's and $g_i' := g_i/g \in R$ for $i = 1,\ldots,t$. We then get $\mathrm{ann}_R(\mathcal{N}) = \langle g\rangle\langle g_1',\ldots,g_t'\rangle$, which shows that $\mathrm{ann}_R(\mathcal{N})$ is principal iff so is

$\langle g'_1, \ldots, g'_t \rangle$, i.e., iff $\langle g'_1, \ldots, g'_t \rangle = R$, i.e., iff there exist $h_i \in R$ for $i = 1, \ldots, t$ such that $\sum_{i=1}^t h_i \, g'_i = 1$. If $\Bbbk = \mathbb{C}$, using *Hilbert's Nullstellensatz* [7,14], this Bézout identity is equivalent to the fact that all the g'_i's have no common zeros in $\mathbb{C}^{d \times 1}$, which can be checked by a *Gröbner basis computation* [7,9]. Now, using Remark 8, $0 \in V_{\mathbb{C}}(\mathrm{ann}_R(\mathcal{N})) = V_{\mathbb{C}}(\langle g \rangle) \bigcup V_{\mathbb{C}}(\langle g'_1, \ldots, g'_t \rangle)$, i.e., $g(0) = 0$ or $g'_i(0) = 0$ for all $i = 1, \ldots, t$. In particular, if $g = 1$, then $\mathrm{ann}_R(\mathcal{N})$ is not a principal ideal. Finally, if $\langle g'_1, \ldots, g'_t \rangle = R$, i.e., $\mathrm{ann}_R(\mathcal{N}) = \langle g \rangle$, then $g(0) = 0$.

The problem of finding the least number of generators $\mu(I)$ of an ideal I is a well-known difficult problem in module theory (see, e.g., [14,15]). In our problem, $\mu(\mathrm{ann}_R(\mathcal{N}))$ is the least number of open sets $D(g_i)$'s which defines a finite open cover of $\mathcal{P}$. Since $\mathrm{ann}_R(\mathcal{N})$ is generated by homogeneous polynomials, it can be proved that $\mu\left(\mathrm{ann}_R(\mathcal{N})\right) = \mu\left(\mathrm{ann}_R(\mathcal{N})/\mathrm{ann}_R(\mathcal{N})^2\right)$ (see Ex. 12 of Chap. V.5 of [14]), where $\mathrm{ann}_R(\mathcal{N})/\mathrm{ann}_R(\mathcal{N})^2$ is the $R/\mathrm{ann}_R(\mathcal{N})$-module *conormal module*.

Example 4. In Example 3, we proved that $g_i = x_i$ for $i = 1, 2, 3$. Hence, if $D(x_i) := \Bbbk^{3 \times 1} \setminus V_{\Bbbk}(\langle x_i \rangle) = \{\psi = (\psi_1 \quad \psi_2 \quad \psi_3)^T \in \Bbbk^{3 \times 1} \mid \psi_i \neq 0\}$ for $i = 1, 2, 3$, then we have $\mathcal{P} = \bigcup_{i=1}^3 D(x_i)$. Moreover, we can check that

$$\begin{aligned}
&\forall\, \psi \in D(x_1): \; E_{x_1}(\psi) := c^{-1} \begin{pmatrix} 0 & 0 & -\dfrac{1}{79\,\psi_1} \end{pmatrix}^T, \\
&\forall\, \psi \in D(x_2): \; E_{x_2}(\psi) := (5956\, c)^{-1} \begin{pmatrix} -\dfrac{77}{\psi_2} & \dfrac{31}{\psi_2} & 0 \end{pmatrix}^T, \\
&\forall\, \psi \in D(x_3): \; E_{x_3}(\psi) := (2978\, c)^{-1} \begin{pmatrix} \dfrac{29}{\psi_3} & \dfrac{27}{\psi_3} & 0 \end{pmatrix}^T,
\end{aligned}$$

are local right inverses of B, i.e., $B\, E_{\psi_i} = 1$, on $D(x_i)$ for $i = 1, 2, 3$. They are computed by the command LocalLeftInverse of the OreModules package [5]. Since $g := \gcd(g_1, g_2, g_3) = 1$, as shown above, $\mathrm{ann}_R(\mathcal{N})$ is not principal, and thus, no global right inverse E of B exists over the whole space $\mathcal{P}$. Using $\mathrm{ann}_R(\mathcal{N}) = \mathfrak{m}_0 = \langle x_1, x_2, x_3 \rangle$, the $R/\mathfrak{m}_0 \cong \Bbbk$-module $\mathfrak{m}_0/\mathfrak{m}_0^2$ is defined by the $\Bbbk$-linear combinations of the generators $\overline{x_i}$'s of $\mathfrak{m}_0/\mathfrak{m}_0^2$, where $\overline{x_i}$ denotes the residue class of x_i in $\mathfrak{m}_0/\mathfrak{m}_0^2$, i.e., $\mathfrak{m}_0/\mathfrak{m}_0^2 \cong \Bbbk^{3 \times 1}$, which shows that $t = \mu(\mathrm{ann}_R(\mathcal{N})) = 3$ is the least number of distinguished open sets of $\Bbbk^{3 \times 1}$ defining a cover of $\mathcal{P}$.

4.2 Existence of a Local/Global Basis C of $\ker_R(B.)$

To study the local/global structure of the solution space (12) of (2), we now investigate the existence of a local/global basis $C(\psi)$ of $\ker(B(\psi).)$ over $\mathcal{P}$.

As explained in Sect. 3, a matrix $C \in R^{r \times s}$ can be computed satisfying $\ker_R(B.) = \mathrm{im}_R(C.)$ (use, e.g., the SyzygyModule command of the OreModules package). By construction, we have the exact sequence of R-modules:

$$0 \longleftarrow \mathcal{N} \xleftarrow{\kappa} R^{l \times 1} \xleftarrow{B.} R^{r \times 1} \xleftarrow{C.} R^{s \times 1}. \tag{23}$$

Let $Q(R) := \Bbbk(x_1, \ldots, x_d)$ be the *field of fractions* of R, i.e., the field of rational functions in the x_i's with coefficients in $\Bbbk$ [7,17]. The *rank* of a finitely generated R-module $\mathcal{L}$ is $\mathrm{rank}_R(\mathcal{L}) := \dim_{Q(R)} (Q(R) \otimes_R \mathcal{L})$. Since $\mathcal{N}$ is a torsion

R-module, $\mathrm{rank}_R(\mathcal{N}) = 0$, the *Euler-Poincaré characteristic* applied to (14) yields $\mathrm{rank}_R(\ker_R(B.)) = r - l$ [7,17], which yields $s \geq r - l$. The equality holds, i.e., $s = r - l$, iff $\ker_R(B.)$ is a *free* R-module, i.e., $\ker_R(B.) \cong R^{r-l}$ [17].

The problem of recognizing whether or not a module is free is an open question in module theory [14,15,17]. It can be effectively solved for $R = \Bbbk[x_1, \ldots, x_d]$ due to the *Quillen-Suslin theorem* [14,15,17]. The Quillen-Suslin theorem is implemented in the QuillenSuslin package [8]. Hence, we can effectively test whether or not $\ker_R(B.)$ is a free R-module and if so, compute a basis of $\ker_R(B.)$, namely, a full column rank matrix $C \in R^{r \times (r-l)}$ such that $\ker_R(B.) = \mathrm{im}_R(C.)$ [8]. We then have $\ker_{\Bbbk}(B(\psi).) = \mathrm{im}_{\Bbbk}(C(\psi).)$ for all $\psi \in \mathcal{P}$, i.e., C is a global basis of $\ker_R(B.)$ on $\mathcal{P}$. In particular, C is a local basis on $D(g_i)$ for all $i = 1, \ldots, t$. Using Theorems 1 and 2, we finally obtain that

$$\forall\, \psi \in D(g_i), \quad \forall\, Y' \in \Bbbk^{(r-l)\times m}, \quad \begin{cases} u = Z\,\psi, \\ v = (E_{g_i}(\psi) \quad C(\psi)) \begin{pmatrix} Y \\ Y' \end{pmatrix}, \end{cases} \tag{24}$$

are solutions of (2) on $D(g_i)$. If $t = 1$, these solutions are globally defined on $\mathcal{P}$.

If $d = 1$, then $R = \Bbbk[x_1]$ is a principal ideal domain, which implies that $\mathrm{ann}_R(\mathcal{N}) = \langle g_1 \rangle$ and $\ker_R(B.)$ is a free R-module of rank $r - l$ [7,17]. Let us show how to compute g_1, $E_{g_1} \in (S_{g_1}^{-1} R)^{r \times l}$ and a basis of $\ker_R(B.)$, i.e., a full column rank matrix $C \in R^{r\times(r-l)}$ satisfying $\ker_R(B.) = \mathrm{im}_R(C.)$. If we note $W := (W_1 \ \ldots \ W_r) \in \Bbbk^{l \times r}$, then we have $B = W\,x_1$. Hence, if $\psi_1 \neq 0$, then we get $\mathrm{rank}_{\Bbbk}(B(\psi_1)) = \mathrm{rank}_{\Bbbk}(W)$, which yields $\mathcal{P} = \emptyset$ if $\mathrm{rank}_{\Bbbk}(W) < l$, i.e., $g_1 = 0$, or $\mathcal{P} = \Bbbk \backslash \{0\}$ if $\mathrm{rank}_{\Bbbk}(W) = l$, i.e., $g_1 = x_1$. In the latter case, if $F \in \Bbbk^{r \times l}$ is a right inverse of W, i.e., $W\,F = I_l$, then $E_{g_1} = x_1^{-1}\,F$ is a right inverse of B. Moreover, let $C \in \Bbbk^{r\times(r-l)}$ be a matrix whose columns define a basis of $\ker_{\Bbbk}(W.)$. Then, we have $\ker_R(B.) = \mathrm{im}_R(C.) \cong R^{r-l}$. We note that E and C can be computed by standard linear algebra methods.

Example 5. Let us consider the following matrices:

$$D_1 = \begin{pmatrix} 1 & 0 & 0 & 0 \\ 0 & 0 & 0 & 0 \\ 0 & 0 & 0 & 0 \\ 0 & 0 & 0 & -1 \end{pmatrix}, \quad D_2 = \begin{pmatrix} 0 & 0 & 0 & 0 \\ 0 & 1 & 0 & 0 \\ 0 & 0 & -1 & 0 \\ 0 & 0 & 0 & 0 \end{pmatrix}, \quad D_3 = \begin{pmatrix} 0 & 0 & 0 & 1 \\ 0 & 0 & 0 & 0 \\ 0 & 0 & 0 & 0 \\ -1 & 0 & 0 & 0 \end{pmatrix},$$

$$D_4 = \begin{pmatrix} 0 & 0 & 0 & 0 \\ 0 & 0 & 1 & 0 \\ 0 & -1 & 0 & 0 \\ 0 & 0 & 0 & 0 \end{pmatrix}, \quad M = \begin{pmatrix} 1 & 0 & 0 & 1 \\ 0 & 1 & -1 & 0 \\ 0 & 1 & 1 & 0 \\ 1 & 0 & 0 & 1 \end{pmatrix}.$$

We can easily check that $l := \mathrm{rank}_{\Bbbk}(M) = 3$, $r = 4$ and:

$$X = \begin{pmatrix} 1 & 0 & 0 \\ 0 & 1 & -1 \\ 0 & 1 & 1 \\ 1 & 0 & 0 \end{pmatrix}, \quad Y = \begin{pmatrix} 1 & 0 & 0 & 1 \\ 0 & 1 & 0 & 0 \\ 0 & 0 & 1 & 0 \end{pmatrix}, \quad L = (1 \quad 0 \quad 0 \quad -1), \quad Z = \begin{pmatrix} -1 & 0 & 0 \\ 0 & 0 & 1 \\ 0 & 1 & 0 \\ 1 & 0 & 0 \end{pmatrix},$$

$$W_1 = \begin{pmatrix} -1 & 0 & 0 \\ 0 & 0 & 0 \\ 0 & 0 & 0 \end{pmatrix}, \; W_2 = -\frac{1}{2} \begin{pmatrix} 0 & 0 & 0 \\ 0 & 1 & -1 \\ 0 & 1 & 1 \end{pmatrix}, \; W_3 = \begin{pmatrix} 1 & 0 & 0 \\ 0 & 0 & 0 \\ 0 & 0 & 0 \end{pmatrix}, \; W_4 = -\frac{1}{2} \begin{pmatrix} 0 & 0 & 0 \\ 0 & -1 & 1 \\ 0 & 1 & 1 \end{pmatrix},$$

$$R = \Bbbk[x_1, x_2, x_3], \quad B = \begin{pmatrix} -x_1 & 0 & x_1 & 0 \\ 0 & -\frac{1}{2}(x_2 - x_3) & 0 & \frac{1}{2}(x_2 - x_3) \\ 0 & -\frac{1}{2}(x_2 + x_3) & 0 & -\frac{1}{2}(x_2 + x_3) \end{pmatrix}.$$

If $g_1 := x_1 (x_2^2 - x_3^2)$, then $\mathrm{ann}_R(\mathcal{N}) = \langle g_1 \rangle$. Hence, $t = 1$ and $\mathcal{P} = \Bbbk^3 \setminus V_{\Bbbk}(\langle g_1 \rangle)$, where $V_{\Bbbk}(\langle g_1 \rangle) = \{x_1 = 0\} \cup \{x_2 - x_3 = 0\} \cup \{x_2 + x_3 = 0\}$. We can check that the R-module $\ker_R(B.)$ is free of rank 1, i.e., $\ker_R(B.) \cong R$. Using [4,8], we get $\ker_R(B.) = \mathrm{im}_R(C.)$, where $C = (1 \quad 0 \quad 1 \quad 0)^T \in R^{4 \times 1}$. Finally, using the OreModules package, we obtain that the following matrix

$$E_{g_1} = \frac{1}{g_1} \begin{pmatrix} 0 & 0 & 0 \\ 0 & -x_1 (x_2 + x_3) & -x_1 (x_2 - x_3) \\ x_2^2 - x_3^2 & 0 & 0 \\ 0 & x_1 (x_2 + x_3) & -x_1 (x_2 - x_3) \end{pmatrix}$$

is a right inverse of B, i.e., $B \, E_{g_1} = I_3$. Hence, all the solutions of (2) with full row rank matrices v can be expressed by a single closed-form given by (24) with $t = 1$ and for all $\psi \in \mathcal{P}$ and for all $Y' = (y_1' \quad y_2' \quad y_3' \quad y_4') \in \Bbbk^{1 \times 4}$ such that:

$$\det((Y^T \quad Y'^T)^T) = y_4' - y_1' \neq 0.$$

Let us now suppose that the R-module $\ker_R(B.)$ is not free. Let us study the module structure of the $S_{g_i}^{-1}R$-module $\ker_{S_{g_i}^{-1}R}(B.)$. Since $S_{g_i}^{-1}R$ is a flat R-module, the functor $S_{g_i}^{-1}R \otimes_R \cdot$ is exact [7,14,17]. Hence, applying $S_{g_i}^{-1}R \otimes_R \cdot$ to (23) and using the fact that $S_{g_i}^{-1}R \otimes_R \mathcal{N} \cong S_{g_i}^{-1}\mathcal{N} = 0$, we get the following split exact sequence of $S_{g_i}^{-1}R$-modules [7,17]:

$$0 \longleftarrow (S_{g_i}^{-1}R)^{l \times 1} \xleftarrow{B.} (S_{g_i}^{-1}R)^{r \times 1} \xleftarrow{C.} (S_{g_i}^{-1}R)^{s \times 1}.$$

See also Remark 9. Hence, we first obtain

$$\ker_{S_{g_i}^{-1}R}(B.) = \mathrm{im}_{S_{g_i}^{-1}R}(C.), \tag{25}$$

and then $(S_{g_i}^{-1}R)^r \cong (S_{g_i}^{-1}R)^l \oplus \ker_{S_{g_i}^{-1}R}(B.)$, which shows that $\ker_{S_{g_i}^{-1}R}(B.)$ is a *stably free* $S_{g_i}^{-1}R$-module of rank $r - l$ [17]. Thus, $\ker_{S_{g_i}^{-1}R}(B.)$ is not necessarily a free $S_{g_i}^{-1}R$-module. Recognizing whether or not a stably free $S_{g_i}^{-1}R$-module is free is an open question in module theory as well as the problem of computing bases of free $S_{g_i}^{-1}R$-modules. For more details, see, e.g., [14,15,17].

If $\ker_{S_{g_i}^{-1}R}(B.)$ is a free $S_{g_i}^{-1}R$-module of rank $r-l$, then there exists a full column rank matrix $C_{g_i} \in (S_{g_i}^{-1}R)^{r \times (r-l)}$ such that

$$\ker_{S_{g_i}^{-1}R}(B.) = \mathrm{im}_{S_{g_i}^{-1}R}(C_{g_i}.) \cong (S_{g_i}^{-1}R)^{(r-l)}, \tag{26}$$

i.e., the $r-l$ columns of the matrix C_{g_i} define a basis of the free $S_{g_i}^{-1}R$-module $\ker_{S_{g_i}^{-1}R}(B.)$. Hence, we obtain $\ker_{\Bbbk}(B(\psi).) = \mathrm{im}_{\Bbbk}(C_{g_i}(\psi).)$ for all $\psi \in D(g_i)$. Thus, C_{g_i} defines a basis of $\ker_R(B.)$ on $D(g_i)$. Theorems 1 and 2 then imply that the solutions of (2) defined on $D(g_i)$ are given by:

$$\forall\, \psi \in D(g_i), \quad \forall\, Y' \in \Bbbk^{(r-l)\times m}, \quad \begin{cases} u = Z\,\psi, \\ v = \begin{pmatrix} E_{g_i}(\psi) & C_{g_i}(\psi) \end{pmatrix} \begin{pmatrix} Y \\ Y' \end{pmatrix}. \end{cases} \tag{27}$$

A stably free module of rank 1 over a commutative ring is free [15]. Hence, (27) holds when $r = \mathrm{rank}_{\Bbbk}(M)+1$. See [8] for the computation of C_{g_i}.

If $\ker_{S_{g_i}^{-1}R}(B.)$ is not a free $S_{g_i}^{-1}R$-module, then no full column rank matrix $C_{g_i} \in (S_{g_i}^{-1}R)^{r \times (r-l)}$ exists such that (26) holds, i.e., such that $\ker_{\Bbbk}(B(\psi).) = C_{g_i}(\psi)\,\Bbbk^{(r-l)\times 1}$ for all $\psi \in D(g_i)$. Hence, no basis of $\ker_{\Bbbk}(B(\psi).)$ exists on $D(g_i)$. But, using (25), we have the following solutions of (2), where $s > r-l$:

$$\forall\, \psi \in D(g_i), \quad \forall\, Y'' \in \Bbbk^{s\times m}, \quad \begin{cases} u = Z\,\psi, \\ v = \begin{pmatrix} E_{g_i}(\psi) & C(\psi) \end{pmatrix} \begin{pmatrix} Y \\ Y'' \end{pmatrix}. \end{cases} \tag{28}$$

Example 6. We consider again Examples 3 and 4. Using [8], we can check that $\ker_R(B.)$ is not a free R-module. Using the OreModules package, we get that

$$C := \begin{pmatrix} -58\,x_2 - 77\,x_3 & -79\,x_1 & 0 \\ -54\,x_2 + 31\,x_3 & 0 & -79\,x_1 \\ 0 & 54\,x_2 - 31\,x_3 & -58\,x_2 - 77\,x_3 \end{pmatrix}$$

is such that $\ker_R(B.) = \mathrm{im}_R(C.)$, i.e., the 3 columns of C generate the R-module $\ker_R(B.)$ of rank $r-l=2$. We get the solutions (28) of (2) on $D(g_i)$ for $i=1,2,3$.

Finally, we study if the solutions of (2) can be written as (27). As explained, the $S_{x_i}^{-1}R$-module $\ker_{S_{x_i}^{-1}R}(B.)$ is stably free of rank 2. Using Corollary 4.10 of [15], i.e., a variant of the Quillen-Suslin theorem for the *generalized Laurent polynomial ring* $S_{x_i}^{-1}R = R[x_i^{\pm 1}, x_j]_{1 \le j \ne i \le 3}$, $\ker_{S_{x_i}^{-1}R}(B.)$ is a free $S_{x_i}^{-1}R$-module of rank 2. Using an implementation of this result in the QuillenSuslin package, a basis of $\ker_{S_{x_i}^{-1}R}(B.)$ is defined by the columns of the matrix C_{x_i} defined by:

$$C_{x_1} = \begin{pmatrix} -79\,x_1 & 0 \\ 0 & -79\,x_1 \\ 54\,x_2 - 31\,x_3 & -58\,x_2 - 77\,x_3 \end{pmatrix},$$

$$C_{x_2} = \begin{pmatrix} -\frac{29\,x_2}{73680} - \frac{77\,x_3}{147360} & -\frac{6083\,x_1}{5956\,x_2} \\ -\frac{9\,x_2}{24560} + \frac{31\,x_3}{147360} & \frac{2449\,x_1}{5956\,x_2} \\ 0 & 1 \end{pmatrix}, \quad C_{x_3} = \begin{pmatrix} -\frac{29\,x_2}{73680} - \frac{77\,x_3}{147360} & \frac{2291\,x_1}{2978\,x_3} \\ -\frac{9\,x_2}{24560} + \frac{31\,x_3}{147360} & \frac{2133\,x_1}{2978\,x_3} \\ 0 & 1 \end{pmatrix}.$$

Hence, we have $\ker_{\mathbb{k}}(B(\psi).) = \mathrm{im}_{\mathbb{k}}(C_{x_i}(\psi).) \cong \mathbb{k}^{2\times 1}$ for all $\psi \in D(g_i)$ and for $i = 1, 2, 3$, and (27) are solutions of (2) defined on the $D(g_i)$'s given in Example 4.

Finally, we emphasize that all the examples were computed with the `Maple` packages OreModules [5] and QuillenSuslin [8]. For more details, see:

https://who.rocq.inria.fr/Alban.Quadrat/MapleConference.

References

1. Barakat, M., Lange-Hegermann, M.: The homalg project. Computeralgebra-Rundbrief **51**, 6–9 (2012). https://github.com/homalg-project/HomalgProject.jl
2. Capdessus, C.: Aide au diagnostic des machines tournantes par traitement du signal. Ph.D. thesis, University of Grenoble (1992)
3. Capdessus, C., Sidahmed, M.: Analyse des vibrations d'un engrenage: cepstre, corrélation, spectre. In: GRETSI, vol. 8, pp. 365–372. Saint Martin d'Hères (1991)
4. Chyzak, F., Quadrat, A., Robertz, R.: Effective algorithms for parametrizing linear control systems over Ore algebras. Appl. Algebra Eng. Commun. Comput. **16**, 319–376 (2005)
5. Chyzak, F., Quadrat, A., Robertz, D.: OreModules: a symbolic package for the study of multidimensional linear systems. In: Chiasson, J., Loiseau, J.J. (eds.) Applications of Time Delay Systems. Lecture Notes in Control and Information Sciences, vol. 352. Springer, Heidelberg (2007). https://doi.org/10.1007/978-3-540-49556-7_15
6. Dagher, R., Hubert, E., Quadrat, A.: On the general solutions of a rank factorization problem arising in vibration analysis (2021)
7. Eisenbud, D.: Commutative Algebra with a View Toward Algebraic Geometry. Springer, Heidelberg (1995). https://doi.org/10.1007/978-1-4612-5350-1
8. Fabiańska, A., Quadrat, A.: Applications of the Quillen-Suslin theorem to multidimensional systems theory. In: Radon Series on Computation and Applied Mathematics, vol. 3, pp. 23–106, de Gruyter publisher (2007)
9. Greuel, G.-M., Pfister, G.: A Singular Introduction to Commutative Algebra. Springer, Heidelberg (2002). https://doi.org/10.1007/978-3-540-73542-7. https://www.singular.uni-kl.de/
10. Hubert, E.: Amplitude and phase demodulation of multi-carrier signals: application to gear vibration signals. Ph.D. thesis, Université de Lyon (2019)
11. Hubert, E., Barrau, A., El Badaoui, M.: New multi-carrier demodulation method applied to gearbox vibration analysis. In: Proceedings of ICASSP 2018 (2018)
12. Hubert, E., Bouzidi, B., Dagher, R., Barrau, A., Quadrat, A.: Algebraic aspects of the exact signal demodulation problem. In: Proceedings of SSSC & TDS (2019)
13. Hubert, E., Bouzidi, B., Dagher, R., Barrau, A., Quadrat, A.: On a rank factorisation problem arising in gearbox vibration analysis. In: Proceedings of IFAC World Congress (2020)
14. Kunz, E.: Introduction to Commutative Algebra and Algebraic Geometry. Birkhäuser, Basel (1985)
15. Lam, T.Y.: Serre's Problem on Projective Modules. Springer, Heidelberg (2006). https://doi.org/10.1007/978-3-540-34575-6
16. Quadrat, A.: An introduction to constructive algebraic analysis and its applications. Les cours du CIRM, 1: Journées Nationales de Calcul Formel **1**, 281–471 (2010)
17. Rotman, J.J.: Introduction to Homological Algebra. Springer, Heidelberg (2009). https://doi.org/10.1007/b98977

Computation of the $\mathcal{L}_\infty$-norm of Finite-Dimensional Linear Systems

Yacine Bouzidi[1], Alban Quadrat[2], Fabrice Rouillier[2], and Grace Younes[2(✉)]

[1] Inria Lille Europe, IMJ – PRG, Sorbonne University, Paris, France
Yacine.Bouzidi@inria.fr

[2] Inria Paris, Ouragan Project, IMJ – PRG, Sorbonne University, Paris, France
{alban.quadrat,fabrice.rouillier,grace.younes}@inria.fr

Abstract. In this paper, we study the problem of computing the $\mathcal{L}_\infty$-norm of finite-dimensional linear time-invariant systems. This problem is first reduced to the computation of the maximal x-projection of the real solutions $(x,\, y)$ of a bivariate polynomial system $\Sigma = \{P, \frac{\partial P}{\partial y}\}$, with $P \in \mathbb{Z}[x, y]$. Then, we use standard computer algebra methods to solve the problem. In this paper, we alternatively study a method based on rational univariate representations, a method based on root separation, and finally a method first based on the sign variation of the leading coefficients of the signed subresultant sequence and then based on the identification of an isolating interval for the maximal x-projection of the real solutions of Σ.

Keywords: $\mathcal{L}_\infty$-norm computation · Real roots · Symbolic computation · Complexity computation · Implementation · Control theory

1 Introduction

An important issue in robust control theory is the computation of the $\mathcal{L}_\infty$-norm of linear systems [11,18]. Contrary to the $\mathcal{L}_2$-norm, no tractable formula is known for the characterization of the $\mathcal{L}_\infty$-norm of finite-dimensional systems (i.e., systems defined either by linear ordinary differential equations or by linear recurrence relations) [11,18]. Hence, the standard methods for the $\mathcal{L}_\infty$-norm computation are numerical (e.g., bisection algorithms, eigenvalues computation of Hamiltonian matrices) [5,7]. In their paper [13], Kano and Smith develop a validated numerical algorithm for the $\mathcal{L}_\infty$-norm computation. They reduce the problem to the localization of the real solutions of a bivariate polynomial and then use Sturm chain tests to guarantee the accuracy of their algorithm. In [8], Chen, Mazza and Xie provide an equivalent study using the theory of border polynomials, which makes the presentation of their solution simpler.

When numerical methods are used, it is worth mentioning that the result is usually obtained within a short time but with a slight error up to a precise accuracy. In contrast, when using symbolic methods, the result usually takes

R. M. Corless et al. (Eds.): MC 2020, CCIS 1414, pp. 119–136, 2021.
https://doi.org/10.1007/978-3-030-81698-8_9

more time to be computed but is exact. In this paper, to compute the $\mathcal{L}_\infty$-norm, we try to develop the right balance between these two approaches.

In this paper, following the approach developed in [8,13], we shall study three different certified symbolic-numeric algorithms for the computation of the $\mathcal{L}_\infty$-norm with the goal of minimizing the drawback of the symbolic part of the computation. This symbolic part consists in computing an isolating interval of the maximal projection of the real solutions of a system of bivariate polynomials. We develop the complexity analysis of each algorithm. Finally, we compare the theoretical complexities of the algorithms and then their performances using an implementation in the computer algebra system `Maple`.

Given two coprime polynomials P and Q in $\mathbb{Z}[x, y]$ of degree bounded by d and of coefficient bitsize bounded by τ, the solving of the system $\Sigma = \{P, Q\}$ can be studied using numerous methods. Typically, isolating boxes of the solutions can be computed either directly from the input system using numerical methods (such as *subdivision* or *homotopy methods*) or indirectly by first computing intermediate symbolic representations such as *triangular sets*, *Gröbner bases*, or *rational parameterizations* [1,6].

Two methods used in the paper require putting the system in a *generic position*, i.e., require to finding a separating linear form $x + a\,y$ that defines a shear of the coordinate system (x, y), i.e., $(x, y) \longmapsto (t - a\,y, y)$, so that no two distinct solutions of the sheared system $\Sigma_a = \{P(t - a\,y, y), Q(t - a\,y, y)\}$ are vertically aligned. This approach has long been used in the literature. For instance, a separating linear form $x + a\,y$ with $a \in \{0, \ldots, 2\,d^4\}$ can be computed as shown in [3,4]. We can then use a *Rational Univariate Representation* (RUR) for the polynomial system Σ_a followed by the computation of isolating boxes for its real solutions. For more details, see [3]. We simply apply this approach (i.e., the so-called *RUR method*) to the polynomial system associated with the $\mathcal{L}_\infty$-norm computation problem and then choose the maximal x-projection of the real solutions of the system. The complexity analysis shows that this algorithm performs $\tilde{O}_B(d_x\,d_y^3\,(d_x^2 + d_x\,d_y + d_y\,\tau))$ bit operations in the worst case, where

$$d_x = \max(\deg_x(P), \deg_x(Q)), \; d_y = \max(\deg_y(P), \deg_y(Q)), \tag{1}$$

and τ is the maximal coefficient bitsize of the polynomials P and Q.

Alternatively, we can also localize the maximal x-projection of the real solutions of the polynomial system Σ by simply applying a linear separating form on the system Σ. The linear separating form $t = x + s\,y$ proposed in [9] preserve the order of the solutions of the sheared system $\Sigma_s = \{P(t - s\,y, y), Q(t - s\,y, y)\}$ with respect to the x-projection of the real solutions of the original system Σ. Thus, the projection of the solutions of Σ_s onto the new separating axis t can be done so that we can simply choose the x-projection corresponding to the maximal t-projection of the real solutions of Σ_s. The drawback of this method lies on the growth of the size of the coefficients of the sheared system for the linear separating form $t = x + s\,y$ due to the large size of s. The complexity analysis shows that this algorithm performs $\tilde{O}_B(d_x^4\,d_y^5\,\tau)$ bit operations in the worst case.

The third method developed in this paper localizes the maximal x-projection of the system real solutions – denoted by $\bar{x}$ – by first isolating the real roots of the univariate *resultant polynomial* $\mathrm{Res}(P, \frac{\partial P}{\partial y}, y)$ and then verifying the existence of a real root of the univariate polynomial $P(\bar{x}, y) \in \mathbb{R}[y]$ as done in [8]. A key point is that we can compute a *Sturm-Habicht sequence* [12] of $P(\bar{x}, y)$ without any consequent overhead. As $P(x, y) = 0$ is bounded in the x-direction, it is then possible to compute the number of real solutions of $P(\bar{x}, y)$ with a good complexity in the worst case, which gives an efficient algorithm as soon as the curve $P(x, y) = 0$ has no isolated real singular points. The complexity analysis shows that this algorithm performs $\tilde{O}_B(d_x^2\, d_y^4\, (d_x + \tau))$ bit operations in the worst case and $\tilde{O}_B(d_x^2\, d_y^4\, \tau)$ when the plane curve $P(x, y) = 0$ has no isolated real singular points.

Finally, we conclude the paper by comparing the bit complexity of those three algorithms and then the experimental time obtained by the implementation of each of these algorithms in `Maple`.

2 Problem Description

Before stating the problem studied in this paper, we first introduce a few standard notations and definitions. If $\mathbb{k}$ is a field and $P \in \mathbb{k}[x, y]$, then $Lc_{var}(P)$ is the *leading coefficient* of P with respect to the variable $var \in \{x, y\}$ and $\deg_{var}(P)$ the *degree* of P in the variable $var \in \{x, y\}$. We also denote by $\deg(P)$ the *total degree* of P. Moreover, let $\pi_x : \mathbb{R}^2 \longrightarrow \mathbb{R}$ be the projection map from the real plane $\mathbb{R}^2$ onto the x-axis, i.e., $\pi_x(x, y) = x$ for all $(x, y) \in \mathbb{R}^2$. For $P, Q \in \mathbb{k}[x, y]$, let $\gcd(P, Q)$ be the greatest common divisor of P and Q, $I := \langle P, Q \rangle$ the ideal of $\mathbb{k}[x, y]$ generated by P and Q, and $V_{\mathbb{K}}(I) := \{(x, y) \in \mathbb{K}^2 \mid \forall\, R \in I : \; R(x, y) = 0\}$, where $\mathbb{K}$ is a field containing $\mathbb{k}$. Finally, let $\mathbb{C}_+ := \{s \in \mathbb{C} \mid \mathrm{Re}(s) > 0\}$ be the *open right-half plane* of $\mathbb{C}$.

Definition 1 ([11,18]). *Let $\mathcal{RH}_\infty$ be the $\mathbb{R}$-algebra of all the proper and stable rational functions with real coefficients, namely:*

$$\mathcal{RH}_\infty := \left\{ \frac{n}{d} \mid n, d \in \mathbb{R}[s],\ \gcd(n, d) = 1,\ \deg_s(n) \leq \deg_s(d), V_{\mathbb{C}}(\langle d \rangle) \cap \mathbb{C}_+ = \emptyset \right\}.$$

An element g of $\mathcal{RH}_\infty$ is holomorphic and bounded on $\mathbb{C}_+$, i.e.,

$$\| \, g \, \|_\infty := \sup_{s \in \mathbb{C}_+} |g(s)| < +\infty,$$

$\mathcal{RH}_\infty$ is a sub-algebra of the *Hardy algebra* $\mathcal{H}_\infty(\mathbb{C}_+)$ of bounded holomorphic functions on $\mathbb{C}_+$. The *maximum modulus principle* of complex analysis yields:

$$\| \, g \, \|_\infty = \sup_{\omega \in \mathbb{R}} |g(i\,\omega)|.$$

Note that this equality shows that the function $g_{|i\,\mathbb{R}} : i\,\omega \in i\,\mathbb{R} \longmapsto g(i\,\omega)$ belongs to the *Lebesgue space* $\mathcal{L}_\infty(i\,\mathbb{R})$ or, more precisely, to the following $\mathbb{R}$-algebra

$$\mathcal{RL}_\infty := \left\{ \frac{n(i\,\omega)}{d(i\,\omega)} \mid n, d \in \mathbb{R}[i\,\omega],\ \gcd(n, d) = 1,\ \deg_\omega(n) \leq \deg_\omega(d),\ V_{\mathbb{R}}(\langle d \rangle) = \emptyset \right\},$$

i.e., the algebra of real rational functions on the imaginary axis $i\,\mathbb{R}$ which are proper and have no poles on $i\,\mathbb{R}$, or simply, the algebra of real rational functions with no poles on $i\,\mathbb{P}^1(\mathbb{R})$, where $\mathbb{P}^1(\mathbb{R}) := \mathbb{R} \cup \infty$.

We can extend the above $\mathcal{L}_\infty$-norms defined on functions of $\mathcal{RH}_\infty$ (resp., $\mathcal{RL}_\infty$) to matrices as follows. Let $G \in \mathcal{RH}_\infty^{u\times v}$ (resp., $G \in \mathcal{RL}_\infty^{u\times v}$, $\mathbb{R}(s)^{u\times v}$), i.e., G is a $u \times v$ matrix with entries in $\mathcal{RH}_\infty$ (resp., $\mathcal{RL}_\infty$, $\mathbb{R}(s)$) and let $\bar{\sigma}\,(\cdot)$ denote the largest *singular value of a complex matrix*. Then, we can define:

$$\| G \|_\infty := \sup_{s\in\mathbb{C}_+} \bar{\sigma}\,(G(s)) \left(\text{resp., } \| G \|_\infty := \sup_{\omega\in\mathbb{R}} \bar{\sigma}\,(G(i\,\omega))\right).$$

If $G \in \mathcal{RH}_\infty^{u\times v}$, then, as above, we have $\| G \|_\infty = \sup_{\omega\in\mathbb{R}} \bar{\sigma}\,(G(i\,\omega))$.

The paper aims at developing certified symbolic-numeric algorithms for the computation of $\| G \|_\infty$ for $G \in \mathbb{R}(s)^{u\times v}$ satisfying $G_{|i\,\mathbb{R}} \in \mathcal{RL}_\infty^{u\times v}$. This problem plays a fundamental role in $\mathcal{H}_\infty$-control theory [11,18].

The *conjugate* $\tilde{G}$ of $G \in \mathbb{R}(s)^{u\times v}$ is defined by $\tilde{G}(s) := G^T(-s)$.

The next proposition gives a first characterization of $\| G \|_\infty$.

Proposition 1 ([13]). *Let $\gamma > 0$, $G \in \mathbb{R}(s)^{u\times v}$ be such that $G_{|i\,\mathbb{R}} \in \mathcal{RL}_\infty^{u\times v}$ and let us consider $\Phi_\gamma(s) = \gamma^2\, I_v - \tilde{G}(s)\, G(s)$. Then, $\gamma > \| G \|_\infty$ if and only if $\gamma > \bar{\sigma}\,(G(i\,\infty))$ and $\det(\Phi_\gamma(i\,\omega)) \neq 0$ for all $\omega \in \mathbb{R}$.*

Let $n(\gamma, \omega)$ and $d(\omega)$ be two coprime polynomials over $\mathbb{R}[\gamma, \omega]$ satisfying:

$$\det(\Phi_\gamma(i\,\omega)) = \frac{n(\gamma,\omega)}{d(\omega)}. \tag{2}$$

Note that $\det(\Phi_\gamma(s))$ is a real function in s^2 and γ^2, and thus, $\det(\Phi_\gamma(i\,\omega))$ is a real function in ω^2 and γ^2. A consequence of Proposition 1 is the next result.

Proposition 2. *Let $G \in \mathcal{RL}_\infty^{u\times v}$ and $n \in \mathbb{R}[\gamma, \omega]$ be defined by (2). We denote by $\bar{n} \in \mathbb{R}[\gamma, \omega]$ the square free part of n. Then, we have:*

$$\| G \|_\infty = \max \left\{ \pi_\gamma \left(V_{\mathbb{R}} \left(\left\langle \bar{n}, \frac{\partial \bar{n}}{\partial \omega} \right\rangle \right) \right) \cup V_{\mathbb{R}}\left(\langle Lc_\omega(\bar{n}) \rangle \right) \right\}.$$

Example 1. If $G \in \mathcal{RL}_\infty$ then, by definition, $\| G \|_\infty$ is the supremum of the continuous function $\omega \in \mathbb{P}^1(\mathbb{R}) := \mathbb{R} \cup \{\infty\} \longmapsto |G(i\,\omega)|$, and thus, we have $\| G \|_\infty = \max_{\omega\in\mathbb{P}^1(\mathbb{R})} |G(i\,\omega)|$, i.e., $\| G \|_\infty = \max\{|G(i\,\infty)|, \gamma_{\max}\}$, where:

$$\gamma_{\max} := \max_{\omega\in\mathbb{R}} |G(i\,\omega)| = \max\left\{\gamma \in \mathbb{R} \mid \exists\, \omega \in \mathbb{R} : \gamma^2 = |G(i\,\omega)|^2\right\}.$$

We find again Proposition 2, i.e., $\gamma > \| G \|_\infty$ iff $\Phi_\gamma(i\,\omega) = \gamma^2 - |G(i\,\omega)|^2 \neq 0$ for all $\omega \in \mathbb{R}$ and $\gamma > |G(i\,\infty)|$. Using $|G(i\omega)|^2 = G(-i\,\omega)\, G(i\,\omega) \in \mathbb{R}(\omega^2)$, a computation of $\| G \|_\infty$ amounts to first computing the zeros of the numerator of $\frac{d|G(i\,\omega)|^2}{d\omega}$, then evaluating $|G(i\,\omega)|$ on these zeros and finally choosing the maximal occurring value, that to say $\bar{\gamma}$, and (iii) $\| G \|_\infty = \max\{|G(i\,\infty)|, \bar{\gamma}\}$.

More explicitly, if we write G as $G(s) = a(s)/b(s)$, where $a, b \in \mathbb{R}[s]$, $\gcd(a,b) = 1$, $q = \deg_s(a) \leq r = \deg_s(b)$, and b does not vanish on $i\,\mathbb{R}$, then $G(i\,\infty) = 0$ if $q < r$ (i.e., G is strictly proper) or $G(i\,\infty) = a_r/b_r$ if $q = r$ (i.e., G is proper), where $a_r = Lc_s(a)$ and $b_r = Lc_s(b)$. Moreover, we have $|G(i\,\omega)|^2 = N(\omega)/D(\omega)$, where $N(\omega) = |a(i\,\omega)|^2$ and $D(\omega) = |b(i\,\omega)|^2 \in \mathbb{R}[\omega^2]$. Since $b(i\,\omega)$ has not real roots, $D(\omega) \neq 0$ for all $\omega \in \mathbb{R}$. Hence, if we note $\mathcal{Z} := \{\omega \in \mathbb{R} \mid N'(\omega)\,D(\omega) - N(\omega)\,D'(\omega) = 0\}$, then we obtain:

$$\| G \|_\infty = \max\{|G(i\,\infty)|, \bar{\gamma}\}, \quad \bar{\gamma} := \max_{\omega \in \mathcal{Z}} \left\{ (N(\omega)/D(\omega))^{1/2} \right\}.$$

Note that if $\mathcal{Z} \cap V_{\mathbb{R}}(\langle D'(\omega)\rangle) = V_{\mathbb{R}}(\langle N'(\omega)\,D(\omega), D'(\omega)\rangle) = \emptyset$, then we also have $\bar{\gamma} = \max_{\omega \in \mathcal{Z}} \left\{ (N'(\omega)/D'(\omega))^{1/2} \right\}$. For instance, if $G(s) = (2\,s+1)/(s+1)$, then $N(\omega) = 4\,\omega^2 + 1$, $D(\omega) = \omega^2 + 1$, $\mathcal{Z} = \{0\}, \mathcal{Z} \cap V_{\mathbb{R}}(\langle D'(\omega)\rangle) = \{0\}$, $\bar{\gamma} = (N(0)/D(0))^{1/2} = 1$, $|G(i\,\infty)| = 2$, and $\| G \|_\infty = \max\{2, \bar{\gamma}\} = 2$.

Finally, according to Proposition 2, we have $n(\gamma, \omega) = D(\omega)\,\gamma^2 - N(\omega)$ and $d(\omega) = D(\omega)$. Now, using $\gcd(N, D) = 1$, $\bar{n} = n$, $Lc_\omega(\bar{n}) = b_r^2\,\gamma^2$ if $q < r$ or $Lc_\omega(\bar{n}) = (b_r^2\,\gamma^2 - a_r^2)$ if $q = r$, which yields:

$$V_{\mathbb{R}}\left(\langle Lc_\omega(\bar{n})\rangle\right) = \begin{cases} 0, & \text{if } q < r, \\ \pm\dfrac{a_r}{b_r}, & \text{if } q = r, \end{cases}$$

and using the fact that $D(\omega) \neq 0$ for all $\omega \in \mathbb{R}$, we have

$$\begin{aligned} \pi_\gamma\left(V_{\mathbb{R}}\left(\left\langle \bar{n}, \frac{\partial \bar{n}}{\partial \omega}\right\rangle\right)\right) &= \pi_\gamma\left(V_{\mathbb{R}}(\langle D(\omega)\,\gamma^2 - N(\omega), D'(\omega)\,\gamma^2 - N'(\omega)\rangle)\right) \\ &= \left\{\gamma \in \mathbb{R} \mid \gamma^2 = \frac{N(\omega)}{D(\omega)}, \omega \in \mathcal{Z}\right\}, \end{aligned}$$

and thus, $\| G \|_\infty = \max\{\bar{\gamma}, 0\} = \bar{\gamma}$ if $q < r$ and $\| G \|_\infty = \max\{\bar{\gamma}, a_r/b_r\}$ if $q = r$.

Corollary 1. *Let $G \in \mathcal{RL}_\infty^{u \times v}$ and $n \in \mathbb{R}[\gamma, \omega]$ be the numerator of* $\det(\gamma^2\,I_v - \tilde{G}(i\,\omega)\,G(i\,\omega))$ *defined by (2). Then, the real γ-projection $\pi_\gamma(V_{\mathbb{R}}(\langle n\rangle))$ of $V_{\mathbb{R}}(\langle n\rangle)$ is bounded by $\| G \|_\infty$.*

According to Proposition 2, given $G \in \mathcal{RL}_\infty^{u \times v}$, the problem of computing $\| G \|_\infty$ can be reduced to the computation of the maximal γ-projection of the real solutions of the following bivariate polynomial system:

$$\Sigma := \left\{\bar{n}(\gamma, \omega), \frac{\partial \bar{n}(\gamma, \omega)}{\partial \omega}\right\}. \tag{3}$$

For studying this problem, we propose three different symbolic-numeric methods – *Rational Univariate Representation method*, *Roots Separation Method* and *Sturm-Habicht method* – and compare them. Without loss of generality, we shall suppose that n is squarefree in $\mathbb{Z}[\gamma, \omega]$.

3 Rational Univariate Representation Method

In this section, we briefly state a straightforward algorithm which computes the maximal γ-projection of the real solutions of (3) based on a *Rational Univariate Representation Method* [3,4,15,16]. This algorithm consists in first computing a rational parametrization (RUR) of the solutions of (3), then isolating the roots of a univariate polynomial p defining the associated field extension, using the intervals obtained to compute isolating boxes for the real solutions of (3) and finally selecting the real solution of (3) with the maximal γ-projection.

If P, $Q \in \mathbb{Q}[x, y]$ are two coprime polynomials, then the computation of the RUR of $V_{\mathbb{K}}(\langle P, Q\rangle)$, $\mathbb{K} = \mathbb{R}, \mathbb{C}$, consists in finding $s \in \mathbb{N}$ such that $x + s\,y$ separates the $\mathbb{K}$-zeros of $\{P, Q\}$ and four polynomials p, q, p_0, $q_0 \in \mathbb{Q}[T]$ which define a 1-1 correspondence between $V_{\mathbb{K}}(\langle \Sigma\rangle)$ and $V_{\mathbb{K}}(\langle p\rangle)$, i.e., the following bijection:

$$\begin{aligned} V_{\mathbb{K}}(\langle P, Q\rangle) &\longrightarrow V_{\mathbb{K}}(\langle p\rangle) \\ (x, y) &\longmapsto \xi = x + s\,y, \\ \left(\frac{p_0(\xi)}{q(\xi)}, \frac{p_1(\xi)}{q(\xi)}\right) &\longleftarrow\!\shortmid\ \xi \end{aligned}$$

Roughly speaking, using the RUR of $V_{\mathbb{K}}(\langle P, Q\rangle)$, we can transform the study of problems on $V_{\mathbb{K}}(\langle P, Q\rangle)$ into corresponding problems on $V_{\mathbb{K}}(\langle p\rangle)$ [3,15].

For the $\mathcal{L}_\infty$-norm computation, the polynomials of $\Sigma \subset \mathbb{Z}[\gamma, \omega]$, defined by (3), are coprime. Hence, to compute $\| G \|_\infty$, we first use the RUR method to obtain isolating boxes for the real solutions (γ, ω) of Σ, choose the maximal γ-projection γ_1, then compute an isolation box γ_2 for the maximal real root of the univariate polynomial $Lc_\omega(n)$, and finally compute $\| G \|_\infty = \max\{\gamma_1, \gamma_2\}$.

Algorithm 1. RUR method

Input: A zero dimensional polynomial system $\{n, \frac{\partial n}{\partial \omega}\} \subset \mathbb{Z}[\gamma, \omega]$.
Output: An isolating interval of $\max\left\{\pi_\gamma\left(V_{\mathbb{R}}\left(n, \frac{\partial n}{\partial \omega}\right)\right) \cup V_{\mathbb{R}}\left(Lc_\omega(n)\right)\right\}$.

1. Apply the RUR function (`Isolate`) for solving the polynomial system $\{n, \frac{\partial n}{\partial \omega}\}$.
2. Let γ_1 be the maximal γ-projection of the system's real solutions.
3. Let γ_2 be the maximal real root of $Lc_\omega(n)$.
4. **Return** the isolating interval of $\max\{\gamma_1, \gamma_2\}$.

Remark 1. *In step* 1 *of Algorithm 1, we obtain isolating boxes* $[a_i, b_i] \times [c_i, d_i]$ *of the system's real solutions* (γ_i, ω_i). *To compare the real values in step* 2 *and then in step* 4, *we can apply a straightforward strategy consisting in computing the resultant polynomial* $R_\gamma = \mathrm{Res}(n, \frac{\partial n}{\partial \omega}, \omega)$ *and then refining the boxes until each interval* $[a_i, b_i]$ *is included in an isolating interval of* R_γ. *Even with this naive approach, the asymptotic complexity of these operations does not exceed the algorithm's overall worst case bit complexity.*

In what follows, $\mathcal{O}_B$ denotes the *bit complexity* and $\tilde{\mathcal{O}}_B$ means that logarithmic factors have been omitted. Given two coprime polynomials $P, Q \in \mathbb{Z}[x, y]$ of degree bounded by d and coefficient bitsize bounded by τ, an algorithm for computing linear separating forms, RUR representations and isolating boxes of the solutions can be obtained in the worst case bit complexity $\tilde{\mathcal{O}}_B(d^6 + d^5\,\tau)$ [3].

Let us compute the complexity of Algorithm 1. We first need the next result.

Lemma 1. *Let $G \in \mathcal{RL}_\infty^{u\times v}$, $\Phi_\gamma(i\,\omega) = \gamma^2 I_v - \tilde{G}(i\,\omega)\,G(i\,\omega)$, $n \in \mathbb{R}[\gamma, \omega]$ be defined by (2), $d_\gamma = \deg_\gamma(n)$, $d_\omega = \deg_\omega(n)$, and τ_n the coefficient bitsize of n. Moreover, let $\alpha = \max\{u, v\}$, $N = \max_{1\le i\le u, 1\le j\le v}\{\deg_\omega(Q_{i,j})\}$, where $G_{jk} = \frac{P_{jk}}{Q_{jk}}$ denotes the $(j,k)^{th}$ entry of G and P_{jk}, $Q_{jk} \in \mathbb{R}[i\,\omega]$ are coprime, and τ_G the maximal coefficient bitsize of $\{P_{jk}, Q_{jk}\}_{1\le j\le u, 1\le k\le v}$. Then, we have:*

$$d_\gamma = \mathcal{O}(\alpha), \quad d_\omega = \mathcal{O}(N\,\alpha^2), \quad \tau_n = \tilde{\mathcal{O}}(\tau_G\,\alpha^2).$$

Proof. Let $G_{jk} = P_{jk}/Q_{jk}$ be the $(j,k)^{\text{th}}$ entry of G, where P_{jk}, $Q_{jk} \in \mathbb{R}[i\,\omega]$ are coprime. Since $G \in \mathcal{RL}_\infty^{u\times v}$, G_{jk} is a proper rational function, and thus, $\deg_\omega(P_{jk}) \le \deg_\omega(Q_{jk}) \le N$, which shows that the degrees in ω of the numerators and the denominator of the entries of $\Phi_\gamma(i\,\omega)$ are bounded by $2\,N\,\alpha$, and thus, d_ω is bounded by $2\,N\,\alpha^2$. Similarly, the maximal coefficient bitsize of the entries of $\Phi_\gamma(i\,\omega)$ is $2\,\tau_n\,\alpha$, which yields τ_n is bounded by $2\,\tau_G\,\alpha^2$. Finally, d_γ is clearly bounded by $2\,\alpha$.

Theorem 1. *With the above notations, the complexity of Algorithm 1 for the computation of $\| G \|_\infty$, where $G \in \mathcal{RL}_\infty^{u\times v}$, is given by:*

$$\tilde{\mathcal{O}}_B(d_\gamma\,d_\omega^3\,(d_\gamma^2 + d_\gamma\,d_\omega + d_\omega\,\tau_n)) = \tilde{\mathcal{O}}_B(\alpha^9\,N^4\,(\alpha + \tau_n)).$$

Proof. According to [3], using the RUR method, the complexity of the resolution of a zero dimensional bivariate polynomial system comes first from the computation of the triangular decomposition of the system after shearing – using a separating linear form $\gamma + s\,\omega$ – then from the root isolation of the univariate polynomial defining the associated field extension, and finally from the computation of the isolating boxes for the solutions.

In the present case, the degrees in ω and γ are not of the same order. Hence, the results of [3] must be adapted.

First, we determine the size and the degree of the sheared system up to the method used in [3]: the degree with respect to the variable ω is $\tilde{\mathcal{O}}(d_\gamma + d_\omega)$ and d_γ with respect to the variable $t = \gamma + s\,\omega$. The size of the sheared system is $\tilde{\mathcal{O}}(\tau_n + d_\gamma)$. From [14], the complexity of the computation of a triangular decomposition of a system over $\mathbb{Z}[x, y]$ costs $\tilde{\mathcal{O}}_B(d_x^3\,d_y^3 + (d_x^2\,d_y^3 + d_x\,d_y^4)\,\tilde{\tau})$, where d_x and d_y are defined by (1) and $\tilde{\tau}$ is its maximal coefficient bitsize. Thus, using Lemma 1, we obtain that the complexity of the computation of the triangular decomposition of the sheared system in $\mathbb{Z}[t, \omega]$ is given by:

$$\begin{aligned}&\tilde{\mathcal{O}}_B(d_\gamma^3\,(d_\gamma + d_\omega)^3 + (d_\gamma^2\,(d_\gamma + d_\omega)^3 + d_\gamma\,(d_\gamma + d_\omega)^4)\,(\tau_n + d_\gamma))\\ &= \tilde{\mathcal{O}}_B(d_\gamma\,d_\omega^3\,(d_\gamma^2 + d_\gamma\,d_\omega + d_\omega\,\tau_n)).\end{aligned}$$

Moreover, by Lemma 1, we obtain $\tilde{O}_B(\alpha^9 N^4 (\alpha + \tau_n))$. This triangular decomposition yields (RUR) polynomials of degree that sum up to $\tilde{O}((d_\gamma + d_\omega) d_\gamma) = \tilde{O}(\alpha^3 N)$ with coefficients of bitsize $\tilde{O}((d_\gamma + d_\omega)(d_\gamma + \tau_n)) = \tilde{O}(\alpha^2 N(\alpha + \tau_n))$.

Finally, according to [3], it is known that the computation of isolating boxes of all the roots of the system can be done in $\tilde{O}_B((\alpha^3 N)^3 + (\alpha^3 N)^2(\alpha^2 N (\alpha + \tau_n))) = \tilde{O}_B(\alpha^8 N^3 (\alpha + \tau_n))$ bit operations.

4 Roots Separation Method

In this section, we localize the maximal γ-projection of the real solutions of the polynomial system Σ by only shearing the system Σ using a special linear separating form [9]. With this linear separating form $t = \gamma + s\,\omega$, we obtain:

$$t_1 = \gamma_1 + s\,\omega_1 < t_2 = \gamma_2 + s\,\omega_2 \implies \gamma_1 \leq \gamma_2.$$

Let $P, Q \in \mathbb{Z}[x, y]$ be coprime and $R_x = \text{Res}(P, Q, y) \in \mathbb{Z}[x]$ be their *resultant.* Moreover, let $x_1 \leq \ldots \leq x_m$ be the real roots of R_x with isolating intervals $[c_1, d_1], \ldots, [c_m, d_m]$. Moreover, let the real numbers δ, M and s be defined by:

$$\delta < \frac{1}{2} \min_{i=1,\ldots,m-1} (x_{i+1} - x_i),\; M > \max\{y \mid (x, y) \in V_{\mathbb{R}}(\langle P, Q\rangle)\},\; 0 < s < \frac{\delta}{M}. \tag{4}$$

We can use the general root bounds for zero dimensional systems to estimate δ and M. In fact, M is the measure of the univariate polynomial $\text{Res}(P, Q, x)$. Note that the resultant computation can be avoided by using the concept of *sleeve functions* studied in [10] and [9, Lemma 3.3].

Let us consider an invertible linear map (a shear) of $\mathbb{R}^2$ to $\mathbb{R}^2$ defined by $\Psi_s : (x, y) \longmapsto (t, y) = (x + s\,y, y)$. Let us also note $\Psi_s(P) = P(t - s\,y, y)$, $\Psi_s(Q) = Q(t - s\,y, y)$, $R_t = \text{Res}(\Psi_s(P), \Psi_s(Q), y)$ and let $t_1 \leq \ldots \leq t_{m'} = t_{\max}$ be the real roots of R_t.

To get a 1-1 correspondence between the zeros of $\{P, Q\}$ and the roots of R_t, $Lc_y(\Psi_s(P))$ and $Lc_y(\Psi_s(Q))$ must not both vanish. It is always possible to choose s such that this condition is satisfied. In what follows, we shall consider this case. For more details for the computation of s up to this condition, see [9].

Remark 2. In Fig. 1, we only draw a part of the plot. In fact, since $n \in \mathbb{Z}[\gamma^2, \omega^2]$, $\Sigma = 0$ is symmetric with respect to the γ and ω axes.

Proposition 3. *With the above notations, let x_m be a real root of R_x with an isolating interval $[c_m, d_m]$ and $t_{\max}$ the maximal real root of R_t. If $t_{\max} \in [c_m - \delta, d_m + \delta]$, then the maximal x-projection of $V_{\mathbb{R}}(\langle P, Q\rangle)$ is equal to x_m.*

Proof. For each real root x_i of R_x with an isolating interval $[c_i, d_i]$, let us denote by $P_{i,j} = (x_i, y_{i,j})$ the real solutions of $\{P, Q\}$ which project onto x_i. Then, $\Psi_s(P_{i,j}) = (x_i + s\,y_{i,j}, y_{i,j})$, where $x_i + s\,y_{i,j}$ is the first coordinate of a real solution of $\{\Psi_s(P), \Psi_s(Q)\}$. Using (4), we obtain that different $\Psi_s(P_{i,j})$ have

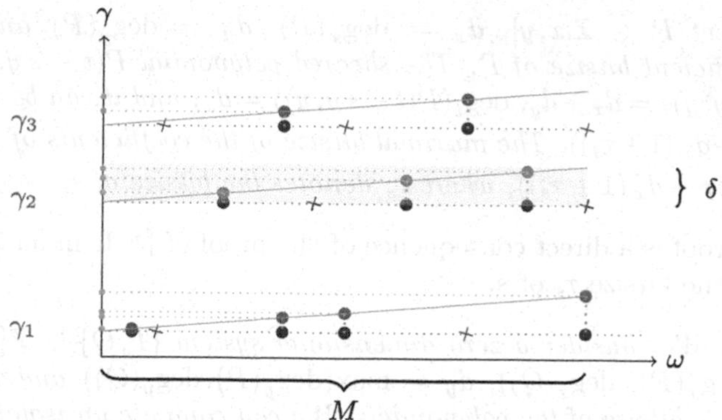

Fig. 1. The blue dots represent the real solutions of $\Sigma = 0$ and the blue crosses are the complex ones; the red dots are the solutions of $\Psi_s(\Sigma) = \{\Psi_s(n), \Psi_s(\frac{\partial n}{\partial \omega})\}$; the orange dots on the γ-axis are the roots of univariate polynomial R_t. (Color figure online)

different first coordinates, and thus, $\{\Psi_s(P, \Psi_s(Q)\}$ is in a *generic position.* Furthermore, we have $|x_i + s\, y_{i,j} - x_i| = |s\, y_{i,j}| < (\frac{\delta}{M})\, M = \delta$. Consequently $\Psi_s(P_{i,j}) \in I_i = [x_i - \delta, x_i + \delta] \times [-M, M]$. In addition, since $\delta < \frac{1}{2}(x_{i+1} - x_i)$, I_i are disjoint for different i. Hence, a real solution $(x,\, y)$ of $\{P,\, Q\}$ is mapped to $(\eta,\, y)$, where $\eta \in [x - \delta, x + \delta]$. Now, since $x_i \in [c_i, d_i]$, the real roots of R_t associated with x_i are in the interval $[c_i - \delta, d_i + \delta]$.

Algorithm 2. Roots Separation method

Input: A zero dimensional system $\{P, Q\} \subset \mathbb{Z}[x, y]$, where $Q = \frac{\partial P}{\partial y}$.
Output: An isolating interval of $\max\{\pi_x(V_{\mathbb{R}}(\langle P, Q\rangle)) \cup V_{\mathbb{R}}(\langle Lc_y(P)\rangle)\}$

1. Isolate $R_x = \mathrm{Res}(P, Q, y)$ up to an accuracy ϵ and let $RI := \{[c_1, d_1], \ldots, [c_m, d_m]\}$ be the isolating intervals of the real roots $\{x_1, \ldots, x_m\}$ of R_x.
2. Compute M and $D = \frac{1}{2}\min_{i=1,\ldots,m-1} |c_{i+1} - d_i|$:
   - if $D > 2\,\epsilon$, let $\epsilon_1 = \epsilon$, $\delta = D - \epsilon_1$ and compute s up to the required conditions.
   - elif $D \leq 2\,\epsilon$, let $\epsilon_1 = D/2$, $\delta = D - \epsilon_1$ and compute s up to the required conditions ;
3. Expand $\{\Psi_s(P), \Psi_s(Q)\}$ and compute $R_t = \mathrm{Res}(\Psi_s(P), \Psi_s(Q), y)$.
4. Isolate R_t up to an accuracy less than ϵ_1 and set $[p_t, q_t]$ to be the isolating interval of its maximal real root t_{max}.
5. for j from 1 to m do:
   - if $[p_t, q_t] \subset [c_j - D, d_j + D]$, then $X_1 = x_j$.
6. Let X_2 be the maximal real root of $Lc_y(P)$.
7. **Return** the isolating interval of $\max\{X_1, X_2\}$.

Lemma 2. *Let $P \in \mathbb{Z}[x,y]$, $d_x = \deg_x(P)$, $d_y = \deg_y(P)$ and τ be the maximal coefficient bitsize of P. The sheared polynomial $P(t - s\,y, y)$ satisfies $\deg_y(P(t-s\,y,y)) = d_x + d_y$, $\deg_t(P(t-s\,y,y)) = d_x$, and it can be expanded in $\tilde{O}_B(d_x\, d_y^2\,(\tau + d_x\,(1+\tau_s))$. The maximal bitsize of the coefficients of $P(t-s\,y,y)$ is equal to $\tilde{O}(\tau + d_x\,(1+\tau_s))$, where τ_s denotes the bitsize of s.*

Proof. The proof is a direct consequence of the proof of [4, Lemma 7] by taking into account the bitsize τ_s of s.

Theorem 2. *We consider a zero dimensional system $\{P, Q\} \subset \mathbb{Z}[x,y]$, where $d_x = \max(\deg_x(P), \deg_x(Q))$, $d_y = \max(\deg_y(P), \deg_y(Q))$ and τ the maximal coefficients bitsize of the polynomials. We can compute an isolating interval for the maximal x-projection of the real solutions of $\{P, Q\}$ in $\tilde{O}_B(d_x^4\, d_y^5\, \tau)$ bit operations, using Algorithm 2.*

Proof. For each root t_j of R_t defined in Algorithm 2 with an isolating interval $[p_j, q_j]$, there exists a unique $i \in \{1, \ldots, m\}$ such that $[p_j, q_j] \subset [c_i - D, d_i + D]$: we know that there exists a unique $[c_i, d_i]$ such that $t_j \in [c_i - \delta, d_i + \delta]$. From step 4, $q_j - p_j < \epsilon_1$ and $D = \delta + \epsilon_1$. Hence, $q_j < t_j + \epsilon_1 < d_i + \delta + \epsilon_1 < d_i + D$. And similarly $p_j > c_i - D$. Consequently, based on Proposition 3, Algorithm 2 outputs an isolating interval for the maximal x-projection of the real solutions of $\{P, Q\}$. As for the complexity, step 1 has worst-case bit complexity $\tilde{O}_B(d_x\, d_y^3\, \tau)$ based on [2, Proposition 8.46]. Step 2 is of worst case bit complexity $\tilde{O}_B(d^3 + d^2\, \tilde{\tau})$, where $d = \deg(R_x)$ and the coefficient size of R_x is equal to $\tilde{\tau}$ [3, Lemma 54]. From [2, Proposition 8.46], $d = \mathcal{O}(d_x\, d_y)$ and $\tilde{\tau} = \tilde{O}(d_y\, \tau)$. Consequently, step 2 is of worst case bit complexity $\tilde{O}_B(d_x^2\, d_y^3\, (d_x + \tau)$. In steps 3 and 4, we get $\delta = 2^{-\tilde{O}(d_x\, d_y^2\, \tau)}$ and $M = 2^{\mathcal{O}(d_x\, \tau)}$ [3]. The bitsize of s is then equal to $\tilde{O}(d_x\, d_y^2\, \tau)$. Consequently, the coefficient bitsize of the sheared system is $\tilde{O}(d_x^2\, d_y^2\, \tau)$, and the worst case bit complexity of step 5 is $\tilde{O}(d_x^3\, d_y^2\, (d_x + d_y)^3\, \tau)$, as computed in Lemma 2. In step 6, we isolate the resultant of the sheared system. Considering the size and degree of the sheared polynomials computed using Lemma 2, the size and degree of the resultant of the sheared system are $\tilde{O}(d_x^2\, d_y^3\, \tau)$ and $\tilde{O}(d_x\,(d_x + d_y))$ respectively. Then, knowing the complexity of the isolation mentioned in [3], we can say that the worst case bit complexity in this line is equal to $\tilde{O}_B((d_x\, d_y)^3 + (d_x\, d_y^2)^2\,(d_x^2\, d_y^3\, \tau)) = \tilde{O}_B(d_x^4\, d_y^5\, \tau)$. Finally, in the step 7, we simply compare two rational numbers. The maximal coefficients bitsize of these rationals is in $\tilde{O}(d_x^3\, d_y^3\,(d_x + d_y)\,\tau)$ and the computation in this step is done in $\tilde{O}(d_x^3\, d_y^3\,(d_x + d_y)\,\tau)$ bit operations. Hence, the overall bit complexity is given by $\tilde{O}_B(d_x^4\, d_y^5\, \tau)$.

Corollary 2. *With the notations of Lemma 1, the worst case bit complexity for the computation of $\| G \|_\infty$ with the separation method (Algorithm 2) is given by $\tilde{O}_B(\alpha^{14}\, N^5\, \tau_n)$.*

From this section, we can conclude that trying to concentrate only on the solution with the maximal γ-projection, after putting the system in a generic position, costs much more than computing isolating boxes for all the real solutions, due to the large size of the separating bound that we must use. Hence, in the next section, using another strategy than shearing the system, we shall try to find the maximal γ-projection of the polynomial solutions without computing isolating boxes for all the real solutions.

5 Sturm-Habicht Method

In this section, as in Sect. 4, we shall concentrate only on the maximal γ-projection $\bar{\gamma}$ of the real solutions of the polynomial system. But instead of shearing the system, we shall verify the existence of a real root of the polynomial system over $\bar{\gamma}$ by studying the sign variation of the leading coefficients of *subresulant polynomials* over $\bar{\gamma}$. Hence, before explaining the third proposed method, we first state again a few standard preliminaries on *subresultants* and *Sturm-Habicht* sequences.

We denote by $\mathcal{K}$ the *unique factorization domain* $\mathbb{Q}[x]$ and we consider $P, Q \in \mathcal{K}[y]$, where $p = \deg_y(P)$ and $q = \deg_y(Q)$. We assume that $p \geq q$. For $0 \leq i \leq \min(q, p-1)$, the ith *subresultant polynomial* of P and Q is denoted by $\mathrm{Sres}_{y,i}(P, p, Q, q)$. When there is no ambiguity on the degrees of the polynomials P and Q, we simply denote it by $\mathrm{Sres}_{y,i}(P, Q)$. It has degree at most i in y and the coefficient of y^i is denoted by $\mathrm{sres}_{y,i}(P, Q)$. It is called the ith *principal subresultant coefficient*. We recall that $\mathrm{sres}_{y,i}(P, Q) = 0$ implies that $\mathrm{Sres}_{y,i}(P, Q)$ vanishes identically. Note that $\mathrm{Sres}_{y,0}(P, Q) = \mathrm{sres}_{y,0}(P, Q)$ is the *resultant* of P and Q with respect to y, also denoted by $\mathrm{Res}(P, Q, y)$. The greatest common divisor $\gcd(P, Q)$ of the polynomials P and Q (uniquely defined up to units of $\mathcal{K}$) is the first non-zero subresultant polynomial $\mathrm{Sres}_{y,i}(P, Q)$ for increasing i.

Letting $v = p + q - 1$ and $\delta_k = (-1)^{\frac{k(k+1)}{2}}$ for $k \in \mathbb{Z}_{\geq 0}$, the jth *polynomial in the Sturm-Habicht sequence* associated to (P, Q), denoted by $\mathrm{StHa}_j(P, Q)$, is then defined by $\delta_{v-j}\, \mathrm{Sres}_{y,j}(P, v+1, P^{'}Q, v)$, where $P^{'}$ denotes the derivative of P with respect to y. The *principal* j^{th} *Sturm-Habicht coefficient* is denoted by $\mathrm{stha}_j(P, Q)$ for $j = 0, \ldots, v+1$. We also denote by **SignVar** the function which maps $\{\mathrm{sign}(\mathrm{stha}_j(P, 1))\}_{j=0,\ldots,v+1}$ to the number of real roots of P. For more details on the function **SignVar**, see [12, Definition 4.1, Theorem 4.1].

As stated above, we aim at computing:

$$\bar{\gamma} = \max\left\{\pi_\gamma\left(V_{\mathbb{R}}\left(\left\langle \bar{n}, \frac{\partial \bar{n}}{\partial \omega}\right\rangle\right)\right) \cup V_{\mathbb{R}}\left(\langle Lc_\omega(\bar{n})\rangle\right)\right\}.$$

Hence, $\bar{\gamma}$ is either the maximal real root of $Lc_\omega(n)$ or an algebraic value over which $\gcd(n(\bar{\gamma}, \omega), \frac{\partial n}{\partial \omega}(\bar{\gamma}, \omega)) \in \mathbb{R}[\omega]$ has at least one real root. We recall that $\gcd(n(\bar{\gamma}, \omega), \frac{\partial n}{\partial \omega}(\bar{\gamma}, \omega))$ is proportional to the first subresultant polynomial $\mathrm{Sres}_{\omega,i}(n, \frac{\partial n}{\partial \omega})$ (for i increasing) that does not identically vanish for $\gamma = \bar{\gamma}$. If $\bar{\gamma}$ is not a real root of $Lc_\omega(n)$, then we can compute the Sturm-Habicht sequence of the univariate polynomial $n(\bar{\gamma}, \omega) \in \mathbb{R}[\omega]$ to check the existence of a real root for $\gcd(n(\bar{\gamma}, \omega), \frac{\partial n}{\partial \omega}(\bar{\gamma}, \omega))$. In what follows, we shall need the next result.

Lemma 3. *Let $P \in \mathbb{Z}[x,y]$ and $\bar{x}$ be a root of* $\mathrm{Res}(P, \frac{\partial P}{\partial y}, y)$. *Moreover, let* $\mathcal{G} = \gcd\left(P(\bar{x},y), \frac{\partial P}{\partial y}(\bar{x},y)\right) \in \mathbb{R}[y]$. *If the x-projection of the points of P is bounded by $\bar{x}$, then we have* $V_{\mathbb{R}}(\langle P(\bar{x},y)\rangle) = V_{\mathbb{R}}(\langle \mathcal{G}(\bar{x},y)\rangle)$.

Proof. If $V_{\mathbb{R}}(\langle \mathcal{G}(\bar{x},y)\rangle) \subsetneq V_{\mathbb{R}}(\langle P(\bar{x},y)\rangle)$, then there exists $y_0 \in \mathbb{R}$ such that $P(\bar{x},y_0) = 0$ and $\mathcal{G}(\bar{x},y_0) \neq 0$. This is equivalent to saying that $P(\bar{x},y_0) = 0$ and $\frac{\partial P}{\partial y}(\bar{x},y_0) \neq 0$. Hence, based on the theorem of implicit functions, there exists a real function φ of class C^p $(p > 0)$, defined on an open interval $V \subset \mathbb{R}$, containing $\bar{x}$, and an open neighborhood Ω of $(\bar{x}, y_0)$ in $\mathbb{R}^2$ such that for all (x, y) in $\mathbb{R}^2$, $\{(x, y) \in \Omega \mid P(x,y) = 0\}$ is equivalent to $\{x \in V \mid y = \varphi(x)\}$. This cannot be true since the x-projection of the points of the curve $P = 0$ is bounded by $\bar{x}$, and thus, an open interval containing $\bar{x}$, such as V, does not exist. Consequently, we obtain $V_{\mathbb{R}}(\langle P(\bar{x},y)\rangle) = V_{\mathbb{R}}(\langle \mathcal{G}(\bar{x},y)\rangle)$.

Algorithm 3. Sturm-Habicht method

Input: A bivariate polynomial $P \in \mathbb{Z}[x,y]$ such that $P = 0$ is bounded in the x-direction.
Output: Isolating interval of $\max\left\{\pi_x\left(V_{\mathbb{R}}\left(\left\langle P, \frac{\partial P}{\partial y}\right\rangle\right)\right) \cup V_{\mathbb{R}}(\langle Lc_y(P)\rangle)\right\}$.

1. Compute $\{\mathrm{Sres}_j(P, \frac{\partial P}{\partial y})\}_{j=0,\ldots,\deg_y(P)}$.
2. Compute $x_1 < \ldots < x_m$ the real roots of sres_0.
3. for i from 1 to m do:
   - if $x_{1-i+m} \in V_{\mathbb{R}}(\langle Lc_y(P)\rangle)$ then **return** the isolating interval of x_{1-i+m};
   - **elif SignVar**$(\{\mathrm{sign}(\mathrm{stha}_{d_y}(x_{1-i+m})), \ldots, \mathrm{sign}(\mathrm{stha}_1(x_{1-i+m}))\}) > 0$, then **return** the isolating interval of x_{1-i+m}.
4. end if end do.

Lemma 4. *Let $P \in \mathbb{Z}[x,y]$, $d_x = \deg_x(P)$, $d_y = \deg_y(P)$ and τ be the maximal coefficients bitsize of P. Let $\{\mathrm{StHa}_j(P(x,y),1)\}_{j=0,\ldots,d_y}$ be the Sturm-Habicht sequence and x_j a real root of* $\mathrm{sres}_{y,0}\left(P, \frac{\partial P}{\partial y}\right)$. *Then,* $\{\mathrm{sign}(\mathrm{stha}_k(x_j))\}_{k=d_y,\ldots,1}$ *can be computed in $\tilde{O}_B(d_x^2\, d_y^4\, (d_x + \tau))$ bit operations.*

Proof. We denote by sres_0 (resp., sres_i) $\mathrm{sres}_{y,0}\left(P, \frac{\partial P}{\partial y}\right)$ (resp., $\mathrm{sres}_{y,i}\left(P, \frac{\partial P}{\partial y}\right)$), where $\mathrm{sres}_i \in \mathbb{Z}[x]$. We first recall that $\mathrm{stha}_i(x_j) = \delta_{d_y-1-i}\,\mathrm{sres}_i(x_j)$. Based on [2, Proposition 8.46], sres_i is of degree $d_x\, d_y$ and of coefficients bitsize $d_y\,\tau$. Thus, the square free part of sres_0 is of coefficients bitsize $\mathcal{O}(d_y\,(d_x + \tau))$ and, based on [4, Lemma 5], can be computed in $\tilde{O}_B(d_x^2\, d_y^3\, \tau)$. By following the proof of [17, Proposition 6], the overall cost for obtaining the list $\{\mathrm{sign}(\mathrm{stha}_{d_y}(x_j)), \ldots, \mathrm{sign}(\mathrm{stha}_1(x_j))\}$ is $\tilde{O}_B(d_x^2\, d_y^4\, (d_x + \tau))$.

Theorem 3. *Let $P \in \mathbb{Z}[x,y]$ be such that $d_x = \deg_x(P)$, $d_y = \deg_y(P)$ and of maximal coefficients bitsize τ. Then, we can compute an isolating interval of the maximal x-projection of the real solutions of $\{P, \frac{\partial P}{\partial y}\}$ (Algorithm 3) in $\tilde{\mathcal{O}}_B(d_x^2\, d_y^4\,(d_x+\tau))$ bit operations in the worst case.*

Proof. The maximal x-projection of the real solutions of $\{P, \frac{\partial P}{\partial y}\}$ is the maximal real root of $\mathrm{sres}_0\left(P, \frac{\partial P}{\partial y}\right)$, say x_m, such that $\gcd\left(P(x_m,y), \frac{\partial P}{\partial y}(x_m,y)\right)$ has at least one real root. If the x-projection of the points of P is bounded by x_m, then, by Lemma 3, the real roots of $\gcd\left(P(x_m,y), \frac{\partial P}{\partial y}(x_m,y)\right)$ are the real roots of $P(x_m,y)$. Consequently, we can compute an isolating interval of x_m using Algorithm 3. According to [2, Proposition 8.46], we can compute the set of principal subresultants in $\tilde{\mathcal{O}}(d_x\, d_y^3\,\tau)$ bit operations and each subresultant polynomial is of degree $\mathcal{O}(d_x\, d_y)$ and of coefficient bit size $\tilde{\mathcal{O}}(d_y\,\tau)$. Thus, step 2, which performs real root isolation of sres_0, is of complexity $\tilde{\mathcal{O}}((d_x, d_y)^3 + (d_x\, d_y)^2\, d_y\,\tau)$ [3, Lemma 54]. Using Lemma 4, step 3 can be done in $\tilde{\mathcal{O}}_B(d_x^2\, d_y^4\,(d_x+\tau))$ operations since its first step is of complexity $\tilde{\mathcal{O}}_B(d_x^3 + d_x^2\,\tau)$. Hence, the overall complexity of this algorithm is $\tilde{\mathcal{O}}_B(d_x^2\, d_y^4\,(d_x+\tau))$.

Considering the notations of Lemma 1, the following result is an immediate consequence of Corollary 1 and Theorem 3.

Corollary 3. *Based on Theorem 3, $\| G \|_\infty$ can be computed by Sturm-Habicht method in the worst case bit complexity $\tilde{\mathcal{O}}_B(\alpha^{10}\, N^4\,(\alpha+\tau_n))$.*

In Algorithm 4, we suppose that there are no real isolated points, and thus, we replace the computation of signs of polynomials at real algebraic numbers by signs of polynomials at rational numbers. Syntactically, these are small modifications but the effect on the computations is consequent in practice, as well as in theory, since the evaluation of signs of polynomials at real algebraic numbers carries the theoretical worst case complexity of Algorithm 3.

Theorem 4. *Let $P \in \mathbb{Z}[x,y]$ be a bivariate polynomial of maximal coefficient bitsize τ and let $d_x = \deg_x(P)$ and $d_y = \deg_y(P)$. Moreover, let us suppose that $V_\mathbb{R}(\langle P\rangle)$ has no isolated singular points. Using Algorithm 4, an isolating interval of the maximal x-projection of the real solutions of $\{P, \frac{\partial P}{\partial y}\}$ can be computed in the worst case bit complexity $\tilde{\mathcal{O}}_B(d_x^2\, d_y^4\,\tau)$.*

Proof. As mentioned in the proof of Theorem 3, we can compute the set of principal subresultants in $\tilde{\mathcal{O}}(d_x\, d_y^3\,\tau)$ bit operations and each subresultant polynomial is of degree $\mathcal{O}(d_x\, d_y)$ and of coefficient bitsize $\tilde{\mathcal{O}}(d_y\,\tau)$ according to [2, Proposition 8.46]. Thus, step 2 of Algorithm 4, which performs the real root isolation of sres_0, is of complexity $\tilde{\mathcal{O}}((d_x\, d_y)^3 + (d_x\, d_y)^2\, d_y\,\tau)$ [3, Lemma 54]. Steps 3 and 4 are of same bit complexity: in these steps, we perform $\mathcal{O}(d_y)$ evaluations of the principal subresultant polynomials over a rational number which is between two real roots of sres_0. This rational number is of

Algorithm 4. Sturm-Habicht method - equidimensional

Input: A bivariate polynomial $P \in \mathbb{Z}[x,y]$ such that the curve $P=0$ is bounded in the x-direction and has not real isolated singular points.
Output: An isolating interval of $\max\left\{\pi_x\left(V_{\mathbb{R}}\left(P, \frac{\partial P}{\partial y}\right)\right) \cup V_{\mathbb{R}}\left(Lc_y(P)\right)\right\}$

1. Compute $\{\mathrm{StHa}_j(P,1)\}_{j=0,\ldots,\deg_y(P)}$.
2. Let $x_1 < \ldots < x_m$ be the roots of sres_0.
3. for i from 1 to m do:
   - if $x_{1-i+m} \in V_{\mathbb{R}}(Lc_y)$, then **return** the isolating interval of x_{1-i+m};
   - **else** let $X' \in \mathbb{Q}$ such that $x_{m-i} < X' < x_{1-i+m}$;
     - if **SignVar**$(\{\mathrm{sign}(\mathrm{stha}_{d_y}(X')),\ldots,\mathrm{sign}(\mathrm{stha}_1(X'))\}) > 0$, then **return** the isolating interval of x_{1-i+m};
     - end if.
   - end if.
4. end do.

---

worst possible coefficient bitsize $\tilde{O}_B(d_x\, d_y^2\, \tau)$, which is equal to the separating bound of sres_0. According to [4, Lemma 6], the d_y evaluations are done in $\tilde{O}_B(d_y\,(d_x\, d_y\,(d_y\,\tau + d_x\, d_y^2\,\tau))) = \tilde{O}_B(d_x^2\, d_y^4\,\tau)$. Hence, the overall cost is given by $\tilde{O}_B(d_x^2\, d_y^4\,\tau)$.

Corollary 4. *Based on Theorem 4,* $\| G \|_\infty$ *can be computed by the Sturm-Habicht method (Algorithm 4) in the worst case bit complexity* $\tilde{O}_B(\alpha^{10}\, N^4\, \tau_n)$.

From the above complexity analysis, we can conclude that RUR method and the Sturm-Habicht method have comparable theoretical complexities since, in our case, we have $\alpha \ll N$.

6 Experiments

6.1 Practical Example

We consider the following transfer matrix:

$$G = \begin{pmatrix} \dfrac{1}{s+1} & \dfrac{1}{s+1} \\ 0 & \dfrac{1}{s+1} \end{pmatrix} \in \mathcal{RH}_\infty^{2\times 2}.$$

Let $\Phi_\gamma(s) = \gamma^2\, I_2 - \tilde{G}(s)\, G(s)$ and $\det(\Phi_\gamma(i\,\omega)) = \dfrac{n(\gamma,\omega)}{d(\omega)}$. We study the real solutions of the polynomial system $\Sigma = \{\bar{n}, \frac{\partial \bar{n}}{\partial \omega}\}$, where:

$$\bar{n}(\gamma,\omega) = \gamma^4\,\omega^4 + \gamma^2\,(2\,\gamma^2 - 3)\,\omega^2 + (\gamma^2+\gamma-1)(\gamma^2-\gamma-1).$$

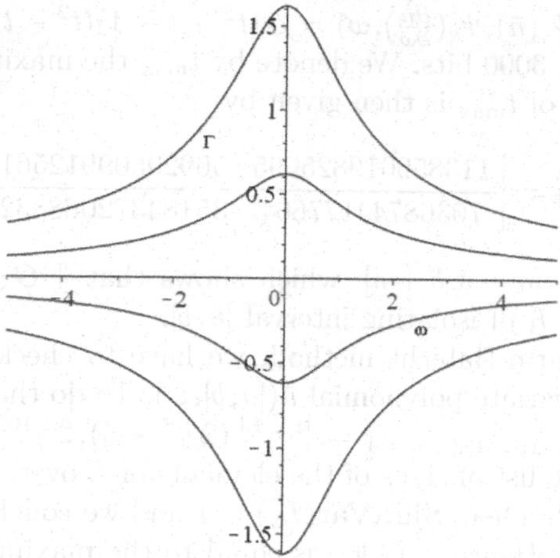

Fig. 2. Plot of $n(\gamma, \omega) = 0$, where ω/γ is in the horizontal/vertical axis.

We first compute $V_{\mathbb{R}}(\mathrm{Lc}_\omega(\bar{n})) = \{0\}$. Then, applying the RUR method, we obtain the following rational univariate representation:

$$\begin{cases} p = (t^2 + t - 1)(t^2 - t - 1), \\ \gamma = \dfrac{3\,t^2 - 2}{t\,(2\,t^2 - 3)}, \\ \omega = 0. \end{cases}$$

Thus, the system's real solutions (γ, ω) are:

$$\left(-\frac{\sqrt{5}}{2} - \frac{1}{2}, 0\right), \left(-\frac{\sqrt{5}}{2} + \frac{1}{2}, 0\right), \left(\frac{\sqrt{5}}{2} - \frac{1}{2}, 0\right), \left(\frac{\sqrt{5}}{2} + \frac{1}{2}, 0\right).$$

Thus, we simply pick their maximal γ-projection to obtain $\frac{\sqrt{5}}{2} + \frac{1}{2}$, which yields:

$$\| G \|_\infty = \max\left\{0, \frac{\sqrt{5}}{2} + \frac{1}{2}\right\} = \frac{\sqrt{5}}{2} + \frac{1}{2}.$$

Following the second approach, which consists in directly focusing on the maximal γ-projection of the system's real solutions, we first compute $\mathrm{Res}(\bar{n}, \frac{\partial \bar{n}}{\partial \omega}, \omega)$ and denote by $R = \gamma\,(\gamma^2 + \gamma - 1)\,(\gamma^2 - \gamma - 1) \in \mathbb{Z}[\gamma]$ its square free part. Then, the maximal real root of R has the following isolating interval:

$$[a, b] = \left[\frac{56929509912547}{35184372088832}, \frac{113859019825121}{70368744177664}\right].$$

Following the Root Separation method, we obtain:

$$\begin{cases} s = \dfrac{12060328540887}{281474976710656}, \\ \delta = \dfrac{43490275647441}{140737488355328}. \end{cases}$$

We have $R_t = \text{Res}(\Psi_s(\bar{n}), \Psi_s(\frac{\partial \bar{n}}{\partial \omega}), \omega) = \alpha\,(t^2+t-1)(t^2-t-1)$, where α is a rational of size about 3000 bits. We denote by $t_{\max}$ the maximal real root of R_t. An isolating interval of $t_{\max}$ is then given by:

$$[c,d] = \left[\frac{113859019825095}{70368744177664}, \frac{56929509912561}{35184372088832}\right].$$

In this case, $[c,d] \subset [a-\delta, b+\delta]$, which shows that $\| G \|_\infty$ is equal to the maximal real root of R of isolating interval $[a,b]$.

Following the Sturm-Habicht method, we have to check the existence of a real root for the univariate polynomial $\bar{n}([a,b],\omega)$. To do that, we first compute $L = [\text{sres}_{\omega,i}(\bar{n}, \frac{\partial \bar{n}}{\partial \omega})]_{i=1,\ldots,\deg_\omega(n)=4} = [5\,\gamma^{14}\,(2\,\gamma^2-3), 2\,\gamma^{10}\,(2\,\gamma^2-3), 4\,\gamma^4, \gamma^4]$. We then compute the list of signs of the elements of L over $[a,b]$. We obtain the list $L_s = [-,-,+,+]$. Then, $\mathbf{SignVar}(L_s) = 1$ and we conclude that $\bar{n}([a,b],\omega)$ admits one real root. Hence, $\| G \|_\infty$ is equal to the maximal real root of R of isolating interval $[a,b]$.

6.2 Experiments

The three proposed methods can be implemented in a few lines of `Maple` but we then have to use implementations at different levels that do not give valuable information about the intrinsic efficiency. For instance, the RUR is implemented in `C` but for general zero dimensional polynomial systems: a variant for bivariate polynomials, the one used for the complexity analysis, is not part of `Maple` and is much more efficient for bivariate systems.

In order to have fair comparisons, we extract the dominating operations and compare them using exactly the same implementations. Namely, resultant computations of sheared/non sheared systems and Root Isolation carry the largest percentage of the computation time. For instance, Algorithm 4 saves time on the resultant computation since it does not perform any shear while it loses time on the root isolation.

For the three methods, the principle subresultant sequence is computed using the routine `SubResultantChain` of the `Maple` package `RegularChain`.

Isolating the real roots of univariate polynomials is another common basic block shared between the three algorithms for which we use `Isolate` provided by the `Maple routine` package `RootFinding`.

In left table of Table 1, we list the main steps of the three algorithms. The check marks mean that the step makes part of the method and the double check marks indicate that this step is the bottleneck of the method. Note that *Res1* stands for the resultant of the original system and *Res2* for the resultant of the sheared system. Keep in mind that the shear done in `Hinf_RUR` is different than the one done in `Hinf_Sep`. Finally, *Iso* means `Isolate`.

In the right table of Table 1, we report the average running time in CPU seconds of the marked steps listed in the table on the left of Table 1 for the three proposed algorithms run on square matrices of size α, with entries given by

Table 1. Left: main steps considered in the implementation of the proposed method. Right: timings for $\mathcal{L}_\infty$-norm for random matrices with $\tau_G = 2$.

	Res1 + Iso	Res2 + Iso	List of signs
Hinf_RUR	✓	✓✓	
Hinf_Srep	✓	✓✓	
Hinf_Sres	✓		✓✓

α	N	Hinf_RUR	Hinf_Sep	Hinf_Sres
	2	0.2	3	0.2
	3	0.5	7	0.5
	4	2.5	25	2
	5	10	83	6
2	6	37	96	10
	7	50	186	47.5
	8	133.5	353	59
	9	236	394	130

random proper rational functions of degree N (degree of the denominators)[1]. It corresponds to a fixed input coefficient bitsize $\tau = 2$, i.e., the rational functions involved in the entries of the matrices have coefficients of magnitude $\mathcal{O}(2^\tau)$.

We finally mention that with these experiments, our goal is not to illustrate the theoretical complexity, but, on the contrary, to show that on practical examples, the results in practice are different than in theory. In theory, the RUR algorithm might asymptotically be the fastest while in practice the Sturm method performs better.

7 Conclusion

In this paper, we have presented three different algorithms for the computation of the $\mathcal{L}_\infty$-norm of the transfer matrix of a finite-dimensional linear control system. By reformulating this problem as the search for the maximal projection of the real solutions of a zero dimensional polynomial system, we have used existing methods such as the rational univariate representation (RUR method). As for the second algorithm, we have only used a special separating linear transformation to shear the polynomial system and put it in a generic position. The last method (Sturm-Habicht method) was based on verifying the existence of a real root for a univariate polynomial having real coefficients.

The complexity analysis has showed that the RUR method has the best theoretical efficiency in comparison with other algorithms. Practically, as we can notice in the tables given in Sect. 6, the practical efficiency is nearly matching with the theoretical efficiency but with a slight advantage for the Sturm-Habicht method probably due to the fact that it is the most adaptive one.

[1] The experiments were conducted on Intel(R) Core(TM) i7-7500U CPU @ 2.70 GHz 2.90 GHz, Installed RAM 8.00 GB under a Windows platform.

References

1. Aubry, P., Maza, M.M.: Triangular sets for solving polynomial systems: a comparative implementation of four methods. J. Symb. Comput. **28**(1–2), 125–154 (1999)
2. Basu, S., Pollack, R., Roy, M.F.: Existential theory of the reals, volume 10 of algorithms and computation in mathematics (2006)
3. Bouzidi, Y., Lazard, S., Moroz, G., Pouget, M., Rouillier, F., Sagraloff, M.: Solving bivariate systems using rational univariate representations. J. Complex. **37**, 34–75 (2016)
4. Bouzidi, Y., Lazard, S., Pouget, M., Rouillier, F.: Separating linear forms and rational univariate representations of bivariate systems. J. Symb. Comput. **68**, 84–119 (2015)
5. Boyd, S., Balakrishnan, V., Kabamba, P.: A bisection method for computing the h_∞ norm of a transfer matrix and related problems. Math. Control Signals Syst. **2**(3), 207–219 (1989)
6. Boztaş, S., Shparlinski, I.E. (eds.): AAECC 2001. LNCS, vol. 2227. Springer, Heidelberg (2001). https://doi.org/10.1007/3-540-45624-4
7. Chen, C., Maza, M.M., Xie, Y.: A fast algorithm to compute compute the h_∞-norm of a transfer function matrix. Syst. Control Lett. **14**, 287–293 (1990)
8. Chen, C., Mazza, M.M., Xie, Y.: Computing the supremum of the real roots of a parametric univariate polynomial (2013)
9. Cheng, J.S., Gao, X.S., Li, J.: Root isolation for bivariate polynomial systems with local generic position method. In: Proceedings of the 2009 International Symposium on Symbolic and Algebraic Computation, pp. 103–110. ACM (2009)
10. Cheng, J.S., Gao, X.S., Yap, C.K.: Complete numerical isolation of real roots in zero-dimensional triangular systems. J. Symb. Comput. **44**(7), 768–785 (2009)
11. Doyle, J.C., Francis, B.A., Tannenbaum, A.R.: Feedback Control Theory. Dover Publications, New York (1992)
12. González-Vega, L., Recio, T., Lombardi, H., Roy, M.F.: Sturm–Habicht sequences, determinants and real roots of univariate polynomials. In: Caviness, B.F., Johnson, J.R. (eds.) Quantifier Elimination and Cylindrical Algebraic Decomposition. Texts and Monographs in Symbolic Computation (A Series of the Research Institute for Symbolic Computation, Johannes-Kepler-University, Linz, Austria). Springer, Vienna (1998). https://doi.org/10.1007/978-3-7091-9459-1_14
13. Kanno, M., Smith, M.C.: Validated numerical computation of the l_∞-norm for linear dynamical systems. J. Symb. Comput. **41**(6), 697–707 (2006)
14. Lazard, S., Pouget, M., Rouillier, F.: Bivariate triangular decompositions in the presence of asymptotes. J. Symb. Comput. **82**, 123–133 (2017)
15. Rouillier, F.: Solving zero-dimensional systems through the rational univariate representation. Appl. Algebra Eng. Commun. Comput. **9**(5), 433–461 (1999)
16. Rouillier, F., Zimmermann, P.: Efficient isolation of polynomial's real roots. J. Comput. Appl. Math. **162**(1), 33–50 (2004)
17. Strzebonski, A., Tsigaridas, E.: Univariate real root isolation in an extension field and applications. J. Symb. Comput. **92**, 31–51 (2019)
18. Zhou, K., Doyle, J.C., Glover, K.: Robust and Optimal Control. Prentice Hall, Upper Saddle River (1996)

A Maple Package to Deal with the Birationality of Curves and Surfaces Parametrizations

Jorge Caravantes[(✉)], Sonia Pérez–Díaz, and J. Rafael Sendra

Universidad de Alcalá, Dpto. Física y Matemáticas, Alcalá de Henares, Madrid, Spain
{jorge.caravantes,sonia.perez,rafael.sendra}@uah.es

Abstract. In this paper we present the Maple package `Luroth` for dealing with the birationality of curves and surfaces parametrizations. The procedures in the package decide whether a given, either curve or surface, parametrization is injective by computing its degree map. In addition, if the parametrization is not injective, it determines a birational reparametrization. For the curve case, the corresponding command always provides an optional answer. For the surface case, not all cases are covered. Nevertheless, we illustrate using Maple some new ideas on how to approach those surface cases not covered in the package.

Keywords: Maple · Symbolic computation · Birational (proper) parametrization · Algebraic curves · Algebraic surfaces

1 Introduction

Algebraic varieties are definable as the zero–set of polynomials. Nevertheless, for some special cases, namely those that are unirational, they can also be represented by means of a tuple of rational functions; see [10,11] for the case of radical parametric representations. It is well-known, and illustrated in the literature (see e.g. [3]), that depending on the particular problem to be approached a different representation might be more advisable. In this paper, we stay within the world of rational parametric representations, that is we deal with unirational varieties, and more specifically with unirational curves and surfaces. Even in this case, different computational and theoretical questions appear that affect to the feasibility of the applications of the parametric representations. More precisely, one may consider the injectivity and/or the surjectivity (see [1,9]) of the parametric representation. We here deal with the injectivity.

The natural question in this context is whether a rationally parametrized variety (i.e. parametrized by means of a tuple of rational functions) can be parametrized birationally (i.e. the map being injective on a non-empty Zariski open subset of the parameter space); we will refer to a birational parametrization as a proper parametrization. This question can be reformulated in terms of field theory by using the field of rational functions of the variety. Then, it holds that

R. M. Corless et al. (Eds.): MC 2020, CCIS 1414, pp. 137–151, 2021.
https://doi.org/10.1007/978-3-030-81698-8_10

any unirationally parametrized curve can be parametrized birationally for any field (see e.g. [12]) because of Lüroth's Theorem:

Theorem 1. *(Lüroth) Consider the inclusion of fields $\mathbb{K} \subset \mathbb{L} \subset \mathbb{K}(x)$. Then, there exists $f(x) \in \mathbb{K}(x)$ such that $\mathbb{L} = \mathbb{K}(x)$.*

Equivalently, let $\mathcal{P} \in \mathbb{K}(t)^n$ a (maybe non proper) curve parametrization. Then, the Zariski closure $\mathcal{C}$ of $\mathcal{P}(\mathbb{K})$ is a rational curve (i.e. there exists a birational parametrization of $\mathcal{C}$).

Example 1. Consider the parametrization

$$t \mapsto (t^6 - 3t^5 - 3t^4 + 11t^3 + 5t^2 - 11t - 6, t^4 - 2t^3 - 3t^2 + 4t + 3). \quad (1)$$

Its image is a curve that, by Lüroth's Theorem, is rational. In fact, it is the nodal cubic $y^3 - x^2 + y^2$, that can be properly (not injectively) parametrized by $t \mapsto (t^3 - t, t^2 - 1)$. It is interesting to have means to find the proper parametrization from the map (1) and the relation between both of them.

However, for the surface case, the characterization is only possible when the field is algebraically closed (see [13]); for other dimensions, the situation is even more restrictive.

In this paper, we deal with the problem of deciding the properness, and computing birational parametrizations, for the case of curves and surfaces. For the curve case, there are constructive proofs of Lüroth's Theorem (see [12] and the references therein), but we will mainly use here the algorithmic approaches in [5] and [6]. For the surface case, one may proceed as follows: compute the implicit equation of the surface, using e.g. [7], and then apply a parametrization algorithm, e.g. Schicho's algorithm (see [8]). However, here, we want to approach the problem without implicitizing, that is by reparametrizing the original parametrization. For this purpose, we will use the algorithm in [6]. Results in [6] provide a wide, but partial, solution to the problem. But, up to our knowledge, there is no complete algorithmic (reparametrizing) answer for the surface case.

More precisely, we present a Maple package, that we call `Luroth` for dealing with the described problems. Furthermore, we also show some ideas on how to approach the cases not covered in [6] that we illustrate with the help of Maple. The paper is structured as follows. In Sect. 2 we briefly recall the results in [5] and [6] that we use in our implementations. In Sect. 3 we give an overview of the package and we show some examples that are additionally illustrated in the Appendix. The Maple package is available (see Sect. 3 for details) at the web site http://www3.uah.es/jorge_caravantes/research.html. In Sect. 4, some on-going working ideas to approach the general case using Maple are presented. Finally, in the appendix (Sect. 5), the Maple executions, corresponding to the examples in the Subsect. 3.2, are shown.

2 Theoretical and Algorithmic Framework

In this section, we briefly recall some theoretical facts and algorithms that will be used in the implementation of the package. Throughout this section, $\mathbb{K}$ is an algebraically close field and $\mathbb{L}$ a subfield of $\mathbb{K}$; usually the ground field where the parametrization is expressed.

2.1 The Curve Case

Let $\mathcal{C} \subset \mathbb{K}^n$ be a curve, rationally parametrized by

$$\mathcal{P}(t) = (p_1(t), \dots, p_n(t)) \in \mathbb{L}(t)^n,$$

where $\mathcal{P}$ is not necessarily proper. By Lüroth's Theorem, we know that there exists a rational function $R \in \mathbb{L}(t)$ and a birational parametrization $\mathcal{Q}(t) \in \mathbb{L}(t)^n$ of $\mathcal{C}$ such that $\mathcal{P}(t) = \mathcal{Q}(R(t))$. In Algorithm 1, we outline the ideas presented in [5] for effectively computing $\mathcal{Q}$ and R. For this purpose, we denote by Res the univariate resultant of two polynomials, by Num the numerator of a rational function expressed in reduced form, and by Den the denominator of a rational function expressed in reduced form.

Algorithm 1. Proper reparametrization of space curves

Input: $\mathcal{P}(t) = (p_1(t), \dots, p_n(t)) \in \mathbb{L}(t)^n \subset \mathbb{K}(t)^n$ a parametrization of $\mathcal{C}$.
Output: A proper parametrization $\mathcal{Q}(t) \in \mathbb{L}(t)^n$ and $R(t) \in \mathbb{L}(t)$ such that $\mathcal{Q}(R(t)) = \mathcal{P}(t)$.

Determine

$$S(s,t) = \gcd(\mathrm{Num}(p_1(s) - p_1(t)), \dots, \mathrm{Num}(p_n(s) - p_n(t))).$$

Let us say that $S(s,t) = C_m(t)s^m + \dots + C_0(t)$.
if $\deg_t(S) = 1$ **then**
  return $\mathcal{Q}(t) = \mathcal{P}(t)$, and $R(t) = t$.
end if
Choose $C, D \in \{C_0, \dots, C_m\}$ such that $\gcd(C, D) = 1$ and $C/D, C \cdot D \notin \mathbb{L}$ (see Section 2 in [5] for further details). Take $R(t) = C/D$.
for $i = 1, \dots, n$ **do**
  Compute

$$L_i(s, x_i) = \mathrm{Res}_t(\mathrm{Num}(x_i - p_i(t)), \mathrm{Num}(s - R(t))).$$

  It holds that L_i is of the form $L_i = (b_i(s)x_i - a_i(s))^{\deg(R)}$.
end for
return the rational function $R(t)$, and the proper parametrization

$$\mathcal{Q}(t) = (a_1(t)/b_1(t), \dots, a_n(t)/b_n(t)) \in \mathbb{L}(t)^n.$$

We illustrate Algorithm 1 by an example.

Example 2. Let $\mathcal{C}$ be a rational space curve over $\mathbb{C}$ defined by the parametrization

$$\mathcal{P}(t) = \left(\frac{(t+1)^2(t^2+1)}{2(t^4+4t^2+1+2t^3+2t)}, \frac{2(t^2+t+1)}{(t+1)^2}, \frac{-(t^2+1)(t^2+1+4t)}{(t+1)^4} \right).$$

In Step 1 of the algorithm, we get

$$\begin{aligned}
\mathrm{Num}(p_1(t)-p_1(s)) &= -(s-t)(ts-1)(ts^2+t^2s+2ts+s+t)\\
\mathrm{Num}(p_2(t)-p_2(s)) &= 2(s-t)(ts-1),\\
\mathrm{Num}(p_3(t)-p_3(s) &= 4(ts-1))(s-t)(ts^2+t^2s+s+4ts+t).
\end{aligned}$$

Thus, we get $S(s,t)=C_0(t)+C_1(t)s+C_2(t)s^2$, where $C_0(t)=t,\ C_1(t)=-t^2-1$, and $C_2(t)=t$. Since $\deg_t(S^{\mathcal{P}})>1$, the parametrization is not proper (observe that, for general t, two values of s vanish S, so give the same image by $\mathcal{P}$), so we go to Step 2 where we choose C as C_1 and D as C_0. Therefore, $R(t)=-(t^2+1)/t$. Note that $\gcd(C_0, C_1)=1$. Now, we compute the polynomials

$$\begin{aligned}
L_1(s,x_1) &= (4x_1+4\,sx_1-2\,s-s^2+2\,s^2x_1)^2,\\
L_2(s,x_2) &= (2x_2-2-2\,s+sx_2)^2,\\
L_3(s,x_3) &= (4x_3+4\,s+4\,sx_3+s^2+s^2x_3)^2.
\end{aligned}$$

Finally, in Step 5, the algorithm outputs the proper parametrization $\mathcal{Q}(t)$, and the rational function $R(t)$:

$$\mathcal{Q}(t)=\left(\frac{t(t+2)}{2(2+2t+t^2)},\frac{2(t+1)}{2+t},\frac{-t(t+4)}{4+4t+t^2}\right)\in\mathbb{C}(t)^3,\quad R(t)=-\frac{t^2+1}{t}\in\mathbb{C}(t).$$

It is not difficult to see that $\mathcal{P}(t)=\mathcal{Q}(R(t))$.

2.2 The Surface Case: A Partial Solution

Let $\mathcal{P}(\overline{t})$ be a rational affine parametrization over $\mathbb{L}$ of an algebraic rational surface $\mathcal{V}$. We express $\mathcal{P}$ as

$$\mathcal{P}(\overline{t})=(p_1(\overline{t}),p_2(\overline{t}),p_3(\overline{t}))\in\mathbb{L}(\overline{t})^3,\quad p_i(\overline{t})=p_{i,1}(\overline{t})/p_{i,2}(\overline{t}),\qquad (2)$$

where $\gcd(p_{i,1},p_{i,2})=1,\ i=1,2,3$, and $\overline{t}=(t_1,t_2)\in\mathbb{K}^2$.

The degree of the rational map induced by $\mathcal{P}$ is denoted by $\mathrm{MapDeg}(\mathcal{P})$; see e.g. [2] pp. 80, or [13] pp. 143 for details. We recall that the properness of $\mathcal{P}(\overline{t})$ is characterized by $\mathrm{MapDeg}(\mathcal{P})$. More precisely, $\mathcal{P}(\overline{t})$ is *proper* if and only if $\mathrm{MapDeg}(\mathcal{P})=1$ (see [2] and [13]). Also, the mapping degree is the cardinality of the fibre of a generic element (see Theorem 7, pp. 76 in [13]). That is,

$$\mathcal{F}_{\mathcal{P}}(P)=\mathcal{P}^{-1}(P)=\{\overline{t}\in\mathbb{K}^2\,|\,\mathcal{P}(\overline{t})=P\},$$

where $\mathcal{F}_{\mathcal{P}}(P)$ is the fibre of a point $P\in\mathcal{V}$. Associated with the parametrization $\mathcal{P}$, we consider the polynomials

$$G_i^{\mathcal{P}}(\overline{t},\overline{s})=p_{i,1}(\overline{t})p_{i,2}(\overline{s})-p_{i,2}(\overline{t})p_{i,1}(\overline{s})\in(\mathbb{K}[\overline{s}])[\overline{t}],\quad i=1,2,3,$$

and

$$\begin{aligned}
S_1^{\mathcal{P}}(t_1,\overline{s}) &= \mathrm{PrimPart}_{\overline{s}}(\mathrm{Content}_Z(\mathrm{Res}_{t_2}(G_1,G_2+ZG_3))),\\
S_2^{\mathcal{P}}(t_2,\overline{s}) &= \mathrm{PrimPart}_{\overline{s}}(\mathrm{Content}_Z(\mathrm{Res}_{t_1}(G_1,G_2+ZG_3))),
\end{aligned}$$

where $\overline{s} = (s_1, s_2) \in \mathbb{K}^2$, Z is an auxiliary variable, and $\text{Content}_x(p)$ and $\text{PrimPart}_x(p)$ are the content and the primitive part of a polynomial p with respect to the variable x. Let $\mathbb{F} = \overline{\mathbb{K}(\overline{s})}$ be the algebraic closure of $\mathbb{K}(\overline{s})$.

The polynomials $S_j^{\mathcal{P}}$ play an important role in deciding the properness of a parametrization $\mathcal{P}$. More precisely, in [4] the following theorem is proved.

Theorem 2. *The following statements hold:*

1. $\mathcal{P}^{-1}(\mathcal{P}(\overline{s})) = \{\overline{t} \in \mathbb{F}^2 \mid G_i^{\mathcal{P}}(\overline{t}, \overline{s}) = 0,\ i = 1, 2, 3\}$ *and, for generic* $\overline{s} \in \mathbb{K}^2$, $\text{MapDeg}(\mathcal{P}) = \text{Card}(\mathcal{P}^{-1}(\mathcal{P}(\overline{s})))$.
2. *The polynomial* $S_i^{\mathcal{P}}$ *defines the* t_i*–coordinates of the points in* $\mathcal{P}^{-1}(\mathcal{P}(\overline{s}))$.
3. *Considering* $\overline{s}$ *as a couple of variables,* $\text{MapDeg}(\mathcal{P}) = \deg_{t_1}(S_1^{\mathcal{P}}(t_1, \overline{s})) = \deg_{t_2}(S_2^{\mathcal{P}}(t_2, \overline{s}))$.

Therefore, by means of resultants and gcd's one can determine the degree map of a surface parametrization, and hence decide whether it is birational or not. We will refer to the algorithms in [4] for this purpose.

Now, let us assume that the given surface parametrization $\mathcal{P}$ is not birational. Then, by Castelnuovo's Theorem, since we are working over an algebraically closed field, there exists a proper parametrization of the same surface. In general, this birational parametrization may require the extension of the ground field $\mathbb{L}$. One possibility, as already mentioned in the introduction, could be an implicitation-parametrization approach. Nevertheless, the idea here is to solve the problem staying within the parametric representation of the variety. For this purpose, we can apply the results in [6] that, although do not provide a complete answer, cover many of the cases.

The method in [6] is based on the application of Algorithm 1 to some *partial* parametrizations associated to $\mathcal{P}$, namely the parametrizations $\mathcal{P}_i(t_j) := \mathcal{P}(\overline{t}) \in (\mathbb{K}(t_i))(t_j)^3$ (that is, $\mathcal{P}$ is seen over $\mathbb{K}(t_i)$), for $i, j \in \{1, 2\}$ and $i \neq j$. Observe that the partial parametrization $\mathcal{P}_i(t_j)$ $(i \neq j)$ defines a space curve over $\mathbb{K}(t_i)$. Hence, the goal of Algorithm 2 is to properly reparametrize the partial parametrizations by applying Algorithm 1. The algorithm outputs a rational parametrization $\mathcal{Q}(\overline{t}) \in \mathbb{K}(\overline{t})^3$ of $\mathcal{V}$, and $R(\overline{t}) \in \mathbb{K}(\overline{t})^2$ such that $\mathcal{P}(\overline{t}) = \mathcal{Q}(R(\overline{t}))$, and $\text{MapDeg}(\mathcal{Q}) < \text{MapDeg}(\mathcal{P})$. In fact, it is proved that $\text{MapDeg}(\mathcal{P}) = \deg_{t_1}(S)\deg_{t_2}(T)\text{MapDeg}(\mathcal{Q})$ (see Theorem 4 in [6]). Furthermore, if some additional properties hold, then $\mathcal{Q}$ is proper. These ideas are described in Algorithm 2 and we illustrate it by an example.

Example 3. Let $\mathcal{V}$ be a rational surface defined over the field of the complex numbers, $\mathbb{C}$, by the parametrization

$$\mathcal{P}(\overline{t}) = (-(3t_2^8t_1^4 + 2t_2^6t_1^6 + 2t_2^{10}t_1^2 + t_2^4t_1^8 + t_2^{12} - t_2^4t_1^2 - t_2^2t_1^4 - t_2^6 + 2t_2^2)/t_2^2,$$
$$-t_1^2t_2^2 - t_1^4 - t_2^4 + 3t_1^4t_2^4 + 2t_1^6t_2^2 + 2t_1^2t_2^6 + t_1^8 + t_2^8 + t_2^2 + t_2^6, 3 + t_1^2t_2^2 + t_1^4 + t_2^4)\,.$$

We apply Algorithm 2. For this purpose, in Step 1, we apply Algorithm 1, and we find that

$$S^{\mathcal{P}_2}(t_1, s_1) = (s_1 - t_1)(s_1 + t_1)(s_1^2 + t_2^2 + t_1^2) \in (\mathbb{C}[t_2])[t_1, s_1]$$

which implies that $\mathcal{P}_2(t_1)$ is not proper; in fact, $\mathrm{MapDeg}(\mathcal{P}_2) = \deg_{t_1}(S^{\mathcal{P}_2}) = 4$. Thus, we go to Step 2 and we apply Algorithm 1 to $\mathcal{P}_2$. We obtain

$$S_2(t_1) = -t_1^2 t_2^2 - t_1^4 \in (\mathbb{C}[t_2])[t_1].$$

Furthermore, we determine the polynomials

$$L_i(s_1, t_2, x_i) = \mathrm{Res}_{t_1}(G_i^{\mathcal{P}_2}(\overline{t}, x_i), s_1 - S_2(t_1)) = (m_{i,2}(s_1, t_2)x_i - m_{i,1}(s_1, t_2))^{\deg_{t_1}(S_2)},$$

where $G_i^{\mathcal{P}_2}(\overline{t}, x_i) = x_i p_{i,2}(\overline{t}) - p_{i,1}(\overline{t})$, for $i = 1, 2, 3$, and we get $\mathcal{M}(\overline{t}) =$

$$(-2 - t_2^{10} + t_2^4 + 2t_1 t_2^6 - t_1 - t_1^2 t_2^2, -t_2^4 + t_2^8 + t_2^2 + t_2^6 + t_1 - 2t_2^4 t_1 + t_1^2, 3 + t_2^4 - t_1)$$

(we rename s_1 as t_1).

Now, in Step 2.2 of the algorithm, we apply Algorithm 1 to $\mathcal{M}_1(t_2) \in (\mathbb{C}(t_1))(t_2)^3$, and we find that

$$S^{\mathcal{M}_1}(t_2, s_2) = (t_2 - s_2)(t_2 + s_2) \in (\mathbb{C}[t_1])[t_2, s_2].$$

Thus, since $\mathrm{MapDeg}(\mathcal{M}_1) = \deg_{t_2}(S^{\mathcal{M}_1}) = 2$, we get that $\mathcal{M}_1$ is not proper. Then, we go to Step 2.3. We apply Algorithm 1 to $\mathcal{M}_1$, and we compute $T_1(t_2) = t_2^2 \in (\mathbb{C}[t_1])[t_2]$, and the polynomials

$$L_i(t_1, s_2, x_i) = \mathrm{Res}_{t_2}(G_i^{\mathcal{M}_1}(\overline{t}, x_i), s_2 - T_1(t_2)) = (q_{i,2}(t_1, s_2)x_i - q_{i,1}(t_1, s_2))^{\deg_{t_2}(T_1)},$$

where $G_i^{\mathcal{M}_1}(\overline{t}, x_i) = x_i m_{i,2}(\overline{t}) - m_{i,1}(\overline{t})$, for $i = 1, 2, 3$. We obtain $\mathcal{Q}(\overline{t}) =$

$$(-2 - t_1 - t_2 t_1^2 + t_2^2 + 2t_1 t_2^3 - t_2^5, t_1 + t_1^2 + t_2 - t_2^2 - 2t_1 t_2^2 + t_2^3 + t_2^4, 3 - t_1 + t_2^2)$$

(we rename s_2 as t_2). Finally, in Step 2.4 of the algorithm, we apply Algorithm 1 to $\mathcal{Q}_2(t_1) \in (\mathbb{C}(t_2))(t_1)^3$. We get that $S^{\mathcal{Q}_2}(t_1, s_1) = s_1 - t_1 \in (\mathbb{C}[t_2])[t_1, s_1]$ which implies that $\mathcal{Q}_2$ is proper. Therefore, Algorithm 2 outputs the parametrization $\mathcal{Q}(\overline{t})$, and

$$R(\overline{t}) = (S(\overline{t}), T(S(\overline{t}), t_2)) = (-t_1^2 t_2^2 - t_1^4, t_2^2) \in \mathbb{C}(\overline{t})^2.$$

One may check that $\mathrm{MapDeg}(\mathcal{P}) = \deg_{t_1}(S)\deg_{t_2}(T) = 8$ and thus $\mathcal{Q}$ is a proper reparametrization (we remind that $\mathrm{MapDeg}(\mathcal{P}) = \deg_{t_1}(S)\deg_{t_2}(T)$ $\mathrm{MapDeg}(\mathcal{Q})$).

Algorithm 2. Proper reparametrization of surfaces (partial case)

Input: A rational parametrization

$$\mathcal{P}(\overline{t}) = (p_1(\overline{t}), p_2(\overline{t}), p_3(\overline{t})) \in \mathbb{K}(\overline{t})^3, \quad p_i(\overline{t}) = p_{i,1}(\overline{t})/p_{i,2}(\overline{t}),$$

$\gcd(p_{i,1}, p_{i,2}) = 1$, $i = 1, 2, 3$, of an algebraic surface $\mathcal{V}$.

Output: A rational parametrization

$$\mathcal{Q}(\overline{t}) = (q_1(\overline{t}), q_2(\overline{t}), q_3(\overline{t})) \in \mathbb{K}(\overline{t})^3, \quad q_i(\overline{t}) = q_{i,1}(\overline{t})/q_{i,2}(\overline{t}),$$

of $\mathcal{V}$, and $R(\overline{t}) \in (\mathbb{K}(\overline{t}) \setminus \mathbb{K})^2$ such that $\mathcal{P}(\overline{t}) = \mathcal{Q}(R(\overline{t}))$, and $1 \leq \mathrm{MapDeg}(\mathcal{Q}) < \mathrm{MapDeg}(\mathcal{P})$.

1: **if** $\mathcal{P}_1$ and $\mathcal{P}_2$ are proper (apply Algorithm 1) **then**
2: **return**
3: **else if** $\mathcal{P}_2$ is not proper **then**
4: Apply Algorithm 1 to $\mathcal{P}_2$. [*It returns a parametrization* $\mathcal{M}(\overline{t}) \in \mathbb{K}(\overline{t})^3$, *and* $S(\overline{t}) \in \mathbb{K}(\overline{t})$ *such that the partial parametrization associated to* $\mathcal{M}$, $\mathcal{M}_2(t_1) \in (\mathbb{K}(t_2))(t_1)^3$, *is proper and* $S_2(t_1) \in (\mathbb{K}(t_2))(t_1)$ *satisfies* $\mathcal{P}_2(t_1) = \mathcal{M}_2(S_2(t_1))$].
5: **if** $\mathcal{M}_1(t_2) \in (\mathbb{K}(t_1))(t_2)^3$ is proper (apply Algorithm 1) **then**
6: **return** $\mathcal{Q} := \mathcal{M}$, and $R(\overline{t}) := (S(\overline{t}), t_2)$.
7: **else**
8: Apply Algorithm 1 to $\mathcal{M}_1(t_2)$. [*It returns a parametrization* $\mathcal{Q}(\overline{t}) \in \mathbb{K}(\overline{t})^3$, *and* $T(\overline{t}) \in \mathbb{K}(\overline{t})$ *such that the partial parametrization associated to* $\mathcal{Q}$, $\mathcal{Q}_1(t_2) \in (\mathbb{K}(t_1))(t_2)^3$, *is proper and* $T_1(t_2) \in (\mathbb{K}(t_1))(t_2)$ *satisfies* $\mathcal{M}_1(t_2) = \mathcal{Q}_1(T_1(t_2))$].
9: **end if**
10: **if** the partial parametrization associated to $\mathcal{Q}$, $\mathcal{Q}_2(t_1) \in (\mathbb{K}(t_2))(t_1)^3$, is proper (apply Algorithm 1) **then**
11: **return** $\mathcal{Q}$, and $R(\overline{t}) := (S(\overline{t}), T(S(\overline{t}), t_2))$.
12: **else**
13: **return** $\mathcal{Q}$, $R(\overline{t}) := (S(\overline{t}), T(S(\overline{t}), t_2))$, and the message "*you may apply the algorithm again (Step 3 and so on) to* $\mathcal{Q}_2$".
14: **end if**
15: **else**
16: Apply Step 8 and the next one to $\mathcal{P}$ and $\mathcal{P}_1$ ($\mathcal{P}_2$ is proper and $\mathcal{P}_1$ is not).
17: **end if**

3 The Package Luroth

In this section, we present the creation of a package in the computer algebra system Maple, that we call **Luroth**. This package consists in several procedures that implement the algorithms described in Sect. 2. More precisely, the package deals with rational parametrizations, either of plane or space curves, or surfaces. It checks the injectivity giving the degree of the map. In addition, it provides birational reparametrizations of a given non-birational parametrization. For the case of curves, all cases are covered. For the case of surfaces, only those treated in [5] and [6] are considered. In Sect. 4, we show how, with the help of Maple, we approach the general case.

3.1 General Description

The Maple package is initialized by the command:

> with(Luroth):

The main procedures in the package are:

- `IsTheCurveProper`:
  i) *Feature:* This procedure checks whether a given rational curve parametrization, non necessarily planar, is birational. Briefly, the input and output of the procedure can be stated as follows:
  ▷ INPUT:
  ∘ A list, of length at least 2, whose entries are univariate rational functions, not all constant.
  ∘ The variable of the rational functions
  ∘ An option $u \in \{\text{probabilistic}, \text{deterministic}\}$.
  ▷ OUTPUT: The command returns either *true* or *false*. *true* means that the parametrization is birational. If *false* then the input parametrization is not birational and the procedure returns also the degree of the map associated to the input parametrization.
  ii) *Calling Sequence:* > IsTheCurveProper(List,variable,option);
  iii) *Mathematical Argumentation:* the procedure is based on the results in [6] and implements the first steps of Algorithm 1.
- `CurveProperReparametrization`:
  i) *Feature:* This procedure computes a birational parametrization of the curve defined by the input parametrization. Briefly, the input and output of the procedure can be stated as follows:
  ▷ INPUT:
  ∘ A list, of length at least 2, whose entries are univariate rational functions, not all constant.
  ∘ The variable of the rational functions
  ▷ OUTPUT: a proper curve parametrization of the input curve.
  ii) *Calling Sequence:* > CurveProperReparametrization(List,variable);
  iii) *Mathematical Foundation:* the procedure is based on the results in [6] and implements the first steps of Algorithm 1
- `IsTheSurfaceProper`:
  i) *Feature:* This procedure checks whether a given rational surface parametrization is birational. Briefly, the input and output of the procedure can be stated as follows:
  ▷ INPUT:
  ∘ A list, of length at least 3, whose entries are bivariate rational functions, which generic Jacobian has rank 2.
  ∘ The variables of the rational functions
  ∘ An option $u \in \{\text{probabilistic}, \text{deterministic}\}$.
  ▷ OUTPUT: The command returns either *true* or *false*. *true* means that the parametrization is birational. If *false* then the input parametrization is not birational and the procedure returns also the degree of the map associated to the input parametrization.

ii) *Calling Sequence:*
> IsTheSurfaceProper(List,variable$_1$, variable$_2$,option);
iii) *Mathematical Foundation:* the procedure is based on the results in [4].

- `SurfaceProperReparametrization`:
  i) *Feature:* This procedure tries to compute a birational parametrization of the surface defined by the input parametrization. Briefly, the input and output of the procedure can be stated as follows:
  ▷ INPUT:
  ∘ A list, of length at least 3, whose entries are univariate rational functions, not all constant.
  ∘ The variables of the rational functions
  ▷ OUTPUT: one of the following possibilities
  ∘ A birational parametrization of the input surface.
  ∘ A non-birational parametrization of the input surface with smaller map degree than the input parametrization.
  ∘ A message informing that no improvement has been possible.
  ii) *Calling Sequence:*
  > SurfaceProperReparametrization(List,variable$_1$,variable$_2$);
  iii) *Mathematical Foundation:* the procedure is based on the results in [6] and implements the first steps of Algorithm 2.

There are other auxiliary procedures in the package that we do not mention here besides `Try`. Given a surface parametrization $\mathcal{P} \in \mathbb{K}(t_1, t_2)^3$, with $\mathbb{K}$ a field of characteristic zero, `Try` decides whether the surface seen as a curve over the algebraic closure of $\mathbb{K}(t_2)(t_1)$, and parametrized by $\mathcal{P}(t_1, t_2)$ as tuple in $\mathbb{K}(t_1)(t_2)$, is birational. In the affirmative case, it looks for a reparametrization function over the ground field $\mathbb{K}(t_1)$. Additionally, `FindR` executes Step 3 in Algorithm 1.

3.2 Illustrative Examples

In this subsection, we illustrate the package by some examples run in Maple. In Example 4, we see the performance of the package for the case of curves. Examples 5, 6 and 7 are devoted to surfaces. In Example 5 the algorithm is able to provide a birational parametrization. Example 6 illustrates the case where the algorithm does not yield to an optimal answer but outputs a parametrization whose map degree has decreased. Finally Example 7 is devoted to the case where the algorithm does not provide any improvement. For additional information on how the execution is performed we will be referring to the tables in the Appendix.

The package and installation instructions can be found at

http://www3.uah.es/jorge_caravantes/research.html

Once the package is installed, after executing the command `with(Luroth)`, the package is ready to be used (see Fig. 1).

Example 4. We consider the curve parametrization

$$\mathcal{P}(t)=\left(\frac{t^6}{(t^3+t+2)^2},\frac{(t^3+t+2)^3}{t^9},\frac{2t^6+2t^4+4t^3+t^2+4t+4}{(t^3+t+2)\,t^3}\right)$$

that parametrizes the space curve defined by

$$\{x^2y+xy-z, xz^2-x^2-2x-1, xyz-z^2+x+1, z^3-xy-xz-y-2z, xy^2-yz+z^2-x-2\}.$$

Using the command `IsTheCurveProper`, one gets that the degree of the map induced by $\mathcal{P}$ is 3 (see Fig. 2). Hence, the curve is traced three times when giving values to the parameter. Since the parametrization is not injective, we apply the command `CurveProperReparametrization` (see Fig. 3) to get the following birational parametrization of the same curve,

$$\mathcal{Q}(t)=\left(\frac{t^2}{(t-2)^2},\frac{t^3-6t^2+12t-8}{t^3},2\,\frac{t^2-2t+2}{t\,(t-2)}\right).$$

Example 5. We consider now the surface parametrization

$$\mathcal{P}(t_1,t_2)=\left(\frac{t_1t_2\,(t_1^2-t_1t_2+t_2^2)}{(t_1+t_2)^2},t_2,\frac{t_2\,(t_1^3-t_2t_1^2+t_2^2t_1+t_1^2+2\,t_1t_2+t_2^2)}{(t_1+t_2)^2}\right).$$

Applying the command `IsTheSurfaceProper`, we get that the degree map is 3, and hence the $\mathcal{P}(t_1,t_2)$ is not birational. We then apply the command

`ProperSurfaceReparametrization`

to get proper parametrization of the surface (see Fig. 4)

$$\mathcal{Q}(t_1,t_2)=\left(-\frac{t_2\,(t_1t_2+1)}{t_1},t_2,-\frac{(1+(t_2-1)\,t_1)\,t_2}{t_1}\right).$$

Example 6. We consider the following surface parametrization

$$\mathcal{P}=\left(-\frac{-93t_1^4-51\,t_1^2t_2^2+90\,t_2^4-22\,t_1^2}{2601\,t_2^4t_1^4-1530t_2^6t_1^2+225t_2^8+2244t_2^2t_1^4-660t_2^4t_1^2+484t_1^4-2},\right.$$
$$-\frac{4743\,t_2^2t_1^6-1395\,t_2^4t_1^4-3825\,t_2^6t_1^2+1125\,t_2^8+2046\,t_1^6-1650\,t_2^4t_1^2-2}{-51\,t_1^2t_2^2+15\,t_2^4-22\,t_1^2},$$
$$\left.-\frac{1}{3}\,\frac{8649\,t_1^8-13950\,t_2^4t_1^4+5625\,t_2^8+51\,t_1^2t_2^2-15\,t_2^4+22\,t_1^2-3}{-31\,t_1^4+25\,t_2^4}\right).$$

In this case, applying the command `IsTheSurfaceProper` we get that the parametrization has map degree 16. The command `SurfacePorper-Reparametrization` does not get a proper parametrization but the procedure returns a new parametrization of the same surface, which is not proper, but where the degree map has decreased from 16 to 4 (see Fig. 5). The output parametrization is

$$\mathcal{Q} = \left(\frac{(22\,t_1 - 93)\,t_2{}^4 - 51\,t_1 t_2{}^3 + 90\,t_1{}^2 t_2{}^2}{(2\,t_1{}^2 - 484)\,t_2{}^4 + 2244\,t_2{}^3 + (-660\,t_1 - 2601)\,t_2{}^2 + 1530\,t_1 t_2 - 225\,t_1{}^2}, \right.$$

$$\frac{2\,t_1{}^3 t_2{}^4 - 1650\,t_1{}^2 t_2{}^2 + 2046\,t_2{}^4 - 1125\,t_1{}^3 + 3825\,t_1{}^2 t_2 + 1395\,t_1 t_2{}^2 - 4743\,t_2{}^3}{t_1{}^2 t_2{}^2\,(22\,t_2{}^2 + 15\,t_1 - 51\,t_2)},$$

$$\left. \frac{(3\,t_2{}^4 + 15\,t_2{}^2 - 5625)\,t_1{}^4 + (22\,t_2{}^4 - 51\,t_2{}^3)\,t_1{}^3 + 13950\,t_1{}^2 t_2{}^2 - 8649\,t_2{}^4}{75\,t_1{}^4 t_2{}^2 - 93\,t_1{}^2 t_2{}^4} \right).$$

Example 7. We consider the following surface parametrization

$$\mathcal{P} = \left(-\frac{4\,t_1{}^4 t_2{}^2 - 4\,t_1{}^2 t_2{}^3 - t_1{}^4 + t_2{}^4 + t_1{}^2 t_2}{(2\,t_1{}^2 - t_2)^2\,(t_1{}^2 - t_2)}, -\frac{t_1 t_2{}^2}{(2\,t_1{}^2 - t_2)\,(t_1{}^2 - t_2)}, \frac{t_2{}^4}{(t_1{}^2 - t_2)^2} \right).$$

The command `IsTheSurfaceProper` ensures that the map degree of $\mathcal{P}$ is 3. However, The command `SurfacePorperReparametrization` does not get any parametrization with smaller map degree (see Fig. 6).

4 Approaching the General Case

As we have already mentioned, Lüroth's Theorem, for the case of surfaces, requires the field to be algebraically closed. This implies that, in general, the ground field $\mathbb{L}$ of the parametrization needs to be extended. Observe that the curve case (see Algorithm 1) does not extend $\mathbb{L}$, and hence the surface partial approach behaves the same (see Algorithm 2). Therefore, a new strategy has to be considered. Here, we present some on-going working ideas to approach the general case. For this purpose, let $\mathcal{P}(\overline{t}) = (p_1(\overline{t}), p_2(\overline{t}), p_3(\overline{t}))$ be as in (2). Let $\mathcal{P}$ be non-birational and $\Phi := \text{MapDeg}(\mathcal{P}) > 1$. In this situation, we know that there exists a birational surface parametrization $\mathcal{Q}$ of the same surface, and a dominant rational map $R : \mathbb{K}^2 \dashrightarrow \mathbb{K}^2$ such that $\mathcal{P} = \mathcal{Q} \circ R$. Therefore, to find $\mathcal{Q}$, it is enough to find R. However, note that the possible pairs $(\mathcal{Q}, R)$, solving the problem, are not unique. Nevertheless, our idea is to observe that an answer can be achieved by looking for rational maps R with the same fiber as $\mathcal{P}$. One may proceed as follows:

1. Let $\mathcal{I}$ be the ideal in $\mathbb{K}[\overline{t}, \overline{s}, w]$ of the fibre of $\mathcal{P}$. That is $\mathcal{I}$ is generated by

$$\{\text{Num}(p_i(\overline{t}) - p_i(\overline{s})), w \cdot \text{lcm}(\text{Den}(p_1), \text{Den}(p_2), \text{Den}(p_3)) - 1\}_{i=1,2}.$$

   Note that $\mathcal{I}$ is zero-dimensional with Φ points.
2. We want R such that the ideal $\mathcal{J}$ in $\mathbb{K}[\overline{t}, \overline{s}, w]$ of its fibre is equal to $\mathcal{I}$. By Bézout's Theorem we know that a linear system of curves defining such $\mathcal{R}$ must have degree greater than or equal to $d = \lceil \sqrt{\Phi} \rceil$.

3. Make an ansatz of unknown coefficients of a linear systems of degree d, and increase the degree by one until a R satisfying $\mathcal{J} = \mathcal{I}$ is found. Since Castelnuovo's Theorem ensures the existence of R, the procedure terminates.

We illustrate these ideas by means of an example.

Example 8. We consider the parametrization $\mathcal{P}$ of Example 7. Using Maple, we compute a Gröbner basis, with respect to the pure lex order with $t_1 < t_2 < w$, of the ideal $\mathcal{I}$. It contains 3 polynomials. Since $\Phi := \text{MapDeg}(\mathcal{P}) = 3$ (see Example 7), a change of variables of degree greater than or equal to 2 is expected.

We start analyzing all the degree-2 transformations of $\mathbb{C}^2$. We consider a generic degree-2 transformation R and we require that $\mathcal{J} \subset \mathcal{I}$. For this purpose, we compute the normal forms w.r.t. to the Gröbner basis above, and we solve the system of equations derived from the vanishing conditions of the normal forms. This provides three different type of expressions for the coordinates of R, namely

$$\left\{\frac{2a_1t_1}{2\,b_4{t_1}^2+2\,b_1t_1-b_4t_2}, \frac{a_5{t_2}^2}{b_4{t_1}^2+b_5{t_2}^2-b_4t_2}, \frac{2\,a_4{t_1}^2+2\,a_1t_1-a_4t_2}{-4\,b_2{t_1}^2+2\,b_1t_1+2\,b_2t_2},\right.$$
$$\left.\frac{-a_4{t_1}^2-a_5{t_2}^2+a_4t_2}{b_2{t_1}^2-b_5{t_2}^2-b_2t_2}\right\}.$$

We now choose from above two shapes to be the entries of the transformation of the plane. For instance, take $b_5 = 0, a_5 + b_4 = 0$ in the last entry, and $a_4 = b_1 = 0, 2a_1 - b_4 = 0$ in the first entry. We get

$$R = (r_1, r_2) = \left(-\frac{{t_2}^2}{{t_1}^2-t_2}, \frac{t_1}{2\,{t_1}^2-t_2}\right).$$

From the equality $\mathcal{P} = \mathcal{Q} \circ R$, we get the parametrization

$$\mathcal{Q}(t_1, t_2) = (t_2^2 + t_1, t_1t_2, t_1^2)\,.$$

Note that one may get $\mathcal{Q}$, for instance, using the idea that $(r_1, r_2, p_i) \in \mathbb{L}(\bar{t})^3$ parametrizes the irreducible polynomial $\text{Num}(q_i(x_1, x_2) - x_3)$ (for $i = 1, 2, 3$), where we denote $\mathcal{Q} = (q_1, q_2, q_3)$, $q_i = q_{i,1}/q_{i,2}$, $\gcd(q_{i,1}, q_{i,2}) = 1$. Thus, one only has to compute the implicit equations of the parametrizations (r_1, r_2, p_i) for $i = 1, 2, 3$ (see e.g. [7]).

Finally, using the command `IsTheSurfaceProper` from the package `Luroth` one checks that $\text{MapDeg}(\mathcal{Q}) = 1$, and hence it is a birational transformation of the surface.

Acknowledgements. This work has been partially supported by FEDER/Ministerio de Ciencia, Innovación y Universidades-Agencia Estatal de Investigación/MTM2017-88796-P (Symbolic Computation: new challenges in Algebra and Geometry together with its applications). Authors belong to the Research Group ASYNACS (Ref. CT-CE2019/683).

5 Appendix

In this appendix, the Maple executions, corresponding to the examples in the Subsect. 3.2, are shown.

```
> with(Luroth);
```

[*Clean*, *CurveProperReparametrization*, *CurveProperReparametrizationAux*, *FindR*, *FindRAux*, *IsTheCurveProper*, *IsTheCurveProperAux*, *IsTheSurfaceProper*, *IsTheSurfaceProperAux*, *IsTheSurfaceProperDeterministic*, *IsTheSurfaceProperProbabilistic*, *SurfaceProperReparametrization*, *Try*]

Fig. 1. Starting the package

```
P:=[t^6/(t^3+t+2)^2, (t^3+t+2)^3/t^9, (2*t^6+2*t^4+4*t^3+t^2+4*t+4)/(
(t^3+t+2)*t^3)];
```

$$P := \left[\frac{t^6}{(t^3+t+2)^2}, \frac{(t^3+t+2)^3}{t^9}, \frac{2t^6+2t^4+4t^3+t^2+4t+4}{(t^3+t+2)\,t^3}\right]$$

```
> IsTheCurveProper(P,t,probabilistic);
```

false

The map degree is

3

Fig. 2. It checks the properness of $\mathcal{P}$ in Example 4. The same result is achieved with the option `deterministic`

```
> Q:=CurveProperReparametrization(P,t);
```

A proper parametrization is

$$Q := \left[\frac{t^2}{(t-2)^2}, \frac{t^3-6t^2+12t-8}{t^3}, \frac{2(t^2-2t+2)}{t(t-2)}\right]$$

Fig. 3. It computes a proper parametrization of the curve in Example 4

```
> IsTheSurfaceProper(P,t[1],t[2],deterministic);
```

false

The map degree is

3

```
> Q:=SurfaceProperReparametrization(P,t[1],t[2]);
```

A proper parametrization is

$$Q := \left[-\frac{t_2 (t_1 t_2 + 1)}{t_1}, t_2, -\frac{(1 + (t_2 - 1) t_1) t_2}{t_1} \right]$$

Fig. 4. It checks the properness of $\mathcal{P}$ in Example 5. The same result is achieved with the option `probabilistic`. Applying the command `SurfacePorperReparametrization` one gets a proper parametrization of the surface.

```
> P:= [-(-51*t[1]^2*t[2]^2+90*t[2]^4-22*t[1]^2-93*t[1]^4)/(2601*t
[2]^4*t[1]^4-1530*t[2]^6*t[1]^2+2244*t[2]^2*t[1]^4+225*t[2]^8
-660*t[2]^4*t[1]^2+484*t[1]^4-2), -(-3825*t[2]^6*t[1]^2+1125*t[2]
^8-1650*t[2]^4*t[1]^2+4743*t[2]^2*t[1]^6-1395*t[2]^4*t[1]^4+2046*
t[1]^6-2)/(-51*t[1]^2*t[2]^2+15*t[2]^4-22*t[1]^2), -1/3*(51*t[1]
^2*t[2]^2-15*t[2]^4+22*t[1]^2+5625*t[2]^8-13950*t[2]^4*t[1]
^4+8649*t[1]^8-3)/(25*t[2]^4-31*t[1]^4)];
```

$$P := \left[-\frac{-93 t_1^4 - 51 t_1^2 t_2^2 + 90 t_2^4 - 22 t_1^2}{2601 t_2^4 t_1^4 - 1530 t_2^6 t_1^2 + 225 t_2^8 + 2244 t_2^2 t_1^4 - 660 t_2^4 t_1^2 + 484 t_1^4 - 2}, -\frac{4743 t_2^2 t_1^6 - 1395 t_2^4 t_1^4 - 3825 t_2^6 t_1^2 + 1125 t_2^8 + 2046 t_1^6 - 1650 t_2^4 t_1^2 - 2}{-51 t_1^2 t_2^2 + 15 t_2^4 - 22 t_1^2}, -\frac{8649 t_1^8 - 13950 t_2^4 t_1^4 + 5625 t_2^8 + 51 t_1^2 t_2^2 - 15 t_2^4 + 22 t_1^2 - 3}{3 (-31 t_1^4 + 25 t_2^4)} \right]$$

```
> IsTheSurfaceProper(P,t[1],t[2],deterministic);
```

false

The map degree is

16

```
> Q:=SurfaceProperReparametrization(P,t[1],t[2]);
```

The algorithm does not get a proper parametrization but it gets

[New Parametrization, Degree Map]

$$Q := \left[\left[\frac{(22 t_1 - 93) t_2^4 - 51 t_1 t_2^3 + 90 t_1^2 t_2^2}{(2 t_1^2 - 484) t_2^4 + 2244 t_2^3 + (-660 t_1 - 2601) t_2^2 + 1530 t_1 t_2 - 225 t_1^2}, \frac{2 t_1^3 t_2^4 - 1650 t_1^2 t_2^2 + 2046 t_2^4 - 1125 t_1^3 + 3825 t_1^2 t_2 + 1395 t_1 t_2^2 - 4743 t_2^3}{t_2^2 t_1^2 (22 t_2^2 + 15 t_1 - 51 t_2)}, \frac{(3 t_2^4 + 15 t_2^2 - 5625) t_1^4 + (22 t_2^4 - 51 t_2^3) t_1^3 + 13950 t_1^2 t_2^2 - 8649 t_2^4}{75 t_2^2 t_1^4 - 93 t_2^4 t_1^2} \right], 4 \right]$$

Fig. 5. It checks the properness of $\mathcal{P}$ in Example 6. The same result is achieved with the option `probabilistic`. Applying the command `SurfacePorperReparametrization` one gets a degree map 4 parametrization of the surface.

```
> P := [-(4*t[1]^4*t[2]^2-4*t[1]^2*t[2]^3-t[1]^4+t[2]^4+t[1]^2*t[2]
)/((2*t[1]^2-t[2])^2*(t[1]^2-t[2])), -t[1]*t[2]^2/((2*t[1]^2-t[2]
)*(t[1]^2-t[2])), t[2]^4/(t[1]^2-t[2])^2];
```

$$P := \left[-\frac{4t_1^4t_2^2-4t_1^2t_2^3-t_1^4+t_2^4+t_1^2t_2}{(2t_1^2-t_2)^2(t_1^2-t_2)}, -\frac{t_1t_2^2}{(2t_1^2-t_2)(t_1^2-t_2)}, \frac{t_2^4}{(t_1^2-t_2)^2}\right]$$

```
> IsTheSurfaceProper(P,t[1],t[2],deterministic);
```

false

The map degree is

3

```
> Q:=SurfaceProperReparametrization(P,t[1],t[2]);
```

$Q :=$ *The algorithm does not get any improvement*

Fig. 6. It checks the properness of $\mathcal{P}$ in Example 7. The same result is achieved with the option `probabilistic`. The command `SurfacePorperReparametrization` does not get any parametrization with smaller map degree.

References

1. Caravantes, J., Sendra, J.R., Sevilla, D., Villarino, C.: On the existence of birational surjective parametrizations of affine surfaces. J. Algebra **501**, 206–214 (2018)
2. Holme, A.: On the dual of a smooth variety. In: Lønsted, K. (ed.) Algebraic Geometry. LNM, vol. 732, pp. 144–156. Springer, Heidelberg (1979). https://doi.org/10.1007/BFb0066642
3. Hoschek, J., Lasser, D.: Fundamentals of Computer Aided Geometric Design. A.K. Peters Ltd., Natick (1993)
4. Pérez-Díaz, S., Sendra, J.R.: Computation of the degree of rational surface parametrizations. J. Pure Appl. Algebra **193**(1–3), 99–121 (2004)
5. Pérez-Díaz, S.: On the problem of proper reparametrization for rational curves and surfaces. Comput. Aided Geom. Des. **23**(4), 307–323 (2006)
6. Pérez-Díaz, S.: A partial solution to the problem of proper reparametrization for rational surfaces. Comput. Aided Geom. Des. **30**(8), 743–759 (2013)
7. Pérez-Díaz, S., Sendra, J.R.: A univariate resultant-based implicitization algorithm for surfaces. J. Symb. Comput. **43**(2), 118–139 (2008)
8. Schicho, J.: Rational parametrization of surfaces. J. Symb. Comput. **26**, 1–29 (1998)
9. Sendra, J.R.: Normal parametrization of algebraic plane curves. J. Symb. Comput. **33**, 863–885 (2002)
10. Sendra, J.R., Sevilla, D.: First steps towards radical parametrization of algebraic surfaces. Comput. Aided Geom. Des. **30**(4), 374–388 (2013)
11. Sendra, J.R., Sevilla, D., Villarino, C.: Algebraic and algorithmic aspects of radical parametrizations. Comput. Aided Geom. Des. **55**, 1–14 (2017)
12. Sendra, J.R., Winkler, F., Pérez-Díaz, S.: Rational Algebraic Curves: A Computer Algebra Approach. Series: Algorithms and Computation in Mathematics, vol. 22. Springer, Heidelberg (2007). https://doi.org/10.1007/978-3-540-73725-4
13. Shafarevich, I.R.: Basic Algebraic Geometry 1 - Varieties in Projective Space. Springer, New York (1994)

A Maple Implementation of the Finite Element Method for Solving Boundary-Value Problems for Systems of Second-Order Ordinary Differential Equations

Galmandakh Chuluunbaatar[1,2], Alexander Gusev[1,3](✉), Vladimir Derbov[4], Sergue Vinitsky[1,2], Ochbadrakh Chuluunbaatar[1,5], Luong Le Hai[6], and Vladimir Gerdt[1,2]

[1] Joint Institute for Nuclear Research, Dubna, Russia
gooseff@jinr.ru
[2] Peoples' Friendship University of Russia (RUDN University), Moscow, Russia
[3] Dubna State University, Dubna, Russia
[4] N.G. Chernyshevsky Saratov National Research State University, Saratov, Russia
[5] Institute of Mathematics and Digital Technology, Mongolian Academy of Sciences, Ulaanbaatar, Mongolia
[6] Ho Chi Minh City University of Education, Ho Chi Minh City, Viet Nam

Abstract. We present a new algorithm of the finite element method (FEM) implemented as KANTBP 5M code in MAPLE for solving boundary-value problems (BVPs) for systems of second-order ordinary differential equations with continuous or piecewise continuous real or complex-valued coefficients. The desired solution in a finite interval of the real-valued independent variable is subject to mixed homogeneous boundary conditions (BCs). To reduce a BVP or a scattering problem with different numbers of asymptotically coupled or entangled open channels in the two asymptotic regions to a BVP on a finite interval, the asymptotic BCs for large absolute values of the independent variable are approximated by homogeneous Robin BCs. The BVP is discretized by means of the FEM using the Hermite interpolation polynomials with arbitrary multiplicity of the nodes, which preserves the continuity of derivatives of the desired solutions. The relevant algebraic problems are solved using the built-in linear algebra procedures. To calculate metastable states with complex eigenvalues of energy or to find bound states with the BCs depending on a spectral parameter, the Newton iteration scheme is implemented. Benchmark examples of the code application to BVPs and scattering problems of quantum mechanics are given.

V. Gerdt—It is painful to think Professor V. Gerdt is no longer among us, and this paper is his last contribution to development of solving BVPs on the base of the FEM which owes remarkable results to him. We are deeply grateful to him for his intuition, insight and support, which were invaluable during our long-standing collaboration.

R. M. Corless et al. (Eds.): MC 2020, CCIS 1414, pp. 152–166, 2021.
https://doi.org/10.1007/978-3-030-81698-8_11

Keywords: Finite element method · Interpolation Hermite polynomials · Boundary-value problem · Scattering problem · System of ordinary differential equations

1 Introduction

Mathematical modeling of quantum-mechanical collisions of molecules, atoms and atomic nuclei, guided propagation of waves (oceanic, optical, electromagnetic), as well as transitions between metastable and bound quantum states using the methods of coupled channels or normal modes reduces to boundary-value problems (BVPs) for systems of N coupled second order ordinary differential equations (ODEs) [1–3].

Mathematical models of the above phenomena, initially formulated as a multidimensional (quantum mechanical) or three-dimensional elliptic BVP [4], reduce to a system of ODEs with variable coefficients (real or complex, tabular or piecewise continuous, or not only continuous, but also having continuous derivatives up to a given order) on a finite interval. The appropriate boundary conditions (BCs) are of the mixed type: Robin (third-type or radiation condition), Neumann and Dirichlet. The procedure implies constructing asymptotes of the desired solution and its expansion in terms of a suitable basis functions, including the calculation of the variable coefficients of the ODE as integrals in the reduction of the original problem in terms of basis functions to be solved by the Kantorovich method or by the incomplete Galerkin method [5,6].

For example, in molecular and nuclear physics, optical waveguides, for the spectrum of beryllium dimer [7], sub-barrier fusion of heavy ions [8] or transverse modes in smoothly irregular optical fibers [9], the proposed approach and the program of its finite element method (FEM) implementation allow the determination of scattering or metastable states in the case of different numbers of asymptotically coupled or entangled open channels [10,11]. The eigenfunctions and the symmetric (or unitary) scattering matrix composed of square matrices of transmission amplitudes and rectangular matrices of reflection amplitudes are found, as well as complex energy eigenvalues and eigenfunctions of metastable states calculated by means of the Newton method [12].

Standard FEM programs with interpolation Lagrange polynomials (ILPs), implemented in FORTRAN and in public domain computer algebra systems like MAPLE and MATHEMATICA solve 3D, 2D and 1D elliptic BVPs [4]. However, they are not applicable to systems of N ODEs of the above general type.

Indeed, in standard public domain FEM programs the desired solution is approximated by ILPs, which do not preserve the continuity of the derivatives of solutions up to a given order, depending on the smoothness of the variable coefficients of the ODE at the boundary points of the finite element mesh subintervals. This can violate the conservation laws inherent in the original problem.

In the present paper we propose new algorithms and software implementation of the FEM for solving BVPs for systems of N ODEs. To approximate the desired solution, the interpolation Hermite polynomials (IHPs) with arbitrary multiplicity of the nodes [6] are used, which preserve the continuity of the derivatives of the desired solution up to a given order, depending on the smoothness of the ODE variable coefficients at the boundary points of the finite element mesh subintervals [13]. The paper continues our previous work presented in the libraries of computer programs of the Computer Physics Communication journal [14–16] implemented in FORTRAN and JINRLIB [17] implemented and executed in MAPLE [18]. In the case of smooth coefficients of the ODE, the approximation by IHPs saves computer resources and provides not only high accuracy, but also the continuity of the solution gradient.

We apply MAPLE to construct and analyze the appropriate FEM schemes, to calculate the IHPs, to approximate the sought solution, to approximate the tabulated ODE coefficients, to implement smooth matching of the FEM solution with its analytical asymptotic extension, to construct the asymptotes of the sought ODE solution necessary for formulating the Robin BCs for the expansion of the desired solution of the original multidimensional BVP in appropriate basis functions. Moreover, using MAPLE we calculate the first derivatives of the basis functions with respect to the parameter – the independent variable of the ODE. The variable coefficients of the ODE – integrals in the reduction of the original multidimensional BVP in terms of basic functions and their first derivatives by the Kantorovich or incomplete Galerkin method are also obtained using MAPLE, as well as a convenient graphical representation of all the items that make up the solution of the BVP.

The structure of the paper is as follows. In Sect. 2, we formulate the BVPs and briefly describe the FEM scheme. Section 3 presents the benchmark examples of using the code to solve bound state problems and scattering problems of quantum mechanics and waveguide physics. In Appendices we present the algorithms of IHPs generation on the standard interval, calculation of the FEM scheme characteristics and FEM generation of an algebraic eigenvalue problem. In Conclusion we summarize the results and prospects of application.

All calculations in this paper were performed by KANTBP 5M code using MAPLE 2019 on PC Intel Pentium 987 2 × 1.5 GHz, 4 Gb, 64bit Windows 8.

2 The Problem Statement

2.1 The Boundary-Value Problems

The proposed approach implemented as program KANTBP 5M is intended for solving BVPs for systems of the ODEs with respect to unknown functions $\boldsymbol{\Phi}(z)=(\Phi_1(z),\ldots,\Phi_N(z))^T$ of independent variable $z\in\Omega(z^{\min},z^{\max})$ numerically using the FEM [10]:

$$(\mathbf{D}-E\,\mathbf{I})\,\boldsymbol{\Phi}(z)\equiv\left(-\frac{1}{f_B(z)}\mathbf{I}\frac{d}{dz}f_A(z)\frac{d}{dz}+\mathbf{V}(z)\right.$$
$$\left.+\frac{f_A(z)}{f_B(z)}\mathbf{Q}(z)\frac{d}{dz}+\frac{1}{f_B(z)}\frac{df_A(z)\mathbf{Q}(z)}{dz}-E\,\mathbf{I}\right)\boldsymbol{\Phi}(z)=0. \qquad (1)$$

Here $f_B(z)>0$ and $f_A(z)>0$ are continuous or piece-wise continuous positive functions, $\mathbf{I}$ is the unit matrix, $\mathbf{V}(z)$ is a symmetric matrix, $V_{ij}(z)=V_{ji}(z)$, and $\mathbf{Q}(z)$ is an antisymmetric $N \times N$ matrix, $Q_{ij}(z) = -Q_{ji}(z)$, of effective potentials. The elements of these matrices are continuous or piecewise continuous real or complex-valued coefficients from the Sobolev space $\mathcal{H}_2^{s\geq 1}(\Omega)$, providing the existence of nontrivial solutions $\boldsymbol{\Phi}(z)$ subjected to homogeneous Dirichlet, Neumann or Robin BCs at the boundary points of the interval $z \in \{z^{\min}, z^{\max}\}$ at given symmetric real or complex-valued $N \times N$ matrix $\mathbf{G}(z) = \mathcal{R}(z) - \mathbf{Q}(z)$

$$\boldsymbol{\Phi}(z^t) = 0, \qquad \lim_{z \to z^t} f_A(z) \left(\mathbf{I}\frac{d}{dz} - \mathbf{Q}(z) \right) \boldsymbol{\Phi}(z) = 0, \tag{2}$$
$$\left(\mathbf{I}\frac{d}{dz} - \mathbf{Q}(z) \right) \boldsymbol{\Phi}(z) \bigg|_{z=z^t} = \mathbf{G}(z^t)\boldsymbol{\Phi}(z^t),$$

where the superscript $t = \min, \max$ labels the boundary points of the interval.

The Scattering Problem at a fixed energy E in the asymptotic form "incident wave + outgoing waves" can be written as:

$$\boldsymbol{\Phi}_{\rightarrow}(z \to \pm\infty) = \begin{cases} \mathbf{X}^{(\rightarrow)}_{\min}(z) + \mathbf{X}^{(\leftarrow)}_{\min}(z)\mathbf{R}_{\rightarrow} + \mathbf{X}^{(c)}_{\min}(z)\mathbf{R}^c_{\rightarrow}, \; z \to -\infty, \\ \mathbf{X}^{(\rightarrow)}_{\max}(z)\mathbf{T}_{\rightarrow} + \mathbf{X}^{(c)}_{\max}(z)\mathbf{T}^c_{\rightarrow}, \; z \to +\infty, \end{cases}$$

$$\boldsymbol{\Phi}_{\leftarrow}(z \to \pm\infty) = \begin{cases} \mathbf{X}^{(\leftarrow)}_{\min}(z)\mathbf{T}_{\leftarrow} + \mathbf{X}^{(c)}_{\min}(z)\mathbf{T}^c_{\leftarrow}, \; z \to -\infty, \\ \mathbf{X}^{(\leftarrow)}_{\max}(z) + \mathbf{X}^{(\rightarrow)}_{\max}(z)\mathbf{R}_{\leftarrow} + \mathbf{X}^{(c)}_{\max}(z)\mathbf{R}^c_{\leftarrow}, \; z \to +\infty. \end{cases}$$

Here $\boldsymbol{\Phi}_{\rightarrow}(z)$, $\boldsymbol{\Phi}_{\leftarrow}(z)$ are matrix solutions with dimensions $N \times N^L_o$, $N \times N^R_o$, where N^L_o, N^R_o are the numbers of open channels, $\mathbf{X}^{(\rightarrow)}_{\min}(z)$, $\mathbf{X}^{(\leftarrow)}_{\min}(z)$ are open channel asymptotic solutions at $z \to -\infty$, dimension $N \times N^L_o$, $\mathbf{X}^{(\rightarrow)}_{\max}(z)$, $\mathbf{X}^{(\leftarrow)}_{\max}(z)$ are open channel asymptotic solutions at $z \to +\infty$, dimension $N \times N^R_o$, $\mathbf{X}^{(c)}_{\min}(z)$, $\mathbf{X}^{(c)}_{\max}(z)$ are closed channel solutions, dimension $N \times (N - N^L_o)$, $N \times (N - N^R_o)$, $\mathbf{R}_{\rightarrow}$, $\mathbf{R}_{\leftarrow}$ are the reflection amplitude square matrices of dimension $N^L_o \times N^L_o$, $N^R_o \times N^R_o$, $\mathbf{T}_{\rightarrow}$, $\mathbf{T}_{\leftarrow}$ are the transmission amplitude rectangular matrices of dimension $N^R_o \times N^L_o$, $N^L_o \times N^R_o$, $\mathbf{R}^c_{\rightarrow}$, $\mathbf{T}^c_{\rightarrow}$, $\mathbf{T}^c_{\leftarrow}$, $\mathbf{R}^c_{\leftarrow}$ are auxiliary matrices. For real-valued potentials $\mathbf{V}(z)$ and $\mathbf{Q}(z)$ the transmission $\mathbf{T}$ and reflection $\mathbf{R}$ amplitudes satisfy the relations

$$\begin{aligned} &\mathbf{T}^{\dagger}_{\rightarrow}\mathbf{T}_{\rightarrow} + \mathbf{R}^{\dagger}_{\rightarrow}\mathbf{R}_{\rightarrow} = \mathbf{I}_{oo}, \quad \mathbf{T}^{\dagger}_{\leftarrow}\mathbf{T}_{\leftarrow} + \mathbf{R}^{\dagger}_{\leftarrow}\mathbf{R}_{\leftarrow} = \mathbf{I}_{oo}, \\ &\mathbf{T}^{\dagger}_{\rightarrow}\mathbf{R}_{\leftarrow} + \mathbf{R}^{\dagger}_{\rightarrow}\mathbf{T}_{\leftarrow} = \mathbf{0}, \quad \mathbf{R}^{\dagger}_{\leftarrow}\mathbf{T}_{\rightarrow} + \mathbf{T}^{\dagger}_{\leftarrow}\mathbf{R}_{\rightarrow} = \mathbf{0}, \\ &\mathbf{T}^{T}_{\rightarrow} = \mathbf{T}_{\leftarrow}, \quad \mathbf{R}^{T}_{\rightarrow} = \mathbf{R}_{\rightarrow}, \quad \mathbf{R}^{T}_{\leftarrow} = \mathbf{R}_{\leftarrow} \end{aligned} \tag{3}$$

ensuring unitarity and symmetry of S-matrix

$$\mathbf{S} = \begin{pmatrix} \mathbf{R}_{\rightarrow} & \mathbf{T}_{\leftarrow} \\ \mathbf{T}_{\rightarrow} & \mathbf{R}_{\leftarrow} \end{pmatrix}, \qquad \mathbf{S}^{\dagger}\mathbf{S} = \mathbf{S}\mathbf{S}^{\dagger} = \mathbf{I}.$$

Here † and ^{T} denote conjugate transpose and transpose of a matrix, respectively. So, for complex potentials $\mathbf{V}(z)$ and $\mathbf{Q}(z)$ the S-matrix is only symmetric $\mathbf{S} = \mathbf{S}^T$ and only the last three conditions of (3) hold.

For set of ODEs (1) with $f_B(z)$=$f_A(z)$=1, $Q_{ij}(z)$=0 and constant effective potentials $V_{ij}(z)$=$V_{ij}^{L,R}$ in the asymptotic region, asymptotic solutions $\mathbf{X}_i^{(*)}(z \to \pm\infty)$ are as follows. The open channel asymptotic solutions $i = i_o = 1, ..., N_o^{L,R}$:

$$\mathbf{X}_{i_o}^{(\rightleftarrows)}(z \to \pm\infty) \to \frac{\exp\left(\pm\imath\sqrt{E - \lambda_{i_o}^{L,R}}z\right)}{\sqrt[4]{E - \lambda_{i_o}^{L,R}}}\boldsymbol{\Psi}_{i_o}^{L,R}, \quad \lambda_{i_o}^{L,R} < E.$$

The closed channels asymptotic solutions $i = i_c = N_o^{L,R} + 1, \ldots, N$:

$$\mathbf{X}_{i_c}^{(c)}(z \to \pm\infty) \to \exp\left(-\sqrt{\lambda_{i_c}^{L,R} - E}|z|\right)\boldsymbol{\Psi}_{i_c}^{L,R}, \quad \lambda_{i_c}^{L,R} \geq E.$$

Here $\lambda_i^{L,R}$ and $\boldsymbol{\Psi}_i^{L,R} = \{\Psi_{1i}^{L,R}, ..., \Psi_{Ni}^{L,R}\}^T$ are solutions of the algebraic eigenvalue problems with matrix $\mathbf{V}^{L,R}$ of dimension $N \times N$ for the entangled channels [11]

$$\mathbf{V}^{L,R}\boldsymbol{\Psi}_i^{L,R} = \lambda_i^{L,R}\boldsymbol{\Psi}_i^{L,R}, \quad (\boldsymbol{\Psi}_i^{L,R})^T\boldsymbol{\Psi}_j^{L,R} = \delta_{ij}. \tag{4}$$

Note that $\lambda_i^{L,R} = V_{ii}^{L,R}$ and $\boldsymbol{\Psi}_i^{L,R} = \delta_{ji}$, if $V_{i\neq j}^{L,R} = 0$, i.e. in the conventional case of orthogonal channels.

Bound or Metastable States. Eigenfunctions $\boldsymbol{\Phi}_m(z)$ obey the normalization and orthogonality conditions

$$(\boldsymbol{\Phi}_m|\boldsymbol{\Phi}_{m'}) = \int_{z^{\min}}^{z^{\max}} f_B(z)(\boldsymbol{\Phi}^{(m)}(z))^T\boldsymbol{\Phi}^{(m')}(z)dz = \delta_{mm'}.$$

For bound states with real eigenvalues E: $E_1 \leq E_2 \leq ...$ the Dirichlet or Neumann BC (2) follow from asymptotic expansions. For metastable states with complex eigenvalues $E = \Re E + \imath\Im E$, $\Im E < 0$: $\Re E_1 \leq \Re E_2 \leq ...$ the Robin BC follow from outgoing wave fundamental asymptotic solutions that correspond to the Siegert outgoing wave BCs [12].

For the set of ODEs (1) with $f_B(z)$=$f_A(z)$=1, $Q_{ij}(z)$=0 and constant effective potentials $V_{ij}(z)$=$V_{ij}^{L,R}$ in the asymptotic region, asymptotic solutions $\mathbf{X}_i^{(*)}(z \to \pm\infty)$ are as follows. For bound states:

$$\mathbf{X}_{i_c}^{(c)}(z \to \pm\infty) \to \exp\left(-\sqrt{\lambda_{i_c}^{L,R} - E_i}|z|\right)\boldsymbol{\Psi}_{i_c}^{L,R}, \quad \lambda_{i_c}^{L,R} \geq E, \quad i_c = 1, \ldots, N,$$

and for metastable states:

$$\mathbf{X}_{i_o}^{(\rightleftarrows)}(z \to \infty) \to \exp\left(+\imath\sqrt{E - \lambda_{i_o}^{L,R}}|z|\right)\boldsymbol{\Psi}_{i_o}^{L,R}, \quad \lambda_{i_o}^{L,R} < \Re E, \quad i_o = 1, ..., N_o^{L,R},$$

$$\mathbf{X}_{i_c}^{(c)}(z \to \infty) \to \exp\left(-\sqrt{\lambda_{i_c}^{L,R} - E}|z|\right)\boldsymbol{\Psi}_{i_c}^{L,R}, \quad \lambda_{i_c}^{L,R} \geq \Re E, \quad i_c = N_o^{L,R} + 1, \ldots, N.$$

In the considered case matrix $\mathcal{R}(z^t)$ of logarithmic derivatives for the corresponding Robin BC takes the form

$$\mathcal{R}(z^t) = \boldsymbol{\Psi}^{L,R}\mathbf{F}^{L,R}\left(\boldsymbol{\Psi}^{L,R}\right)^{-1},$$

where $\mathbf{F}^{L,R} = \mathrm{diag}(..., \pm\sqrt{\lambda_{i_c}^{L,R} - E}, ..., \mp\imath\sqrt{E - \lambda_{i_o}^{L,R}}, ...)$ and $\boldsymbol{\Psi}^{L,R}$ is the matrix, composed from solutions $\boldsymbol{\Psi}_j^{L,R}$ of algebraic eigenvalue problem (4).

2.2 Finite Element Scheme

Finding solution $\boldsymbol{\Phi}(z) \in \mathcal{H}_2^{s\geq 1}(\bar{\Omega})$ of BPVs (1)–(2) reduces to the FEM calculation of stationary points of symmetric quadratic functional

$$\begin{aligned}
&\boldsymbol{\Xi}(\boldsymbol{\Phi}, E, z^{\min}, z^{\max}) \equiv \int_{z^{\min}}^{z^{\max}} f_B(z)\boldsymbol{\Phi}^\bullet(z)\left(\mathbf{D} - E\mathbf{I}\right)\boldsymbol{\Phi}(z)dz = \boldsymbol{\Pi}(\boldsymbol{\Phi}, E, z^{\min}, z^{\max})\\
&-f_A(z^{\max})\boldsymbol{\Phi}^\bullet(z^{\max})\mathbf{G}(z^{\max})\boldsymbol{\Phi}(z^{\max}) + f_A(z^{\min})\boldsymbol{\Phi}^\bullet(z^{\min})\mathbf{G}(z^{\min})\boldsymbol{\Phi}(z^{\min}),\\
&\boldsymbol{\Pi}(\boldsymbol{\Phi}, E, z^{\min}, z^{\max}) = \int_{z^{\min}}^{z^{\max}} \left[f_A(z)\frac{d\boldsymbol{\Phi}^\bullet(z)}{dz}\frac{d\boldsymbol{\Phi}(z)}{dz} + f_B(z)\boldsymbol{\Phi}^\bullet(z)\mathbf{V}(z)\boldsymbol{\Phi}(z)\right.\\
&\left.+f_A(z)\boldsymbol{\Phi}^\bullet(z)\mathbf{Q}(z)\frac{d\boldsymbol{\Phi}(z)}{dz} - f_A(z)\frac{d\boldsymbol{\Phi}(z)^\bullet}{dz}\mathbf{Q}(z)\boldsymbol{\Phi}(z) - f_B(z)E\boldsymbol{\Phi}^\bullet(z)\boldsymbol{\Phi}(z)\right]dz, \qquad (5)
\end{aligned}$$

where $\mathbf{G}(z) = \mathcal{R}(z) - \mathbf{Q}(z)$ is a symmetric $N \times N$ matrix and $^\bullet$ stands for ^{T} or † depending on the problem considered.

High-accuracy computational schemes for solving BVP (1)–(2) are derived from variational functional (5) basing on the FEM. The general idea of FEM in a one-dimensional space is to divide the interval $[z^{\min}, z^{\max}]$ into many small subintervals referred to as elements. The choice of subintervals size (length) is free enough to account for physical properties or qualitative behavior of the sought solutions, such as smoothness.

The interval $\Delta = [z^{\min}, z^{\max}]$ is covered by a set of n subintervals $\Delta_j = [z_{(j-1)}, z_{(j)}]$, $z_{(0)} = z^{\min}$, $z_{(n)} = z^{\max}$ in such a way that $\Delta = \bigcup_{j=1}^n \Delta_j$. On each subinterval $\Delta_j = [z_{(j-1)}, z_{(j)}]$ of a length $h_j = z_{(j)} - z_{(j-1)}$ we introduce a set of local functions given by the IHPs [6]: $\varphi_r^\kappa(z)$, $r = 0, ..., p$, $\kappa = 0, \ldots, \kappa_r^{\max} - 1$, where $\kappa_r^{\max}$ is referred to as the multiplicity of the nodes $z_r \in \Delta_j$, $z_0 = z_{(j-1)}$, $z_p = z_{(j)}$. The values of functions $\varphi_r^\kappa(z)$ of the order $p' = \sum_{r=0}^p \kappa_r^{\max} - 1$ with their derivatives up to the order $(\kappa_r^{\max} - 1)$ are determined by expressions

$$\varphi_r^\kappa(z_{r'}) = \delta_{rr'}\delta_{\kappa 0}, \quad \left.\frac{d^{\kappa'}\varphi_r^\kappa(z)}{dz^{\kappa'}}\right|_{z=z_{r'}} = \delta_{rr'}\delta_{\kappa\kappa'}. \qquad (6)$$

IHPs are calculated using analytical formulas [13] implemented in the algorithm of Appendix A. The numerical solution $\boldsymbol{\Phi}^h(z) \approx \boldsymbol{\Phi}(z)$ is sought in the form of a finite sum over the basis of local functions $N_s(z)$ at each nodal point $z = z_\rho$ of the grid $\Omega^p_{h_j(z)}[z^{\min}, z^{\max}]$ on interval $z \in \Delta = [z^{\min}, z^{\max}]$:

Table 1. Interrelation of subscripts μ, ν, i, r, ρ, j, S, s and κ by the example of BVP (1) with $N=2$ solved by FEM with the IHPs of the order $p'=4$ ($p'=\sum_{r=0}^{p}\kappa_r^{\max}-1$) with multiplicities $(\kappa_1^{\max},\kappa_2^{\max},\kappa_3^{\max})=(2,1,2)$ on $n=4$ finite elements. Here (D) means using the Dirichlet conditions at $z=z^{\min}$ and $z=z^{\max}$.

odd j	1												3															
even j							2												4									
ρ	0	0	0	0	1	1	2	2	2	2	3	3	4	4	4	4	5	5	6	6	6	6	7	7	8	8	8	8
κ	0	0	1	1	0	0	0	0	1	1	0	0	0	0	1	1	0	0	0	0	1	1	0	0	0	0	1	1
i	1	2	1	2	1	2	1	2	1	2	1	2	1	2	1	2	1	2	1	2	1	2	1	2	1	2	1	2
r (odd j)	0	0	0	0	1	1	2	2	2	2			0	0	0	0	1	1	2	2	2	2						
r (even j)							0	0	0	0	1	1	2	2	2	2			0	0	0	0	1	1	2	2	2	2
S (odd j)	1	1	2	2	3	3	4	4	5	5			11	11	12	12	13	13	14	14	15	15						
S (even j)							6	6	7	7	8	8	9	9	10	10			16	16	17	17	18	18	19	19	20	20
s	0	0	1	1	2	2	3	3	4	4	5	5	6	6	7	7	8	8	9	9	10	10	11	11	12	12	13	13
ν	1	2																							1	2		
μ	1	2	3	4	5	6	7	8	9	10	11	12	13	14	15	16	17	18	19	20	21	22	23	24	25	26	27	28
μ(D)			1	2	3	4	5	6	7	8	9	10	11	12	13	14	15	16	17	18	19	20	21	22			23	24

$$\boldsymbol{\Phi}^h(z)=\sum_{s=0}^{L-1}\boldsymbol{\Phi}_s^h N_s(z),\quad \boldsymbol{\Phi}_s^h=(\Phi_{s1}^h,\ldots,\Phi_{sN}^h)^T,\quad \left.\frac{d^\kappa N_s(z)}{dz^\kappa}\right|_{z=z_\rho}=\delta_{ss'(\kappa,\rho)}. \quad (7)$$

Basis functions $N_s(z)$ are piecewise polynomials calculated by IHP matching (see for the details [13,17]). The substitution of expansion (7) into variational functional (5) reduces problem (1)–(2) to an algebraic problem for the unknown eigenvalues E or S-matrix and vector $\boldsymbol{\Phi}^h=\{\Phi_\mu^h\}_{\mu=1}^{LN}=\{\{\Phi_{si}^h\}_{i=1}^{N}\}_{s=0}^{L-1}$:

$$(\mathbf{A}-E\mathbf{B})\boldsymbol{\Phi}^h=0. \quad (8)$$

Here $\mathbf{A}=\mathbf{A}^{(2)}+\mathbf{A}^{(1)}+\mathbf{V}+\mathbf{M}^{\min}-\mathbf{M}^{\max}$ and positive definite $\mathbf{B}$ are symmetric $NL\times NL$ matrices of stiffness and mass, respectively:

$$\begin{aligned}
A^{(2)}_{\mu_1,\mu_2}&=\int_\Delta \frac{dN_{s_1}(z)}{dz}\delta_{i_1 i_2}\frac{dN_{s_2}(z)}{dz}f_A(z)dz,\\
A^{(1)}_{\mu_1,\mu_2}&=\int_\Delta\left(N_{s_1}(z)Q_{i_1 i_2}(z)\frac{dN_{s_2}(z)}{dz}-\frac{dN_{s_1}(z)}{dz}Q_{i_1 i_2}(z)N_{s_2}(z)\right)f_A(z)dz,\\
V_{\mu_1,\mu_2}&=\int_\Delta N_{s_1}(z)V_{i_1 i_2}(z)N_{s_2}(z)f_B(z)dz,\\
B_{\mu_1,\mu_2}&=\int_\Delta N_{s_1}(z)\delta_{i_1 i_2}N_{s_2}(z)f_B(z)dz. \quad (9)
\end{aligned}$$

According to the definition of local function $N_s(z)$, the integrals in (9) are calculated only on subinterval Δ_j in which both $N_{s_1}(z)$ and $N_{s_2}(z)$ are localized. $NL\times NL$ matrices $\mathbf{M}^{\max}$ and $\mathbf{M}^{\min}$ have only one nonzero $N\times N$ submatrix:

$$M^{\min}_{\nu_1,\nu_2}=f_A(z^{\min})R_{\nu_1,\nu_2}(z^{\min}),\ M^{\max}_{\nu_0+\nu_1,\nu_0+\nu_2}=f_A(z^{\max})R_{\nu_1,\nu_2}(z^{\max}), \quad (10)$$

where $\nu_0=N(L-\kappa_r^{\max})$, respectively. Each element of the eigenvector $\boldsymbol{\Phi}^h$ is marked by the multi-index notation μ. The dependence of multi-index μ on

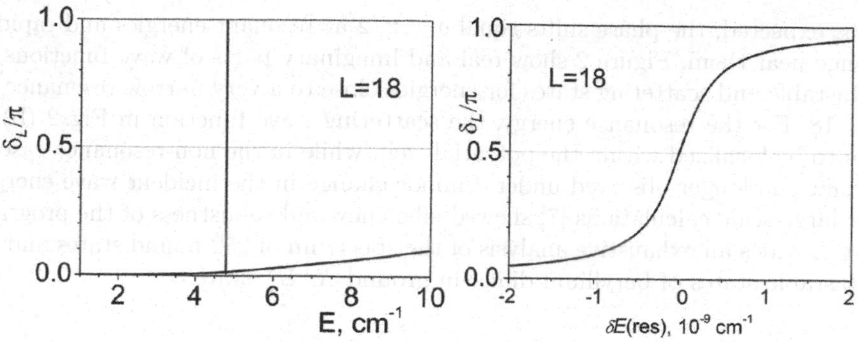

Fig. 1. Phase shift δ vs scattering energy E at $L = 18$. Here $\delta E(res) = E - E(res)$.

indices ν, i, r, ρ, j, s, κ defined above and index S used for global enumeration of local functions φ_r^κ on each of finite elements Δ_j (see Appendix B) is illustrated in Table 1 by an example of BVP (1) with $N = 2$, solved using FEM with fourth-order IHPs $(\kappa_1^{\max}, \kappa_2^{\max}, \kappa_3^{\max}) = (2, 1, 2)$ on $n = 4$ finite elements.

The algorithms for calculating characteristics of FEM scheme and generating an algebraic eigenvalue problem are given in Appendixes B and C. The algebraic eigenvalue problem is solved using either built-in linear algebra procedures, or the continuous analog of Newton method [12].

3 Benchmark Calculations

3.1 ODE with Potential Calculated by Quantum Chemistry

In quantum chemical calculations, effective potentials of interatomic interaction are presented in the form of numerical tables calculated with a limited accuracy and defined on a nonuniform mesh of nodes in a finite range of interatomic distances. It is important that the proposed FEM scheme with IHPs ensures smooth matching of the tabulated potential with its analytical asymptotic expression, as well as high-quality smooth approximation of eigenfuntions [7]. Consider, e.g., the Schrödinger equation for a diatomic beryllium molecule in the adiabatic approximation, commonly referred to as Born–Oppenheimer approximation

$$\left(-\frac{1}{r^2}\frac{d}{dr}s_2 r^2\frac{d}{dr}+V(r)+\frac{L(L+1)}{r^2}s_2-E_{vL}\right)\Phi_{vL}(r)=0,$$

where $s_2 = 1/0.2672973729$, r is the distance between the atoms in angstroms, E_{vL} is the energy in cm^{-1} and L is the total angular momentum quantum number. The potential $V(r)$ is defined by its values on a grid and an asymptotic expansion beyond it (see for details [7]). For $L = 18$ there are 7 bound states $E_{vL} = (-600.3, -392.4, -240.7, -150.4, -96.6, -54.4, -20.3)$ cm^{-1} and 1 metastable state $E_{1L}^M = (4.788 - 4 \cdot 10^{-10}\imath)$ cm^{-1}. The computation time does not exceed 20 s. Figure 1 shows phase shifts δ as functions of scattering energy

E; as expected, the phase shifts equal $\delta = \pi/2$ at resonant energies and rapidly change near them. Figure 2 show real and imaginary parts of wave functions of metastable and scattering states for energies close to a very narrow resonance at $L = 18$. For the resonance energy, the scattering wave function in Fig. 2 (b) is seen to be localized within the potential well, while in the non-resonance case it becomes no longer observed under a minor change in the incident wave energy. Our large-scale calculations [7] showed efficiency and robustness of the program that provides an exhaustive analysis of the spectrum of 252 bound states and 58 metastable states of beryllium dimer in ground $X^1\Sigma_g^+$ state.

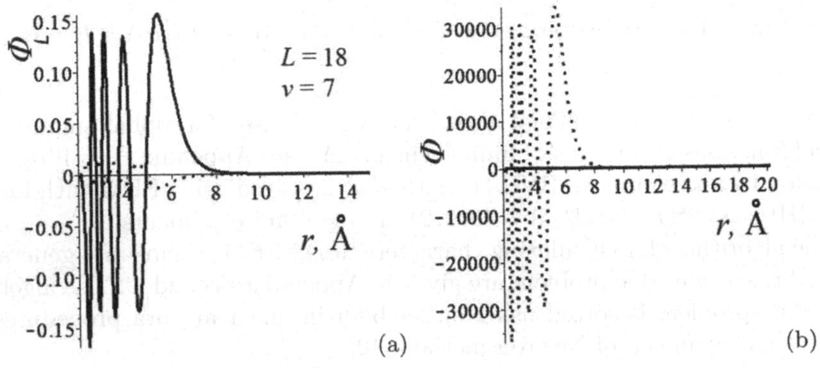

Fig. 2. Plots of the real (solid curves) and imaginary (dashed curves) parts of metastable state wave function $\Phi_{vL}(r)$ (a) and scattering functions $\Phi_L(r)$ in the vicinity of resonance energy $E(res) \approx 4.788\,\mathrm{cm}^{-1}$ (b). Here $L = 18$ and $v = 7$.

3.2 ODE System with Piecewise Constant Potentials

Consider, e.g., a BVP similar to that of Ref. [11] for the Schrödinger equation in 2D domain $\Omega_{yz}=\{y\in(0,\pi), z\in(-\infty,+\infty)\}$, with potential

$$V(y,z)=\{0, z<2; -2y, |z|\le 2; 2y, z>2\}.$$

We seek the solution in the form of expansion $\Psi(y,z)=\sum_{i=1}^{N} B_i(y)\Phi_i(z)$ in a set of basis functions $B_i(y)=\frac{\sqrt{2}}{\sqrt{\pi}}\sin(iy)$, which leads to Eqs. (1) with $f_B(z)=f_A(z)=1$, $Q_{ij}(z)=0$ and effective potentials

$$V_{ij}(z)= i^2\delta_{ij}+\left\{0, z<-2; -2, |z|\le 2; 2, z>2\right\} \times\left\{\pi/2, i=j; 0, \text{ even } i-j; \frac{-8ij}{\pi(i^2-j^2)^2}, \text{ odd } i-j\right\}.$$

For example, let us choose $N = 6$. The considered system has sets of threshold energies that differ in the left and right asymptotic regions of the z-axis: $\lambda_i^{(L)} = \{1, 4, 9, 16, 25, 36\}$ and $\lambda_i^{(R)} = \{3.742260, 7.242058, 12.216485, 19.188688,$

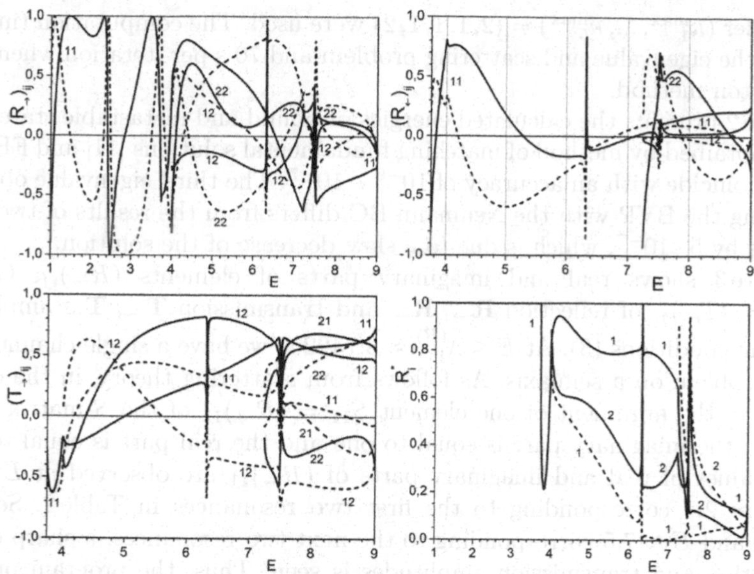

Fig. 3. Real (solid curves) and imaginary (dashed curves) parts of elements $(R_{\rightarrow})_{ij}$, $(R_{\leftarrow})_{ij}$, $(T_{\rightarrow})_{ij} = (T_{\leftarrow})_{ji}$ of reflection $\mathbf{R}_{\rightarrow}$, $\mathbf{R}_{\leftarrow}$ and transmission $\mathbf{T}_{\rightarrow}$, $\mathbf{T}_{\leftarrow}$ amplitudes, and reflection coefficients $R_i^* = (\mathbf{R}_*^{\dagger}\mathbf{R}_*)_{ii}$ at $* = \rightarrow$ (solid curves) and $* = \leftarrow$ (dashed curves) as functions of scattering energy E.

Table 2. Eigenvalues E_i, $i = 1, 2, 3$ of bound states and E_i^M, $i = 1, ..., 4$ of metastable states obtained by solving the eigenvalue problem with Neumann BC (E) by Newton method (N) and method of matching fundamental solutions (M).

E	−2.12846503065	−0.925565889437	0.835126562953
N	−2.12846503036	−0.925565881437	0.835126980234
M	−2.12846503156	−0.925565883542	0.835126979072
N	1.35989392876−$\imath$0.00016253897		2.43040517408−$\imath$0.0789059067115
M	1.35989392695−$\imath$0.00016253895		2.43040517183−$\imath$0.0789059070893
N	6.32021061134−$\imath$0.00326071312		7.50608788873−$\imath$0.0194121454599
M	6.32021060910−$\imath$0.00326071319		7.50608789245−$\imath$0.0194121442796

28.173689,39.286376}, respectively. So, we have different numbers of open channels and entangled channels (4) in the right-hand asymptotic region.

The bound states were calculated on a grid [−25.78125, −18.1875, −13.125, −9.75, −7.5, −6(1)6] built up a geometric progression of steps in accordance with a slow exponential decay of solutions at $z < -6$ subject to the Neumann BC. The metastable states were found by Newton method on a grid [−4(1)4] with the Robin BC dependent on the eigenvalue. As initial data, the solution obtained on a grid [−2(1)2] with the Neumann BC was taken. The same grid [−4(1)4] was used to solve the scattering problem. In both cases, IHPs of the sixth order $(\kappa_1^{\max},...,\kappa_5^{\max}) = (2,1,1,1,2)$ were used. The computation time was 62 s for the eigenvalue and scattering problem and 70 s per iteration when using the Newton method.

Table 2 presents the calculated energies of bound and metastable states. The results obtained by method of matching fundamental solutions [11] and FEM are seen to coincide with an accuracy of $10^{-9} \div 10^{-7}$. The third eigenvalue obtained by solving the BVP with the Neumann BC differs from the results of two other methods by $5 \cdot 10^{-7}$, which is due to a slow decrease of the solution.

Figure 3 shows real and imaginary parts of elements $(R_{\rightarrow})_{ij}$, $(R_{\leftarrow})_{ij}$, $(T_{\rightarrow})_{ij} = (T_{\leftarrow})_{ji}$ of reflection $\mathbf{R}_{\rightarrow}$, $\mathbf{R}_{\leftarrow}$ and transmission $\mathbf{T}_{\rightarrow}$, $\mathbf{T}_{\leftarrow}$ amplitudes satisfying conditions (3). At $E \leq \lambda_1^{(R)} \approx 3.742260$ we have a single-channel scattering problem on a semiaxis. As follows from scattering theory, in the case of resonance, the argument of one element $S_{11} = (R_{\rightarrow})_{11}$ of the S-matrix equals $\pi/2$, i.e., the imaginary part is equal to one and the real part is equal to zero. These values of real and imaginary parts of $(R_{\rightarrow})_{11}$ are observed at $E \approx 1.1$ and $E \approx 2.4$ corresponding to the first two resonances in Table 2. So, near $E \approx 6.3$ and $E \approx 7.5$ corresponding to the next two resonances, a sharp change of reflection and transmission amplitudes is seen. Thus, the program provides an exhaustive analysis of the scattering problem with a different number of *open entangled channels* similar to the one present in sub-barrier fusion reactions [8].

4 Conclusion

We presented the FEM scheme and showed its efficiency by benchmark examples of using the KANTBP 5M program (an upgrade of KANTBP 4M [17] containing 1484 lines) implemented and executed in MAPLE. We showed that the program provides a suitable tool for solving multichannel scattering and eigenvalue problems for systems of second-order ODEs with continuous or piecewise continuous real or complex-valued coefficients with a given accuracy. The new type of FEM discretization is implemented using IHPs *with an arbitrary multiplicity of IHPs nodes*, determined by Eqs. (6) and (7) given in Appendices A and B and Gauss quadratures given in Appendix C, whereas only a fixed multiplicity of the nodes and analytical integration of polynomial approximants were available in KANTBP 4M, which preserves the continuity of derivatives of the sought solutions. To reduce the new type of a scattering problem with a different number of *open entangled channels* (whereas in KANTBP 4M only non-entangled open channels could be considered) in the left and right asymptotic regions to a BVP on a finite interval, the *new type of entangled asymptotic BCs determined by Eq. (4)* are approximated by the homogeneous third-type (Robin) conditions. To calculate metastable states with complex eigenvalues, or to solve a bound state problem with Robin BC depending on the spectral parameter, the Newtonian iteration scheme is implemented. The open code of the KANTBP 5M and test examples including *INPUT* and *OUTPUT* both implemented and executed in MAPLE of solving eigenvalue problems and scattering problems of quantum mechanics [7,8,10–12] and adiabatic waveguide modes [9] will be presented in JINRLIB program library.

The authors thank Prof. E. Zima for useful discussions. The work was partially supported by the RFBR and MECSS, project number 20-51-44001, the Ministry of Science and Higher Education of the Russian Federation, grant number 075-10-2020-117, the Bogoliubov-Infeld program, the Hulubei-Meshcheryakov program, the RUDN University Strategic Academic Leadership Program, grant of Plenipotentiary of the Republic of Kazakhstan in JINR, the Foundation of Science and Technology of Mongolia, grant number SST_18/2018 and the Ho Chi Minh City University of Education (Grant CS.2020.19.47).

A Generation of IHPs on the Standard Interval

This appendix presents an algorithm for constructing IHPs according to their characteristics: p is the number of partitions of a finite element, z_r are the IHP nodes with multiplicities $\kappa_r^{\max}$. They are applied to construct IHPs in the FEM scheme, then the conditions $z_r \in [0,1]$, $z_0 = 0$, $z_p = 1$, $\kappa_0^{\max} = \kappa_p^{\max}$ are to be satisfied. For further implementation it is convenient to number IHPs with n''.

Input: $r = 0, ..., p$ is the number of the node,
$\kappa_r^{\max}$ is the multiplicity of node z_r,
Output: n' is the number of last IHP, IHP(0),...,IHP(n') the set of IHPs,
$r(n')$ and $\kappa(n')$ are values of r and κ vs n',
p' is the degree of IHPs.

1.1.: $n' := -1$;
  for r:=0 to p do
    $w_r := \prod_{r'=0, r'\neq r}^{p} \left(\frac{z - z_{r'}}{z_r - z_{r'}}\right)^{\kappa_{r'}^{\max}}$;
1.2.: $g_r^0 := 1$; $g_r^1 := \sum_{r'=0, r'\neq r}^{p} \frac{\kappa_{r'}^{\max}}{z - z_{r'}}$;
    for κ:=2 to $\kappa_r^{\max} - 1$ do
      $g_r^\kappa := \frac{d g_r^{\kappa-1}}{dz} + g_r^1 g_r^{\kappa-1}$;
    end for;
    $g_r^\kappa := g_r^\kappa(z \to z_r), \quad \kappa := 1, ..., \kappa_r^{\max} - 1$
  end for;
1.3.: for r:=0 to p do
  $a_0 = H_0$;
  for r':=1 to $\kappa_r^{\max} - 1$ do
    $a_{r'} := H_{r'}/r'! - \sum_{r''=0}^{r'-1} a_{r''} g_{r'-r''}^{\kappa}/(r' - r'')!$
  end for;
  for r':=1 to $\kappa_r^{\max} - 1$ do
    $n' := n' + 1$; $r(n') := r$; $\kappa(n') := r'$;
    $\mathrm{IHP}(n') = w_r(z) + \sum_{r'=0}^{\kappa_{r'}^{\max}-1} a_{r'}(H_{r''} \to \delta_{rr''})(z - z_r)^{r'}$
  end for;
 end for;
 $p' = \sum_{r=0}^{p} \kappa_r^{\max} - 1$

B Calculation of the FEM Scheme Characteristics

Note that when calculating the matrices (9) of the algebraic problem (8), we do it without explicitly calculating $N_s(z)$ from (7) by introducing global numbering φ_r^κ on each of the finite elements Δ_j, i.e. $\varphi_S \equiv \varphi_{n''}(z \in \Delta_j) \equiv \varphi_r^\kappa(z \in \Delta_j)$. In our implementation, the FEM IHP schemes are numbered so that S increases with an increase in j, or with a constant j and an increase in n'', or with constant j and n'' and an increase in i. For convenience, arrays of length $n \times 3$ are introduced: $E(j,1)$ is the minimum S at which φ_S is defined on Δ_j, $E(j,2)$ is the minimum S for which $r = p$ and φ_S is defined on Δ_j, $E(j,3)$ is the maximum S at which φ_S is defined on Δ_j and a two-dimensional array C with dimension $S^{\max} \times 3$, where depending on S, $C(S,1)$, $C(S,2)$, $C(S,3)$ correspond to μ (the number of element of eigenvector $\boldsymbol{\Phi}^h$), n'' (the number of IHP) and i (the number of equation in the system of ODEs from Eq. (1)).

Input: n is the number of finite elements $\Delta_j = [z_{j-1}, z_j]$, $\Delta = \cup_{j=1}^n \Delta_j$
n' is the number of last IHP,
IHP(0),...,IHP(n') the set of IHPs,
$r(n'')$ and $\kappa(n'')$ are values of r and κ vs n''.
Output: $E(n, 1:3)$ and $C(S^{\max}, 3)$ are the FEM scheme characteristics

```
for j from 1 to n do
E(j,1) := 0;
E(j,2) := 0;
for n'' from 0 to n' do
    if (not
      ((Dirichlet BC on z^min and j = 1 and r(n'') = 0 and κ(n'') = 0)
      or
      (Dirichlet BC on z^max and j = n and r(n'') = p and κ(n'') = 0 ))
   ) then
     for i from 1 to N do
        S:=S+1;
        if (E(j,1) = 0) then E(j,1) := S; fi;
        if (E(j,2) = 0 and r(i2) = p) then E(j,2) := S; fi;
        E(j,2) := S;
        C(S,2) := n'';
        C(S,3) := i;
        if (r(n'') = 0 and j > 1) then
          if ∃S' ∈ {E(j-1,2),...,E(j-1,3)}:
                C(S,3) = C(S',3) and κ(C(S,2)) = κ(C(S',2))):
           then C(S,1) := C(S',1);
            else increase μ and C(S,1) := μ
End of all cycles and conditions
```

C FEM generation of Algebraic Eigenvalue Problem

Input: $E(n, 1 : 3)$ and $C(S^{\max}, 3)$ are FEM scheme characteristics from Appendix B. $\mathrm{IHP}_g(C(S,2))$ and $\mathrm{IHP}'_g(C(S,2))$ are the values of IHPs and their derivatives in Gaussian nodes $\bar{z}_g$ (in local coordinates)
w_g are the Gaussian weights
Output: A_{μ_1,μ_2}, B_{μ_1,μ_2} are matrix elements of **A** and **B**

$\mathbf{A} = \mathbf{0}$, $\mathbf{B} = \mathbf{0}$
for j from 1 to n do
  $\Delta z_j := z_{(j)} - z_{(j-1)}$;
  for S, S' from $E(j, 1)$ to $E(j, 3)$ do
   if $C(S, 3) = C(S', 3)$ then

$$B_{C(S,1),C(S',1)} := B_{C(S,1),C(S',1)} + \sum_g w_g \mathrm{IHP}_g(C(S,2))\mathrm{IHP}_g(C(S',2)) \times (\Delta z_j)^{1+\kappa(C(S,2))+\kappa(C(S',2))} f_b(z_{(j-1)} + \Delta z_j \bar{z}_g);$$

$$A_{C(S,1),C(S',1)} := A_{C(S,1),C(S',1)} + \sum_g w_g \mathrm{IHP}'_g(C(S,2))\mathrm{IHP}'_g(C(S',2)) \times (\Delta z_j)^{-1+\kappa(C(S,2))+\kappa(C(S',2))} f_a(z_{(j-1)} + \Delta z_j \bar{z}_g);$$

  else

$$A_{C(S,1),C(S',1)} := A_{C(S,1),C(S',1)} + \sum_g w_g (\mathrm{IHP}_g(C(S,2))\mathrm{IHP}'_g(C(S',2)) - \mathrm{IHP}'_g(C(S,2))\mathrm{IHP}_g(C(S',2))) \times (\Delta z_j)^{\kappa(C(S,2))+\kappa(C(S',2))} f_a(z_{(j-1)} + \Delta z_j \bar{z}_g) \times Q_{C(S,3),C(S',3)}(z_{(j-1)} + \Delta z_j \bar{z}_g);$$

  fi;

$$A_{C(S,1),C(S',1)} := A_{C(S,1),C(S',1)} + \sum_g w_g \mathrm{IHP}_g(C(S,2))\mathrm{IHP}_g(C(S',2)) \times (\Delta z_j)^{1+\kappa(C(S,2))+\kappa(C(S',2))} f_b(z_{(j-1)} + \Delta z_j \bar{z}_g) \times V_{C(S,3),C(S',3)}(z_{(j-1)} + \Delta z_j \bar{z}_g);$$

End of all cycles

References

1. Alder, B., Fernbach, S., Rotenberg, M. (eds.): Atomic and Molecular Scattering: Methods In Computational Physics. Academic Press, New York, London (1971)
2. Katsenelenbaum, B.Z., Mercader del Rio, L., Pereyaslavets, M., Sorolla Ayza, M., Thumm, M.: Theory of Nonuniform Waveguides the Cross-Section Method. The Institution of Electrical Engineers (1998)
3. Brekhovskikh, L.M., Lysanov, Y.P.: Fundamentals of Ocean Acoustics. Springer-Verlag, Berlin (2003)
4. Ramdas Ram-Mohan, L.: Finite Element and Boundary Element Applications in Quantum Mechanics. Oxford University Press, New York (2002)
5. Kantorovich, L.V., Krylov, V.I.: Approximate Methods of Higher Analysis. Wiley, New York (1964)
6. Berezin, I.S., Zhidkov, N.P.: Computing Methods. Pergamon Press, Oxford (1965)
7. Derbov, V.L., et al.: Spectrum of beryllium dimer in ground $X^1\Sigma_g^+$ state. J. Quant. Spectrosc. Radiat. Transf. **262**, 107529–1-10 (2021). https://doi.org/10.1016/j.jqsrt.2021.107529

8. Wen, P.W., et al.: Near-barrier heavy-ion fusion: role of boundary conditions in coupling of channels. Phys. Rev. C **101**, 014618-1-10 (2020). https://doi.org/10.1103/PhysRevC.101.014618
9. Divakov, D.V., Tiutiunnik, A.A., Sevastianov, A.L.: Symbolic-numeric study of geometric properties of adiabatic waveguide modes. In: Boulier, F., England, M., Sadykov, T.M., Vorozhtsov, E.V. (eds.) CASC 2020. LNCS, vol. 12291, pp. 228–244. Springer, Cham (2020). https://doi.org/10.1007/978-3-030-60026-6_13
10. Gusev, A.A., Gerdt, V.P., Hai, L.L., Derbov, V.L., Vinitsky, S.I., Chuluunbaatar, O.: Symbolic-numeric algorithms for solving BVPs for a system of ODEs of the second order: multichannel scattering and eigenvalue problems. In: Gerdt, V.P., Koepf, W., Seiler, W.M., Vorozhtsov, E.V. (eds.) CASC 2016. LNCS, vol. 9890, pp. 212–227. Springer, Cham (2016). https://doi.org/10.1007/978-3-319-45641-6_14
11. Chuluunbaatar, G., Gusev, A.A., Chuluunbaatar, O., Vinitsky, S.I., Hai, L.L.: KANTBP 4M: Program for solving the scattering problem for a system of ordinary second-order differential equations. EPJ Web Conf. **226**, 02008 (2020). https://doi.org/10.1051/epjconf/202022602008
12. Gusev, A.A., et al.: Symbolic-numeric solution of boundary-value problems for the Schrödinger equation using the finite element method: scattering problem and resonance states. In: Gerdt, V.P., Koepf, W., Seiler, W.M., Vorozhtsov, E.V. (eds.) CASC 2015. LNCS, vol. 9301, pp. 182–197. Springer, Cham (2015). https://doi.org/10.1007/978-3-319-24021-3_14
13. Gusev, A.A., et al.: Symbolic-numerical solution of boundary-value problems with self-adjoint second-order differential equation using the finite element method with interpolation hermite polynomials. In: Gerdt, V.P., Koepf, W., Seiler, W.M., Vorozhtsov, E.V. (eds.) CASC 2014. LNCS, vol. 8660, pp. 138–154. Springer, Cham (2014). https://doi.org/10.1007/978-3-319-10515-4_11
14. Chuluunbaatar, O., et al.: KANTBP: A program for computing energy levels, reaction matrix and radial wave functions in the coupled-channel hyperspherical adiabatic approach. Comput. Phys. Commun. **177**, 649–675 (2007). https://doi.org/10.1016/j.cpc.2007.05.016
15. Chuluunbaatar, O., Gusev, A.A., Abrashkevich, A.G., Vinitsky, S.I.: KANTBP 2.0: new version of a program for computing energy levels, reaction matrix and radial wave functions in the coupled-channel hyperspherical adiabatic approach. Comput. Phys. Commun. **179**, 685–693 (2008). https://doi.org/10.1016/j.cpc.2008.06.005
16. Chuluunbaatar, O., Gusev, A.A., Abrashkevich, A.G., Vinitsky, S.I.: KANTBP 3.0: new version of a program for computing energy levels, reflection and transmission matrices, and corresponding wave functions in the coupled-channel adiabatic approach. Comput. Phys. Commun. **185**, 3341–3343 (2014). https://doi.org/10.1016/j.cpc.2014.08.002
17. Gusev A.A., Hai L.L., Chuluunbaatar O., Vinitsky S.I.: KANTBP 4M: program for solving boundary problems of the system of ordinary second order differential equations. JINRLIB (2015). http://wwwinfo.jinr.ru/programs/jinrlib/kantbp4m/indexe.html
18. http://www.maplesoft.com

Blends in MAPLE

Robert M. Corless[1,2(✉)] and Erik J. Postma[3]

[1] School of Mathematical and Statistical Sciences, Western University, London, Ontario, Canada

[2] David R. Cheriton School of Computer Science, University of Waterloo, Waterloo, Ontario, Canada
rcorless@uwo.ca

[3] Maplesoft, Waterloo, Ontario, Canada
epostma@maplesoft.com

Abstract. A *blend* of two Taylor series for the same smooth real- or complex-valued function of a single variable can be useful for approximation of said function. We use an explicit formula for a two-point Hermite interpolational polynomial to construct such blends. We show a robust MAPLE implementation that can stably and efficiently evaluate blends using linear-cost Horner form, evaluate their derivatives to arbitrary order at the same time, or integrate a blend exactly. The implementation is suited for use with **evalhf**. We provide a top-level user interface and efficient module exports for programmatic use.

Keywords: Two-point Hermite interpolants · Blends · MAPLE · Stable and efficient implementation

1 Introduction

Taylor series are one of the basic tools of analysis and of computation for functions of a single variable. However, outside of specialist circles it is not widely appreciated that two Taylor series can be rapidly and stably combined to give what is usually a much better approximation than either one alone. In this paper we only consider blending Taylor series at two points, say $z = a$ and $z = b$. We convert to the unit interval by introducing a new variable s with $z = a + s(b - a)$. Most examples in this paper will just use s, but it is a straightforward matter to adjust back to the original variables, and we will give examples of how to do so.

1.1 The Basic Formula

Consider the following formula, known already to Hermite, which states that the grade $m+n+1$ polynomial

$$H_{m,n}(s) = \sum_{j=0}^{m} \left[\sum_{k=0}^{m-j} \binom{n+k}{k} s^{k+j} (1-s)^{n+1} \right] p_j + \sum_{j=0}^{n} \left[\sum_{k=0}^{n-j} \binom{m+k}{k} s^{m+1} (1-s)^{k+j} \right] (-1)^j q_j \qquad (1)$$

This work partially funded by NSERC.

R. M. Corless et al. (Eds.): MC 2020, CCIS 1414, pp. 167–184, 2021.
https://doi.org/10.1007/978-3-030-81698-8_12

has a Taylor series matching the given $m+1$ values $p_j = f^{(j)}(0)/j!$ at $s=0$ and another Taylor series matching the given $n+1$ values $q_j = f^{(j)}(1)/j!$ at $s=1$. Putting this in symbolic terms and using a superscript (j) to mean the jth derivative with respect to s, we have

$$\frac{H_{m,n}^{(j)}(0)}{j!} = p_j\,, \qquad 0 \le j \le m\,, \qquad \text{and} \qquad \frac{H_{m,n}^{(j)}(1)}{j!} = q_j\,, \qquad 0 \le j \le n\,.$$

This is a kind of interpolation, indeed a special case of what is called *Hermite* interpolation. As with Lagrange interpolation, where for instance two points give a grade one polynomial, that is, a line, here $m+n+2$ pieces of information gives a grade $m+n+1$ polynomial. We will see that this formula can be evaluated in $O(m)+O(n)$ arithmetic operations.

We use the word *grade* to mean "degree at most". That is, a polynomial of grade (say) 5 is of degree at most 5, but because here the leading coefficient is not visible, we don't know the exact degree, which could be lower. Typically, with a blend we will not know the degree unless we compute it. This use of the word "grade" is common in the literature of matrix polynomial eigenvalue problems.

1.2 Applications

Our initial motivation was in writing code in MAPLE to solve the Mathieu differential equation in [5] using a Hermite-Obreschkoff method [17–19], which uses Taylor series at either end of each numerical step; this implicit high-order method is especially suited to differential equations for which the Taylor series at any point may be computed quickly, such as so-called D-finite or *holonomic* functions [10, 14, 15]. Blends can also be used for quadrature (numerical evaluation of definite integrals), or for approximation of functions. We will see examples of that last, in Sect. 3.

Using a companion matrix discussed in [13], we can also find approximate zeros of nonlinear functions from Taylor series data at either end of an interval. This can be turned into an efficient iterative method of order $1+\sqrt{3} > 2$ with the same cost as Newton's method, by using *reversed* Taylor series: see [7]. This method was used, as were blends, in [6] to compute several graphs of nodal lines in an elliptical drum, according to the Mathieu equation.

There are also interesting pedagogical applications. Using blends can strengthen the notion of convergence in students' minds; this can be done at an elementary calculus level or at a real analysis level. The companion matrix mentioned earlier can be used in Linear Algebra as a topic in computing eigenvalues. The derivation of the method is a lovely exercise in contour integration for a course in complex variables. Of course, it provides a topic in approximation theory and in numerical analysis: a proof that the method is numerically stable will be given elsewhere. Interestingly, blends generate an infinite number of different quadrature rules, unusual members of which can be used as unique tools of student assessment.

1.3 Initial Examples

We show an example in Fig. 1. We take the function $f(s) = 1/\Gamma(s-3)$. For a reference on the Gamma function, see [4]. This function has known series at both ends of the interval: at $s = 0$ we have

$$\frac{1}{\Gamma(s-3)} = -6s + (-6\gamma + 11)s^2 + \left(\frac{\pi^2}{2} - 3\gamma^2 + 11\gamma - 6\right)s^3 + O\left(s^4\right)$$

and at $s = 1$ we have

$$\frac{1}{\Gamma(s-3)} = 2(s-1) + (2\gamma - 3)(s-1)^2 + \left(-\frac{\pi^2}{6} + \gamma^2 - 3\gamma + 1\right)(s-1)^3 + O\left((s-1)^4\right),$$

as computed by MAPLE's **series** command[1]. The series coefficients get complicated as the degree increases, so we suppress printing them. We compute them up to degrees $m = 9$ and $n = 9$ and make a blend for this function. This gives a grade $9+9+1 = 19$ approximation (and indeed the blend turns out to be actually degree 19; the lead coefficient does not, in fact, cancel). In the figure, we plot the error $H_{9,9}(s) - f(s)$ and the derivative error $H'_{9,9}(s) - f'(s)$, first in the top row computing the blend in 15 Digits (which takes a third of a second on a 2018 Surface Pro to compute the blend and three of its derivatives at 2021 points, so 8084 values) and then comparing against MAPLE's built-in evaluator (computed at higher precision, in fact 60 Digits because of the apparent end-point singularity, and then rounded correctly to 15 digits). In the second row we compare the blend computed at 30 Digits, which takes 3.14 seconds, about ten times longer than the 15 Digit computation. We see in the second row of the figure that the *truncation error*—that is, the error in approximation by taking a degree 19 polynomial—is smaller than $6 \cdot 10^{-16}$; MAPLE's hardware floats use IEEE double precision with a unit roundoff of $2^{-53} \approx 10^{-16}$. We therefore expect rounding error to dominate if we do computation in only 15 Digits, and that is indeed what we see in the top row—and moreover we see that the rounding error is not apparently amplified very much, if at all: the errors plotted are all modest multiples of the unit roundoff. The unit roundoff itself can be seen in the apparent horizontal lines, in fact. This will be indicative of the general behaviour of a blend: when carefully implemented, rounding errors do not affect it much. Since, as we will see, balanced blends are quite well-conditioned, this will result in usually accurate answers.

To compare with Taylor series and other methods of approximating this particular $f(s)$, an equivalent cost Taylor series would be degree 19. The Taylor series of degree 19 at $s = 0$ has an error at $s = 1^-$ of about $3.5 \cdot 10^{-7}$, many orders of magnitude greater than the error in the blend; the series at $s = 1$ has a similar-sized error at $s = 0^+$. This is well-known: Taylor series are really good at their point of expansion, but will be bad at the other end of the interval. On the other hand, the "best" polynomial approximation to this function, best in the minimax sense and found by the Remez algorithm, is of course better than the blend we produce here. Similarly a Chebyshev approximation to this function, as would be produced by Chebfun [1], is also better: either cheaper for the same accuracy, or more accurate for the same effort. And then there is the new AAA algorithm, which is better still [16], which we do not pursue further here. But where does a blend fit in on this scale of best-to-Taylor? The Chebyshev approximation

[1] In fact, MAPLE can compute the series for $1/\Gamma(z)$ at $z = -n$ where n is a symbol, assumed to be a nonnegative integer. This example will be discussed further in a separate paper.

accurate to $6 \cdot 10^{-16}$ (as computed by `numapprox[chebyshev]` is of degree 16, not 19, so it is about 20% cheaper to evaluate[2]. For the rational Remez best approximation, by `numapprox[minimax]` the degree [7,8] gets an error $3 \cdot 10^{-16}$ and is cheaper yet to evaluate. Conversely, when using a single Taylor series, experiments at high precision show that we need to use degree 29 to get an error strictly less than the error of the $(9,9)$ blend everywhere in the interval, and a degree 28 Taylor series is strictly worse. Therefore both the best approximation and the Chebyshev series are better than a blend—but in this case not by that much, while a blend beats a single Taylor series by a considerable margin. There are other examples where Chebyshev series beat blends by a similarly large margin, but because blends are relatively simple to compute and to understand, being "sometimes in the ballpark" of the best kinds of approximation is likely good enough to make these objects interesting. We are especially interested in situations where Taylor series at either end of an interval are known or very cheap to compute, e.g. for so-called *holonomic* or *D-finite* functions [10, 14, 15]. Note that even there we find that sometimes Chebyshev series are worth the extra effort [3].

2 Truncation Error and Rounding Error

The error in Hermite interpolation is known, see for instance [12]. Here, the general real results simplify to

$$f(s) - H_{m,n}(s) = \frac{f^{(m+n+2)}(\theta)}{(m+n+2)!} s^{m+1}(s-1)^{n+1} \tag{2}$$

for some $\theta = \theta(s)$ between 0 and 1. This is quite reminiscent of the Lagrange form of the remainder of Taylor series, and indeed it reduces exactly to that if we have an $(m,-1)$ or $(-1,n)$ blend—that is, without using any information from the other point. We saw in the high-precision graphs in Fig. 1 that the actual error curve really does flatten out at both ends, when information is known at both ends.

The errors in derivatives have a similar form, and as shown in [12] essentially lose only one order of accuracy for each derivative taken.

The numbers $\binom{n+k}{k}$, for $0 \le k \le m$, and identically $\binom{m+k}{k}$, $0 \le k \le n$, which appear in formula (1), grow large rather quickly. If we made a table of $\binom{m+k}{k}$, the largest entries would be on the diagonal, and indeed

$$\binom{2m}{m} \sim \frac{4^m}{\sqrt{\pi m}}\left(1 + O\left(\frac{1}{m}\right)\right) \tag{3}$$

as we find out from the MAPLE command

```
asympt( binomial(2*m,m), m )
```

[2] This is harder to judge than we are saying, here. Optimal evaluation of Chebyshev polynomials via preprocessing is not usually done; the Clenshaw algorithm is backward stable (see e.g. [8]) and usually used because it is $O(n)$ in cost. Similarly, evaluation of a blend is $O(m+n)$. So this figure of 20% is likely not very true, but rather merely indicative.

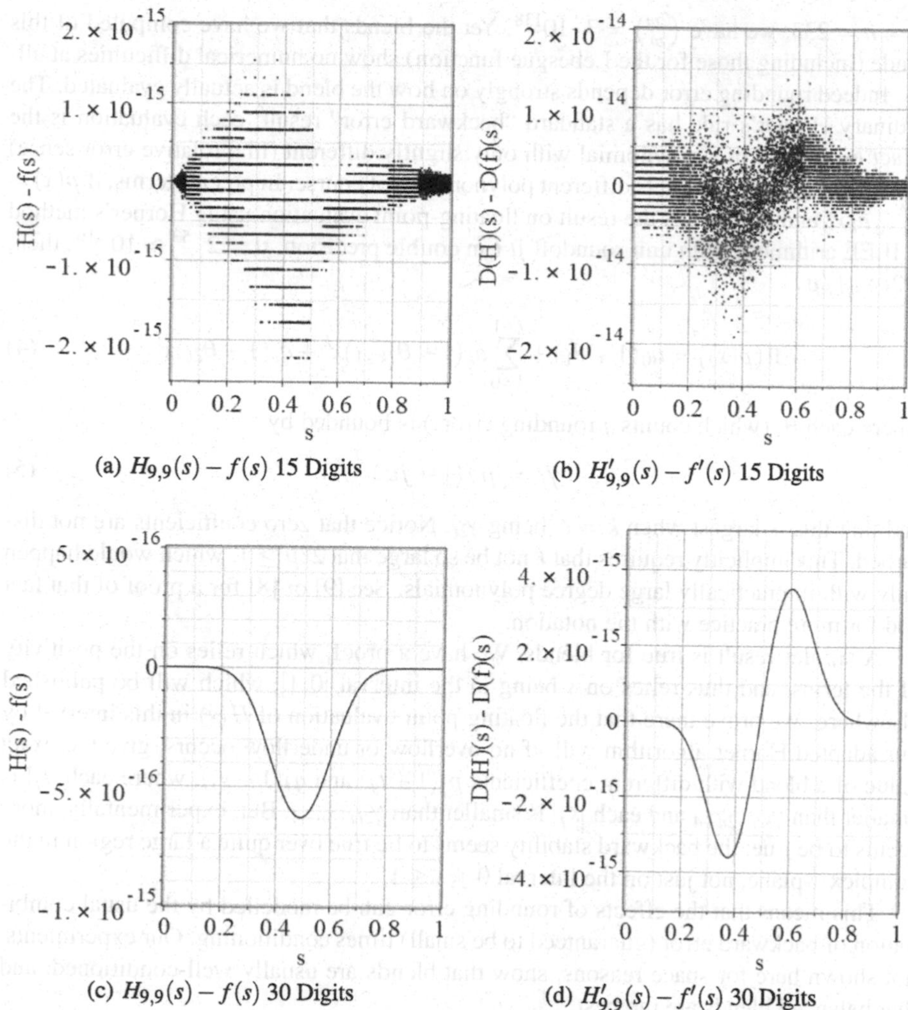

(a) $H_{9,9}(s) - f(s)$ 15 Digits

(b) $H'_{9,9}(s) - f'(s)$ 15 Digits

(c) $H_{9,9}(s) - f(s)$ 30 Digits

(d) $H'_{9,9}(s) - f'(s)$ 30 Digits

Fig. 1. The error in an $(9,9)$ blend for $f(s) = 1/\Gamma(s-3)$. This grade of blend produces an approximation that is nearly accurate to full double precision; when the correctly rounded reference result is subtracted from the computed result, the truncation error is obscured by rounding errors. As is shown in the second row of graphs, recomputing these approximations at higher precision gives smoother error curves of about the same size and showing the theoretical $s^{10}(1-s)^{10}$ behaviour. As usual with approximation methods, the accuracy degrades as the derivative order increases.

and some simplification. One worries about the numerical effect of these large numbers for high-degree blends. These *do* have some bad numerical effects sometimes, such as in the companion matrices of [13], but we will also see by the example of high-degree blends that their influence is not as bad as it might have been feared. For instance, when

$m = n = 233$, we have $\binom{2m}{m} = 7 \cdot 10^{138}$. Yet the blends that we have computed of this grade (including those for the Lebesgue function) show no numerical difficulties at all.

Indeed rounding error depends strongly on how the blend is actually evaluated. The ordinary Horner's rule has a standard "backward error" result: each evaluation is the *exact* evaluation of a polynomial with only slightly different (in a relative error sense) coefficients. Different x gets different polynomials, of course. In precise terms, if $p(x) = \sum_{k=0}^{\ell} a_k x^k$ and "fl" means the result on floating-point evaluation using Horner's method in IEEE arithmetic with unit roundoff μ (in double precision, $\mu = 2^{-53} \approx 10^{-16}$) then, if $2\ell < 1/\mu$,

$$\mathrm{fl}\,(p(x)) = a_0(1+\theta_1) + \sum_{k=0}^{\ell-1} a_k(1+\theta_{2k+1})x^k + a_\ell(1+\theta_{2\ell})x^\ell \tag{4}$$

where each θ_j (which counts j rounding errors) is bounded by

$$\gamma_j = j\mu/(1-j\mu) \tag{5}$$

and that this is largest when $k = \ell$, being $\gamma_{2\ell}$. Notice that zero coefficients are not disturbed. This implicitly requires that ℓ not be so large that $2\ell\mu \geq 1$, which would happen only with impractically large degree polynomials. See [9] or [8] for a proof of that fact and for more practice with the notation.

A similar result is true for blends. We have a proof, which relies on the positivity of the terms, and thus relies on s being in the interval $[0,1]$, which will be published elsewhere. We prove there that the floating-point evaluation of $H(s)$ in this interval by our adapted Horner algorithm will, if no overflow or underflow occurs, give the *exact* value of a blend with different coefficients $p_j(1+t_j)$ and $q_j(1+s_j)$, where each $|t_j|$ is smaller than $\gamma_{3m+2n+4}$ and each $|s_j|$ is smaller than $\gamma_{2m+3n+4}$. But, experimentally, more seems to be true: the backward stability seems to be true over quite a large region in the complex s-plane, not just on the interval $0 \leq s \leq 1$.

This means that the effects of rounding error can be modelled by the usual combination of backward error (guaranteed to be small) times conditioning. Our experiments, not shown here for space reasons, show that blends are usually well-conditioned, and that balanced blends are the best.

2.1 Conditioning and the Lebesgue Function

One common measure of the numerical behaviour of a polynomial expression in a given basis is the so-called *Lebesgue* function of the basis: this is defined to be what you would get if absolute values are taken of each term multiplying a coefficient, and moreover all coefficients are also replaced by 1. More formally, if we expand $f(z)$ using the basis $\phi_k(z)$ for $0 \leq k \leq n$, so that

$$f(z) = \sum_{k=0}^{n} a_k \phi_k(z), \tag{6}$$

then

$$|f(z)| = \left|\sum_{k=0}^{n} a_k \phi_k(z)\right| \leq \max_{0\leq j\leq n} |a_j| \sum_{k=0}^{n} |\phi_k(z)| \tag{7}$$

and if we write $L(z) = \sum_{k=0}^{n} |\phi_k(z)|$ and call it the Lebesgue function for the basis $\phi_k(z)$ then the absolute value of $f(z)$ is bounded by the infinity norm of the vector of coefficients of $f(z)$ times the Lebesgue function. This is simply the Hölder inequality applied to the expression for $f(z)$.

In our case we can see that on the interval $0 \le s \le 1$ all terms in the first series, with p_j, are positive anyway; in the second term, we may make everything positive by choosing $q_j = (-1)^j$. Outside that interval, the absolute values are needed.

$$L_{m,n}(s) = \sum_{j=0}^{m} \left| \sum_{k=0}^{m-j} \binom{n+k}{k} s^{k+j} (1-s)^{n+1} \right| + \sum_{j=0}^{n} \left| \sum_{k=0}^{n-j} \binom{m+k}{k} s^{m+1} (1-s)^{k+j} \right| \tag{8}$$

Thus the Lebesgue function for our blend is (inside $0 \le s \le 1$) a blend itself, for a function with the same Taylor series at $s=0$ as $1/(1-s)$, and the same Taylor series at $s=1$ as $1/s = 1/(1+(s-1))$. There is no function analytic everywhere with those properties, of course, but nonetheless these polynomials are useful. Having a small size of L is a guarantee of good numerical behaviour, if one implements things carefully. Here, for the balanced case $m=n$, one can show that *inside the interval* $1 \le L_{m,m}(s) \le 2$, no matter how large m is. If m and n are large but not balanced, then we can have $L_{m,n}(s)$ as large as the maximum of m and n. See Fig. 2.

Outside the interval $0 \le s \le 1$ the Lebesgue function grows extremely rapidly: not exponentially fast, but like a degree $m+n+1$ polynomial in $|s|$. This essentially guarantees that blends are typically useful numerically only between the two endpoints and in a small region in the complex plane surrounding that line segment; that is, where $L(s)$ remains of moderate size. By refining this argument somewhat, we may do better for certain polynomials by taking better account of the polynomial coefficients through the theory of conditioning, see [8]. We do not pursue this further here.

3 Integration of a Blend

We will now see that the *definite* integral of a blend over the entire interval will allow us to construct a new blend whose value at any point is the *indefinite* integral of the original blend up to that point, $F(x) = \int_{s=0}^{x} H_{m,n}(s)\, ds$.

Direct integration over the entire interval $0 \le s \le 1$ and use of the formula

$$\int_{s=0}^{1} s^a (1-s)^b \, ds = \frac{a!b!}{(a+b+1)!}$$

gets us a formula for $F(1)$ involving the symbolic sum

$$\sum_{k=0}^{m-j} \frac{\binom{n+k}{k} (j+k)!\,(n+1)!}{(n+2+j+k)!} \tag{9}$$

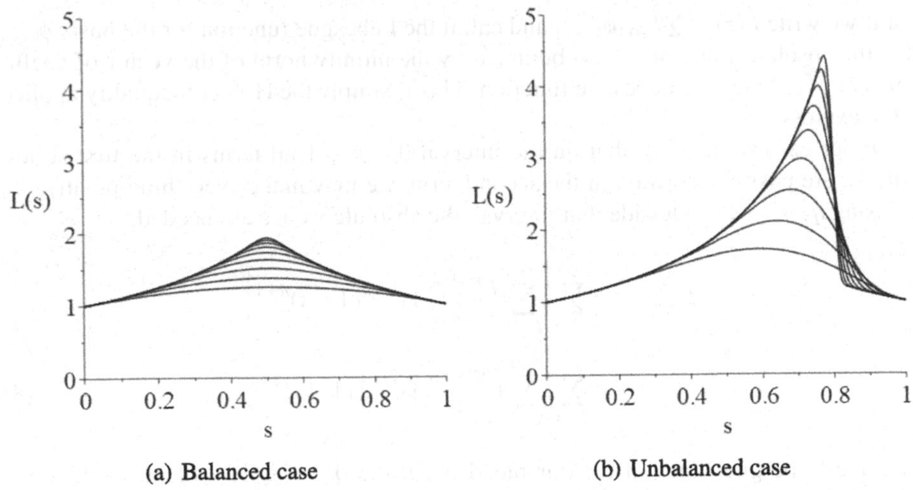

(a) Balanced case (b) Unbalanced case

Fig. 2. Lebesgue functions for $(m,n) = 2^k$, $0 \le k \le 8$ (balanced case) and $(m,n) = (4\cdot 2^k, 2^k)$, $0 \le k \le 8$ (unbalanced case). We see that in the balanced case, errors will be amplified by at most a factor of two; in the unbalanced case, it can be more, depending on the degree of unbalancing, but never more than the maximum of m and n.

and a similar one interchanging m and n. MAPLE can evaluate both those sums:

```
sm := sum( binomial(n+k,k)*(n+1)!*(k+j)!/(j+k+n+2)!, k=0..m-j ):
simplify( sm );
```

yields the right-hand side of the equation below:

$$\sum_{k=0}^{m-j} \frac{\binom{n+k}{k}(j+k)!\,(n+1)!}{(n+2+j+k)!} = \frac{(n+m-j+1)!\,(1+m)!}{(j+1)\,(n+2+m)!\,(m-j)!}. \tag{10}$$

Similarly we find the other sum, and finally we get

$$\int_{s=0}^{1} H_{m,n}(s)\,ds = \frac{(m+1)!}{(m+n+2)!}\sum_{j=0}^{m}\frac{(n+m-j+1)!}{(j+1)\,(m-j)!}\,p_j + \frac{(n+1)!}{(m+n+2)!}\sum_{j=0}^{n}\frac{(n+m-j+1)!}{(j+1)\,(n-j)!}\,(-1)^j q_j\,. \tag{11}$$

The numbers showing up in this formula turn out to be smaller for the higher-order Taylor coefficients, as one would expect. We emphasize that the above formula gives (in exact arithmetic) the *exact* integral of the blend over the whole interval. If the blend is approximating a function $f(s)$, then integrating Eq. (2) gives us

$$\int_{s=0}^{1} f(s)\,ds - F(1) = (-1)^{n+1}\frac{(m+1)!(n+1)!}{(m+n+3)!}\frac{f^{(m+n+2)}(c)}{(m+n+2)!} \tag{12}$$

where, using the Mean Value Theorem for integrals and the fact that $s^{m+1}(1-s)^{n+1}$ is of one sign on the interval, we replace the evaluation of the derivative at one unknown point θ with another unknown point c on the interval.

Once we have the value $F(1)$, we can construct a new blend from the old one as follows. First, we put a value of 0 for the new $F(0)$ at the left end (in a string of blends, we would accumulate integrals; for now, we are just integrating from the left end). Then we adjust all the Taylor coefficients at the left: the old $f(0)/0!$ becomes the new $F'(0)/1!$, the old $f'(0)/1!$ becomes the new $F''(0)/2!$ and so we have to divide the old p_1 by 2; the old $f''(0)/2!$ becomes the new $F'''(0)/3!$ so we have to divide by 3, and so on until the old $f^{(m)}(0)/m!$ becomes the new $F^{(m+1)}(0)/(m+1)!$; the new blend will have $m+2$ Taylor coefficients on the left (indexing starts at 0).

Now we make $F(1)$ = the integral given above. We then shift all the old $q_j = f^{(j)}(1)/j!$ into the new $F^{(j+1)}(1)/(j+1)!$ for $j = 0, \ldots, n$.

We now have a type $(m+1, n+1)$ blend $H_{m+1,n+1}(s)$. Its Taylor coefficients on the left are the same as the Taylor coefficients of $F(x) = \int_{s=0}^{x} H_{m,n}(s)\,ds$ as a function of x. Its Taylor coefficients on the right are also the same as those of $F(x)$ at $x = 1$. Thus we have a blend for the integral. Its grade is $m+1+n+1+1$ which is $m+n+3$, not $m+n+2$. However, in exact arithmetic, the result is actually of degree at most $m+n+2$, because the value is the exact integral of a polynomial, and thus we see that the blend we have is actually using more information than it needs. We *could* throw one of the highest derivatives away (it's natural to do so at the right end) but there is no real need unless we expect to do this process repeatedly to a single blend.

To use this formula on integration from $z = a$ to $z = b$ one must incorporate the change of variable from z to $s = (z-a)/(b-a)$. Putting $h = (b-a)$ then we must (as always) scale the Taylor coefficients p_j and q_j by multiplying each by h^j, and then finally the integral is just

$$\int_{z=a}^{b} H_{m,n}\left(\frac{z-a}{b-a}\right) dz = h \int_{s=0}^{1} H_{m,n}(s)\,ds\,. \tag{13}$$

The case $m = n = 0$ just gives the trapezoidal rule, which is right because the blend is just a straight line; if instead $m = n = 1$ then we get what is called the "corrected trapezoidal rule"

$$\int_{z=a}^{b} H_{1,1}\left(\frac{z-a}{b-a}\right) dz = \frac{h}{2}\left(f(a)+f(b)\right) + \frac{h^2}{12}\left(f'(b)-f'(a)\right). \tag{14}$$

A $(4,4)$ blend gives the rule

$$\begin{aligned}\int_{s=0}^{1} H_{4,4}(s)\,ds = & \frac{p_0}{2} + \frac{p_1}{9} + \frac{p_2}{36} + \frac{p_3}{168} + \frac{p_4}{1260} \\ & + \frac{q_0}{2} - \frac{q_1}{9} + \frac{q_2}{36} - \frac{q_3}{168} + \frac{q_4}{1260}\end{aligned} \tag{15}$$

To get a valid rule on an interval of width h, one needs powers of h in the Taylor series. We see that this balanced blend gives coefficients that will telescope at odd orders for composite rules on equally-spaced intervals. [This is well-known.] See also [20] for optimal formulas of this balanced type.

If we have more than one blend lined up in a row, which we call a "string of blends" (this is quite natural, as can be seen from the fact that blends are joined at what are termed "knots" in the spline and piecewise polynomial literature), then this formula can be used to generate composite quadrature rules. For example, in the repository https://github.com/rcorless/Puiseux-series-Mathieu-double-points you can find the workbook `MathieuTalk.maple` where the following integral is computed:

$$\int_0^{\pi} w_1^2(z)\,dz \doteq -2.9\times 10^{-15} + 5.780\times 10^{-15} i .$$

Here $w_1(z)$ is a complex-valued eigenfunction of the Mathieu differential equation, at a *double eigenvalue*. The function $w_1(z)$ is represented as a string of nine blends, with Taylor series of order 9 at each end (so the blends are of grade 17). We multiply the blend for $w_2(z)$ with itself by using the Cauchy product of the Taylor series at each knot. When we integrate the result by the method outlined above, we get zero (to numerical accuracy), which is correct.

4 Horner Form

If we look at Eq. (1) with a programmer's eye, we see a lot of room for economization. First, the sums are polynomials in s and in $1-s$. Because $0 \le s \le 1$, both of these terms are positive, so we do not want to expand powers of $(1-s)$, for instance; introducing subtraction means potentially revealing rounding errors made earlier. But as a first step we may put the sums in Horner form. We remind you that the *Horner form* of a polynomial $f(x) = f_0 + f_1 x + f_2 x^2 + f_3 x^3$ is a rewriting so that no powers occur, only multiplication: $f(x) = f_0 + x(f_1 + x(f_2 + x f_3))$. The form can be programmed in a simple loop:

```
p := f[n];
for j from n-1 by -1 to 0 do
  p := f[j] + x*p;
end do;
```

Here we have a double sum, and in each sum we may write in Horner form; that is, where the loop above has a simple `f[j]` we would have an inner Horner loop to compute it.

But the inner sum is simply $\sum_{k=0}^{m-j}\binom{n+k}{k}s^k$ once the $s^j(1-s)^{n+1}$ is factored out of it. These inner sums should be precomputed by the simple recurrence (adding the next term to the previous sum), outside of the innermost loop, so that the cost is proportional to either n or m, and not their product.

The numbers $\binom{m+k}{k}$ and $\binom{n+k}{k}$ occur frequently, and perhaps they should be precomputed. Except that they, too, can be split in a Horner-like fashion, because for $k \ge 1$

$$s^k\binom{m+k}{k} = s\frac{m+k}{k}\cdot s^{k-1}\binom{m+k-1}{k-1} .$$

While this is actually more expensive than precomputing the numbers, by keeping s involved, the loop keeps the size of the numbers occurring in the formula small (remember $0 \le s \le 1$), and this contributes to numerical stability. This is best seen by example.

In the $m = n = 3$ case, one of the terms is

$$1 + 4s + 10s^2 + 20s^3$$

which rewritten in Horner form is just $1 + s(4 + s(10 + s \cdot 20))$. But if we factor out the binomial coefficient factors using the rule above, it becomes

$$1 + 4s\left(1 + \frac{5}{2}(1 + 2s)s\right).$$

It might be better to keep only integers in the rewritten form; we do not know how to do that in general, although it is simple enough for this example.

A final and important efficiency is to realize that the sum for the left-hand terms and the sum for the right-hand terms is invariant under a symmetry: exchange m and n, exchange s and $1 - s$, and account for sign changes in the second sum by absorbing them into the q_j, and the sums can be executed by the same program. This leads to later *programmer* efficiency as well, if one thinks of a further improvement to the code: then it only has to happen in one place. [This actually happened here.] We give the algorithm for this half-sum in Algorithm 1. To compute the blend, this algorithm is called once with m, n, and s and the coefficient vector p_j, and once with n, m (note the reverse order), $1 - s$, and the coefficient vector $(-1)^j q_j$, and the results are added.

The goal is to make the innermost loop as efficient as is reasonably possible. We expect that these blends will be evaluated with hundreds of points (routinely) and on occasion with tens of thousands of points (for a tensor product grid of a bivariate function, for instance). In MAPLE, we want to be able to use **evalhf** or even the compiler. This provides significant speedup.

4.1 "Automatic" Differentiation

The Horner loop above can be rewritten to provide not only the value of $p(x)$ but also of $p'(x)$, the derivative with respect to x. This is also called program differentiation. MAPLE's **D** operator can differentiate simple programs such as that. Supposing we define

```
Horner := proc(x, f, n)
  local i, p;
  p := f[n];
  for i from n-1 by -1 to 0 do
    p := f[i] + x*p;
  end do;
  return p;
end proc:
```

Then the command `D[1](Horner)` produces a procedure returns just the derivative, not the derivative and the polynomial value. If one wishes that, one may use instead `codegen[GRADIENT]`, with the syntax

```
codegen[GRADIENT](Horner, [x], function_value = true)
```

Algorithm 1. Horner's algorithm adapted for one of the two sums of the blend.

procedure HSF(m, n, σ, w)
  $a_0 \leftarrow 1$
  for $k \leftarrow 1 \ldots m$ **do**
    $a_k \leftarrow (n+k)\sigma a_{k-1}/k$
  end for
  for $k \leftarrow 1$ to m **do**
    $a_k \leftarrow a_{k-1} + a_k$
  end for
  $u \leftarrow 0$
  for $j \leftarrow m$ by -1 to 0 **do**
    $u \leftarrow a_{m-j} w_j + \sigma u$
  end for
  $c \leftarrow 1$
  for $j \leftarrow 1$ to $n+1$ **do**
    $c \leftarrow (1-\sigma)c$
  end for
  $e \leftarrow cu$
  return e
end procedure

Procedures for evaluating higher-order derivatives may be computed in a similar way.

For our purposes, though, it is better to allow an *arbitrary* number `nder` of derivatives. This means not adding one or more statements to the Horner loop, but rather writing a loop to evaluate all the derivatives. Here is this idea applied to the Horner program above.

```
Horner := proc(x, f, n, nder)
  local i, ell, p;
  p := Array(0..nder,0);
  p[0] := f[n];
  for i from n-1 by -1 to 0 do
    for ell from nder by -1 to 1 do
      p[ell] := p[ell]*x + ell*p[ell-1];
    end do;
    p[0] := f[i] + x*p[0];
  end do;
  return p;
end proc:
```

But the strength of this technique is not for symbolic use, but rather for numeric use. When calling the modified program with a numeric x then the loop just performs (reasonably efficient) numerical computation; this program can be translated into other languages, as well.

For the code for our blends, we simply wrote all the loops ourselves as above. We have not yet tried to translate the resulting code (which is more complicated than the simple Horner loop above) into any other languages.

4.2 User Interface Considerations

There is similar code in the `Interpolation` package and in the `CurveFitting` package, namely `Spline` and `ArrayInterpolation`. The interface to this code should not be too much different to those. Consideration of the various possible kinds of inputs demonstrates that a front-end that dispatches to the most appropriate routine would be helpful; if the input s is a symbol, then there is no point in calling **`evalhf`**, for instance. If the input is an **`Array`** of complex floating-point numbers, then depending on Digits it might indeed be appropriate to try the hardware float routine.

For that reason we chose a **`module`** with `ModuleApply` as being most convenient; this would allow the user to be relatively carefree. We also allowed the module to export the basic 'fast' routines so that if the user wanted to look after the headaches of working storage of hardware floating point datatypes then the user could use blends in their own code without a significant performance penalty.

The minimum information that the routine needs is z, a, b, and the Arrays p and q of Taylor coefficients. If the user does not request a number of derivatives, it can be safely assumed that only $H(z)$ is wanted. The grade (m,n) of the blend can be deduced from the input Arrays p and q. As a convenience to the user we allow the ability to specify m or n even if the input Arrays are larger.

The types of data input can vary considerably. We allow rationals, exact numerics, software floats, hardware floats, and complex versions of all of those. We do not provide for finite fields (the binomial coefficients would in some cases then possibly be zero—and we don't even know if formula (1) is even true over finite fields—in that case) or for matrix values although for that latter case the concept is well-defined.

The data type of the output can vary, as well: when there is an Array of inputs, and only function values are wanted and no derivatives, the user would surely expect an Array of outputs of the same dimension. If derivatives are wanted, though, then there will be a higher-dimensional Array output; sometimes the special case of an index 0 for such a higher-dimensional output would fit the user's expectations so we allow an option to specify such. The default is just to be sensible: scalar in, scalar out; Vector in, Vector out.

Currently several operations take place outside the code, in "main MAPLE". This includes series manipulations and the construction of the companion matrix pair. Construction of the integrated blend is also currently left in the user's hands.

5 Testing and Timing

Tests show the computing time depends linearly on $m+n$. We first used the MAPLE **`rand`** function to generate coefficients for the maximum m and n. Subsequent calls to `Blend` used subsets of those data. The blends were evaluated in 15 Digit precision at 2021 points equally-spaced on $0 \le s \le 1$ including the endpoints. The code was asked to compute derivatives up to order 3. That is, four quantities were computed at each point: $H_{m,m}(s)$, $H'_{m,m}(s)$, $H''_{m,m}(s)$, and $H'''_{m,m}(s)$. The computing time was modest and showed linear growth, with a fit of $0.023m$ to its data (in seconds). Thus the computing time seems, as expected, linear in the degree of the balanced blend. We ran a further test with the same coefficients but this time without asking for derivatives; the cost (not shown) was a factor 4.2 less. In both cases the main call was used, so these times include the times for preparation and dispatch to **evalhf**.

For testing stability and accuracy, we first looked at very smooth functions. In Fig. 3 we see error curves at 15 Digits for $(8,8)$ blends for $f(s) = \cos \pi s$ and its derivatives. This shows the effects, scaled with the appropriate power of π, of taking the derivative. This function has known Taylor series at each end (indeed the coefficients are just the negatives of each other): $\cos \pi s = 1 - (\pi s)^2/2! + (\pi s)^4/4! - \cdots$ and at the other end $\cos \pi s =$

$$-1 + \frac{\pi^2}{2}(s-1)^2 - \frac{\pi^4}{24}(s-1)^4 + \frac{\pi^6}{720}(s-1)^6 - \frac{\pi^8}{40320}(s-1)^8 + O\left((s-1)^{10}\right).$$

The $(9,9)$ blends are better—and use essentially the same information because the Taylor coefficients of degree 9 at either end are zero—but these curves are informative about the numerical stability and efficiency of these blends.

We then chose a harder example. We consider the results of a "stress test", namely a blend for the function $f(s) = \exp(-1/s)$. This has all its right-sided derivatives at $s = 0^+$ being zero, but the function is not analytic there, and indeed has an essential singularity there. At the other end, $s = 1$, we use MAPLE's symbolic-order differentiation capability [2] `diff(exp(-1/x),x$k)` to find that, for $k \geq 1$,

$$\left.\frac{d^{(k)}f}{dx^k}\right|_{x=1} = (-1)^{k+1} e^{-1/2} \text{WhittakerM}\left(k, \frac{1}{2}, 1\right) = (-1)^{k+1} e^{-1} F\left(\left.\begin{matrix} 1-k \\ 2 \end{matrix}\right| 1\right). \quad (16)$$

Here F represents `hypergeom([1-k], [2], 1)`. For $k = 0$ one uses the same formula but adds 1. MAPLE knows how to evaluate these; they are rational multiples of $\exp(-1)$. It is amusing to note that apart from sign, the first 5 are just $\exp(-1)/k!$, but the degree 5 term is $-19\exp(-1)/5!$: only computing to degree 4 could have led to a false experimental conclusion! This formula allows us to compute as many series coefficients at $s = 1$ as we could wish. We take $n = 900$, and $m = 100$, giving a grade 1001 blend. Indeed only the q part of formula (1) is present, so the blend is actually degree 1001 not just grade 1001. The largest binomial coefficient appearing is $\binom{1000}{100}$ which is about $6.4 \cdot 10^{139}$ which suggests that numerical difficulties are to be expected. None, however, appear. The blend is entirely smooth, and the difference between the blend and $f(s)$ is no more than 10^{-5} at its greatest. One expects that the blends will converge as (m,n) go to infinity, for any fixed ratio of m and n. Here because the ratio was $9/10$ we find the maximum error occurring near $s = 0.1$ (about $s = 0.095$).

It is natural to compare with the pure Taylor series at $s = 1$, both of degree 900 and of 1001. The errors at $s = 0$ are, naturally, far larger, because that series diverges there. The degree 900 polynomial has error -0.0558, while the degree 1001 polynomial has error 0.0576. The blend wins very handily.

We now give another stress test, this one (finally) showing some numerical failure (overflow and underflow, resulting in NaNs, or floating-point (Not A Number)s). We blend the step function $f(s) = -1$ at $s = 0$ with all derivatives zero and $f(s) = 1$ at $s = 1$ with all derivatives zero. Depending on the ratio of m and n, the step will be located somewhere between; near $s = (m+1)/(m+n+2)$ in fact. The Lebesgue function is

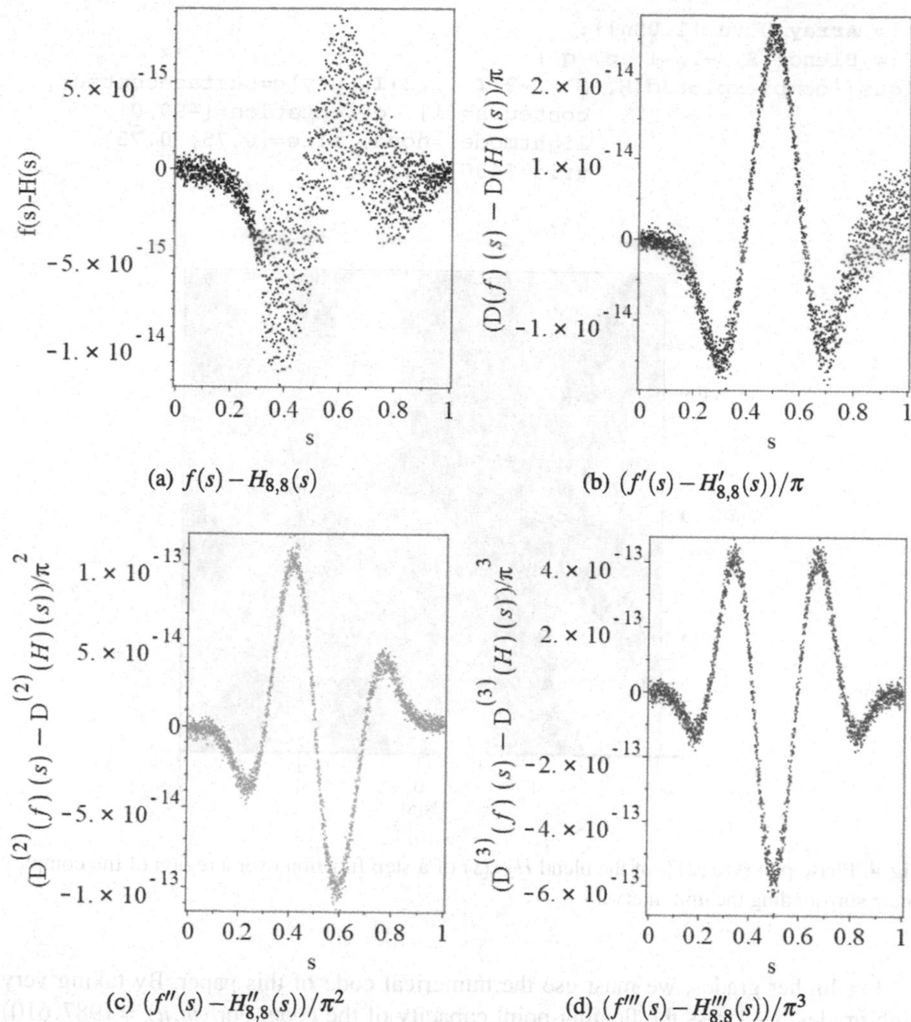

(a) $f(s) - H_{8,8}(s)$

(b) $(f'(s) - H'_{8,8}(s))/\pi$

(c) $(f''(s) - H''_{8,8}(s))/\pi^2$

(d) $(f'''(s) - H'''_{8,8}(s))/\pi^3$

Fig. 3. The error in an $(8,8)$ blend for $f(s) = \cos \pi s$. This grade of blend produces an approximation that is nearly accurate to full double precision; the truncation error, proportional to $s^9(1-s)^9$, is beginning to be obscured by rounding errors. Recomputing these errors at higher precision gives smoother curves of about the same size. As usual with approximation methods, the accuracy degrades as the derivative order increases.

maximal at that point, with value $(\max(m,n)+1)/(\min(m,n)+1)$. In Fig. 4 we see a phase plot in the complex plane of a modest $(3,5)$ blend for this function; this was symbolically computed and plotted with no difficulty using the code below.

```
Digits := 15:
(m,n)  := (3,5):
p := Array(0..m,[-1,0$m]):
```

```
q := Array(0..n,[1,0$n]):
H := Blend( S, -1, 1, p, q ):
plots:-complexplot3d(H, S = -2-I .. 2+I, style=surfacecontour,
                     contours=[1], orientation=[-90,0],
                     lightmodel=none, size=[0.75, 0.75],
                     grid=[400,400] );
```

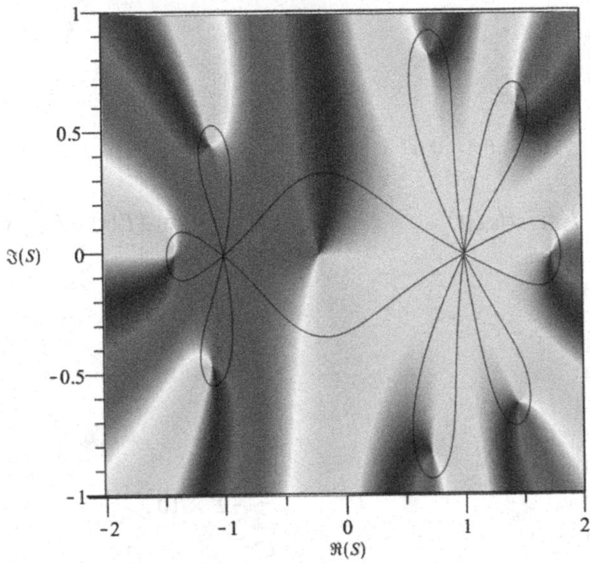

Fig. 4. Phase plot (see [21]) of the blend $H_{3,5}(s)$ of a step function over a region of the complex plane surrounding the unit interval.

For higher grades, we must use the numerical code of this paper. By taking very high grades, we stress the floating-point capacity of the code. For $(m,n) = (987,610)$ the largest binomial coefficient is about $3.5 \cdot 10^{459}$ which must *overflow* in IEEE double precision, which is used by **evalhf**. The corresponding powers of s and $1-s$ may underflow. In spite of that, the blend correctly computes (taking 7.7s CPU time) the step portion of the plot: overflow and underflow causing NaNs only happen in the flat portions of the blend. Computing instead in 30 Digits (which takes about 70 seconds on the same machine) does not suffer from overflow or underflow because software floats in MAPLE have a greater range. At this precision, MAPLE computes the complete figure (not shown). Moreover, comparing the numerical values computed at 15 Digits to the values computed at 30 Digits, we find that the largest difference is smaller than $7 \cdot 10^{-14}$. Working at 15 Digits, the blend was able to correctly compute the interesting part of the curve, even though overflow/underflow prevented it from computing the flat parts. The derivative was computed in 15 Digits correct to 10^{-11} in the same region the function was computed correctly.

We conclude that blends of degrees high enough to produce binomial coefficients $\binom{m+n}{n}$ that overflow will cause numerical difficulty. What seems surprising is that this is the only case where we have seen numerical difficulty. We have also looked at cases where the function has nearby complex poles and so the series cannot converge, and while the blends do have unexpected features in the regions between the two points of expansion of the series, in all cases they behaved smoothly. Even when we tried series for functions whose Taylor series are known to be ill-conditioned (such as $\exp(-50s)$ blending with $\exp(150(s-1)/\pi)$ the resulting behaviour was explainable.

6 Future Work

The idea of blending two Taylor series is quite old, and people have tried to do it in several different ways. The Hermite interpolation idea is one of the oldest, but we think that not enough attention has been paid to it. There are other ways in the literature. For example, there is the *very* similar work in [11], which makes a kind of blend with a variable upper limit and uses that to construct *rational* approximations.

The next step of course is to combine different blends into what we call a string of blends, joined at "knots" where the same local Taylor series are reused. This is a kind of *piecewise polynomial*, similar to splines which are another kind of piecewise polynomial. Of course having a single blend use more than two truncated Taylor series is just Hermite interpolation. There are a great many other similar ideas in the literature.

We would like to extend this code to vector and matrix blends. We would also like to blend Laurent and Puiseux series (Laurent series seem, in fact, very simple: just do a blend of the Taylor series for $(z-a)^{\alpha}(z-b)^{\beta}f(z)$—but we haven't tried this yet). Creating an environment where one can add, subtract, multiply, and apply other operations to blends and produce new blends, might be of interest, in a way similar to Chebfun (www.chebfun.org). It is in such an environment where the companion pair and the integration of blends would fit most naturally.

Extending this work to the multivariable case (aside from the use of tensor product grids) may also be of interest. However, the residue argument breaks down for finding formulas; other approaches will have to be used. There is a significant literature on low-degree multivariate Hermite interpolation, and in the context of finite element methods there is interest in higher-degree methods (and considerable published work) as well.

Acknowledgements. We thank Jürgen Gerhard for coordinating the reviewing, and the referees for their comments.

References

1. Battles, Z., Trefethen, L.: An extension of Matlab to continuous functions and operators. SIAM J. Sci. Comput. **25**(5), 1743–1770 (2004)
2. Benghorbal, M., Corless, R.M.: The *n*th derivative. ACM SIGSAM Bull. **36**(1), 10–14 (2002)
3. Benoit, A., Joldeş, M., Mezzarobba, M.: Rigorous uniform approximation of D-finite functions using Chebyshev expansions. Math. Comput. **86**(305), 1303–1341 (2017)
4. Borwein, J.M., Corless, R.M.: Gamma and factorial in the monthly. Am. Math. Mon. **125**(5), 400–424 (2018)

5. Brimacombe, C., Corless, R.M., Zamir, M.: Computation and applications of Mathieu functions: a historical perspective. https://arxiv.org/abs/2008.01812 (2020)
6. Corless, R.M.: Pure tone modes for a 5:3 elliptic drum. https://arxiv.org/abs/2008.06936 (2020)
7. Corless, R.M.: Sobre la iteración cúbica inversa. La Gaceta de la RSME **24**, 147–159 (2021). English version at https://arxiv.org/abs/2007.06571
8. Corless, R.M., Fillion, N.: A Graduate Introduction to Numerical Methods. Springer, New York (2013). https://doi.org/10.1007/978-1-4614-8453-0
9. Higham, N.J.: Accuracy and Stability of Numerical Algorithms, 2nd edn. SIAM, Philadelphia (2002)
10. van der Hoeven, J.: Fast evaluation of holonomic functions near and in regular singularities. J. Symb. Comput. **31**(6), 717–744 (2001)
11. Hummel, P., Seebeck, C., Jr.: A generalization of Taylor's expansion. Am. Math. Mon. **56**(4), 243–247 (1949)
12. Kansy, K.: Elementare fehlerdarstellung für ableitungen bei der Hermite-interpolation. Numerische Mathematik **21**(4), 350–354 (1973)
13. Lawrence, P.W., Corless, R.M.: Numerical stability of barycentric Hermite root-finding. In: Proceedings of the 2011 International Workshop on Symbolic-Numeric Computation (SNC 2011), pp. 147–148. ACM, New York (2011). https://doi.org/10.1145/2331684.2331706
14. Mezzarobba, M.: NumGfun: a package for numerical and analytic computation with D-finite functions. In: Proceedings of the 2010 International Symposium on Symbolic and Algebraic Computation, pp. 139–145 (2010)
15. Mezzarobba, M.: A note on the space complexity of fast D-finite function evaluation. In: Gerdt, V.P., Koepf, W., Mayr, E.W., Vorozhtsov, E.V. (eds.) CASC 2012. LNCS, vol. 7442, pp. 212–223. Springer, Heidelberg (2012). https://doi.org/10.1007/978-3-642-32973-9_18
16. Nakatsukasa, Y., Sète, O., Trefethen, L.N.: The AAA algorithm for rational approximation. SIAM J. Sci. Comput. **40**(3), A1494–A1522 (2018)
17. Nedialkov, N., Jackson, K.: An interval Hermite-Obreschkoff method for computing rigorous bounds on the solution of an initial value problem for an ordinary differential equation. Reliable Comput. **5**(3), 289–310 (1999)
18. Nedialkov, N., Pryce, J.: Solving differential-algebraic equations by Taylor series (I): computing Taylor coefficients. BIT Numer. Math. **45**(3), 561–591 (2005)
19. Nedialkov, N., Pryce, J.: Solving differential-algebraic equations by Taylor series (II): computing the system Jacobian. BIT Numer. Math. **47**(1), 121–135 (2007)
20. Talvila, E.: Higher order corrected trapezoidal rules in Lebesgue and Alexiewicz spaces. J. Class. Anal. **8**(1), 77–90 (2016)
21. Wegert, E.: Visual Complex Functions: An Introduction with Phase Portraits. Springer, Basel (2012). https://doi.org/10.1007/978-3-0348-0180-5

Certified Evaluations of Hölder Continuous Functions at Roots of Polynomials

Parker B. Edwards[1], Jonathan D. Hauenstein[1(✉)], and Clifford D. Smyth[2]

[1] Department of Applied and Computational Mathematics and Statistics, University of Notre Dame, Notre Dame, IN 46556, USA
{parker.edwards,hauenstein}@nd.edu

[2] Department of Mathematics and Statistics, University of North Carolina at Greensboro, Greensboro, NC 27402, USA
cdsmyth@uncg.edu

https://sites.nd.edu/parker-edwards, https://www.nd.edu/~jhauenst, https://www.uncg.edu/~cdsmyth/

Abstract. Various methods can obtain certified estimates for roots of polynomials. Many applications in science and engineering additionally utilize the value of functions evaluated at roots. For example, critical values are obtained by evaluating an objective function at critical points. For analytic evaluation functions, Newton's method naturally applies to yield certified estimates. These estimates no longer apply, however, for Hölder continuous functions, which are a generalization of Lipschitz continuous functions where continuous derivatives need not exist. This work develops and analyzes an alternative approach for certified estimates of evaluating locally Hölder continuous functions at roots of polynomials. An implementation of the method in Maple demonstrates efficacy and efficiency.

Keywords: Roots of polynomials · Hölder continuous functions · Certified evaluations

1 Introduction

For a univariate polynomial $p(x)$, the Abel-Ruffini theorem posits that the roots cannot be expressed in terms of radicals for general polynomials of degree at least 5. A simple illustration of this is that the solutions of the quintic equation

$$p(x) = x^5 - x - 1 = 0 \tag{1}$$

cannot be expressed in radicals. Thus, a common technique is to compute numerical approximations with certified bounds for the roots of a polynomial. Some approaches based on Newton's method are the Kantorovich theorem [7] and Smale's α-theory [19]. Kantorovich's approach is based on bounds for a twice-differentiable function in an open set while Smale's approach only uses local

R. M. Corless et al. (Eds.): MC 2020, CCIS 1414, pp. 185–203, 2021.
https://doi.org/10.1007/978-3-030-81698-8_13

estimates at one point coupled with the analyticity of the function. Certified approximations of roots of polynomials can also be obtained using interval methods such as [4,8,13,18] along with the Krawczyk operator [10,12].

Although computing certified estimates for roots of polynomials is important, many applications in science and engineering utilize the roots in further computations. As an illustrative example, consider the optimization problem

$$\min\{21x^8 - 42x^4 - 56x^3 + 3 \; : \; x \in \mathbb{R}\}. \tag{2}$$

For this problem, the global minimum is the minimum of the critical values which are obtained by evaluating the objective function $g(x) = 21x^8 - 42x^4 - 56x^3 + 3$ at its critical points, i.e. at the real roots of $g'(x) = 168x^2(x^5 - x - 1)$. Since the quintic in (1) is a factor of $g'(x)$, only approximations of the roots of g' can be computed. One must translate these approximate roots to certified evaluations of the objective function $g(x)$ evaluated at the roots of the polynomial $g'(x)$ to obtain certified bounds on the global minimum of (2).

One approach for computing a certified evaluation of $f(x)$ at roots of a polynomial $p(x)$ is via certified estimates of solutions to the multivariate system $p(x) = y - f(x) = 0$. For sufficiently smooth f, approaches based on Newton's method generate certified estimates. When f is not differentiable, one can alternatively follow a two-stage procedure: first, certifiably estimate a root of $p(x)$ to error at most ϵ and then use interval evaluation methods, e.g. see [13, Chap. 5], to compute a certified estimate of $f(x)$ evaluated at the root. Such an approach provides direct control on the approximation error of a root of $p(x)$ but not on the output evaluation error of $f(x)$ which will typically be larger than ϵ.

The approach in this paper considers certified evaluations of locally Hölder continuous functions at roots of polynomials and links the desired output of the certified evaluation with the error in the approximation of the root. Hölder continuous functions are a generalization of Lipschitz functions which are indeed continuous, but need not be differentiable anywhere, e.g., see Sect. 5.3. Moreover, satisfying the local Hölder continuity condition does not guarantee that a function can be evaluated exactly for, say, rational input. Therefore, our approach also incorporates numerical evaluation error into the certified bounds.

The rest of the paper is organized as follows. Section 2 describes the necessary analysis of locally Hölder continuous functions, with a particular focus on polynomials and rational functions. Section 3 summarizes the approach used for developing certified bounds on roots of polynomials. Section 4 combines the certification of roots and evaluation bounds on Hölder continuous functions yielding our approach for computing certified evaluations. Section 5 presents information regarding the implementation in `Maple` along with several examples demonstrating its efficacy and efficiency. Section 6 applies the techniques developed for certified evaluations to prove non-negativity of coefficients arising in a series expansion of a rational function. The paper concludes in Sect. 7.

2 Hölder Continuous Functions

The following describes the collection of functions under consideration.

Definition 1. *A function* $f : \mathbb{C} \to \mathbb{C}$ *is* locally Hölder continuous *at a point* $x^* \in \mathbb{C}$ *if there exist positive real constants* ϵ, C, α *such that*

$$|f(x^*) - f(y)| \leq C \cdot |x^* - y|^\alpha \leq C \cdot \epsilon^\alpha \tag{3}$$

for all $y \in B(x^*, \epsilon)$ *where* $B(x^*, \epsilon) = \{z \in \mathbb{C} : |z - x^*| \leq \epsilon\}$. *In this case,* $f(x)$ *is said to have* Hölder constant C *and* Hölder exponent α *at* x^*. *Moreover, if* $\alpha = 1$, *then* $f(x)$ *is said to be* Lipschitz continuous *at* x^* *with* Lipschitz constant C.

Functions which are locally Hölder continuous at a point are clearly continuous at that point and the error bound provided in (3) will be exploited in Sect. 4 to provide certified evaluations. Every function $f(x)$ which is continuously differentiable in a neighborhood of x^* is locally Hölder continuous with $\alpha = 1$, i.e., locally Lipschitz continuous. For $n \geq 1$, $f(x) = \sqrt[n]{|x|}$ is continuous but not differentiable at $x^* = 0$. It is locally Hölder continuous at $x^* = 0$ with Hölder constant $C = 1$ and Hölder exponent $\alpha = 1/n$.

A computational challenge is to determine a Hölder constant C and Hölder exponent α for $f(x)$ on $B(x^*, \epsilon)$ given $f(x)$, x^*, and $\epsilon > 0$. Sections 2.1 and 2.2 describe a strategy for polynomials and rational functions, respectively.

2.1 Polynomials

Since every polynomial $f(x)$ is continuously differentiable, we can take the Hölder exponent to be $\alpha = 1$ at any point x^*. However, the Hölder constant C depends upon x^* and ϵ. The Fundamental Theorem of Calculus shows that one just needs

$$C \geq \max_{y \in B(x^*, \epsilon)} |f'(y)|. \tag{4}$$

Although one may attempt to compute this maximum directly, the Taylor series expansion of $f'(x)$ at x^* provides an easy to compute upper bound. If $d = \deg f$, $f'(x) = \sum_{i=1}^{d} \frac{f^{(i)}(x^*)}{(i-1)!}(x - x^*)^{i-1}$ so that the triangle inequality yields the bound

$$C := \sum_{i=1}^{d} \frac{|f^{(i)}(x^*)|}{(i-1)!} \epsilon^{i-1} \geq \max_{y \in B(x^*, \epsilon)} |f'(y)|. \tag{5}$$

2.2 Rational Functions

The added challenge with a rational function $f(x)$ is to ensure that it is defined on $B(x^*, \epsilon)$. One may attempt to compute the poles of $f(x)$ and ensure that none are in $B(x^*, \epsilon)$. However, the implementation in Sect. 5 is based on a local approach that also enables computing local upper bounds on $|f'(x)|$. For $f(x) = a(x)/b(x)$, one can prove $b(y) \neq 0$ for all $y \in B(x^*, \epsilon)$ by showing that $|b(x^*)| > |b(y) - b(x^*)|$ for all $y \in B(x^*, \epsilon)$. If $d_b = \deg b$, then

$$|b(y) - b(x^*)| = \left| \sum_{i=1}^{d_b} \frac{b^{(i)}(x^*)}{i!}(y - x^*)^i \right| \leq \sum_{i=1}^{d_b} \frac{|b^{(i)}(x^*)|}{i!} \epsilon^i.$$

Therefore, a certificate that $f(x)$ is continuously differentiable on $B(x^*,\epsilon)$ is

$$|b(x^*)| > \sum_{i=1}^{d_b} \frac{|b^{(i)}(x^*)|}{i!}\epsilon^i$$

yielding

$$\min_{y\in B(x^*,\epsilon)} |b(y)| \geq |b(x^*)| - \sum_{i=1}^{d_b} \frac{|b^{(i)}(x^*)|}{i!}\epsilon^i > 0. \tag{6}$$

When $b(x^*) \neq 0$, it is clear that one can always take ϵ small enough to satisfy (6).

When $f(x)$ is continuously differentiable on $B(x^*,\epsilon)$, then one can take the Hölder exponent $\alpha = 1$ and the Hölder constant C as in (4). Hence,

$$\max_{y\in B(x^*,\epsilon)} |f'(x)| \leq \frac{\max_{y\in B(x^*,\epsilon)} |a'(y)|}{\min_{y\in B(x^*,\epsilon)} |b(y)|} + \frac{\max_{y\in B(x^*,\epsilon)} |a(y)| \cdot \max_{y\in B(x^*,\epsilon)} |b'(y)|}{\min_{y\in B(x^*,\epsilon)} |b(y)|^2}$$

where the maxima can be upper bounded similar to (5) and the minimum can be lower bounded using (6).

3 Certification of Roots

The initial task of determining certified evaluation bounds at roots of a given polynomial is to compute certified bounds of the roots. From a theoretical perspective, we assume that we know the polynomial $p(x)$ exactly. From a computational perspective, we assume that $p(x)$ has rational coefficients, i.e., $p(x) \in \mathbb{Q}[x]$. The certification of roots of $p(x)$ can thus be performed using `RealRootIsolate` based on [3,15,20–22] in `Maple` as follows.

Since $p(x)$ is known exactly, we can first reduce down to the irreducible case with multiplicity 1 roots by computing an irreducible factorization of $p(x)$, say

$$p(x) = p_1(x)^{r_1} \cdots p_s(x)^{r_s}$$

where $p_1, \ldots, p_s$ are irreducible with corresponding multiplicities $r_1, \ldots, r_s \in \mathbb{N}$. For $p(x) \in \mathbb{Q}[x]$, `factor` in `Maple` computes the irreducible factors in $\mathbb{Q}[x]$, i.e., each $p_i(x) \in \mathbb{Q}[x]$. If $z \in \mathbb{C}$ is a root of $p_j(x)$, then z has multiplicity 1 with respect to $p_j(x)$, i.e., z is a simple root of $p_j(x)$ with $p_j(z) = 0$ and $p_j'(z) \neq 0$. In contrast, z has multiplicity r_j with respect to $p(x)$. Note that one could alternatively use a squarefree factorization with appropriate modifications.

For each irreducible factor $q := p_j$, one computes certified approximations of each root. Although methods over $\mathbb{C}$ are more efficient [2,14], we utilize the `RealRootIsolate` function in `Maple` by transforming the domain $\mathbb{C}$ into $\mathbb{R}^2$ via

$$q(x+iy) = q_r(x,y) + i \cdot q_i(x,y) \qquad \text{where } i = \sqrt{-1}. \tag{7}$$

Therefore, solving $q = 0$ on $\mathbb{C}$ corresponds with solving $q_r = q_i = 0$ on $\mathbb{R}^2$. Applying `RealRootIsolate` with an optional absolute error bound `abserr` that will be utilized later guarantees as output isolating boxes for every real solution to $q_r = q_i = 0$ on $\mathbb{R}^2$. Therefore, looping over the irreducible factors of p, one obtains certified bounds for every root z of $p(x)$ in $\mathbb{C}$ of the form $a_1 \leq \text{real}(z) \leq a_2$ and $b_1 \leq \text{imag}(z) \leq b_2$ where $a_1, a_2, b_1, b_2 \in \mathbb{Q}$.

4 Certified Evaluations

Combining information on Hölder continuous functions from Sect. 2 and certification of roots of polynomials from Sect. 3 yields the following approach to develop certified evaluations. With input a polynomial $p(x)$, a Hölder continuous function $f(x)$ which is defined at each root of $p(x)$, and an error bound $\epsilon > 0$, the goal is to develop an approach that computes an approximation of $f(z)$ within ϵ for each root z of $p(x)$. Since one may not be able to evaluate $f(x)$ exactly, we incorporate an evaluation error of $\delta \in (0, \epsilon)$. Typically, δ can be decreased by utilizing higher precision computations. For rational input, rounding to produce finite decimal representations constitutes the only source of representation error in `Maple`. Our implementation utilizes enough digits to have $\delta = \epsilon/10$.

The first step is to utilize Sect. 3 to determine initial certified bounds for each root of $p(x)$. For an initial error bound on the roots, we start with $\gamma = \epsilon/2$ and approximate each root z by x^* with $|z - x^*| < \gamma$. Certified evaluations are then obtained root by root since the Hölder constants are dependent upon local information near each root. In particular, the next step is to compute a Hölder exponent α and Hölder constant C that is valid on the ball $B(x^*, 2\gamma)$. If this is not possible, e.g., if $f(x)$ cannot be certified to be defined on $B(x^*, 2\gamma)$, one can simply reduce γ, e.g., by replacing γ by $\gamma/2$ and repeating the process using a newly computed certified approximation of z. Since $f(x)$ is defined at each root of $p(x)$, such a loop must terminate.

The final step is to utilize local information to compute a new approximation of root z that will produce a certified evaluation within ϵ. Consider μ such that

$$0 < \mu \leq \min\left\{\gamma, \sqrt[\alpha]{\frac{\epsilon - \delta}{C}}\right\}$$

and z^* an approximation of z such that $z \in B(z^*, \mu)$. Since $|x^* - z^*| \leq 2\gamma$, we have $B(z^*, \mu) \subset B(x^*, 2\gamma)$ so that all of the Hölder constants are valid on $B(z^*, \mu)$. Hence, all that remains is to compute a certified approximation of $f(z^*)$, say f^*, within the evaluation error of δ since

$$|f^* - f(z)| \leq |f^* - f(z^*)| + |f(z^*) - f(z)| \leq \delta + C \cdot |z^* - z|^\alpha \leq \delta + C \cdot \mu^\alpha \leq \epsilon.$$

Remark 1. Software packages that deal with numerical approximations express numerical estimates either in terms of absolute or relative error. The choice is typically made based on the intended application. For example, relative error is more easily associated with memory requirements for storing approximations regardless of the magnitude of a value being approximated so that control over relative error in approximations is often easier than controlling absolute error.

We have chosen to bound the absolute error here since the application of our Hölder function based approach presented in Sect. 6 is simpler to state in terms of absolute error.

Remark 2. When f is polynomial, one could use the built-in `Maple` function `RootFinding[Isolate]` which implements various root isolation strategies based on [1,9,16,17,20]. The `Maple` function call is:

```
RootFinding[Isolate]([p_r,p_i],[x,y],constraints=[f],
                     digits=ceil(-log[10](eps)))
```

where p_r and p_i are the real and imaginary parts of $p(x+iy)$, respectively. The outputs estimate both the roots of p and evaluations of f at those roots. One principal difference between our approach and the built-in functionality is that `RootFinding[Isolate]` is only implemented for polynomial evaluation functions. See Sect. 5.1 for a polynomial example using `RootFinding[Isolate]`.

Example 1. As an illustration, consider evaluating the Cantor ternary function

$$f\left(\sum_{j=1}^{\infty}\frac{a_j}{3^j}\right)=\frac{1}{2^N}+\frac{1}{2}\sum_{j=1}^{N-1}\frac{a_j}{2^j} \text{ where } \begin{cases} a_j\in\{0,1,2\} \text{ for } j=1,2,\ldots, \\ N=\min\{j \mid a_j \text{ is odd}\}\in\mathbb{Z}_{>0}\cup\{\infty\}. \end{cases}$$

at the unique root $z\in[0,1]$ of the polynomial

$$p(x)=\left(\frac{3}{2}\right)^{101}x^5+17\left(\frac{3}{2}\right)^{101}x-1$$

with error $\epsilon=10^{-16}$. Figure 1 plots the Cantor ternary function on the domain $[0,1]$ along with the point $(z,f(z))$. Clearly, the Cantor ternary function $f(x)$ is not polynomial so that `RootFinding[Isolate]` can not be utilized. Since $f(x)$ can be evaluated exactly at points with a finite ternary expansion, we can take $\delta=0$. Moreover, $f(x)$ is Hölder continuous on $[0,1]$ with Hölder exponent $\alpha=\log 2/\log 3$ and Hölder constant $C=2$ so that we can simply take

$$\mu=10^{-26}<\sqrt[\alpha]{\frac{\epsilon-\delta}{C}}.$$

Hence,

$$z^*=\frac{2}{3^{40}}+\frac{2}{3^{41}}+\frac{1}{3^{42}}+\frac{1}{3^{43}}+\frac{2}{3^{44}}+\frac{2}{3^{45}}+\frac{1}{3^{48}}+\frac{1}{3^{49}}+\frac{1}{3^{50}}+\frac{1}{3^{51}}+\frac{2}{3^{52}}+\frac{1}{3^{53}}+\frac{1}{3^{54}}+\frac{2}{3^{55}}+\frac{1}{3^{56}}$$

satisfies $|z-z^*|<\mu$ so that

$$f(z^*)=\frac{1}{2^{42}}+\frac{1}{2}\left(\frac{2}{2^{40}}+\frac{2}{2^{41}}\right)=\frac{7}{2^{42}}$$

is certifiably within ϵ of $f(z)$.

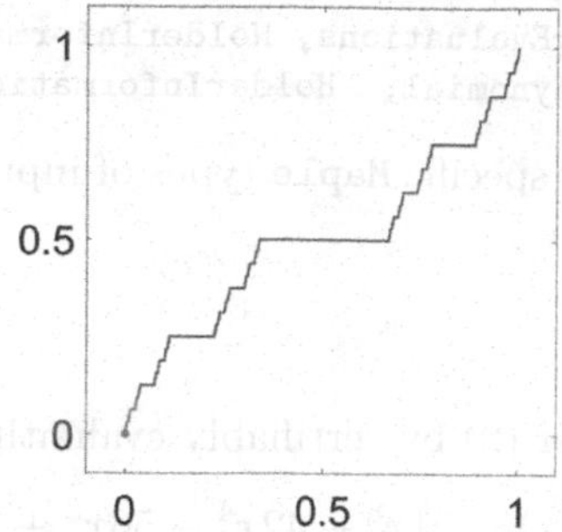

Fig. 1. Plot of the Cantor ternary function f with evaluation at a root of p.

5 Implementation and Examples

The certified evaluation procedure has been implemented as a `Maple` package entitled `EvalCertification` available at https://github.com/P-Edwards/EvalCertification along with `Maple` notebooks for the examples. The main export is the procedure `EstimateRootsAndCertifyEvaluations` which has the following high level signature:

Input:

- Univariate polynomial $p \in \mathbb{Q}[x]$.
- List of locally Hölder continuous functions $f_1, \ldots, f_m$ with which to certifiably estimate evaluations at the roots of $p(x)$.
- List of procedures specifying how to compute local Hölder constants and exponents for $f_1, \ldots, f_m$. (See Sect. 5.3 for example of the syntax).
- Desired accuracy $\epsilon \in \mathbb{Q}_{>0}$.

Main output:

- Complex rational root approximations $z_1^*, \ldots, z_s^*$, one for each of the distinct roots $z_1, \ldots, z_s$ of $p(x)$, such that $|z_j - z_j^*| \leq \epsilon$.
- For each f_i and x_j, a complex decimal number f_{ij}^* with $|f_i(x_j) - f_{ij}^*| \leq \epsilon$.

The `EvalCertification` package is formatted in a .mpl file which can be read into a notebook with:

```
read("EvalCertification.mpl")
with(EvalCertification)
```

This lists the package's following four exports: the main function and three built in procedures for determining local Hölder constants and exponents for common classes of Hölder functions.

EstimateRootsAndCertifyEvaluations, HolderInformationForExponential, HolderInformationForPolynomial, HolderInformationForRationalPolynomial

The following highlight specific Maple types of inputs and outputs as well as other interface details.

5.1 Critical Values

As a first example, consider (2) by certifiably evaluating

$$f(x) = 21x^8 - 42x^4 - 56x^3 + 3$$

at the roots of $p(x) = f'(x) = 168x^2(x^5 - x - 1)$ with error $\epsilon = 10^{-14}$.

```
f_polynomial := 21x^8 - 42x^4 - 56x^3 + 3;
f_derivative := diff(f_polynomial, x);
EstimationPrecision := 1/10^14;
```

The main call to EstimateRootsAndCertifyEvaluations is subsequently:

```
solutions_information :=
EstimateRootsAndCertifyEvaluations(f_derivative,
                                   [f_polynomial, f_derivative],
                                   HolderInformationForPolynomial,
                                   EstimationPrecision);
```

The first argument provides the polynomial to solve and the second is a list of polynomials to evaluate. For illustration, we include evaluating the polynomial to solve in the evaluation list. The third argument is a procedure for computing Hölder constants which, in this case, uses the procedure that implements the estimates in Sect. 2.1 for polynomials. Notice that we need only provide the procedure once since all functions for evaluation fall into the same class of Hölder functions, namely polynomials. The last argument is the final error bound.

The output solutions_information is formatted as a Record. Certifiably estimated roots are stored in a list as illustrated.

```
solutions_information:-root_values =
[0,
2691619717901426047/2305843009213693952,
26745188167908553113/147573952589676412928 -
19995423894655642147*I/18446744073709551616, ...]
```

Evaluations are also stored in lists, one list for each function to evaluate with one entry for each root of p. Estimates are ordered so that the estimate at index i in its list corresponds to the root at index i in the roots list.

```
solutions_information:-evaluations_functions_1 =
[3., -91.6600084778015707, ...];
solutions_information:-evaluations_functions_2 =
[0, -6.692143197043304*10^(-16), ...];
```

Therefore, the solution to (2) is -91.6600084778015707 which is certifiably correct within an error of 10^{-14}.

Since $f(x)$ is polynomial, we can compare with `RootFinding[Isolate]` as discussed in Remark 2. Since evaluations at only the real critical points are of interest, one can simply utilize

```
Isolate(f_derivative, constraints = [f_polynomial], digits = 14);
```

which yields

```
[x = 0., x = 1.1673039782614],
[[21*x^8 - 42*x^4 - 56*x^3 + 3 = 3.],
[21*x^8 - 42*x^4 - 56*x^3 + 3 = -91.660008477802]]
```

The 14 digits of 1.1673039782614 are indeed correct, but the result has an absolute error of approximately $1.87 \cdot 10^{-14}$ while the evaluation -91.660008477802 has an absolute error of approximately $4.29 \cdot 10^{-13}$. Nonetheless, the relative error for both the root and evaluation estimates is less than the requested 10^{-14}, both of which are controlled by the `digits` argument. This fulfills the documented specifications for `Isolate`, though the `digits` argument's control over the evaluation's relative error is undocumented [11]. An additional function call with `digits` set to 16 would be necessary to decrease the absolute error below 10^{-14} for both the root and evaluation.

5.2 Comparison with Ball Arithmetic

Interval and ball arithmetic methods can provide similar certification functionality as `EvalCertification` by first isolating each root and then evaluating a interval extension of the function. As mentioned in the Introduction, this two-step procedure does not provide direct control on the size of the evaluation error and thus one may need to perform several loops to refine the isolation of each root to have sufficiently small evaluation error. In contrast, for evaluating functions without poles, `EvalCertification` always performs root estimation exactly twice: once to obtain local Hölder information and then a second time to guarantee small evaluation error.

For illustration, consider evaluating the function $f(x) = 50^x$ at the roots of the polynomial $p(x) = (x^7 + x - 1)(x - 1000)$ with an error at most $\epsilon = 2^{-1000}$. The library `Arb` [6] required a root estimation error of at most $0.91 \cdot 2^{-8176}$ to provide the requisite evaluation error. In `Arb`, relative error is input as a number of bits of precision available to computations. Thus, by supplying additional bits of precision, one lowers the relative and absolute error. For our computation, the precision in `Arb` was initialized at $2^{10} = 1024$ bits of precision which is enough to accurately store the desired error of 2^{-1000} exactly. We then utilized the two-stage procedure which loops back to refine the root if the output evaluation error is unacceptably large. If we simply double the number of bits of precision used in each loop, then three iterations are required to yield 8192 bits of precision which is sufficient to perform root estimation accurately enough for the function evaluation to yield the desired evaluation error.

As mentioned in Sect. 3, transforming complex root isolation into bivariate real root isolation is a costly maneuver. However, such an approach was used in `EvalCertification` to take advantage of the already existing `RealRootIsolate` in `Maple`. Since `Arb` implements a faster univariate solver that only allows relative error bounds on the estimates as input, this accounts for the drastic difference in computing time on this problem using `Arb` (0.52 s) and using `RealRootIsolate` in `Maple` via `EvalCertification` (153 s).

5.3 Extending with Custom Hölder Information Procedures

Polynomial and rational functions can utilize the built-in procedures for computing local Hölder constants. One more feature of `EvalCertification` is the ability to extend the certification procedures to new classes of functions by specifying how to compute local Hölder constants. To illustrate, consider the Weierstrass function $f : \mathbb{R} \to \mathbb{R}$ given by

$$f(x) = \sum_{n=0}^{\infty} 7^{-\frac{n}{3}} \cos(7^n \pi x)$$

which we aim to evaluate at the unique real root z of $p(x) = x^7 + x - 1$. Figure 2 plots $f(x)$ and $p(x)$ along with the point $(z, f(z))$. The Weierstrass function $f(x)$ is nowhere differentiable but is globally Hölder continuous [5] with exponent $\alpha = 1/3$ and constant $C \leq 4.73$. The following is the format for defining a new procedure to supply the Hölder information:

```
WeierInfo := proc(f, point, radius, domain_estimate := false)
return Record('exponent' = 1/3,
              'constant' = 4.73,
              'avoid_roots' = false); end proc;
```

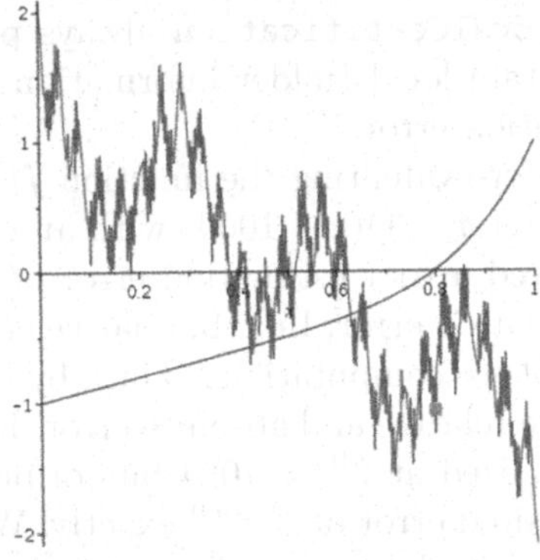

Fig. 2. Plot of the Weierstrass function $f(x)$, the polynomial $p(x)$, and point $(z, f(z))$ where z is the unique real root of $p(x)$.

All custom Hölder information procedures must follow the same signature as this example. The `exponent` α and `constant` C in the output `Record` should satisfy the Hölder conditions for `InputFunction` on the ball $B($`point`, `radius`$)$. For this example, the Hölder information is independent of the `point` and `radius` since the Weierstrass function is globally Hölder continuous. The entry `avoid_roots` lists estimates within `radius` of points missing from the input function's domain or false if defined everywhere.

Since $f(x)$ is an infinite series, we must evaluate a finite truncation of it, say

$$f_N(x) = \sum_{n=0}^{N} 7^{-\frac{n}{3}} \cos(7^n \pi x) \text{ with } |f(x) - f_N(x)| \leq \sum_{n=N+1}^{\infty} 7^{-\frac{n}{3}} = \frac{7^{-\frac{N+1}{3}}}{1 - 7^{-\frac{1}{3}}} =: E_N.$$

Therefore, to approximate $f(z)$, one has three sources of error: approximation error in z, finite truncation error E_N, and numerical error when evaluating f_N. After selecting N such that $E_N < \epsilon$, one can simply replace ϵ by $\epsilon - E_N$ with the other two errors already accounted for in our approach. The following commands produce certified evaluations of f to precision 10^{-14} at z utilizing $N = 51$ so that $E_{51} < 4.71 \cdot 10^{-15}$:

```
p := x^7 + x - 1;
MaxErr := 1/10^14-E_N;
solutions_information :=
EstimateRootsAndCertifyEvaluations(p,[F_N],WeierInfo,MaxErr);
```

This yields $z^* = 0.79654435412846$ and $f^* = -1.06659590869988$.

5.4 Benchmarking

As mentioned in Sect. 5.2, the dominant computational cost is in estimating roots with the next largest cost associated with computing local Hölder constants. Suppose that $R(p, \epsilon)$ is the complexity of approximating roots of p within ϵ, $H(f_1, \ldots, f_n, p, \epsilon)$ is the minimum complexity of computing Hölder constants at one root, and $A(p, f_1, \ldots, f_n, \epsilon)$ is the number of repetitions required to find an accuracy $\gamma \leq \epsilon$ where local Hölder constants can be calculated. Then,

$$A(p, f_1, \ldots, f_n, \epsilon)(R(p, \epsilon) + n \deg(p) H(f_1, \ldots, f_n, p, \epsilon)) + R(p, \epsilon)$$

is a lower bound on the complexity. The number of repetitions A is 1 for functions without poles and otherwise depends on the input in a complicated way which we do not attempt to characterize here.

We benchmarked `EvalCertification` using random polynomials generated by the command `randpoly` in `Maple`. All tests computed roots of a random polynomial $p(x)$ with integer coefficients between -10^{10} and 10^{10} and evaluated rational functions where the numerator and denominator were polynomials of degree D. The average was taken over 50 random selections. Figure 3 shows the results of the benchmarking tests, which were performed on Ubuntu 18.04 running `Maple` 2020 with an Intel Core i7-8565U processor. They were based on the degree d of $p(x)$, the value of D, the number of functions n to evaluate, and the size of the output error ϵ.

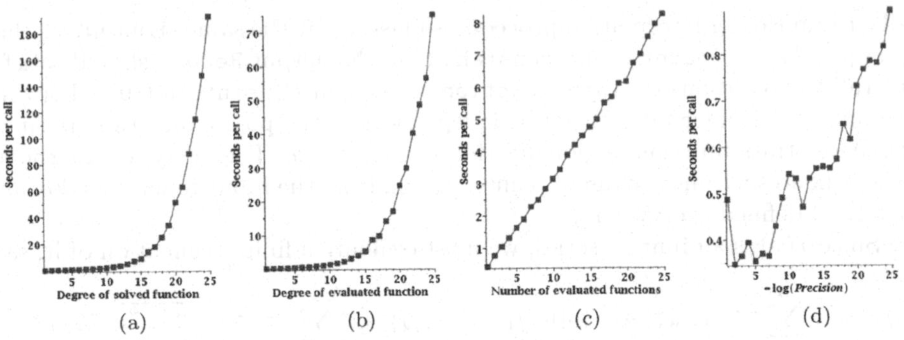

Fig. 3. Results of tests with (a) $d \in \{1, \ldots, 25\}$, $D = 5$, $n = 1$, and $\epsilon = 10^{-14}$; (b) $d = 5$, $D \in \{1, \ldots, 25\}$, $n = 1$, and $\epsilon = 10^{-14}$; (c) $d = 5$, $D = 5$, $n \in \{1, \ldots, 25\}$, and $\epsilon = 10^{-14}$; (d) $d = 5$, $D = 5$, $n = 1$, and $\epsilon \in \{1, 10^{-1}, \ldots, 10^{-25}\}$.

6 Application to Prove Non-negativity

One application of our approach for computing certified evaluations is to certifiably decide whether or not all coefficients of the Taylor series expansion centered at the origin are non-negative for a given real rational function $r(x)$. We focus on non-negativity since non-positivity is equivalent to non-negativity for $-r(x)$ and alternating in sign is equivalent to non-negativity for $r(-x)$. The following method uses certified evaluations to obtain information about the coefficients in the tail of the Taylor series expansion reducing the problem to only needing to inspect finitely many coefficients. This approach assumes that the function does not have a pole at the origin, its denominator has only simple roots, and its denominator has a real positive root that is strictly smallest in modulus amongst all its roots. This approach can be extended to more general settings, but will not considered here due to space considerations.

We will make use of the following standard theorem.

Theorem 1. *Let $p(x), q(x) \in \mathbb{R}[x]$ such that $p(x)$ and $q(x)$ have no common root, $q(0) \neq 0$ and $\deg(p(x)) < \deg(q(x)) = d$. If $q(x)$ has only simple roots say $\alpha_1, \ldots, \alpha_d \in \mathbb{C}$, then $r(x) = p(x)/q(x)$ has a Taylor series expansion of the form $r(x) = \sum_{n=0}^{\infty} r_n x^n$ converging for all $x \in \mathbb{C}$ with $|x| < \min\{|\alpha_1|, \ldots, |\alpha_d|\}$. Furthermore, for all $n \geq 0$,*

$$r_n = -\sum_{i=1}^{d} \frac{p(\alpha_i)}{\alpha_i q'(\alpha_i)} \alpha_i^{-n}. \tag{8}$$

Theorem 1 follows from partial fraction decomposition of rational functions or using linear recurrences. For completeness, we provide a proof in the Appendix. Using Theorem 1, we obtain the following result on the eventual behavior of the coefficients of the Taylor series of certain rational functions.

Theorem 2. *With the setup from Theorem 1, define $C_i = -p(\alpha_i)/(\alpha_i q'(\alpha_i))$ for $i = 1, \ldots, d$. If $\alpha_1 \in \mathbb{R}$ is such that $|\alpha_1| < \min\{|\alpha_2|, \ldots, |\alpha_d|\}$, then there exists N after which exactly one of the following conditions on r_n holds:*

1. *If $\alpha_1 > 0$ and $C_1 > 0$, then $r_n > 0$ for all $n > N$.*
2. *If $\alpha_1 > 0$ and $C_1 < 0$, then $r_n < 0$ for all $n > N$.*
3. *If $\alpha_1 < 0$, then r_n is alternating in sign for all $n > N$, i.e., $(-1)^n \cdot r_n > 0$ for all $n > N$ or $(-1)^n \cdot r_n < 0$ for all $n > N$.*

Moreover, one may take $N = \log(K)/\log(M/m)$ where $K = \sum_{i=2}^{d} |C_i|/|C_1|$, $m = |\alpha_1|$, and $M = \min\{|\alpha_2|, \ldots, |\alpha_d|\}$.

A proof of Theorem 2 is provided in the Appendix. Theorems 1 and 2 yield the following.

Corollary 1. *Suppose that $f(x), q(x) \in \mathbb{R}[x]$ have no common root, $q(0) \neq 0$, and $q(x)$ has only simple roots, namely $\alpha_1, \ldots, \alpha_d \in \mathbb{C}$, such that $\alpha_1 \in \mathbb{R}$ and $|\alpha_1| < \min\{|\alpha_2|, \ldots, |\alpha_d|\}$. Let $g(x), p(x) \in \mathbb{R}[x]$ be the unique polynomials such that $f(x) = q(x) \cdot g(x) + p(x)$ with $\deg(p(x)) < \deg(q(x))$. Define $C_i = -p(\alpha_i)/(\alpha_i q'(\alpha_i))$ for $i = 1, \ldots, d$. Then, $f(x)/q(x)$ has a Taylor series expansion $f(x)/q(x) = \sum_{n=0}^{\infty} R_n x^n$ converging for all $x \in \mathbb{C}$ with $|x| < \min\{|\alpha_1|, \ldots, |\alpha_d|\}$ and there is a threshold N_0 so that exactly one of the following conditions on R_n holds:*

1. *If $\alpha_1 > 0$ and $C_1 > 0$, then $R_n > 0$ for all $n > N_0$.*
2. *If $\alpha_1 > 0$ and $C_1 < 0$, then $R_n < 0$ for all $n > N_0$.*
3. *If $\alpha_1 < 0$, then R_n is alternating in sign for all $n > N_0$, i.e. $(-1)^n \cdot R_n > 0$ for all $n > N_0$ or $(-1)^n \cdot R_n < 0$ for all $n > N_0$.*

One can take $N_0 = \max\{\deg(f(x)) - \deg(q(x)) + 1, \log(K)/\log(M/m)\}$ where $K = \sum_{i=2}^{d} |C_i|/|C_1|$, $m = |\alpha_1|$, and $M = \min\{|\alpha_2|, \ldots, |\alpha_d|\}$.

Proof. Since $f(x)/q(x) = g(x) + p(x)/q(x)$, applying Theorem 1 yields the first part. Since the Taylor series coefficients of $f(x)/q(x)$ and $p(x)/q(x)$ are same for $n > \deg(f(x)) - \deg(g(x))$, the second part immediately follows from Theorem 2.

One key to utilizing Theorem 2 and Corollary 1 is to certify that $q(x)$ satisfies the requisite assumptions. Validating that $q(x)$ has only simple roots follows from computing an irreducible factorization as in Sect. 3 and checking if every factor has multiplicity 1. Section 6.1 describes a certified approach to verify the remaining conditions on $q(x)$. Section 6.2 yields a complete algorithm for certifiably deciding non-negativity of all Taylor series coefficients which is demonstrated on two examples.

6.1 Classification of Roots

Given a polynomial $q(x) \in \mathbb{R}[x]$ with only simple roots and $q(0) \neq 0$, the following describes a method to certifiably determine if $q(x)$ has a positive root that is strictly smallest in modulus amongst all its roots. This method uses the ability to certifiably approximate all real points in zero-dimensional semi-algebraic sets. Computationally, this can be accomplished using the command `RealRootIsolate` in `Maple`. Note that some of these computations could also be accomplished using `RootFinding[Isolate]` following Remark 2.

The first step is to certifiably determine if $q(x)$ has a positive root via

$$\mathcal{P} = \{p \in \mathbb{R} \ : \ q(p) = 0, p > 0\}.$$

If $\mathcal{P} = \emptyset$, then one returns that $q(x)$ does not have a positive root. Otherwise, one proceeds to test the modulus condition for $\alpha_1 = \min \mathcal{P}$.

The modulus condition needs to be tested against negative roots and non-real roots. For negative roots, consider

$$\mathcal{N} = \{n \in \mathbb{R} \ : \ q(-n) = 0, n > 0\} \ \text{ and } \ \mathcal{B} = \{b \in \mathbb{R} \ : \ q(b) = q(-b) = 0, b > 0\}.$$

By using certified approximations of α_1 and points in $\mathcal{N}$ and $\mathcal{B}$ of decreasing error, one can certifiably determine which of the following holds: $\alpha_1 < \min \mathcal{N}$, $\alpha_1 > \min \mathcal{N}$, or $\alpha_1 \in \mathcal{B} \subset \mathcal{P}$. If $\alpha_1 > \min \mathcal{N}$ or $\alpha_1 \in \mathcal{B}$, then one returns that $q(x)$ does has not a positive root that is strictly smallest in modulus amongst all its roots. Otherwise, one proceeds to the non-real roots by considering

$$\begin{aligned} \mathcal{L} &= \{(r,a,b) \in \mathbb{R}^3 \ : \ q(r) = 0, r > 0, q(a+ib) = 0, b > 0, a^2 + b^2 < r^2\} \text{ and} \\ \mathcal{E} &= \{(r,a,b) \in \mathbb{R}^3 \ : \ q(r) = 0, r > 0, q(a+ib) = 0, b > 0, a^2 + b^2 = r^2\}. \end{aligned}$$

Note that $q(a + ib) = 0$ provides two real polynomial conditions on $(a, b) \in \mathbb{R}^2$ via the real and imaginary parts as in (7) so that $\mathcal{L}$ and $\mathcal{E}$ are clearly zero-dimensional semi-algebraic sets. Moreover, for the projection map $\pi_1(r, a, b) = r$, $\pi_1(\mathcal{L} \cup \mathcal{E}) \subset \mathcal{P}$. By using certified approximations of α_1 and points in $\mathcal{L}$ and $\mathcal{E}$ of decreasing error, one can certifiably determine if $\alpha_1 \in \pi_1(\mathcal{L} \cup \mathcal{E})$ or $\alpha_1 \notin \pi_1(\mathcal{L} \cup \mathcal{E})$. If the former holds, then one returns that $q(x)$ does has not a positive root that is strictly smallest in modulus amongst all its roots. If the later holds, then one returns that $q(x)$ does indeed have a positive root that is strictly smallest in modulus amongst all its roots.

6.2 Certification of Non-negativity

Suppose that $f(x), q(x) \in \mathbb{R}[x]$ which satisfy the assumptions in Corollary 1. The following describes a method to certifiably determine if all of the coefficients R_n of the Taylor series expansion for $f(x)/q(x)$ centered at the origin are non-negative or provides an integer n_0 such that $R_{n_0} < 0$.

First, the Euclidean algorithm is utilized to determine $g(x), p(x) \in \mathbb{R}[x]$ with $\deg(p(x)) < \deg(q(x))$ such that $f(x) = q(x) \cdot g(x) + p(x)$. Define $h(x) = x \cdot q'(x)$ and $C(x) = -p(x)/h(x)$. Hence, $d = \deg(q(x)) = \deg(h(x))$ such that $q(x)$ and $h(x)$ have no common roots. As in Corollary 1, let $\alpha_1, \ldots, \alpha_d$ be the roots of $q(x)$ with $\alpha_1 \in \mathbb{R}_{>0}$ such that $\alpha_1 < \min\{\alpha_2, \ldots, \alpha_d\}$. Let $\beta_1, \ldots, \beta_d \in \mathbb{C}$ (not necessarily all distinct) be the roots of $h(x)$.

Certified evaluations at the roots of $q(x)$ and $h(x)$ with error bound $\epsilon_k = 2^{-k}$ for $k = 1, 2, \ldots$ can be used until the following termination conditions are met:

1. α_i^* and β_j^* are such that $\alpha_1^* \in \mathbb{R}$, $|\alpha_i^* - \alpha_i| < \epsilon_k$, and $|\beta_j^* - \beta_j| < \epsilon_k$
2. the set $\{0, \alpha_1^*, \ldots, \alpha_d^*\}$ is $2 \cdot \epsilon_k$ separated, i.e., $|s-t|^2 \geq (2\epsilon_k)^2$ for all distinct s, t in this set,
3. $\gamma^* \leq \min\{|\alpha_i^* - \beta_j^*| \ : \ 1 \leq i, j \leq d\}$ such that $\gamma^* > 2 \cdot \epsilon_k + \epsilon_k^{1/(4d)}$,
4. for $m^* = \alpha_1^* + \epsilon_k$ and $M^* \leq \min\{|\alpha_2^*|, \ldots, |\alpha_d^*|\} - \epsilon_k$, one has $m^* < M^*$,
5. L_i^* such that $L_i^* \geq |c_d|^{-2} \sum_{\ell=0}^{2d-1} |u^{(\ell)}(\alpha_i^*)|\epsilon_k^\ell / \ell!$ where c_d is the leading coefficient of $q(x)$ and $u(x) = -p'(x)h(x) + p(x)h'(x)$, and
6. either (a) $C_1^* + L_1^*\sqrt{\epsilon_k} < 0$ or (b) $C_1^* - L_1^*\sqrt{\epsilon_k} > 0$.

Note that starred quantities in the termination conditions above, or in the further discussion below, are either exact rational approximations to real constants or complex numbers with rational real and imaginary parts that approximate roots.

Before proving that such a termination condition can be met, we describe the last steps which are justified by Corollary 1. Let $\epsilon = \epsilon_k$ be the value where the termination conditions are met and calculate $0 < b^* \leq \min\{|C_1^* \pm L_1^*\sqrt{\epsilon}|\}$,

$$K^* \geq (1/b^*) \sum_{i=2}^d (|C_i^*| + L_i^* \sqrt{\epsilon}), \ A^* \geq \log(K^*)/\log(M^*/m^*),$$
$$N_0^* = \max\{\deg(f(x)) - \deg(q(x)), \lceil A^* \rceil\}.$$

The inequalities in Items 3, 4, and 5 and these values above are meant to signify the rounding direction in machine precision used to compute the values. With this, if $C_1^* + L_1^*\sqrt{\epsilon_k} < 0$, then return $n_0 = N_0^* + 1$ for which $R_{n_0} < 0$. Otherwise, the non-negativity of all R_n is equivalent to the non-negativity of $R_0, \ldots, R_{N_0^*}$ which can be computed by explicit computation. If there exists $n_0 \in \{0, \ldots, N_0^*\}$ such that $R_{n_0} < 0$, return n_0. Otherwise, return that all R_n are non-negative.

First, note that Item 1 is obtained by real root certification. We have

$$\delta = \min\{|\alpha_i - \alpha_j|, |\alpha_i| \ : \ 1 \leq i < j \leq d\} > 0.$$

By Item 1, $|\alpha_i^* - \alpha_j^*| \geq |\alpha_i - \alpha_j| - 2\epsilon_k$ and $|\alpha_i^*| \geq |\alpha_i| - \epsilon_k$. Thus,

$$\delta^* = \min\{|\alpha_i^* - \alpha_j^*|, |\alpha_i^*| \ : \ 1 \leq i < j \leq d\} \geq \delta - 2\epsilon_k - \eta$$

where $\eta > 0$ is the machine precision on the lower bounds on the quantities in δ^*. Since $\epsilon_k \to 0$ and η can be made arbitrarily small, eventually $\delta^* - 2\epsilon_k - \eta > 2\epsilon_k$ and Item 2 will be met.

Since $q(x)$ and $h(x)$ have no roots in common, consider

$$\gamma = \min\{|\alpha_i - \beta_j| \ : \ 1 \leq i, j \leq d\} > 0.$$

We have $\gamma^* \geq \gamma - 2\epsilon_k - \eta$ where $\eta > 0$ is the machine precision on the lower bounds of the quantities in γ^*. Eventually, $\gamma^* \geq \gamma - 2\epsilon_k - \eta > 2\epsilon_k + \nu + \epsilon_k^{1/(4d)}$ where $\nu > 0$ is the machine precision on the upper bound for $\epsilon_k^{1/(4d)}$ and Item 3 will be met.

Since $\Delta = M - m > 0$, we have $M^* > M - \epsilon_k - \eta$ where $\eta > 0$ is the machine precision on the lower estimates in M^* and also $m^* = \alpha_1^* + \epsilon_k < \alpha_1 + 2\epsilon_k$. Thus, $M^* - m^* \geq \Delta - 3\epsilon_k - \eta$ so that eventually Item 4 will be met.

Since $C(x) = p(x)/h(x)$, we have $C'(x) = u(x)/h^2(x)$. Assuming the previous items have all been met, we have

$$|\beta_j - \alpha_i^*| \geq |\beta_j^* - \alpha_i^*| - |\beta_j - \beta_J|^* > 2\epsilon_k + \epsilon_k^{1/(4d)} - \epsilon_k > \epsilon_k.$$

Hence, $h(x)$ has no roots in $B(\alpha_i^*, \epsilon_k)$ and $C'(x)$ is continuous on $B(\alpha_i^*, \epsilon_k)$. Fix $z \in B(\alpha_i^*, \epsilon_k)$ and let ζ be the straight line segment contour from α_i^* to z in $B(\alpha_i^*, \epsilon_k)$. Since $C'(x)$ exists on $B(\alpha_i^*, \epsilon_k)$, $C(z) - C(\alpha_i^*) = \int_\zeta C'(x)dx$. Thus, $|C(z) - C(\alpha_i^*)| \leq P \cdot |z - \alpha_i^*|$ where $P = \max\{|C'(x)| \ : \ x \in B(\alpha_i^*, \epsilon_k)\}$. Therefore, we have $P \leq P_1/P_2^2$ where $P_1 = \max\{|u(x)| \ : \ x \in B(\alpha_i^*, \epsilon_k)\}$ and $P_2 = \min\{|h(x)| \ : \ x \in B(\alpha_i^*, \epsilon_k)\}$. Since $u(z) = \sum_{\ell=0}^{2d-1} u^{(\ell)}(\alpha_i^*)(z - \alpha_i^*)^\ell/\ell!$, we have $P_1 \leq \sum_{\ell=0}^{2d-1} |u^{(\ell)}(\alpha_i^*)|\epsilon_k^\ell/\ell!(1+\eta)$ where $\eta > 0$ is the machine precision that results from the upper bound on the quantities $|u^{(\ell)}(\alpha_i^*)|$. Since

$$h(z) = c_d \prod_{j=1}^{d}(z - \beta_j) \ \text{ and } \ |z - \beta_j| \geq |\alpha_i^* - \beta_j^*| - |z - \alpha_i^*| - |\beta_j^* - \beta_j| > \epsilon_k^{1/(4d)},$$

we have $|h(z)| \geq |c_d|\epsilon^{1/4}$. Thus, $P_2^2 \geq |c_d|^2\epsilon^{1/2}$ and $|C(z) - C(\alpha_i^*)| \leq L_i^*\sqrt{\epsilon}$. Thus, $|C_i^* - C_i| \leq L_i^*\sqrt{\epsilon}$ where $C_i = C(\alpha_i)$ and $C_i^* = C(\alpha_i^*)$. By the inclusions $B(\alpha_i^*, \epsilon_k) \subset B(\alpha_i, 2\epsilon_k) \subset B(\alpha_i, 1)$, all estimates $|u^{(\ell)}(\alpha_i^*)|$ will be bounded above by the corresponding maximum values of $|u^{(\ell)}(z)|$ for $z \in B(\alpha_1, 1)$. Thus, even though P_1 varies in each step k, P_1 and L_i^* will be uniformly bounded above for all k. Since L_1^* is uniformly bounded above and $C_1 \neq 0$ by the proof of Theorem 2, Item 6(a) will eventually be met if $C_1 < 0$ while Item 6(b) will eventually be met if $C_1 > 0$.

Having proved termination, we note that in order to get a value of N_0^* which is reasonably close to the value of N_0 in Corollary 1, one may continue to decrease ϵ_k past the point where all termination conditions are first met. The reason for this is to separate m^* and M^* as far all possible, i.e., to match the actual gap between m and M as closely as possible. This could be wise especially when m^* is very close to M^* in which case $\log(M^*/m^*)$ will be very close to 0 so that N_0^* will be very large.

Example 2. To demonstrate the approach, consider the rational functions

$$(1 - x^3 - x^7 + x^{18})^{-1} \qquad \text{and} \qquad (1 - x^3 - x^7 + x^{21})^{-1}.$$

The implementation of this approach in `Maple` certifies that both rational functions have Taylor series expansions centered at the origin where all of the coefficients are non-negative. The value of N_0 from Corollary 1 which could be certified by the method described above was $N_0^* = 204$ and $N_0^* = 55$, respectively. Thus, it was easy to utilize `series` in `Maple` to check the non-negativity of the Taylor series coefficients up to N_0^* combined with Corollary 1 for the tail.

7 Conclusion

This manuscript developed techniques for certified evaluations of locally Hölder continuous functions at roots of polynomials along with an implementation in `Maple`. These techniques were demonstrated on several problems including certified bounds on critical values and proving non-negativity of coefficients in Taylor series expansions. Although this paper focused on roots of univariate polynomials, it is natural to extend to multivariate polynomial systems in the future.

Acknowledgments. JDH was supported in part by NSF CCF 1812746. CDS was supported in part by Simons Foundation grant 360486.

A Appendix

Proof of Theorem 1. Suppose that $C \neq 0$ such that $q(x) = C \cdot \prod_{i=1}^{d}(x-\alpha_i)$. Thus, we know $q'(x) = C \cdot \sum_{i=1}^{d} \prod_{j\neq i}(x-\alpha_j)$ and $q'(\alpha_i) = C \cdot \prod_{j\neq i}(\alpha_i - \alpha_j) \neq 0$ for all i. Let $p_i(x) = q(x)/(x-\alpha_i) = C \cdot \prod_{j\neq i}(x-\alpha_j)$. Hence, $p_i(\alpha_i) = q'(\alpha_i)$ and $p_i(\alpha_j) = 0$ if $j \neq i$. The polynomials $p_1, \ldots, p_d$ are linearly independent since, if $\sum_{i=1}^{d} a_i p_i(x) = 0$, then evaluating at $x = \alpha_j$ yields $a_j \cdot q'(\alpha_j) = 0$ which implies $a_j = 0$. Thus, they must form a basis for the d-dimensional vector space of polynomials of degree at most $d-1$.

Since $p(x)$ has degree at most $d-1$, there are unique constants a_i so that $\sum_{i=1}^{d} a_i p_i(x) = p(x)$. Evaluating at $x = \alpha_j$ yields $a_j q'(\alpha_j) = p(\alpha_j)$ so that $a_j = p(\alpha_j)/q'(\alpha_j)$. Therefore, for all $x \in \mathbb{C} \setminus \{\alpha_1, \ldots, \alpha_d\}$,

$$\frac{p(x)}{q(x)} = \sum_{i=1}^{d} \frac{p(\alpha_i)}{q'(\alpha_i)} \frac{1}{x-\alpha_i} = \sum_{i=1}^{d} -\frac{p(\alpha_i)}{\alpha_i q'(\alpha_i)} \frac{1}{1-x/\alpha_i}. \tag{9}$$

The terms in (9) have a Taylor series expansion centered at the origin that converge for all x with $|x| < \min\{|\alpha_1|, \ldots, |\alpha_d|\}$ such that, as (8) claims,

$$\frac{p(x)}{q(x)} = \sum_{i=1}^{d} -\frac{p(\alpha_i)}{\alpha_i q'(\alpha_i)} \sum_{n=0}^{\infty} \alpha_i^{-n} x^n = \sum_{n=0}^{\infty} \left(-\sum_{i=1}^{d} \frac{p(\alpha_i)}{\alpha_i q'(\alpha_i)} \alpha_i^{-n} \right) x^n.$$

Proof of Theorem 2. Clearly, one has $r_n = \left.\frac{d^n}{dz^n} \frac{p(z)}{q(z)}\right|_{z=0}$. Since $p(x)$ and $q(x)$ have real coefficients, r_n is real for all $n \geq 0$. For $i \in \{1, \ldots, d\}$, let $t_n^i = C_i \alpha_i^{-n}$ so that (8) reduces to $r_n = \sum_{i=1}^{d} t_n^i$. Moreover, $\alpha_1 \in \mathbb{R} \setminus \{0\}$ implies $C_1 \in \mathbb{R} \setminus \{0\}$. Clearly, if $\alpha_1 < 0$, then t_n^1 is alternating in sign.

Consider the case when $\alpha_1 > 0$. First, note that t_n^1 and C_1 always have the same sign. The following derives a threshold N such that $|r_n - t_n^1| < |t_n^1|$ for all $n > N$. Given such an N, r_n will have the same sign as t_n^1 and C_1 for $n > N$ and the theorem will be proved. To that end, since $(r_n - t_n^1)/t_n^1 = \sum_{i=2}^{d} t_n^i/t_n^1$,

$$\frac{|r_n - t_n^1|}{|t_n^1|} \leq \sum_{i=2}^{d} \frac{|C_i|}{|C_1|} \frac{|\alpha_1|^n}{|\alpha_i|^n} \leq K \left(\frac{m}{M}\right)^n$$

for all n. Since, by assumption, $m/M < 1$, there is a threshold N so that $K(m/M)^n < 1$ and $|r_n - t_n^1| < |t_n^1|$ for all $n > N$. We may take N so that $K(m/M)^N = 1$ or $N = \log(K)/\log(M/m)$ as claimed.

References

1. Aubry, P., Lazard, D., Moreno Maza, M.: On the theories of triangular sets. J. Symbolic Comput. **28**(1–2), 105–124 (1999)
2. Becker, R., Sagraloff, M., Sharma, V., Yap, C.: A near-optimal subdivision algorithm for complex root isolation based on the Pellet test and Newton iteration. J. Symbolic Comput. **86**, 51–96 (2018)
3. Boulier, F., Chen, C., Lemaire, F., Maza, M.M.: Real root isolation of regular chains. In: The Joint Conference of ASCM 2009 and MACIS 2009, COE Lect. Note, vol. 22, pp. 15–29. Kyushu Univ. Fac. Math, Fukuoka (2009)
4. Gargantini, I., Henrici, P.: Circular arithmetic and the determination of polynomial zeros. Numer. Math. **18**, 305–320 (1971/72)
5. Hardy, G.H.: Weierstrass's non-differentiable function. Trans. Amer. Math. Soc. **17**(3), 301–325 (1916)
6. Johansson, F.: Arb: efficient arbitrary-precision midpoint-radius interval arithmetic. IEEE Trans. Comput. **66**, 1281–1292 (2017)
7. Kantorovich, L.V.: On Newton's method for functional equations. Doklady Akad. Nauk SSSR (N.S.) **59**, 1237–1240 (1948)
8. Kearfott, R.B.: Rigorous global search: continuous problems, Nonconvex Optimization and its Applications, vol. 13. Kluwer Academic Publishers, Dordrecht (1996)
9. Kobel, A., Rouillier, F., Sagraloff, M.: Computing real roots of real polynomials ... and now for real! In: Proceedings of the 2016 ACM International Symposium on Symbolic and Algebraic Computation, pp. 303–310. ACM, New York (2016)
10. Krawczyk, R.: Newton-Algorithmen zur Bestimmung von Nullstellen mit Fehlerschranken. Computing (Arch. Elektron. Rechnen) **4**, 187–201 (1969)
11. Maple 2020 Program Committee Chairs: Private Communication
12. Moore, R.E.: A test for existence of solutions to nonlinear systems. SIAM J. Numer. Anal. **14**(4), 611–615 (1977)
13. Moore, R.E., Kearfott, R.B., Cloud, M.J.: Introduction Interval Analysis. Society for Industrial and Applied Mathematics (SIAM), Philadelphia, PA (2009)
14. Pan, V.Y.: Old and new nearly optimal polynomial root-finders. In: England, M., Koepf, W., Sadykov, T.M., Seiler, W.M., Vorozhtsov, E.V. (eds.) CASC 2019. LNCS, vol. 11661, pp. 393–411. Springer, Cham (2019). https://doi.org/10.1007/978-3-030-26831-2_26
15. Rioboo, R.: Real algebraic closure of an ordered field: implementation in axiom. In: Papers from the International Symposium on Symbolic and Algebraic Computation. ISSAC 1992, pp. 206–215, New York, NY, USA. Association for Computing Machinery (1992)
16. Rouillier, F.: Solving zero-dimensional systems through the rational univariate representation. Appl. Algebra Engrg. Comm. Comput. **9**(5), 433–461 (1999)
17. Rouillier, F., Zimmermann, P.: Efficient isolation of polynomial's real roots. J. Comput. Appl. Math. **162**(1), 33–50 (2003)
18. Rump, S.M.: Verification methods: rigorous results using floating-point arithmetic. Acta Numer. **19**, 287–449 (2010)

19. Smale, S.: Newton's method estimates from data at one point. In: Ewing, R.E., Gross, K.I., Martin, C.F. (eds.) The Merging of Disciplines: New Directions in Pure, Applied, and Computational Mathematics (Laramie, Wyo., 1985), pp. 185–196. Springer, New York (1986). Doi: https://doi.org/10.1007/978-1-4612-4984-9_13
20. Xia, B., Yang, L.: An algorithm for isolating the real solutions of semi-algebraic systems. J. Symbolic Comput. **34**(5), 461–477 (2002)
21. Xia, B., Zhang, T.: Real solution isolation using interval arithmetic. Comput. Math. Appl. **52**(6–7), 853–860 (2006)
22. Yang, L., Hou, X., Xia, B.: A complete algorithm for automated discovering of a class of inequality-type theorems. Sci. China Ser. F **44**(1), 33–49 (2001)

Maple for Distance Education in Secondary Schools During the COVID-19 Emergency

Cecilia Fissore[1](✉), Francesco Floris[2], Marina Marchisio[2], and Matteo Sacchet[2]

[1] Department of Foreign Languages, Literatures and Modern Cultures, University of Turin, Via Giuseppe Verdi 41, 10124 Turin, Italy
cecilia.fissore@unito.it

[2] Department of Molecular Biotechnology and Health Sciences, University of Turin, Via Nizza 52, 10126 Turin, Italy
{francesco.floris,marina.marchisio,matteo.sacchet}@unito.it

Abstract. Many educational institutions have been closed to contain the spread of COVID-19 and teachers took action to continue education through remote learning with the use of technologies. The context of this research is a community of Italian STEM teachers from different secondary schools, within the Ministerial Project PP&S-Problem Posing and Solving. The project involves the use of a Digital Learning Environment (DLE), a Moodle platform integrated with Maple. The research question is: what kind of support can Maple integrated in a DLE give for Distance Education? To answer this question, we considered teachers who used Maple through the DLE during this school year. We carried out an analysis on 74 courses to understand how much they used this type of resource; for what purpose it was used; and whether this type of resource is related to the participation of the students in the course. Analysis shows that there are teachers who have conducted the entire course with this teaching material. They have used Maple in a meaningful way to propose problem-solving activities, theoretical explanations, interactive resources, explanation of Maple commands and text of exercises and resolution. Representative examples of each category are shown. Student views of worksheets increase when teachers use at least three categories of worksheets within a course. In particular, student participation increases when using the problem solving methodology. This type of resources is very effective for online teaching, where immediate feedback and interactivity are essential to involve students more.

Keywords: Digital learning environment · Distance Education · Interactive resources · Maple · Secondary schools · STEM Education

1 Introduction

As UNESCO reports [1], in 2020 most governments around the world have temporarily closed educational institutions to contain the spread of COVID-19 pandemic. In Italy, schools were closed on March 5, 2020 until the end of the school year in June. After

R. M. Corless et al. (Eds.): MC 2020, CCIS 1414, pp. 204–218, 2021.
https://doi.org/10.1007/978-3-030-81698-8_14

the lockdown of the schools and the suspension of face-to-face lessons, it was necessary to switch to "distance learning", in order to not interrupt the didactic continuity and to guarantee the right to education for all the students. All teachers took action to facilitate continuity of education through remote learning with the use of technological methods and tools. At the end of September, the lessons of the new school year began face to face but, due to the uncertainty of the evolution of the pandemic, teachers were asked to contemplate blended scenarios with their students. In some cases, in fact, single infected students or entire classes were isolated for few weeks and for these students the teachers had to carry out the lessons online. In December, a new closure of all upper secondary schools was necessary. All teachers must therefore be ready for hybrid teaching. At the beginning of the emergency, not many schools were prepared for this type of change. For example, since they did not adopt online teaching in regular teaching, teachers did not have the adequate digital skills and adequate devices, so teachers and students did not get used to work on a Digital Learning Environment (DLE), a shared virtual space in which teachers deliver activities and students consult educational resources. The Italian Ministry of Education created a site dedicated to distance learning, to globally distribute instructions to teachers and schools who had to activate types of distance learning. The context of this research is a community of teachers in disciplines like Science, Technology, Engineering and Mathematics (STEM) from different Italian secondary schools, within the Ministerial Project PP&S-Problem Posing and Solving, one of the proposed initiatives [2–4].

The PP&S - "Problem Posing and Solving" - project (available at www.progettop ps.it), headed by the Italian Ministry of Education, promotes since 2012 the training of teachers of secondary schools on innovative teaching methods, through the use of digital technologies, and on the creation of a culture of problem posing and problem solving, with the use of Information and Communication Technology (ICT). Teachers involved in the project learn how to use different kinds of digital tools and new methodologies, in order to enhance their daily teaching. The project proposes innovative methodologies like problem posing and solving using Maple, automatic formative assessment [5, 6] and collaborative learning among teachers and students [7].

In the PP&S community, teachers exchange materials, ideas, and useful advices, they participate in training activities and have the constant support of expert tutors. The project involves the use of a DLE for STEM, a Moodle platform integrated with Maple for the creation of interactive materials, which help the exploration of mathematical concepts and the developing of problem solving skills [8]. Maple allows numerical and symbolic calculations, static and animated graphical representations in 2 and 3 dimensions, writing procedures in simple language, programming and connecting all these different representation registers in a single worksheet using verbal language, too [9].

The term "problem-solving" includes all mathematical tasks that have the potential to provide intellectual challenges for enhancing students' mathematical understanding and development [10]. The ability to solve problems in everyday situations includes the ability to understand the problem, devise a mathematical model, develop the solving process and interpret the obtained solution [11]. The use of Maple in problem solving activities offers a precious diversity of ways to represent and explore the tasks and makes teachers and learners active participants in the learning process [8]. A very important aspect of Maple for problem solving is the design and programming of interactive components (such as math container, text area, slider, tables, graphs, etc.). They allow to visualize the variation of the output when the input parameters change and therefore they allow to generalize the solving process of a problem. Generalizing is an important process by which the specifics of a solution are examined and questions as to why it worked are investigated [12].

Teachers enrolled in the project can work inside an integrated DLE available for each class of students they wish. In addition to using a DLE integrated with Maple, teachers and students receive the software in equipped laboratories thanks to the enrollment in the Project. The choice of the ACE Maple integrated inside the DLE was influenced by different factors. The most important one is the close connection with the Automatic Assessment System Möbius Courseware. Maple engine allows students to analyze, explore, visualize and solve complex mathematical problems. Moreover, Maple suite for education has an attractive, easy-to-use, well designed interface, with tutoring commands and math applications for learning. Finally, Maple is very close to academic research. On the project platform, throughout the year, multiple online synchronous training activities are offered to teachers, about the use of the integrated digital learning environment and about the use of Maple to create interactive files and to design problem posing and solving activities [13, 14]. Teachers who were part of the PP&S before the pandemic and the closure of schools, already used the integrated DLE in their daily teaching and they continued to use it during the emergency. As a result, it was much easier for them to switch to distance learning. For online teaching, but not only, the interactive nature of the resources within the DLE is fundamental. Compared to static resources, interactive resources allow to enter numbers and expressions, to click on buttons, to move sliders, etc. and above all to receive immediate feedback on the exploration. This aspect is very important because it increases student engagement, where by engagement we mean "students' dynamic participation and coparticipation in recognition of opportunity and purpose in completing a specific learning task" [15]. Peculiarity of this definition is the characterization of engagement as an interactive and purposive process; it allows to examine how it may change over time and vary according to situations and contexts. When students participate eagerly in a specific learning activity, they deploy appropriate strategies, regulate processes and monitor their actions [16]. In addition, interactive files provide students with immediate feedback on the exploration and allow teachers to implement formative assessment strategies [17] such as: stimulating discussions, feedback that advance students, and activating students and peers as the protagonists of their learning [18]. These strategies act on student involvement. Maple allows the creation of interactive files and these materials, thanks to the integration of the DLE with the software, maintain interactivity even on the platform (they are called Maple worksheets) and students can actively explore the resource even if they do not have the software on their device.

Before the emergency, teachers used Maple in different ways: carrying out problem solving activities in a computer lab also in groups; having students submit their works (problem solving, exercises or research) created with Maple; projecting the interactive files in class through the IWB (Interactive WhiteBoard) for a theoretical explanation or for an exploratory activity; uploading interactive materials on the DLE for asynchronous activities. After the school closure, teachers and students worked exclusively online via the platform. The teachers had to get used to this new type of teaching and they had to develop strategies to involve students even remotely, through synchronous online lessons but also and above all through asynchronous activities and interactive resources to explore and study. PP&S teachers have the great advantage of being part of a large community of teachers, some of whom have been part of the project for many years and therefore more experienced, and some who are less experienced. Even the teachers who joined the project during the emergency had the opportunity to access the Community database, in which many materials created with Maple for STEM didactics have been inserted and cataloged. Therefore, even teachers who are not experts in using Maple can immediately use the proposed methodologies. Furthermore, within the Community, teachers can exchange ideas and opinions on the design and creation of teaching materials and on the activities carried out with students.

The research conducted fits into this context, in particular we want to study what kind of support can Maple, integrated in a DLE, give for Distance Education. The analyzes conducted and the results obtained show how the use of Maple has facilitated teaching and increased student participation.

2 Methodologies

To answer this question, we considered PP&S teachers who used Maple online through the DLE during the school year 2019/20, uploading a Maple worksheet resource within one of their courses with students. We carried out an analysis on all the courses of the platform used by the teachers in the school year 2019/20 to understand how much they had used this type of resource.

Through the creation of a configurable report, a plugin of the Moodle platform that allows you to create customized reports on all courses using the SQL language, the following information has been automatically reported for each course opened in the 2019/20 school year:

- name and ID (Identification code) of the course;
- teacher data (name, surname, email, institute);
- total number of activities and resources in the course;
- number of Maple worksheets in the course.

The 74 courses in which there were at least five activities and resources (in which at least one of these resources was represented by a Maple worksheet) were selected. For each course, the ratio between the number of Maple worksheet resources and the total number of activities and resources present in the course was also calculated, to analyze how much this resource was used by teachers in relation to other types of online teaching tools. An analysis was then conducted to study for what purpose this educational resource was used, to understand the potential and effectiveness that Maple resources can give to STEM Education. All the worksheets present in the various courses were then classified into one of the following categories:

- Contextualized problem-solving activities - Materials in which it is proposed: a contextualized problem, its resolution with the use of Maple and an active exploration of the generalization of the problem through a system of interactive components.
- In-depth theoretical explanations - Materials for theoretical explanations with texts and formulas but also graphics (static and animated), tables and interactive components for the exploration of theoretical concepts.
- Interactive resources for mathematical exploration - Materials characterized exclusively by interactive explorations for the exploration of mathematical concepts (such as MathApps).
- Explanation of Maple commands - Materials for explaining the use of Maple: the basic commands, the commands in the various packages and the design and creation of interactive components.
- Text of exercises and resolution - Materials in which Mathematics exercises are presented (for example taken from textbooks) and their resolution.

By classifying the worksheets into these categories, for each course it was indicated how many worksheets of each type are present. In this way, it was possible to study:

- if teachers use more types of worksheets, and therefore more teaching methods;
- which are the most used types;
- if there are patterns in the use of the various types (for example if teachers who use worksheets for theoretical explanations tend to assign exercises with Maple as well).

Another goal was to see if, in the courses where the teachers used Maple worksheet resources, the amount of resources of this type is related to the participation of the students in the activities and resources of the course, and if it also reflects/affects their design by the teachers. To do this, two pieces of information have been added to the previously created configurable report to analyze student participation. For each course it was reported:

- number of views of course worksheets (total student logs to worksheets in the course);
- course logs (total student logs to all course activities).

Using the software R, the correlation matrix between the following variables was calculated:

- course logs;
- total number of activities and resources present in the course;
- number of course worksheets;
- number of views of course worksheets.

The last goal was to test if and how using worksheets of different categories influences student participation in the online course. In this case, we have used the subdivision of courses based on the number of categories of worksheet used (courses with one category of worksheets, two categories, etc.) with respect to the variables inherent to student logs.

Finally, a specific analysis was carried out to study the impact of using the "contextualized problem-solving activities" category of worksheet on student participation.

3 Results

3.1 Descriptive Analysis

The first analysis was carried out on the 74 selected courses, in which there were at least five activities and resources (in which at least one of these resources was represented by a Maple worksheet). The 74 courses analyzed were held by 40 teachers. 62.5% of teachers teach upper secondary school (divided equally between High Schools and Technical Institutes) and 37.5% lower secondary school. The subjects taught by the teachers in the various courses are: Physics (1%), Mathematics (70%), Mathematics and Physics (15%), Mathematics and Computer Science (14%).

A total of 1289 Maple worksheets were uploaded into 74 courses. The number of worksheets in a course varies from a minimum of three to a maximum of 138, and the median is 10.5. For each course, the ratio between the number of maple worksheets and the total number of activities and resources in the course was calculated. On average, 27% of course materials are worksheets, with a range from 5% to 84%. Therefore, there are teachers who conducted the entire online course using Maple worksheet and teachers who rarely used this type of resource. The graph shown in Fig. 1 shows the various percentages of worksheets available in the courses compared to the total number of activities and resources, divided into four percentage classes: 5%–24%, 25%–44%, 45%–64%, 65%–85%.

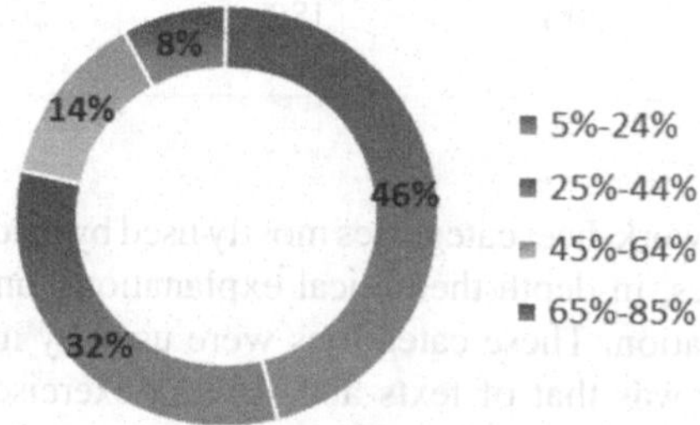

Fig. 1. Percentages of the worksheets available in the courses compared to the total number of activities and resources.

As shown in the graph, the most populated class is the "5%–24%" class (46%): this class is represented by teachers who have used this type of resource less than other types of activities and resources available within the DLE. However, it can be seen that for more than half of the courses, more than a quarter of the materials are worksheet resources. A percentage of teachers (8%) made almost exclusive use of this type of resource when teaching their subject.

3.2 Classification of Worksheets into Five Categories

For each course, all the Maple worksheets were analyzed and each resource was classified into one of the five categories: contextualized problem solving activities; in-depth theoretical explanations; interactive resources for mathematical exploration; explanation of Maple commands; and text of exercises and resolution. The number of worksheets of each type was then reported for each course. Table 1 shows, for each category of worksheet:

- the number of teachers who used worksheets of this category;
- the number of worksheets in this category;
- the percentage of worksheets in this category with respect to the total number of worksheets (1289).

Table 1. Classification of worksheets in the five categories and use by teachers.

	Problem solving activities	Theoretical explanations	Maple explanations	Interactive resources	Exercises
Number of teachers who used worksheets of this type	31	34	27	34	23
Number of worksheet	348	259	229	339	114
Percentage of worksheets	27%	20%	18%	26%	9%

This table shows that the worksheet categories mostly used by teachers are contextualized problem solving activities, in-depth theoretical explanations, and interactive resources for mathematical exploration. These categories were used by most of the 40 teachers. The least used category was that of texts and solving exercises, probably because it is a static resource, very similar to the textbooks that all students possess. The great advantage of using Maple is precisely the possibility of creating interactive contents for the active exploration of mathematical concepts and problems by students. In fact, of the 1289 worksheets, more than half are problem solving and interactive exploration activities. Figure 2 shows the number of teachers who used one, two, three, four or five types of Maple worksheets within the same course.

The graph shows that 66% of teachers used more than one category of Maple worksheet, adopting different teaching methods depending on the purpose of the activity. 34% of teachers who used only one type of Maple worksheet chose more problem solving activities (36%) and interactive resources (32%), but fewer Maple explanations (4%). This last result may be due to the fact that this type of file is mainly designed for students who use Maple alone even at home, and can be more supportive when combined with other types of worksheets. The first result is in line with the previous results.

A final analysis on the various categories of worksheet used by teachers concerned the search for patterns in the use of the various types. Table 2 shows the numerous combinations of use of different types of worksheets in the 74 courses.

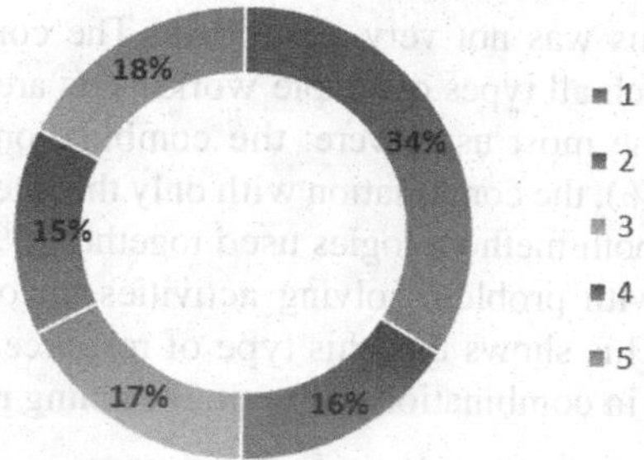

Fig. 2. Number of Maple worksheet categories used by teachers.

Table 2. Combination of the types of worksheets used by teachers looking for patterns.

Problem solving activities	Theoretical explanations	Maple explanations	Interactive resources	Exercises	
X	X	X	X	X	13
X					9
			X		8
X	X			X	6
X	X	X			4
X			X		4
	X				4
	X	X	X	X	3
					3
X	X		X	X	2
X	X		X		2
X	X				2
X		X			2
X			X	X	2
X				X	2
	X	X	X		2
X		X		X	1
	X		X		1
		X	X		1
		X		X	1
		X			1
			X	X	1

The search for patterns was not very significant. The combination used by most teachers is the one in which all types of Maple worksheets are used (18%). According to the previous results, the most used were: the combination with only the problem solving methodology (12%), the combination with only the interactive resources (11%), or the combination with both methodologies used together (5%). Another widely used combination is the one with problem solving activities, theoretical explanations and solving exercises (8%). This shows that this type of resource, even if more static and ordinary, can be effective in combination with other teaching methodologies.

3.3 Representative Examples of the Various Categories of Worksheet

Contextualized Problem Solving Activities

The first category of worksheet includes problem solving activities. All materials of this type, according to the methodology proposed within the project, are characterized by: the presentation of a contextualized problem (Example in Fig. 3), a resolution proposal with the use of Maple and the generalization of the solution process through a system of interactive components. The contextualization of the problem in a real context or in one that is familiar to the students is one of the main factors of the proposed methodology. This allows students to approach the world of mathematics and to understand its importance in everyday life. The solution of the problem is proposed using different registers (calculations, formulas, tables, graphs, etc.). Another very important aspect is the generalization of the problem. This is an activity that would not be proposed to students during regular lessons with pen and paper, but which is fundamental for developing computational thinking and abstraction skills. Through the interactive components, students can change the data of the problem and see how the solution of the problem varies. In this way students can focus on the solution process and its abstraction and modeling, and not just on the calculations.

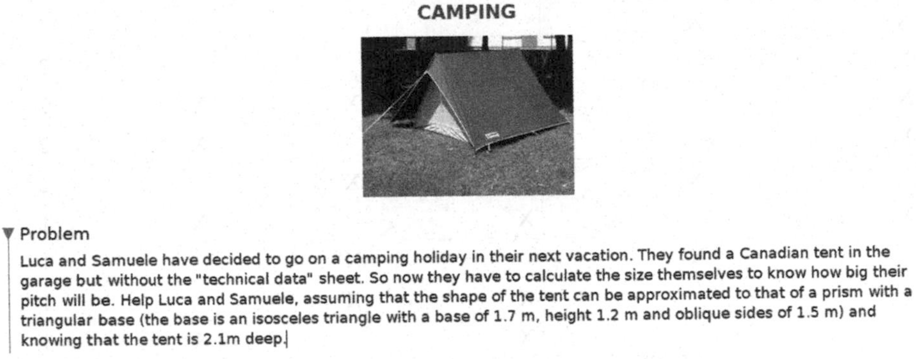

CAMPING

Problem

Luca and Samuele have decided to go on a camping holiday in their next vacation. They found a Canadian tent in the garage but without the "technical data" sheet. So now they have to calculate the size themselves to know how big their pitch will be. Help Luca and Samuele, assuming that the shape of the tent can be approximated to that of a prism with a triangular base (the base is an isosceles triangle with a base of 1.7 m, height 1.2 m and oblique sides of 1.5 m) and knowing that the tent is 2.1m deep.

Fig. 3. Example of contextualized problem created by teachers

In-Depth Theoretical Explanations

The second category of worksheet is the one dedicated to theoretical explanations.

Through this type of resource, the teachers' goal is to enrich the theoretical explanation of the textbooks, which all students have and use, with animated graphics and interactions. In the example in Fig. 4, the teacher proposes an introduction to the concept of limit of a function in an accumulation point and enriches the explanation with an animated graph that clarifies the explanation and provides an example. Within the integrated DLE the animations within the worksheets are displayed as GIFs.

INTRODUCTION TO THE CONCEPT OF FUNCTION LIMIT IN ONE POINT

To calculate the limit of a function, at an accumulation point, we need to study the behavior of the function around that point.

For example, consider the graph of the function $Y = \frac{k}{x}$

We want to calculate the limit for x tending to 2 of the function: we focus our attention on the values that the function assumes in a neighborhood of 2

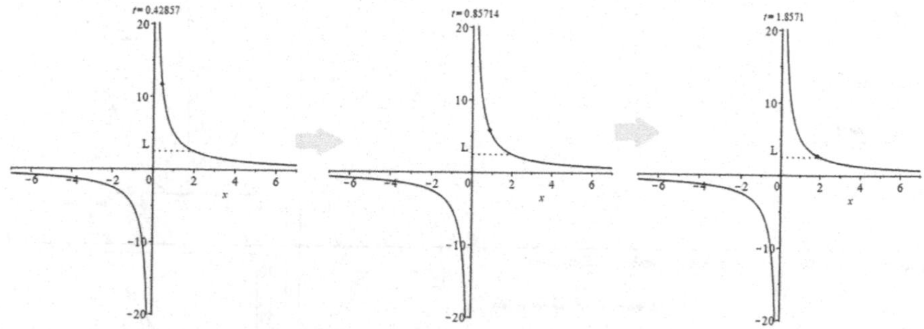

Fig. 4. Example of worksheet for a theoretical explanation enriched by animated graphics

Interactive Resources for Mathematical Exploration

The third category of worksheet is "interactive resources for mathematical exploration". These types of resources are very similar to MathApps (available in Maple) but are created by the teachers themselves. Through a system of interactive components, mathematical explorations are proposed to study mathematical concepts, for modeling activities, to support the study or for insights. The example shown in Fig. 5 concerns the problem of calculating the areas. Through a step-by-step procedure (of which only the first steps are shown), the student is guided in the reasoning of calculating the area of a flat region. In the first part, the student can choose the function to study and write it inside the math container, after which he can visualize it graphically by clicking the button. In the second part, the student must choose the extremes in which to draw and calculate the area of the function and how many parts to subdivide the interval into, and again visualize the function in the chosen interval. In the third part, by clicking the two buttons, the student can view the inscribed or circumscribed rectangles. This type of material is very effective for formative evaluation because it provides an example of good practice and gives immediate feedback. Students are active and engaged in learning.

Explanation of Maple Commands

This category of worksheet includes all materials dedicated exclusively to the explanation of how to use Maple (commands, particular packages, procedures, interactive components, etc.). Usually, teachers teach students to use the software face-to-face in a computer lab but, having to do it remotely, they have created explanatory files. One of the most interesting examples in this category is the interactive guided explanation of how to create an interactive component system. To enrich the explanation, the teacher also created a short video commenting the Maple file.

Text of Exercises and Resolution
This category of worksheet differs from the previous ones because it is not interactive. This type of resource is similar to the exercises found in a textbook. In this case the teachers use Maple as a text editor to write the texts of the exercises and comment on the resolutions and commands (also maple tools like Tasks and Tutors) to calculate the solutions.

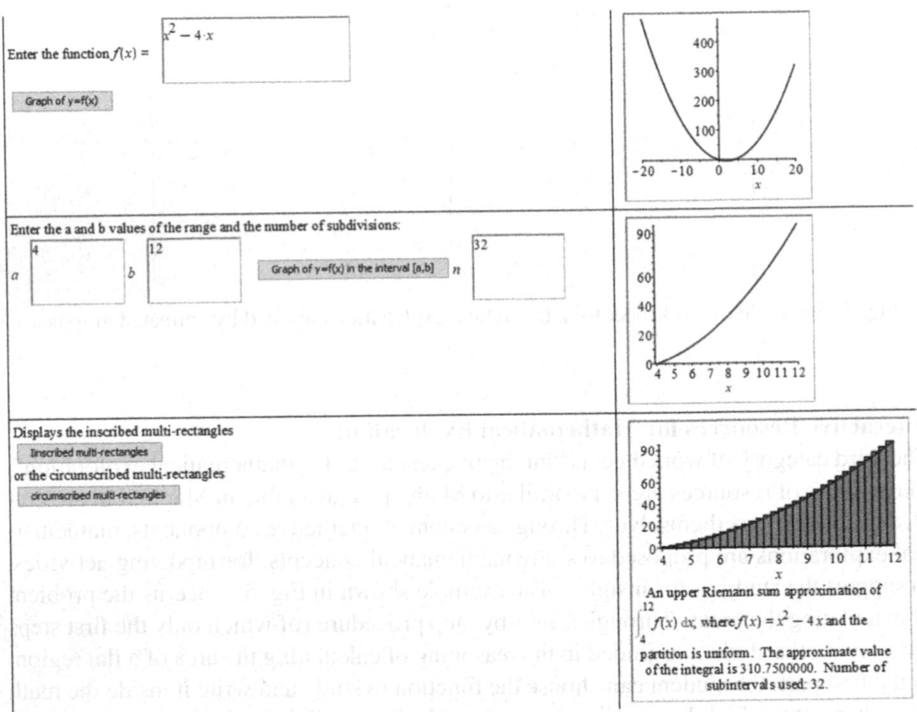

Fig. 5. Example of an interactive resource on the problem of calculating areas

3.4 Analysis of Student Participation

As seen in Table 3, the linear correlation index showed that there is a relationship between the two variables "total number of activities" and "course logs" (index value 0.727). This result is not particularly surprising, since the logs are automatic recordings made by the DLE at every user interaction with the resources or activities of the course. Therefore, as their number increases within the course, the course logs increase in turn.

The correlation index between the variables "Maple worksheet views" and the "number of Maple worksheet" highlights how the increase in views is moderately correlated with the number of Maple worksheets (index value 0.444). This may be due to the fact that, as a teaching methodology, it is not enough to just insert many worksheets into the course for students to use. Especially in online teaching, students need to be guided within the DLE to use the resources.

There is also a moderate correlation between the total number of course activities and the number of Maple worksheets (index value 0.696). This result may seem obvious since we considered the worksheets within the course activities. However, on average, worksheets represent approximately 30% of course activities. This means that other activities and resources are used in parallel with the use of worksheets, such as, for example, material for explanations on the use of DLE, for formative assessment activities, for learning tests, for in-depth studies, etc.

Table 3. Correlation matrix

	Total number of activities	Number of Maple worksheet	Maple worksheet views	Course logs
Total number of activities	1	0.696	0.389	0.727
Number of Maple worksheet	0.696	1	0.444	0.17
Maple worksheet views	0.389	0.444	1	0.37
Course logs	0.727	0.17	0.37	1

The latest analysis concerned the impact of the number of worksheet categories used by teachers on student participation in the course. The courses were divided into five classes according to the number of worksheet categories used in the course (as in Fig. 2). Since the "course logs" and "worksheet views" variables are not normally distributed, we used the Kruskal – Wallis test by ranks. The first result is that there is no significant difference between the medians of the course logs in the various classes (p-value = 0.2). Since having used multiple types of worksheets has no influence on the course logs, the influence on the number of views of worksheets within the course was studied. This analysis shows that there is a significant difference among the medians of the worksheet views in the course classes (p-value < 0.0001). Figure 6 shows how the averages of the views appear to be higher in the last three course classes (those in which teachers have used at least three categories of worksheet).

By dividing the courses into two macroclasses (class 1 and 2 on the one hand and classes 3, 4 and 5 on the other), the test shows (Fig. 7) that the median of the views is greater for the second macroclass (p-value < 0.0001).

These results show that using multiple types of worksheets affects students' participation in this type of resource and not the entire course. Therefore, using this resource for different methodologies engages the students. Student participation increases when

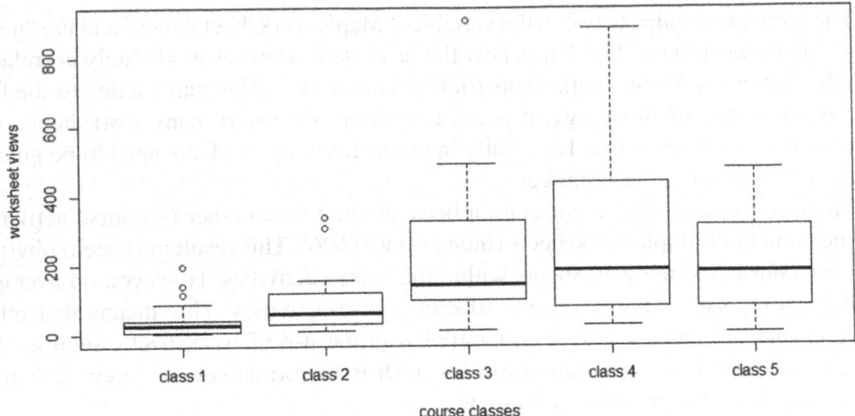

Fig. 6. Boxplot of the averages of the worksheet views in the course classes

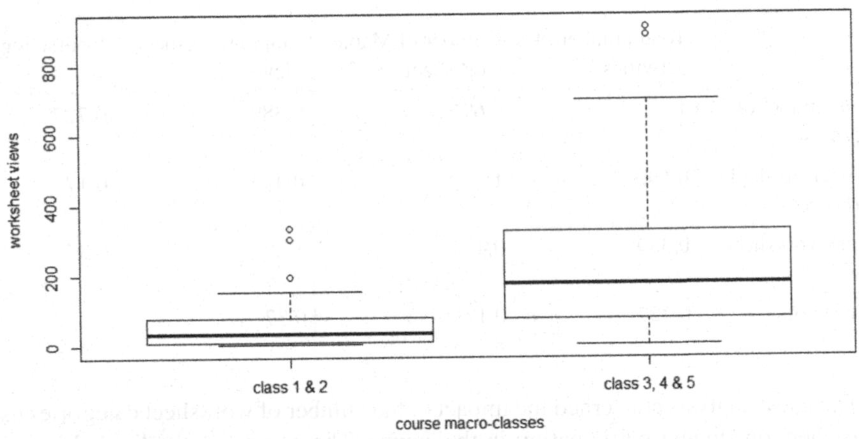

Fig. 7. Boxplot of the averages of the worksheet views in the two macro-classes

teachers use at least three types of worksheets. From this type of analysis, it does not emerge directly which type of worksheet has the greatest influence on student participation (which will be the subject of future analysis). However, the test shows that in courses in which worksheets of the "contextualized problem-solving activities" category were inserted, student views are on average higher than in courses in which this category of worksheet is not present (p-value = 0.04).

4 Conclusion

This paper investigates Maple software support for online teaching during the pandemic emergency in the 2019/20 school year. The context of the research is the Italian community of teachers of the PP&S ministerial project of STEM disciplines. Teachers enrolled in the project can have a DLE, Moodle platform integrated with Maple, available for each class of students. 74 courses, in which the Maple worksheet resource was used, were analyzed: 1289 Maple worksheets were uploaded into the courses. Analysis shows that there are teachers who have conducted the entire course with these Maple worksheet resources and teachers who have rarely used them.

Teachers have used Maple in a meaningful way to propose problem-solving activities, in-depth theoretical explanations, interactive resources for mathematical exploration, explanation of Maple commands, and text of exercises and resolution. Some significant examples have been shown to explain how this type of resource can enrich STEM teaching and learning. Not being able to teach Maple in the school computer lab, some teachers have designed files with step-by-step explanations for students to allow them to use the software independently at home. Most of the teachers used more than one type of Maple worksheet, adopting different teaching methods depending on the purpose of the activity.

The worksheet categories most used by teachers are contextualized problem solving activities, in-depth theoretical explanations, and interactive resources for mathematical exploration. The most appreciated feature by teachers is the interactive Maple files (which is maintained thanks to the integrated platform) used for interactive theoretical explanations, interactive explorations of mathematical concepts and for the generalization of solution processes in problem solving activities.

Analysis shows that it is not enough to insert many worksheets into the course for students to view and interact with them. Particularly in online teaching, students have to be guided within the DLE to use the resources. In addition, student views of worksheets increase when teachers use at least three categories of worksheets within a course with different teaching methodologies. In particular, student participation increases when using the problem solving methodology.

This type of resource, more interactive and effective than classic paper textbooks, is very effective for online teaching, where immediate feedback and interactivity are essential to involve students more. During the emergency, Maple not only proved to be useful for dealing with forced distance teaching more easily, but it also showed to be a very convenient tool, flexible for new modalities and new purposes like synchronous online lessons, formative online assessment, adaptive activities for students with difficulties. This resource can enrich STEM teaching and learning at any level of education.

References

1. UNESCO: Education - From disruption to recovery. https://en.unesco.org/covid19/educationresponse. Accessed 04 Jan 2021
2. Fissore, C., Marchisio, M., Rabellino, S.: Secondary school teacher support and training for online teaching during the COVID-19 pandemic. In: European Distance and E-Learning Network (EDEN) Proceedings, Timisoara, pp. 311–320. European Distance and E-Learning Network (2020)
3. Barana, A., Fissore, C., Marchisio, M., Pulvirenti, M.: Teacher training for the development of computational thinking and problem posing and solving skills with technologies. In: Proceeding of eLearning Sustainment for Never-Ending Learning. Proceedings of the 16th International Scientific Conference ELearning and Software for Education, vol. 2, pp. 136–144 (2020).

4. Brancaccio, A., Demartini, C.G., Marchisio, M., Pardini, C., Patrucco, A.: The PP&S computer science project in school. Mondo Digitale **13**, 565–574 (2014)
5. Barana, A., Fissore, C., Marchisio, M.: From standardized assessment to automatic formative assessment for adaptive teaching. In: Proceedings of the 12th International Conference on Computer Supported Education, Prague, Czech Republic, pp. 285–296. SCITEPRESS - Science and Technology Publications (2020). https://doi.org/10.5220/0009577302850296
6. Barana, A., Conte, A., Fissore, C., Marchisio, M., Rabellino, S.: Learning analytics to improve formative assessment strategies. J. e-Learning Knowl. Soc. **15**, 75–88 (2019). https://doi.org/10.20368/1971-8829/1135057
7. Barana, A., Marchisio, M.: Dall'esperienza di digital mate training all'attività di alternanza scuola lavoro. Mondo Digitale **15**, 63–82 (2016)
8. Barana, A., et al.: The role of an advanced computing environment in teaching and learning mathematics through problem posing and solving. In: Proceedings of the 15th International Scientific Conference eLearning and Software for Education, Bucharest, pp. 11–18 (2019). https://doi.org/10.12753/2066-026X-19-070
9. Barana, A., Conte, A., Fissore, C., Floris, F., Marchisio, M., Sacchet, M.: The creation of animated graphs to develop computational thinking and support STEM education. In: Gerhard, J., Kotsireas, I. (eds.) MC 2019. CCIS, vol. 1125, pp. 189–204. Springer, Cham (2020). https://doi.org/10.1007/978-3-030-41258-6_14
10. National Council of Teachers of Mathematics: Executive Summary Principles and Standards for School Mathematics (2000)
11. Samo, D.D., Darhim, D., Kartasasmita, B.: Culture-based contextual learning to increase problem-solving ability of first year university student. J. Math. Educ. **9**(1) (2017). https://doi.org/10.22342/jme.9.1.4125.81-94
12. Liljedahl, P., Santos-Trigo, M., Malaspina, U., Bruder, R.: Problem Solving in Mathematics Education. Springer, Cham (2016). https://doi.org/10.1007/978-3-319-40730-2
13. Barana, A., Pulvirenti, M., Fissore, C., Marchisio, M.: An online math path to foster the transition of students between lower and upper secondary school. In: eLearning Sustainment for Never-Ending Learning. Proceedings of the 16th International Scientific Conference ELearning and Software for Education, Bucharest, pp. 568–575. Carol I National Defence University Publishing House (2020)
14. Barana, A., Marchisio, M., Miori, R.: MATE-BOOSTER: design of an e-learning course to boost mathematical competence. In: Proceedings of the 11th International Conference on Computer Supported Education (CSEDU 2019), pp. 280–291 (2019)
15. Ng, C., Bartlett, B., Elliott, S.N.: Empowering Engagement: Creating Learning Opportunities for Students from Challenging Backgrounds. Springer, Cham (2018). https://doi.org/10.1007/978-3-319-94652-8
16. Barana, A., Marchisio, M., Rabellino, S.: Empowering engagement through automatic formative assessment. In: 2019 IEEE 43rd Annual Computer Software and Applications Conference (COMPSAC), Milwaukee, WI, USA, pp. 216–225. IEEE (2019). https://doi.org/10.1109/COMPSAC.2019.00040
17. Hattie, J., Timperley, H.: The power of feedback. Rev. Educ. Res. **77**, 81–112 (2007). https://doi.org/10.3102/003465430298487
18. Black, P., Wiliam, D.: Developing the theory of formative assessment. Educ. Assess. Eval. Acc. **21**, 5–31 (2009). https://doi.org/10.1007/s11092-008-9068-5

Development of Problem Solving Skills with Maple in Higher Education

Cecilia Fissore[1], Marina Marchisio[2], Fabio Roman[2], and Matteo Sacchet[2](✉)

[1] Department of Foreign Languages and Literatures and Modern Cultures, University of Turin, Via Giuseppe Verdi 41, 10124 Turin, Italy
cecilia.fissore@unito.it

[2] Department of Molecular Biotechnology and Health Sciences, University of Turin, Via Nizza 52, 10126 Turin, Italy
{marina.marchisio,fabio.roman,matteo.sacchet}@unito.it

Abstract. Problem solving is the ability to understand the environment, identify complex problems, and review related information to develop, evaluate strategies and implement solutions to build the desired outcome. Mathematics boosts problem solving skills and, in Higher Education, all scientific degree programs deliver at least one module in Mathematics that should develop students' problem solving skills. Mathematics Modules of the Biotechnology Bachelor Degree and of the Strategic Science Bachelor and Master Degrees at the University of Turin use innovative digital technologies, like the Advanced Computing Environment Maple, and methodologies to facilitate the learning of Mathematics and the development of problem solving skills. At the beginning of the courses, students must learn how to use Maple through dedicated lab sessions to solve contextualized problems related to their future careers. Moreover, for the final examination, students must study, present and discuss a science-based problem solved with Maple. In this paper, we investigated how the use of Maple enabled students to develop problem solving skills. We examined 110 students' submissions through a rubric that analyzes different dimensions: comprehension, resolution strategy identified, solution process, representation, argument, use of Maple. Dimensions are correlated with module attendance, involvement, exam marks. A qualitative analysis was also performed. The research shows that the adopted approach is useful and effective: students' scores are high and submissions indicate the presence of problem solving skills. Problem solving labs with Maple should be introduced, in connection with other disciplines, to facilitate analysis of data, visualization, communication, and deep understanding of concepts.

Keywords: Higher education · Maple · Mathematics education · Problem solving

1 Introduction

Humans have always been fascinated and entertained by solving problems, games, mysteries, and puzzles. Even in ancient history, there were some famous problems, halfway

R. M. Corless et al. (Eds.): MC 2020, CCIS 1414, pp. 219–233, 2021.
https://doi.org/10.1007/978-3-030-81698-8_15

between legend and reality, such as the measurement of the Great Pyramid performed by Thales and the mythological riddle of the Sphinx, asked to the tragic hero Oedipus. With the evolving of human history, societies became complex systems, with a great variety of professionals and citizens who now have to face practical problems in their daily routine. The challenge to tackle such complexity and such practical problems nowadays mainly resides on Higher Education that needs to foster creative problem solvers [1]. Solving a problem requires creativity, but at the same time problems help in developing creativity. Novelty in creativity does not necessarily mean that something is new for the whole humanity, but it can be such also just for the individual who needs to develop his capacities and to focus on relevance and effectiveness.

More specifically, Problem Solving (PS) can be defined as the ability to understand the environment, identify complex problems, review related information to develop, evaluate strategies and implement solutions to build the desired outcome. It is the basis for creative thinking, new inventions, evolution, continuous improvement, communication, and learning. PS skills are then essential for every citizen in the world, not only for their career but also for their nonprofessional, everyday life. There is indeed scientific research on PS skills for adults, especially in connection with technology-rich environments [2] and teacher training [3, 4]. PS competences are essential inside the world of work, in [5] PS appears twice in the list of the Top 15 skills for 2025, and those skills are also very important in the framework of Digital Competences [6]. Higher Education plays a pivotal role: findings indicate that there is a tendency to have high problem-solving skills among adults with Higher Education degrees [2].

The best area to foster PS skills is Mathematics, the queen of sciences. In Higher Education, all scientific degree programs deliver at least one module in Mathematics that should develop students' PS skills beyond notions. However, this is not always the case, since the curricula contain a lot of theory and exercises that do not leave much space to PS, which looks more like a side effect of education. That is why at the University of Turin we are addressing PS with different actions, involving students, teachers at secondary schools [7], and university students [8].

1.1 Outline of the Research

At the University of Turin, technology is used to enhance the development of PS skills. In this research, we are going to address a specific way to face this challenge adopted in Mathematics Modules of the Biotechnology Bachelor Degree, of the Earth Sciences Bachelor Degree, and of the Strategic Science Bachelor and Master Degrees at the University of Turin. The approach makes use of innovative digital technologies, like the Advanced Computing Environment (ACE) Maple, and methodologies, like contextualized problem solving and formative assessment, in order to facilitate the learning of Mathematics and, more in detail, the development of PS skills [9, 10].

The research question of this paper is the following: can Maple be a valuable support for students to develop problem solving skills, to learn Mathematics and to understand its applications?

The research does not aim at stating the importance of the software, which is indubitable, but it considers a variety of actions (activities, labs, teaching strategies, online education) that are devoted to the development of learners' PS skills. These actions must provide the basis in which the use of the ACE Maple gives an enhancement to the development of students' skills.

Section 2 delineates the state of the art about PS skills development, its close relation with Mathematics and its relation with Higher Education. Section 3 illustrates the experience and the approach about PS skills development inside modules at the University of Turin. Section 4 is devoted to the presentation of the research methodology. Section 5 shows results of the experience, measuring scores and the point of view of students. Section 6 provides a showcase of some elements of the work of students that exhibit the presence of PS skills. In Sect. 7, conclusions are drawn.

2 State of the Art

One of the first authors to describe methods of PS was the Hungarian mathematician George Pólya in the book "How to solve it" [11]. In its list, Pólya suggests the following steps when solving a mathematical problem.

- First step: you have to understand the problem and determine unknowns, data and conditions.
- Second step: make a plan, find connections between data and unknowns, according also to previous experience, eventually split into auxiliary problems.
- Third step: carry out the plan, carefully checking each step of the solution process.
- Fourth step: look back on your work, examine the solution obtained and see if something better could have been done.

Pólya stressed the importance of the teacher: this role requires time, practice, devotion and sound principles. Students need to develop PS skills independently, but they should not be alone, because in this case they may make no progress at all. There is a balance in these two aspects. This happens at all stages of education.

To improve PS skills, students should regularly work to solve problems according to their own attitude [12]. Using non-routine problems and open problems, students can face a wide range of possibilities and good or bad performance is an indicator of the presence of PS skills. On the other side, learned PS strategies tend to be ignored at higher levels, possibly because of the routine nature of examinations [12].

Modellization should not take place exclusively into the students' mind. Words, numbers, symbols, graphics, simulations, technologies and many others must mediate contextualized real-world situations [13]. Representation is at the core of PS, and different computerized systems enable and enhance different approaches to the problem. Mathematics skills are also reinforced after implementation in the systems [14].

Different activities carried out at the University of Turin aim at enhancing PS skills using technology. The target of these activities are secondary school students and teacher [15–17]. As an example, Digital Math Training (DMT) wants to develop and strengthen Mathematics and Computer Science skills through problem solving activities using the ACE Maple [18]. This activity is meant for secondary schools, but in many cases the performance required by students is quite challenging, close to university level. This activity also promotes Collaborative Problem Solving.

Approach in Collaborative Problem Solving is affected by gender, too. In [19], the authors studied the effects of gender pairings on CPS performances, processes, and attitudes in a social learning context. Single-gender groups had more focused discussions; male-male groups tended to develop and test their solutions directly without spending significant time on problem identification; female-female groups were more attentive to the benefits of social learning.

Collaboration can be promoted through Digital Learning Environments (DLE), which are suitable places to share, assess, communicate and interact between peers, tutors and machine agents [20–22].

There are evidences that the participation to online collaborative activities and discussions promotes critical thinking [23]. A close connection brings together Information and Communication Technology (ICT), Mathematics and PS, too. ICT and PS work on problem-based software development [24] and allow students to develop Computational thinking competences [10, 25]. In fact, programming is not only about writing code but also about the ability to analyze a scenario, identify key components, model data and processes, and refine a program through an agile design-thinking approach [26]. In educational settings, programming can be used as a tool to develop PS skills and to engage participants in creative PS activities.

3 Problem Solving Inside University Modules

Regular teaching at the university is an opportunity to experiment different approaches. The main degrees in which we adopted new methodologies are the Biotechnology Bachelor Degree, in the academic year 2019/2020, the Strategic Science Bachelor and Master Degrees, in the academic years 2017/2018, 2018/2019, 2019/2020 and the Earth Sciences Bachelor Degree, in the academic years 2017/2018, 2018/2019.

Students of these modules must learn how to use the ACE Maple. The choice of the ACE Maple integrated inside the DLE was influenced by different factors. The most important one is the close connection with the Automatic Assessment System Möbius Courseware. Maple engine allows students to analyze, explore, visualize and solve complex mathematical problems. Moreover, Maple suite for education has an attractive, easy-to-use, well designed interface, with tutoring commands and math applications for learning. Finally, Maple is very close to academic research.

Students learn Maple through dedicated lab sessions and they must use it alone and in groups with different modalities: during in-person lessons and lab sessions, inside a DLE integrated with the ACE. The aim of these activities is to develop PS skills by solving contextualized problems connected with possible challenges that students will face in their future job. Students attend 3 to 12 h labs solving problems inside the ACE Maple, meanwhile learning how to use the software. During labs, tutors briefly introduce a mathematical topic, its applications and the basic commands and operations that can be performed inside the ACE Maple; afterwards students can practice with a contextualized exercise. Students practice with PS, with examples of applications and with Maple even at home with their personal device.

As part of the final examination, students must study, present and discuss a science-based problem solved with Maple. In order to get a high score, they must show mastery in Mathematics underlying the problem, critical and computational thinking, proper use of the ACE Maple, ability to generalize, to justify and to make arguments for the provided solutions. Students do not only solve a problem, but also pose it, according to their experience, to the study of other scientific topics and, more importantly, based on examples seen during classes and inside the DLE.

4 Methodology

To answer the research question stated in the introduction, in this work we examined the effectiveness of the approach adopted through the analysis of worksheets that students submitted in order to take the exam. From the various submission activities, we collected 110 distinct files, which correspond to 127 students since some of the submissions came from a group work. It is worth mentioning that this sample does not cover all students that took an exam with this approach of submitting a science-based problem with the ACE Maple. Going back to previous academic years, we could collect more submissions, but this approach has been refined over the years, thus we decided to only take into account students from the academic years 2017/2018, 2018/2019, 2019/2020. Moreover, this research considers only students that submitted the work through the official submission activity: other methods of submission (email, hard drive, cloud sharing) lie outside this research. Students come from three different degree courses: the Biotechnology Bachelor Degree, the Earth Sciences Bachelor Degree, and the Strategic Science Bachelor and Master Degrees at the University of Turin. They concern different disciplines, but for each of them it is important to develop PS skills. Figure 1 shows the percentage of students for each degree course, while Table 1 provides the precise number of students divided by degree course and academic year.

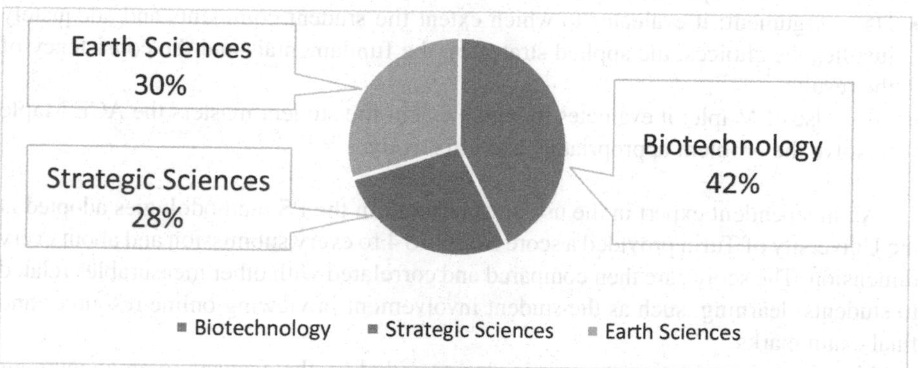

Fig. 1. Percentage of students that submitted a science-based problem divided by degrees: biotechnology, earth sciences and strategic sciences.

The evaluation is based on a rubric. The rubric has been prepared in accordance to the literature on PS, the Italian Mathematics assessment grid, and to previous experience of DMT at the University of Turin. The specific rubric has been developed to assess the students' competence in PS. This rubric is quite transversal among education levels and

Table 1. Number of students divided by degree courses and academic years (AY)

Degree course	AY	Students	Total per AY
Biotechnology Bachelor Degree	2019/2020	54	63
Strategic Sciences Master Degree	2019/2020	9	
Earth Sciences Bachelor Degree	2018/2019	16	30
Strategic Sciences Master Degree	2018/2019	14	
Earth Sciences Bachelor Degree	2017/2018	22	34
Strategic Sciences Master Degree	2017/2018	12	
Total		**127**	

it is the basis for this paper's evaluations of submissions. In this research, the rubric evaluates six different dimensions:

- D1 – Comprehension: it evaluates to which extent the student is able to analyze the situation, to represent and translate the data into mathematical language
- D2 – Resolution strategy: it evaluates to which extent the student implements and identifies the most suitable solution strategies through modeling.
- D3 – Solution process: it evaluates to which extent the student solves the problem in a coherent, complete and correct way, applying the rules and performing the necessary calculations.
- D4 – Representation: it evaluates to which extent the student graphically represents the problem and the results and communicates effectively to the reader with proper diagrams, tables, plots, animations.
- D5 – Argument: it evaluates to which extent the student comments and adequately justifies the choices, the applied strategies, the fundamentals and the consistency of the results.
- D6 – Use of Maple: it evaluates to which extent the student masters the ACE Maple to solve the problem appropriately and effectively.

An independent expert in the use of Maple and in the PS methodologies adopted at the University of Turin provided a score from 1 to 4 to every submission and about every dimension. The scores are then compared and correlated with other measurables related to students' learning, such as the student involvement in viewing online resources and final exam marks.

Moreover, a qualitative analysis is also provided by the answers to an anonymous questionnaire that student had to submit after attending the Biotechnology Bachelor Degree, which is the course with the highest attendance among the considered courses. We had 87 respondents. The questionnaire covered different topics, but we extracted only the relevant items, related to PS skills and the ACE Maple. Different kinds of questions compose the questionnaire, from Likert scale to open-ended answers. For scales, basic statistics with median and IQR (Inter Quartile Range) are provided.

5 Results

Each of the 110 submissions received a score from 1 to 4 on the various dimensions that were considered in the rubric. Figure 2 shows the distribution of scores according to these six dimensions.

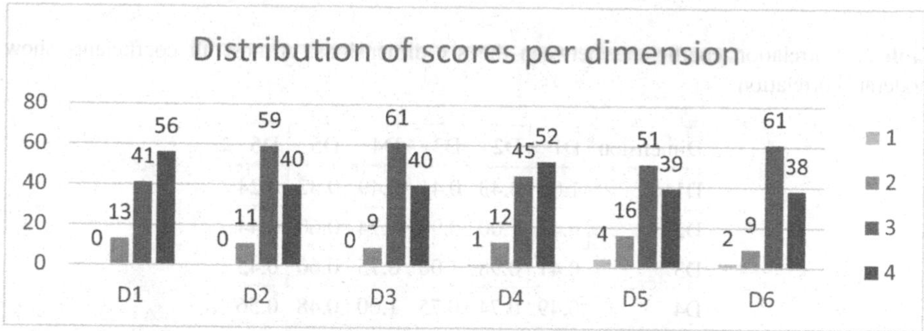

Fig. 2. Distribution of scores per dimensions: D1 Comprehension, D2 Resolution strategy, D3 Solution process, D4 Representation, D5 Argument, D6 Use of Maple

From Fig. 2, we can see that students' scores are in general quite high. The median score in every dimension is 3, except for the first dimension D1, in which the median score is 4, the highest value. This can have two possible interpretations. On one side, without comprehension it is very hard to make progress on the resolution of the problem, thus it is the expected behavior. On the other side, students were not only solving the problem, they were also posing the problem, so students may even have created ill-conditioned problems.

Considering the average mark per student across the six dimensions, we can see, as shown in Fig. 3, that marks are shifted towards the highest values, 61% of submissions being between 3 (not included) and 4 (included). Categories are open on the left and closed to the right: the higher extremum is included in the category.

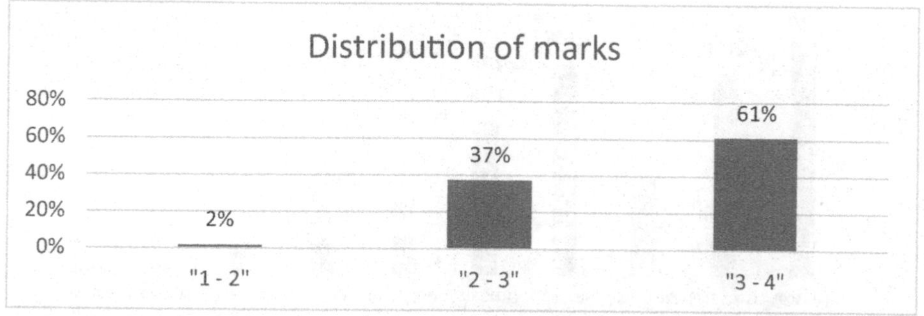

Fig. 3. Distribution of marks, calculated as average across the six dimensions. Categories are open on the left: the higher extremum is included in the category.

Correlations among the various dimensions are very different, but there is generally a moderate correlation (Pearson coefficient between 0.3 and 0.6), stating a good balance between independence of the dimensions and overall quality of the submission. Just one Pearson coefficient between D1 and D6 is quite low (0.24), and one Pearson coefficient between D2 and D3 is very high (0.98). Table 2 reports all the correlation coefficients between the six dimensions.

Table 2. Correlation coefficients between the six dimensions. Almost all coefficients show moderate correlation.

Dimension	D1	D2	D3	D4	D5	D6
D1	1.00	0.43	0.41	0.49	0.32	0.24
D2	0.43	1.00	0.98	0.74	0.60	0.44
D3	0.41	0.98	1.00	0.75	0.60	0.45
D4	0.49	0.74	0.75	1.00	0.48	0.56
D5	0.32	0.60	0.60	0.48	1.00	0.34
D6	0.24	0.44	0.45	0.56	0.34	1.00

One possible reason for the high correlation between D2 and D3 may rely on the formulation of indicators of good or bad performance. Both dimensions are related to the resolution, in D2 concerning strategy and in D3 concerning process. The dimensions are different, but very close.

As we already mentioned, students can attend and use a parallel online course, a different one for every degree course, in which they find suitable materials and examples. Collecting all together data from course usage, we can see that, on average, each student made 282 views. The distribution of views is shown in Fig. 4.

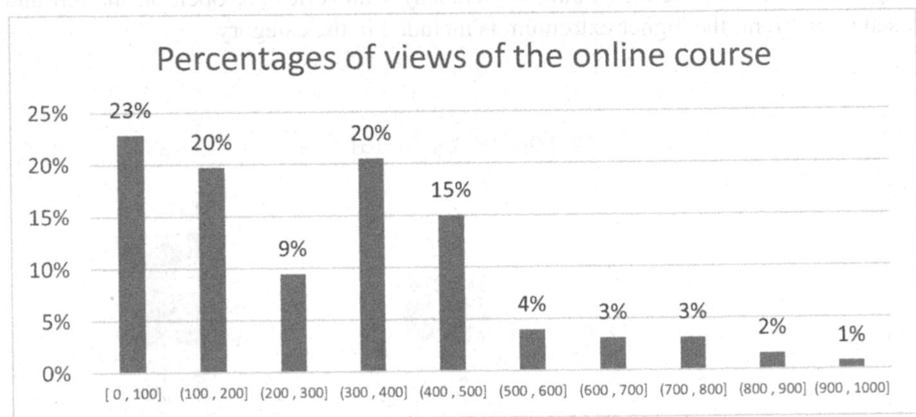

Fig. 4. Percentages of views of the online course divided into categories open to the left. 77% of students made more than 100 views.

Similar data can be collected focusing mainly on visualizations of Maple Worksheets. On average, students made 70 views each, but in this case the distribution, shown in Fig. 5, is denser on low numbers, since 76% of students made less than 50 views.

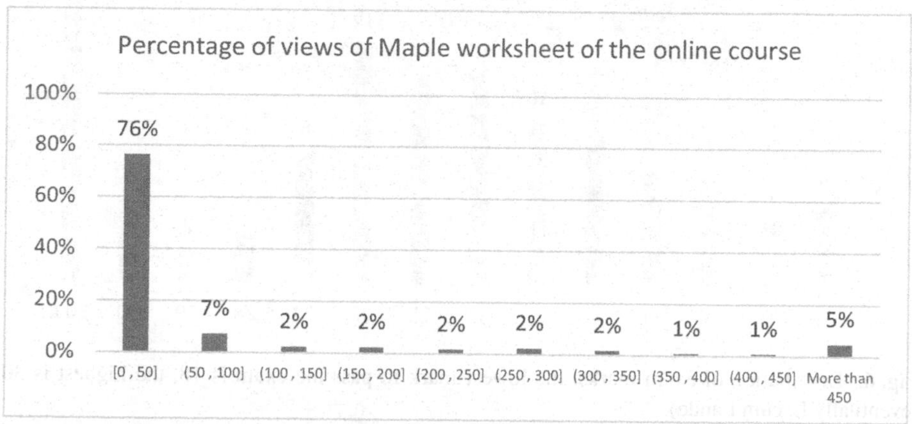

Fig. 5. Percentages of views of Maple worksheets of the online course divided into categories open to the left. 76% of students viewed less than 50 times.

The different behavior with respect to general views of the course can be explained in different ways. On one side, by the presence of different kinds of resources, as an example a consistent portion of formative assessment activities. On the other side, the number of available Maple Worksheets varies depending on the degree course and on the size of the program.

Since the submission is part of the exam, we considered the exam marks of the involved students. 120 students compose the subsample of students with exam marks since few of them still have to take the exam. The average mark is 24.9, standard deviation 3.6, with 30 cum Laude counting as 30. The median is 25. Figure 6 shows the distribution of exam marks.

Since the submission was part of the exam, we evaluated the correlation between submission score and exam mark. The Pearson coefficient gives 0.41, meaning there is correlation, but not a strong one. A possible explanation relies on the different evaluators that intervened in the research (an independent evaluator and the professors, the module leader). Another possible explanation relies on the difference between exercises that students usually do on paper and practical applications that students must deeply understand and develop.

Another source of useful data comes from the questionnaire that students submit before the examination. The questionnaire concerns the students in Biotechnology, academic year 2018/2019. The questionnaire is anonymous, thus there is no chance to consider relations with submission scores and exam marks. 87 students responded to the questionnaire, which touched numerous aspects. In the following, we are going to focus on the questions related to the adopted approach and to PS.

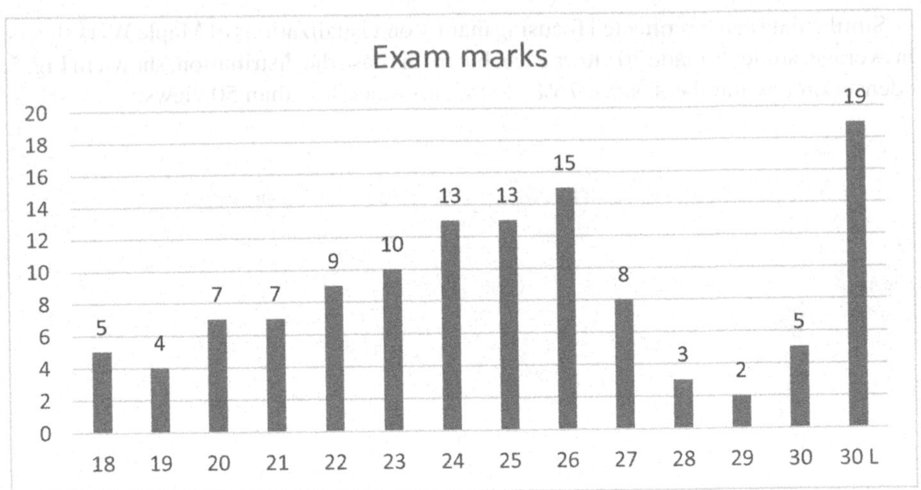

Fig. 6. Distribution of exam marks, the lowest mark to pass the exam is 18, the highest is 30 (eventually L, cum Laude).

After attending the module, students stated their agreement from 1 (disagree) to 5 (agree) to the following statements on PS:

- "Solving contextualized problems helps me learn theory better"
- "Studying mathematics through contextualized problems helps me to better face my university career"
- "Studying mathematics through contextualized problems helps me to better face my career path"

We can state that students found the adopted methodology useful in order to understand theory and to face their future study or job (median 4, over a 5-point Likert scale, IQR 1).

Concerning the use of Maple, students are more skeptical. Students had to state how much they used Maple, their confidence after the course and its usefulness over a 5-point scale, in which 1 means not at all and 5 means very much. Results are illustrated in Table 3.

The first concern is that they did not use Maple too much and, when you do not use an instrument, you do not understand its full potential. During classes, three labs were dedicated to learn PS with Maple. A total amount of 12 h over a total of 80 h for the whole course. However, many students declared that they would like to have more time to deepen and understand all potentialities of Maple.

A last sentence, with which 54% of the respondents agree, is "Mathematics can provide reasoning methods that are valid in many areas", underlying the importance of the discipline in first year scientific courses.

Table 3. Median of students' answers over the usage, confidence and usefulness of Maple.

Question	Median	IQR
How much have you used ACE Maple?	3	1
How much has the use of ACE Maple contributed to your training?	2	1
How competent do you feel in using the ACE Maple?	2	1
In your opinion, how useful will the skills acquired using ACE Maple be for your future professional career?	3	1

6 Examples

In this section, we are going to highlight some parts of the submissions to show the presence of indicators of good performance and the presence of problem solving skills.

The contexts of the various submissions were quite different and so were the topics related to the course of study that students attend. Thus, problems vary from health and natural sciences to applications about military, as shown in Fig. 7. Students had the chance to watch and study applications and problems that teachers prepared as example of good submissions.

SITUATION

The Lt. col. Alessandro Affinito commands the "Terrestrial Armaments" section of the Multifunctional Experimentation Center, located in Montelibretti (RM). He is *tasked* directly by the army logistic commander Gen. Giuseppe Roggio to research on the ballistic capabilities of the new weapon system "Hitrole light 2.0 remote turret", which will be installed on the modern VBM "Freccia" and VCC "Dardo" equipment. In this case, the task is to calculate, by varying the reference system, the angular speed with which the 25 mm cannon rotates, engaging, on his radar, the objective. Moreover, the results must be compared with the previous versions already in use by other apparatus of the armed force.

Fig. 7. Contextualization of a problem in the military field: the student is going to calculate the velocity of a specific rotating cannon.

Various types of representation enable students to solve the problem. Students adopted and combined several modalities, such as words, graphics, symbolic calculations, numerical approximations and experiments through computerized simulations. Students are free to choose the modalities, so they need to constantly validate and carefully justify every step of their resolution process and of their solving strategy. The ACE Maple allows users to use all these modalities in a single environment, helping clearness and comprehension. Figure 8 shows explanations of a basic analysis in a simple context.

One of the most known PS strategies consists in considering the same problem with smaller size or less degrees of freedom.

How many kilograms of contaminant would remain after 3 years?

We start calculating the contaminant variations as a function of time knowing that, annually, 60 Kg are added and 15% are biodegraded by nature.

$AddedContaminants := 60; RemovedContaminants := \frac{15}{100} y(t);$

$$60$$

$$\frac{3}{20} y(t) \quad (1.1)$$

Thus we have the relation to solve

$ode := \frac{d}{dt} y(t) = AddedContaminants - RemovedContaminants; dsolve(ode, y(t));$

$$\frac{d}{dt} y(t) = 60 - \frac{3}{20} y(t)$$

$$y(t) = 400 + e^{-\frac{3}{20} t} \_C1 \quad (1.2)$$

in which $\_C1$ is determined by initial condition at $t = 0$

$InitialCondition := y(0) = 200$

$$y(0) = 200 \quad (1.3)$$

obtaining as a result the function that describes the quantity of contaminant with respect to time:

$dsolve(\{ode, InitialCondition\}, y(t))$

$$y(t) = 400 - 200\, e^{-\frac{3}{20} t} \quad (1.4)$$

Fig. 8. Students explained the problem with examples and calculations. In this case, the student was considering the quantity of a contaminant in a lake after three years.

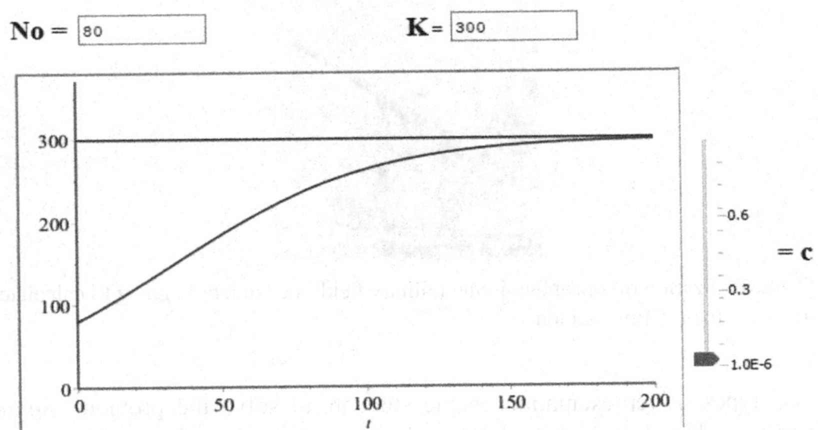

Fig. 9. Students prepared custom interactive components to generalize the studied context, letting the user choose initial conditions and parameters and watch the resulting plot.

After studying the problem, students were strongly encouraged to create a generalization of the situation using a standard or custom interactive interface, using embedded components such as buttons, sliders, input and output areas. In this part of the submission, the teacher, when evaluating, is able to change parameters and input variables to check the situation with a different perspective and to check possibly different behaviors according to the initial conditions. Figure 9 shows an example of interactive component.

7 Conclusion

This research provided a useful insight on possible approaches to the development of PS skills inside Mathematics university modules. The approach make use of contextualized PS with the help of the ACE Maple. The collected data show the presence of PS skills, detected by the six dimensions of Comprehension, Resolution strategy, Solution process, Representation, Argument, and Use of Maple. The use of the technology is not essential, but we can state that it facilitated students in the development of PS skills, in learning Mathematics and in understanding its applications, thanks to the opportunity of symbolic and numerical calculation, graphical representations and interactivity. Students emphasized this, stating that studying mathematics through contextualized problems will help better face future challenges, even if they still do not fully comprehend the mediation of the ACE Maple.

Future work will be carried out at the University of Turin in these areas. For sure, future modules and future examination led by our research group will maintain and boost this approach. In the Academic Year 2020/2021, we have already experimented guided problem posing, in which students choose a problem from a list of available ones. In this way, given the precise formulation of the problem, there is a benefit in Comprehension and, consequently, in the other dimensions.

Other work needs to be done also on institutional policies. PS skills development is left to Mathematics modules and in the work of teachers who adopt new teaching approaches and methodologies. Moreover, PS dedicated labs with the use of Maple should be introduced, in connection also with other disciplines, not only scientific, to facilitate analysis of data, visualization, communication, intersection with other disciplines and deep understanding of concepts.

References

1. Zhou, C.: Handbook of Research on Creative Problem-Solving Skill Development in Higher Education. IGI Global (2016)
2. Hämäläinen, R., De Wever, B., Nissinen, K., Cincinnato, S.: What makes the difference – PIAAC as a resource for understanding the problem-solving skills of Europe's higher-education adults. Comput. Educ. **129**, 27–36 (2019). https://doi.org/10.1016/j.compedu.2018.10.013
3. Barana, A., et al.: The role of an advanced computing environment in teaching and learning mathematics through problem posing and solving. In: Proceeding of the 15th International Scientific Conference eLearning and Software Education, vol. 2, pp. 11–18 (2019). https://doi.org/10.12753/2066-026X-19-070
4. Barana, A., Fissore, C., Marchisio, M., Pulvirenti, M.: Teacher training for the development of computational thinking and problem posing & solving skills with technologies. In: Proceedings of the 16th eLearning and Software for Education Conference (eLSE 2020), pp. 136–144 (2020). https://doi.org/10.12753/2066-026X-20-103

5. World Economic Forum: The Future of Jobs Report 2020 (2020).
6. Carretero, S., Vuorikari, R., Punie, Y.: DigComp 2.1: the digital competence framework for citizens with eight proficiency levels and examples of use, EUR 28558 EN (2017). https://doi.org/10.2760/38842
7. Barana, A., Marchisio, M.: Dall'esperienza di Digital Mate Training all'attività di Alternanza Scuola Lavoro. MONDO DIGITALE **15**(64), 63–82 (2016). http://mondodigitale.aicanet.net/2016-3/DidamaticaSessioni/Alternanza/paper_100.pdf
8. Marchisio, M., Remogna, S., Roman, F., Sacchet, M.: Teaching mathematics in scientific bachelor degrees using a blended approach. In: 2020 IEEE 44th Annual Computers, Software, and Applications Conference (COMPSAC), pp. 190–195 (2020). https://doi.org/10.1109/COMPSAC48688.2020.00034
9. Barana, A., Marchisio, M., Miori, R.: MATE-BOOSTER: design of tasks for automatic formative assessment to boost mathematical competence. In: Lane, H.C., Zvacek, S., Uhomoibhi, J. (eds.) CSEDU 2019. CCIS, vol. 1220, pp. 418–441. Springer, Cham (2020). https://doi.org/10.1007/978-3-030-58459-7_20
10. Barana, A., Conte, A., Fissore, C., Floris, F., Marchisio, M., Sacchet, M.: The creation of animated graphs to develop computational thinking and support STEM education. In: Gerhard, J., Kotsireas, I. (eds.) MC 2019. CCIS, vol. 1125, pp. 189–204. Springer, Cham (2020). https://doi.org/10.1007/978-3-030-41258-6_14
11. Polya, G.: How to Solve It. Princeton University Press, Princeton, New Jersy (1957)
12. Leong, Y.H., Janjaruporn, R.: Teaching of problem solving in school mathematics classrooms. In: Cho, S.J. (ed.) The Proceedings of the 12th International Congress on Mathematical Education, pp. 645–648. Springer, Cham (2015). https://doi.org/10.1007/978-3-319-12688-3_79
13. Lesh, R., Leher, R.: Models and modeling perspectives on the development of students and teachers. Math. Think. Learn. **5**(2–3), 109–129 (2009). https://doi.org/10.1080/10986065.2003.9679996
14. Shoenfeld, A.H.: Learning to think mathematically. In: Grouws, D.A. (ed.) Handbook of Research on Mathematics Teaching and Learning, pp. 334–370. Macmillan, New York, (1992)
15. Fissore, C., Floris, F., Marchisio, M., Rabellino, S., Sacchet, M.: Digital competences for educators in the Italian secondary school: a comparison between DigCompEdu reference framework and the PP&S project experience. Proc. Int. Conf. E-Learn. **2020**, 47–54 (2020)
16. Barana, A., Fissore, C., Marchisio, M., Pulvirenti, M.: An online math path to foster the transition of students between lower and upper secondary school. In: Proceedings of the 16th eLearning and Software for Education Conference (eLSE 2020), pp. 568–575 (2020)
17. Fissore, C., Marchisio, M., Rabellino, S.: Secondary school teacher support and training for online teaching during the Covid-19 pandemic. In: Proceedings of EDEN 2020 - Human and Artificial Intelligence for the Society of the Future, pp. 311–320 (2020)
18. Barana, A., Fioravera, M., Marchisio, M.: Developing problem solving competences through the resolution of contextualized problems with an advanced computing environment. In: Proceedings of the 3rd International Conference on Higher Education Advances, HEAd 2017 Universitat Politecnica de Valencia, Valencia, pp. 1015–1023 (2017). https://doi.org/10.4995/HEAd17.2017.5505
19. Lin, Y.-T., Wu, C.-C., Chen, Z.-H., Ku, P.-Y.: How gender pairings affect collaborative problem solving in social-learning context: the effects on performance, behaviors, and attitudes. Educ. Technol. Soc. **23**(4), 30–44 (2020)
20. Barana, A., Conte, A., Fissore, C., Marchisio, M., Rabellino, S.: Learning analytics to improve formative assessment strategies. J. E-Learn. Knowl. Soc. **15**(3), 75–88 (2019). https://doi.org/10.20368/1971-8829/1135057

21. Marchisio, M., Rabellino, S., Roman, F., Sacchet, M., Salusso, D.: Boosting up data collection and analysis to learning analytics in open online contexts: an assessment methodology. J. E-Learn. Knowl. Soc. **15**(3), 49–59 (2019). https://doi.org/10.20368/1971-8829/1135048
22. Marchisio, M., Rabellino, S., Spinello, E., Torbidone, G.: Advanced e-learning for IT-army officers through virtual learning environments. J. E-Learn. Knowl. Soc. **13**(3) (2017). https://doi.org/10.20368/1971-8829/1382
23. Abdul Razzak, N.: Strategies for effective faculty involvement in online activities aimed at promoting critical thinking and deep learning. Educ. Inf. Technol. **21**(4), 881–896 (2014). https://doi.org/10.1007/s10639-014-9359-z
24. Lamprecht, A., Margaria, T., Neubauer, J.: On the use of XMDD in software development education. In: Proceedings of the 2015 IEEE 39th Annual Computer Software and Applications Conference, Taichung, pp. 835–844 (2015). https://doi.org/10.1109/COMPSAC.2015.178
25. Gossen, F., Kühn, T., Margaria, T., Lamprecht, A.: Computational thinking: learning by doing with the Cinco adventure game tool. In: Proceedings of the 2018 IEEE 42nd Annual Computer Software and Applications Conference (COMPSAC), Tokyo, pp. 990–999 (2018). https://doi.org/10.1109/COMPSAC.2018.00175
26. Romero, M., Lepage, A., Lille, B.: Computational thinking development through creative programming in higher education. Int. J. Educ. Technol. High Educ. **14**, 42 (2017). https://doi.org/10.1186/s41239-017-0080-z

Modelling and Sensitivity Analysis of Nonlinear Firefighting Systems Using Maple

Flóra Hajdu(✉), Győző Molnárka, and Rajmund Kuti

Széchenyi István University, Győr 9026, Hungary
{hajdfl,molnarka,kuti.rajmund}@sze.hu
https://mgt.sze.hu/

Abstract. Mathematical modelling and numerical simulations have greatly contributed to the development of technical sciences in the recent decades. With powerful tools, like Maple, the examination of ever newer engineering applications in simulation environment was made possible. This paper gives an overview of mathematical modelling and numerical examination of nonlinear fire truck suspension systems using Maple. The examined models are the suspension system of a heavy-duty fire truck with different degrees of freedom and a special double-cabin fire truck suspension system with a crew compartment. The construction of mathematical models, their implementation to Maple and the numerical simulation results are explained. Detailed One-at-a-Time sensitivity analysis results using a novel fuzzy logic based evaluation method developed in Maple are also presented. With the proposed method an extended parameter range can be examined and the parameters can be easily compared. From the sensitivity analysis it was concluded that the spring characteristics and the road models greatly affect simulation results.

Keywords: Maple · Fire truck · Nonlinear system modelling · Sensitivity study

1 Introduction

In Hungary there has been only a limited amount of research about operation of fire trucks and firefighting systems so far. In special purpose vehicles, such as fire trucks, high value equipment is built in. This can be damaged by various adverse effects, like harmful vibrations during firefighting. A search of the available literature shows that different purpose of trucks [1,2] or passenger cars [3,4] have been examined in simulation environment, but there are only a few publications dealing with fire trucks [5,6]. The aim of this study was therefore the mathematical modelling and numerical simulation of nonlinear fire truck suspension systems.

For the purpose of mathematical modelling different tools are available from script based ones to computer algebra systems [7]. Maple and MapleSim are

R. M. Corless et al. (Eds.): MC 2020, CCIS 1414, pp. 234–251, 2021.
https://doi.org/10.1007/978-3-030-81698-8_16

powerful tools for system modelling and numerical simulation with a wide range of engineering applications [8]. Moreover, Maplesoft products have already been successfully and effectively used to solve simulation tasks of different trucks [9–11]. The main advantages of Maple include the effective combination of symbolic and numeric computations, a user friendly, clear interface, an intuitive syntax easy to understand and debug, and the availability of a lot of useful extensions by the Application Center [12]. Maple also offers a lot of packages, which can be effectively used for solving engineering tasks [13]. With regard to all these advantages, Maple was chosen was chosen for our research.

An effective method to study nonlinear systems is their sensitivity study, which can be used to detect the weak points of a system and to develop new models effectively. It is primarily used to examine how changes in parameters affect system behavior [14]. Using it parameter identification [15] and inverse simulation tasks [16] can also be solved and the uncertainty of the system can also be calculated [17]. Sensitivity study is also widely used in robust control tasks [18]. There are different approaches to carry out a sensitivity analysis ranging from partial differential techniques to statistical methods [19]. In this study OaT (One-at-a-Time) sensitivity study method was chosen, because it is easy to implement and on fast computers a lot of parameter combinations can be examined in a short time. OaT sensitivity study is, however, mainly used in a short examination range and only the most sensitive parameters with the highest sensitivity index are determined. To use the method in a wider examination range, to obtain the degree of sensitivity and to compare the parameters more precisely a fuzzy-logic evaluation method developed in Maple is proposed.

This paper is organized as follows: first a novel fuzzy logic based OaT sensitivity study method developed in Maple is presented, followed by the description of the modelling, the numerical analysis and the sensitivity study of 2 different fire trucks. The paper concludes with further research tasks.

2 One-at-a-Time Sensitivity Study Using Fuzzy Logic

During our research OaT sensitivity study of different fire truck suspension systems was carried out. Using a fuzzy logic based [20,21] evaluation method developed with Maple's FuzzySets[RealDomain] package [22] a finer, broader analysis can be carried out with an extended parameter range. Moreover, with defuzzification a sensitivity number can be calculated, with which the parameters could be easily compared to each other. The aim of using fuzzy logic based sensitivity analysis is, to analyze the models in detail in order to select the most appropriate one and to compare the applicability of different models. The proposed method is shown in Fig. 1.

After selecting the output variable, sensitivity functions were created. On a sensitivity function a selected output variable versus the change in the selected parameter is shown. The parameter is sensitive where the slope of the sensitivity function is large, therefore a small change in the parameter changes the output parameter significantly (see Fig. 2).

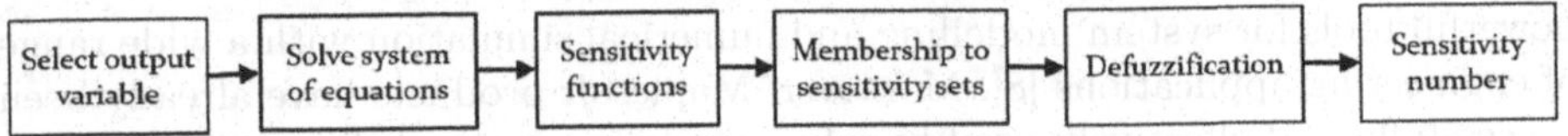

Fig. 1. Proposed OaT sensitivity study with fuzzy logic

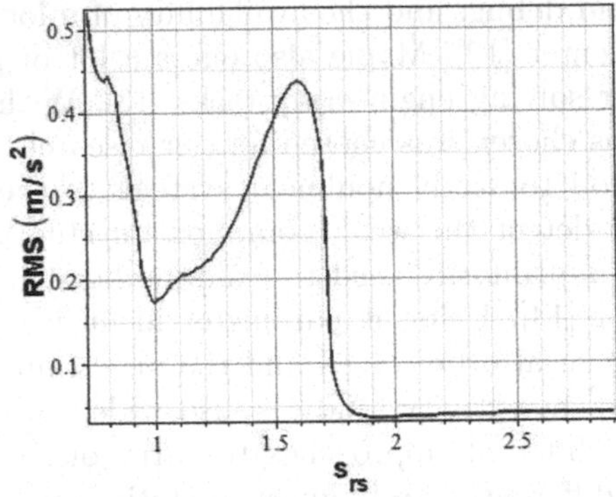

Fig. 2. Example of sensitivity function. The parameter is sensitive from 1 to 1.8 and is not sensitive above 1.8

From the sensitivity function sensitivity index can be calculated. Sensitivity index is the ratio of the relative change in the output variable and the relative change in the selected parameter:

$$SI = \frac{\Delta v}{\Delta p} \tag{1}$$

From sensitivity functions and the calculated sensitivity index the membership to the sensitivity sets can be calculated [21]. Sensitivity sets were established as fuzzy sets and are shown in Fig. 3.

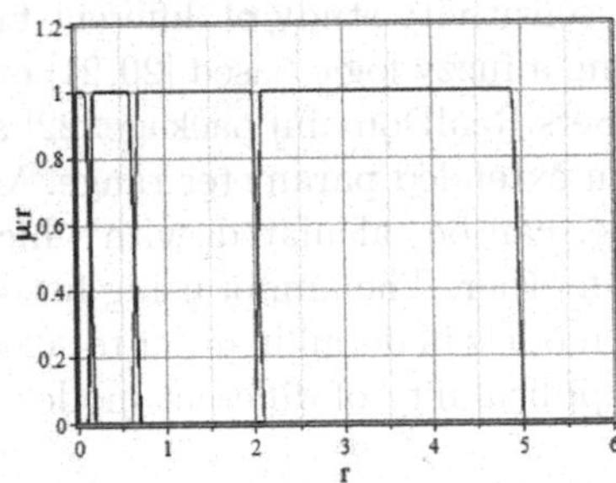

Fig. 3. Fuzzy sensitivity sets

To calculate the membership to the sensitivity sets the sensitivity function was divided into segments based on the sensitivity index. After division, the membership is calculated with the following formula:

$$\mu_y = \sum \frac{R_i}{R} \tag{2}$$

where R_i is the length of the examined segment and R is the length of the entire examination range.

From the membership the centroids for defuzzification can be established [20]. The functions for defuzzification and a centroid is shown is Fig. 4. For defuzzification Maple's command defuzzify was used, which provides the center of gravity or the center of area [22]. After defuzzification the sensitivity number (SN) is obtained.

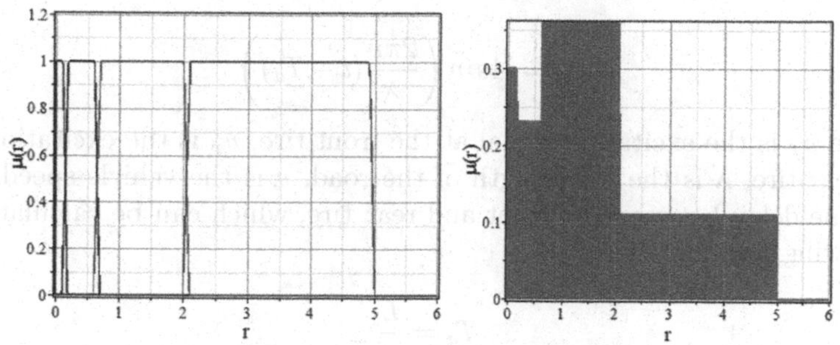

Fig. 4. Rules for defuzzification (left) and example for centroid (right)

3 Modelling and Sensitivity Study of Nonlinear Fire Truck Suspension Models

According to the literature, in case of system modelling, taking into account nonlinear effects is important. Linear models do not always reflect certain essential properties and behavior of the system, therefore fire truck suspensions were modelled as mass-spring-damper systems with nonlinear spring and damper characteristics [23]. Mass parameters were taken from manufacturing catalogs and spring and damper coefficients were taken from literature of similar truck models [24,25]. Our starting model was based on the half-vehicle model because it is detailed and suitable for many tests, easily expandable with new elements but does not require as much computing capacity as a full vehicle model. A simple, easily modifiable model was selected to include the nonlinear effects. The nonlinearity of the spring, the damper and the tire were included with the following formulas [26,27]:

$$F_k = k_i sgn(\Delta x_i)|\Delta x_i|^{s_i} \tag{3}$$

$$F_c = c_i sgn(\Delta\dot{x}_i)|\Delta\dot{x}_i|^{d_i} \tag{4}$$

where F_k is the spring force, F_c is the damping force k_i is the spring stiffness, Δx_i is the relative displacement, c_i is the damping coefficient, s_i and d_i are the nonlinear coefficients. The tire was modelled as a spring-damper pair with nonlinear spring characteristics and a linear damper. In case of other mass-spring-damper parts (e.g. cabin mount) linear spring and damper were used.

A theoretical sinusoidal road profile, which corresponds to the profile of an undulated road [28], was used to develop and test the simulation models more easily. The road was included as an input excitation signal at the tires with the following formulas:

$$u_f(t) = Asin\left(\frac{2\pi v}{\lambda}t\right) \tag{5}$$

$$u_r(t) = Asin\left(\frac{2\pi v}{\lambda}(t - T_d)\right) \tag{6}$$

where u_f is the excitation signal at the front tire, u_r is the excitation signal at the rear tire, λ is the wavelength of the road, v is the vehicle speed and T_d is the time delay between the front and rear tire, which can be calculated with the following formula:

$$T_d = \frac{L}{v} \tag{7}$$

A Maple script was written for each suspension model. In the script first the simulation parameters were specified, then the system of equations was given and was solved with the rkf45 numerical algorithm with 0.01 s step size. After solving the system of equations, the results were evaluated with time-dependent diagrams, phase-plane diagrams, Poincaré-sections, frequency diagrams and sensitivity diagrams.

3.1 Heavy-Duty Fire Truck

The model was based on a Hungarian Csepel SCD-755-10 heavy-duty fire truck. The vehicle has a ladder chassis with rigid bridges. Suspension is provided by leaf springs for both bridges, and an auxiliary leaf spring has also been installed at the rear to support the constant load. Hydraulic shock absorbers are installed on both sides of the bridge. The vehicle and its model are shown in Fig. 5.

The equations describing the system behavior are the following:

$$m_c\ddot{x}_c = -(k_c\Delta x_c + c_c\Delta\dot{x}_c) \tag{8}$$

$$m_e\ddot{x}_e = k_e\Delta x_e + c_e\Delta\dot{x}_e \tag{9}$$

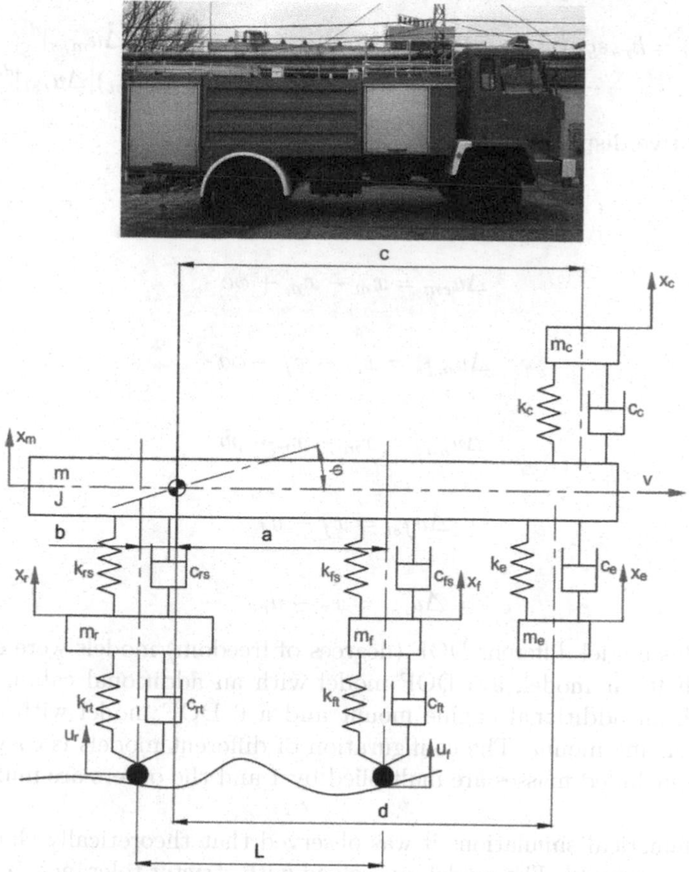

Fig. 5. Csepel CSD-755-10 heavy-duty fire truck and its suspension model

$$\begin{aligned} m\ddot{x}_m = {} & k_c\Delta x_c + c_c\Delta\dot{x}_c - k_e\Delta x_e(t) - c_e\Delta\dot{x}_e \\ & - k_{fs}sgn(\Delta u_{m_{fs}})|\Delta u_{m_{fs}}|^{s_{fs}} - c_{fs}sgn(\Delta\dot{u}_{m_{fs}})|\Delta\dot{u}_{m_{fs}}|^{d_{fs}} \\ & - k_{rs}sgn(\Delta u_{m_{rs}})|\Delta u_{m_{fs}}|^{s_{rs}} - c_{rs}sgn(\Delta\dot{u}_{m_{fs}})|\Delta\dot{u}_{m_{rs}}|^{d_{rs}} \end{aligned} \tag{10}$$

$$\begin{aligned} J\ddot{\phi} = {} & -(k_c\Delta x_c + c_c\Delta\dot{x}_c)c - (k_e\Delta x_e - c_e\Delta\dot{x}_e)d \\ & + (k_{fs}sgn(\Delta u_{m_{fs}})|\Delta u_{m_{fs}}|^{s_{fs}} - c_{fs}sgn(\Delta\dot{u}_{m_{fs}})|\Delta\dot{u}_{m_{fs}}|^{d_{fs}})a \\ & - (k_r ssgn(\Delta u_{m_{rs}})|\Delta u_{m_{fs}}|^{s_{rs}} - c_{rs}sgn(\Delta\dot{u}_{m_{fs}})|\Delta\dot{u}_{m_{rs}}|^{d_{rs}})b \end{aligned} \tag{11}$$

$$\begin{aligned} m_f\ddot{x}_f(t) = {} & k_{fs}sgn(\Delta u_{m_{fs}})|\Delta u_{m_{fs}}|^{s_{fs}} - c_{fs}sgn(\Delta\dot{u}_{m_{fs}})|\Delta\dot{u}_{m_{fs}}|^{d_{fs}} \\ & - k_{ft}sgn(\Delta u_{f_{st}})|\Delta u_{f_{st}}|^{s_{rs}} - c_{rs}sgn(\Delta\dot{u}_{f_{st}})|\Delta\dot{u}_{f_{st}}|^{d_{rs}} \end{aligned} \tag{12}$$

$$m_r\ddot{x}_r(t) = k_{rs}sgn(\Delta u_{m_{rs}})|\Delta u_{mrs}|^{s_{fs}} - c_{rs}sgn(\Delta u_{\dot{m}}rs)|\Delta u_{\dot{m}_{rs}}|^{d_{rs}}$$
$$- k_{rt}sgn(\Delta u_{rst})|\Delta u_{rst}|^{s_{rs}} - c_{rs}sgn(\Delta u_{\dot{r}st})|\Delta u_{\dot{r}st}|^{d_{rs}} \quad (13)$$

The relative displacements are:

$$\Delta u_{cm} = x_c - x_m + \phi c \quad (14)$$

$$\Delta u_{em} = x_m - x_m + \phi c \quad (15)$$

$$\Delta u_{mfs} = x_m - x_f - \phi a \quad (16)$$

$$\Delta u_{mrs} = x_m - x_r + \phi b \quad (17)$$

$$\Delta u_{fst} = x_f - u_f \quad (18)$$

$$\Delta u_{rst} = x_r - u_r \quad (19)$$

With this model different DOF (degrees of freedom) models were examined: a 4 DOF half car model, a 5 DOF model with an additional cabin, a 5 DOF model with an additional engine mount and a 6 DOF model with additional cabin and engine mount. The configuration of different models is easy to set in Maple: the included masses are multiplied by 1 and the others are multiplied by 0 (Fig. 6).

With numerical simulations it was observed that theoretically chaotic oscillations can occur [29]. The model was solved with stricter tolerances and smaller step size, but the same chaotic behavior occurred. Consequently to be a numerical chaos was concluded to be an unlikely cause [30]. To prove the chaotic behavior of the model, further work with a more detailed backward error analysis is planned [31]. It was also observed that by adding new elements to the model (e.g. cabin, engine) the behavior of the system can completely change. For example, new attractors may appear, which can be seen on the Poincaré-sections (Fig. 7). The same tests were also carried out on completely linear models, which, however showed that the results in these only varied in the amplitude of the output signals, while the Poincaré sections remained the same regardless of the DOF of the model [32]. Hence, nonlinear models were the better choice for our purposes.

```
> #tömegek, tehetetlenségi nyomatékok
mf := 310 : mr := 740 : mv := 5750 : mc := 250 : me := 1200 : J
     := 28600 :
m := mv + 1 · mc + 1 · me :
#felfüggesztés rugó és csillapítás
kfs := 300000 : krs := 400000 : cfs := 20000 : crs := 40000 :
#kerék rugó és csillapítás
kft := 1000000 : krt := 1800000 : cft := 500 : crt := 1000 :
#kabin rugó és csillapítás
kc := 0·75000 : cc := 0·7500 :
#motor rugó és csillapítás
> ke := 0·3500000 : ce := 0·8000 :
```

```
> #tömegek, tehetetlenségi nyomatékok
mf := 310 : mr := 740 : mv := 5750 : mc := 250 : me := 1200 : J
     := 28600 :
m := mv + 0 · mc + 0 · me :
#felfüggesztés rugó és csillapítás
kfs := 300000 : krs := 400000 : cfs := 20000 : crs := 40000 :
#kerék rugó és csillapítás
kft := 1000000 : krt := 1800000 : cft := 500 : crt := 1000 :
#kabin rugó és csillapítás
kc := 1·75000 : cc := 1·7500 :
#motor rugó és csillapítás
> ke := 1·3500000 : ce := 1·8000 :
```

Fig. 6. Maple script to change the DOF of the suspension model

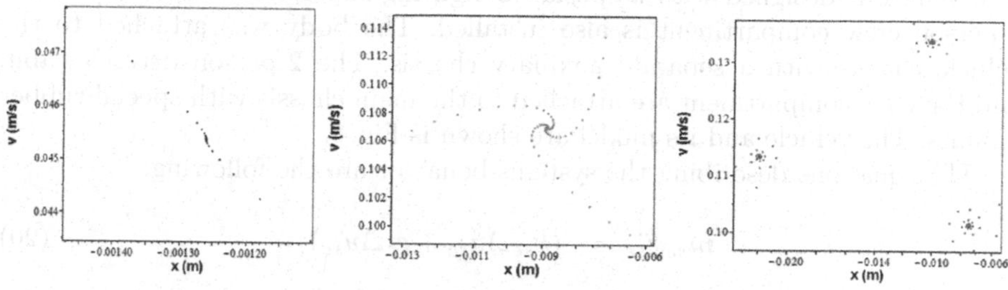

Fig. 7. Poincaré sections (left: 4 DOF model, middle: 5 DOF model with engine, right: 6 DOF model with engine and cabin)

The sensitivity analysis was carried out with the 4 DOF model. The output variable was the RMS (root mean square) of the acceleration at the center of mass of the vehicle. The detailed OaT sensitivity study of the 4 DOF system was carried out in a previous research [33]. In this research only the sensitivity numbers with defuzzification were calculated, which were respectively:

$SN(m) = 1.633; SN(m_f) = 0.08; SN(m_r) = 0.08; SN(J) = 0.08;$
$SN(k_{fs}) = 0.175; SN(k_{rs}) = 0.89; SN(k_{ft}) = 0.2; SN(k_{rt}) = 0.08;$
$SN(c_{fs}) = 0.08; SN(c_{rs}) = 0.232; SN(c_{ft}) = 0.08; SN(c_{rt}) = 0.08;$
$SN(a) = 0.231; SN(b) = 0.4; SN(s_{fs}) = 2.652; SN(s_{rs}) = 2.894;$
$SN(d_{fs}) = 0.08; SN(d_{rs}) = 0.247; SN(s_{ft}) = 0.08; SN(s_{rt}) = 0.737;$
$SN(\lambda) = 2.091; SN(A) = 1.137; SN(v) = 1.889.$

The most sensitive parameter was the nonlinear coefficient of the rear spring (s_{rs}), as it has the highest sensitivity number value. It was 1.09 times more sensitive than the nonlinear coefficient of the front spring suspension (s_{fs}), 1.38 times more sensitive than the wavelength of the road (λ) and 1.53 times more sensitive than the vehicle speed (v). It is 36.18 times more sensitive than the non-sensitive parameters (the parameters with SN = 0.8). It can be concluded that the nonlinear coefficients of the springs were amongst the most sensitive parameters, therefore accurate spring models are essential in vehicle simulations. The parameters, which are connected to the time delay (wavelength of the road, vehicle speed) were also sensitive. This corresponds to the findings of other studies in the literature, in which it was showed that chaotic behavior can be caused by the time delay between the tires [34].

3.2 Double Cabin Fire Truck

The model was based on a mid-size fire truck with an Austrian Rosenbauer superstructure built on a Mercedes-Benz 1124 chassis. The vehicle has a ladder chassis with rigid bridges. Suspension is provided by leaf springs for both bridges, and an auxiliary leaf spring has also been installed at the rear to support the constant load. Hydraulic shock absorbers are installed on both sides of the bridge. The vehicle is designed with a special firefighting superstructure, in which the 4-person crew compartment is also installed. The body was attached to the vehicle chassis with a separate auxiliary chassis. The 2-person driver's cabin and the crew compartment are attached to the main chassis with special rubber mounts. The vehicle and its model are shown is Fig. 8.

The equations describing the systems behavior are the following:

$$m_{se}\ddot{x_{se}} = -(k_{se}\Delta u_{sc} + c_c\Delta \dot{u_{sc}}) \tag{20}$$

$$m_c\ddot{x_c}(t) = -(k_c\Delta u_{chc} + c_c\Delta \dot{u_{chc}}(t)) + k_{se}\Delta u_{sc} + c_c\Delta \dot{u_{sc}}) \tag{21}$$

$$m_b\ddot{x_b} = k_b\Delta u_{chb} + c_e\Delta \dot{u_{chb}}(t) \tag{22}$$

$$\begin{aligned} m_{ch}\ddot{x_{ch}} = {} & k_c\Delta u_{chc} + c_c\Delta \dot{u_{chc}} + (k_e\Delta u_{chb} + c_e\Delta \dot{u_{chb}} \\ & - k_{fs}sgn(\Delta u_{chfs})|\Delta u_{chfs}|^{s_{fs}} - c_{fs}sgn(\Delta \dot{u_{chfs}})|\Delta \dot{u_{chfs}}|^{d_{fs}} \\ & - k_{rs}sgn(\Delta u_{chrs})|\Delta u_{chrs}|^{s_{rs}} - c_{rs}sgn(\Delta \dot{u_{chrs}})|\Delta \dot{u_{chrs}}|^{d_{rs}} \end{aligned} \tag{23}$$

$$\begin{aligned} J\ddot{\phi} = {} & -(k_c\Delta u_{chc} + c_c\Delta \dot{u_{chc}})c \\ & + (k_{fs}sgn(\Delta u_{chfs})|\Delta u_{chfs}|^{s_{fs}} - c_{fs}sgn(\Delta \dot{u_{chfs}})|\Delta \dot{u_{chfs}}|^{d_{fs}})a \\ & - (k_{rs}sgn(\Delta u_{chrs})|\Delta u_{chrs}|^{s_{rs}} + c_{rs}sgn(\Delta \dot{u_{chrs}})|\Delta \dot{u_{chrs}}|^{d_{rs}})b \end{aligned} \tag{24}$$

$$\begin{aligned} m_f\ddot{x_f}(t) = {} & k_{fs}sgn(\Delta u_{chfs})|\Delta u_{chfs}|^{s_{fs}} - c_{fs}sgn(\Delta \dot{u_{chfs}})|\Delta \dot{u_{chfs}}|^{d_{fs}} \\ & - k_{ft}sgn(\Delta u_{fst})|\Delta u_{fst}|^{s_{rs}} - c_{rs}sgn(\Delta \dot{u_{fst}})|\Delta \dot{u_{fst}}|^{d_{rs}} \end{aligned} \tag{25}$$

$$\begin{aligned} m_r\ddot{x_r}(t) = {} & k_{fs}sgn(\Delta u_{chrs})|\Delta u_{chrs}|^{s_{fs}} - c_{rs}sgn(\Delta \dot{u_{chrs}})|\Delta \dot{u_{chrs}}|^{d_{fs}} \\ & - k_{ft}sgn(\Delta u_{rst})|\Delta u_{rst}|^{s_{rs}} - c_{rs}sgn(\Delta \dot{u_{rst}})|\Delta \dot{u_{rst}}|^{d_{rs}} \end{aligned} \tag{26}$$

The relative displacement are:

$$\Delta u_{sc} = x_s - x_c \tag{27}$$

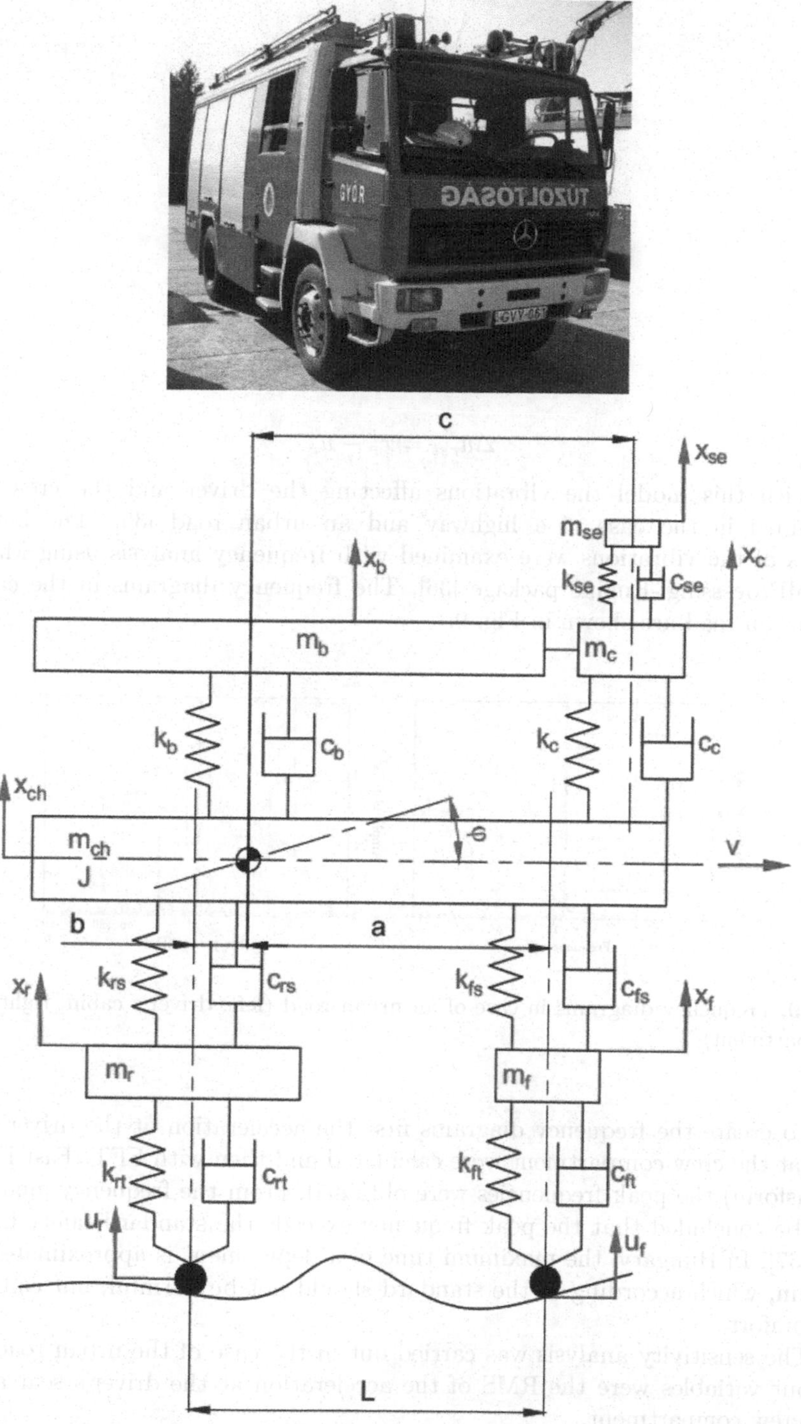

Fig. 8. MB RB TLF 2000 double-cabin fire truck and its suspension model

$$\Delta u_{chb} = x_b - x_{ch} \tag{28}$$

$$\Delta u_{chc} = x_c - x_m + \phi c \tag{29}$$

$$\Delta u_{chfs} = x_m - x_f - \phi a \tag{30}$$

$$\Delta u_{chrs} = x_b - x_r + \phi b \tag{31}$$

$$\Delta u_{fst} = x_f - u_f \tag{32}$$

$$\Delta u_{rst} = x_r - u_r \tag{33}$$

With this model the vibrations affecting the driver and the crew were examined in the case of a highway and an urban road [35]. The harmful effects of the vibrations were examined with frequency analysis using Maple's SignalProcessing:-Engine package [36]. The frequency diagrams in the case of the urban road are shown in Fig. 9.

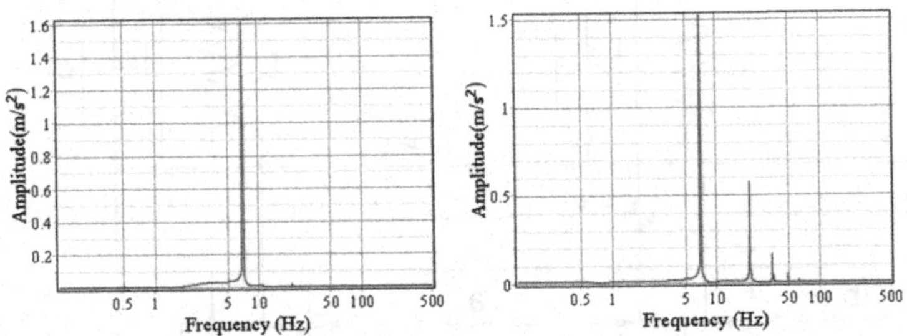

Fig. 9. Frequency diagrams in case of an urban road (left: driver's cabin, right: crew compartment)

To create the frequency diagrams first the acceleration at the driver's seat and at the crew compartment were calculated and then with FFT (Fast Fourier Transform) the peak frequencies were obtained. From the frequency analysis it can be concluded that the peak frequency exceeds the standard safety limit of 8 h [37]. In Hungary, the maximum time of a deployment is approximately 25–30 min, which according to the standard should not be harmful, but can cause discomfort.

The sensitivity analysis was carried out in the case of the urban road. The output variables were the RMS of the acceleration at the driver's seat and at the crew compartment.

The results of the sensitivity study are summarized in Table 1.

Table 1. Sensitivity study of the double cabin fire truck (upper row: driver's seat, lower row: crew compartment)

Parameter	Initial value	SImax	S1	S2	S3	S4	SN
m_{ch}	3515 kg	7.823	0.000	0.321	0.482	0.197	2.113
		0.589	0.363	0.637	0.000	0.000	0.363
m_b	205 kg	0.027	1.000	0.000	0.000	0.000	0.008
		3.079	0.000	0.448	0.427	0.125	1,825
m_c	205 kg	1.364	0.000	0.000	1.000	0.000	1.350
		0.689	0.109	0.811	0.079	0.000	0.596
m_f	225 kg	0.539	0.338	0.662	0.000	0.000	0.366
		1.606	0.000	0.000	1.000	0.000	1.350
m_r	335 kg	0.645	0.006	0.949	0.044	0.000	0.514
		0.190	0.702	0.298	0.000	0.000	0.284
m_{se}	100 kg	8.675	0.000	0.000	0.408	0.592	2.962
		0.059	1.000	0.000	0.000	0.000	0.080
J	26130 kgm^2	0.828	0.000	0.780	0.220	0.000	0.826
		5.107	0.000	0.000	0.462	0.538	2.869
k_{fs}	300000 N/m	0.020	1.000	0.000	0.000	0.000	0.080
		0.048	1.000	0.000	0.000	0.000	0.080
k_{rs}	400000 N/m	0.028	1.000	0.000	0.000	0.000	0.080
		0.023	1.000	0.000	0.000	0.000	0.080
k_{ft}	1000000 N/m	2.716	0.000	0.241	0.591	0.167	1.998
		1.913	0.000	0.102	0.898	0.000	1.312
k_{rt}	1800000 N/m	1.184	0.047	0.487	0.466	0.000	1.087
		0.709	0.082	0.812	0.106	0.000	0.651
k_c	75000 N/m	0.650	0.000	0.926	0.074	0.000	0.580
		0.105	0.994	0.006	0.000	0.000	0.087
k_e	75000 N/m	0.042	1.000	0.000	0.000	0.000	0.080
		0.061	1.000	0.000	0.000	0.000	0.080
k_{se}	20000 N/m	1.299	0.000	0.082	0.918	0.000	1.319
		0.039	1.000	0.000	0.000	0.000	0.080
c_{fs}	20000 Ns/m	0.316	0.818	0.182	0.000	0.000	0.233
		1.148	0.426	0.416	0.158	0.000	0.801
c_{rs}	40000 Ns/m	0.579	0.225	0.775	0.000	0.000	0.380
		0.341	0.633	0.367	0.000	0.000	0.306
c_{ft}	500 Ns/m	0.032	1.000	0.000	0.000	0.000	0.080
		0.123	0.804	0.196	0.000	0.000	0.240
c_{rt}	1000 Ns/m	0.053	1.000	0.000	0.000	0.000	0.080
		0.026	1.000	0.000	0.000	0.000	0.080

(continued)

Table 1. (*continued*)

Parameter	Initial value	SImax	S1	S2	S3	S4	SN
c_c	7500 Ns/m	1.306	0.000	0.483	0.517	0.000	1.112
		0.731	0.407	0.538	0.054	0.000	0.543
c_e	7500 Ns/m	0.059	1.000	0.000	0.000	0.000	0.080
		0.289	0.000	1.000	0.000	0.000	0.400
c_{se}	1000 Ns/m	1.764	0.000	0.000	1.000	0.000	1.350
		0.023	1.000	0.000	0.000	0.000	0.080
a	3.44 m	10.850	0.010	0.032	0.098	0.860	3.366
		11.161	0.019	0.041	0.163	0.777	3.274
b	0.2 m	0.374	0.108	0.892	0.000	0.000	0.392
		0.200	0.764	0.236	0.000	0.000	0.259
c	3.5 m	0.383	0.078	0.922	0.000	0.000	0.394
		0.736	0.000	0.019	0.981	0.000	1.343
r	2.5 m kg	–	–	–	–	–	–
		2.631	0.000	0.000	0.000	1.000	3.500
s_{fs}	1.3	3.027	0.655	0.175	0.137	0.033	1.362
		7.878	0.478	0.225	0.168	0.129	2.127
s_{rs}	1.45	0.338	0.591	0.379	0.030	0.000	0.291
		0.467	0.565	0.435	0.000	0.000	0.324
s_{ft}	1.1	0.467	0.565	0.435	0.000	0.000	0.324
		5.032	0.040	0.104	0.339	0.517	2.909
s_{rt}	1.1	4.533	0.011	0.135	0.604	0.250	2.262
		3.271	0.043	0.240	0.584	0.133	1.883
d_{fs}	2.2	0.048	1.000	0.000	0.000	0.000	0.080
		0.382	0.004	0.996	0.000	0.000	0.400
d_{rs}	2.2	0.325	0.683	0.317	0.000	0.000	0.291
		0.128	0.964	0.036	0.000	0.000	0.123
λ	2 m	11.342	0.000	0.044	0.057	0.899	3.410
		34.565	0.012	0.055	0.222	0.711	3.182
v	50 km/h	3.385	0.017	0.063	0.667	0.253	2.262
		5.532	0.048	0.127	0.206	0.619	3.110
A	0.02 m	2.270	0.000	0.000	0.191	0.809	3.278
		2.665	0.000	0.000	0.000	1.000	3.500

It can be seen that using the RMS of the acceleration at the driver's seat the wavelength of the road (λ) was the most sensitive parameter, because it had the highest sensitivity number (SN = 3.410). The distance between the center of mass and the front tire (a) was almost as a sensitive parameter, since their ratio was only 1.01. The amplitude of the road (A) was also sensitive, the ratio compared to the wavelength of the road was 1.04. The membership value of these parameters to the extremely sensitive set was high, which means that they were extremely sensitive over almost the entire examination range. The nonlinear coefficient of the rear tire (s_{rt}) and the vehicle speed (v) were also sensitive parameters, their ratio to the wavelength of the road was 1.15. The parameters mass of the chassis (m_{ch}), mass of the seat (m_{se}) and the nonlinear coefficient of the front tire (s_{ft}) were also sensitive parameters with sensitivity numbers above 2.

In case of selecting the RMS of the acceleration at the crew compartment, the most sensitive parameters were the amplitude of the road (A) and, not surprisingly, the distance between the crew compartment and the center of mass (r), as they had the highest sensitivity number (SN = 3.5). The distance between the center of mass and the front tire (a) was almost as a sensitive parameter as the amplitude of the road, their ratio was 1.06. The next most sensitive parameters were the wavelength of the road (λ) with a ratio of 1.1 and the speed of the vehicle (v) with a ratio of 1.12 compared to the amplitude of the road. The nonlinear coefficient of the front tire (s_{ft}) and the moment of inertia (J) were also sensitive parameters their ratio to the amplitude was 1.12. These parameters had a membership value to the extremely sensitive set above 0.5, which means that they were extremely sensitive over more than half of the examination range. The nonlinear coefficient of the front suspension spring (s_{fs}) was also sensitive with sensitivity number above 2.

It can be concluded that the road parameters and the vehicle speed were amongst the most sensitive parameters in both output variables. The road parameters had similar SN values (their ratio was 1.07), but the vehicle speed was 1.3 more sensitive, when the RMS of acceleration at the crew compartment was the output variable. Therefore it can be concluded that road model is particularly important in vehicle simulations. The nonlinear coefficients of the tires were also amongst the most sensitive parameters. Surprisingly the nonlinear coefficient of the front tire was more sensitive, when the output variable was the RMS of the acceleration at the crew compartment. On the other hand the nonlinear coefficient of the rear tire was more sensitive, when the output variable was the RMS of the acceleration at the driver's seat. Therefore, the nonlinear effects of the tire cannot be neglected. The distance between the center of mass and the front tire were also amongst the more sensitive parameters. The other parameters, such as spring stiffnesses, damping coefficients and most of the masses were moderately sensitive or not sensitive parameters with SN below 2. Also, the nonlinear coefficients of the dampers were not sensitive parameters. According to the results of the presented sensitivity study in the future more accurate spring and road models and simpler damper models can be used to improve the suspension models further.

4 Conclusions and Further Research Tasks

In this study the mathematical modelling and numerical simulation of a heavy-duty and a double-cabin fire truck were carried out using Maple scripts. During the simulation of the heavy-duty fire truck it was observed that theoretically chaotic oscillations can occur in the case of undulated roads. It was also observed, that by adding a new element to the system, its behavior can be changed completely, for example new attractors can appear on the phase plane. During the simulation of the double cabin fire truck frequency analysis using Maple's SignalProcessing:-Engine package showed that in the case of a theoretical undulated road the vibrations affecting the crew should not be harmful, but might cause discomfort. A novel fuzzy logic based OaT sensitivity study method developed in Maple was also tested with these systems. The advantage of the proposed method is that an extended parameter range can be examined and the parameters can easily be compared. From the sensitivity study of the heavy-duty fire truck it was observed that the nonlinear coefficient of the rear spring was the most sensitive parameter. Therefore, in vehicle simulations it is particularly important to use accurate spring models. From the sensitivity study of the double-cabin fire truck it was observed that parameters which affect the time delay between the front and rear tire (like the wavelength of the road, the distance between the tires and the vehicle speed) were sensitive parameters. The nonlinear coefficients of the tires were also sensitive parameters. Therefore, the nonlinear effects of the tire also cannot be neglected in vehicle simulations. The parameters of the dampers were not sensitive, therefore, in the future simpler damping models can be used.

In this study it was also shown, that Maple can be effectively used to develop and carry out numerical simulations in case of nonlinear systems. Using different packages of Maple, several engineering tasks can be solved fast. For example, with SignalProcessing:-Engine package, the vibrations affecting people travelling in a vehicle can be examined and a more comfortable seat can be developed. FuzzySets[RealDomain] package can be used in several fields of engineering, like the design of a controller or of a robotic system. Moreover, the scripts can be easily modified and therefore a lot of parameter combinations can be examined as well as new models can be developed fast. The metods and the models presented in this paper can be used with little modification in other similar studies as well, for example to carry out numerical simulations of similar trucks or other special firefighting vehicles. The presented sensitivity study can be used to develop new mathematical models and simulation of other nonlinear systems in different fields of engineering in an effective, convenient way. It might also be used for model callibration and to select the most appropriate model for a specific tasks.

The models are planned to be developed further according to the results of the presented sensitivity study. For example more accurate and realistic road profiles are planned to be used in simulations. To create more accurate models, laboratory and field measurements are planned with parameter identification using Maple's Optimization package. For establishing stochastic road profiles, Maple's Statistics package is planned to be used. To carry out an OaT sensitivity study a large number of calculations are necessary as there are many different parameter combinations. To speedup the calculations parallelization is an important task, for which Maple's Grid package is planned to be used. Another future research goal is to model and carry out the sensitivity study of other nonlinear firefighting systems using Maple and MapleSim.

References

1. Pidl, R.: Analytical approach to determine vertical dynamics of a semi-trailer truck from the point of view of goods protection. In: Podgorski, J., et al. (eds.) AIP Conference Proceedings 2018, vol. 1922, Paper No. 120003 (2008). https://doi.org/10.1063/1.5019118
2. Wu, W., Zhang S., Zhang Z.: athematical simulations and on-road experimentations of the vibration energy harvesting from mining dump truck hydro-pneumatic suspension. Shock and Vibration (2019). Article ID: 4814072
3. Danko, J., Milesich, T., Bucha, J.: Nonlinear model of the passenger car seat suspension system. Strojnícky časopis - J. Mech. Eng. **67**(1), 23–28 (2017)
4. Avesh, Mohd, Srivastava, Rajeev: Passenger car active suspension system model for better dynamic characteristics. Natl. Acad. Sci. Lett. **43**(1), 37–41 (2019). https://doi.org/10.1007/s40009-019-00807-z
5. Kovtun, V., Korotkevich, S., Mirchev, Y., Lodnya, V.: Optimization of fire truck's tanks on the chassis MAZ-6317 by the method of computer simulation. Int. J. "NDT Days" **2**(4), 495–500 (2019)
6. Kovtun, V.A., Korotkevich, S.G., Zharanov, V.A.: Computer simulation and research of the stress-strain state of fire tank truck construction. Vestnik Universiteta grazhdanskoy zashchity MCHS Belarusi (Bulletin of the University of Civil Protection of the MES of Belarus) **2017**(1), 81–90 (2017). (in Russian)
7. Sund, S. M., Plouvier, M., Lie, B.: Comparison of simulation tools for dynamic models. In: Proceedings of The 59th Conference on Simulation and Modelling (SIMS 59), 26–28 September 2018, Oslo Metropolitan University, Norway, pp. 177–184 (2018)
8. Molnárka, Gy., Gergó, L., Wettl, F., Horváth, A., Kallós, G.: Maple V and its applications. Springer Hungarica, Budapest (1996)
9. Zhao, X.X., Ing, A.H., Azad, N.L., McPhee, J.: An optimal gear-shifting strategy for heavy trucks with trade-off study between trip time and fuel economy. Int. J. Heavy Veh. Syst. **22**(4), 356–374 (2015)
10. Zhao, X., Yang, J.: An optimal gear-shifting strategy for heavy trucks with trade-off study between trip time and fuel economy. Int. J. Heavy Veh. Syst. **22**(4), 356–374 (2015)
11. Maplesoft: Engineering Solutions for Heavy Vehicles. https://www.maplesoft.com/solutions/engineering/IndustrySolutions/HeavyVehicles.aspx. Accessed 10 Dec 2020
12. Chonacky, N., Winch, D.: Maple, mathematica, and matlab: the 3M'S without the tape. Comput. Sci. Eng. **7**, 8–16 (2005)
13. GarVan, F.: The Maple Book. CRC Press, Boca Raton (2001)
14. Saltelli, A., Chan, K., Scott, E.M.: Sensitivity Analysis. Wiley, Hoboken (2009)
15. Iwaniec, J.: Sensitivity analysis of an identification method dedicated to nonlinear systems working under operational loads. J. Theorethical Appl. Mech. **49**, 419–438 (2011)

16. Murray-Smith, D.J.: The application of parameter sensitivity analysis methods to inverse simulation models. Math. Comput. Model. Dyn. Syst. **19**(1), 1–24 (2012)
17. Zádor, J., Zsély, I. Gy., Turányi, T., Ratto, M., Tarantola, S., Saltelli, A.: Local and global uncertainty analyses of a methane flame model. J. Phys. Chem. A. **109**(43), 9795–9807 (2005)
18. Rauh, A., Minisini, J., Hofer, E.P.: Verification techniques for sensitivity analysis and design of controllers for nonlinear dynamic systems with uncertainties. Int. J. Appl. Math. Comput. Sci. **19**(3), 425–439 (2009). https://doi.org/10.2478/v10006-009-0035-1. https://sciendo.com/downloadpdf/journals/amcs/19/3/article-p425.pdf
19. Hamby, D.M.: A review of techniques for parameter sensitivity analysis of environmental models. Environ. Monit. Assess. **32**(2), 135–154 (1994)
20. Kóczy, L.T., Tikk, D.: Fuzzy Systems (in Hungarian), Typotex (2012)
21. Fogel, D.B., Liu, D., Keller, J.M.: Fundamentals of Computational Intelligence: Neural Networks, Fuzzy Systems, and Evolutionary Computation, IEEE Press Series on Computationa Intelligence. Wiley (2016)
22. Maple FuzzySets. https://16.www.maplesoft.com/applications/view.aspx?SID=96899. Accessed 15 Oct 2020
23. Ahmad, I., Khan, A.: A comparative analysis of linear and nonlinear semi-active suspension system. Mehran Univ. Res. J. Engi. Technol. **37**, 233–240 (2018)
24. Jiao, L.: Vehicle model for tyre-ground contact force evaluation, M.Sc thesis, KTH Royal Institute of Technology (2013)
25. Evers, W.J., Besselin, I., Teerhuism, A., Oomen, T., Nijmeijer, H.: Experimental validation of a quarter truck model using asynchronous measurements with low signal-to-noise ratios. In: 10th AVEC conference, Loughborough, UK. 22–26. August, 2010, pp. 177–182, Loughborough University Department of Aeronautical & Automotive Engineering & Transport Studies (2010)
26. Zhu, Q., Ishitobi, M.: Chaotic vibration of a nonlinear full-vehicle model. Int. J. Solids Struct. **43**(3–4), 747–759 (2015)
27. Cui, Y., Kurfess, T. R., Messman, M.: Testing and modeling of nonlinear properties of shock absorbers for vehicle dynamics studies. In: Proceedings of the World Congress on Engineering and Computer Science 2010 Vol II WCECS 2010, San Francisco, USA, pp. 949–954. Newswood Limited (2010)
28. Dixon, J.C.: Suspension Geometry and Computation. Wiley, Chichester (2009)
29. Hajdu, F., Kuti, R.: Examination of chaotic vibrations during operation of a fire truck. In: Proceedings of MAC 2018 in Prague Prag, Czech Republic, 25–27 May 2018, pp. 163–170. MAC Prague consulting Ltd., Prague, Czech Republic (2018)
30. Corless, R.M.: What good are numerical simulations of chaotic dynamical systems? Comput. Math. Appl. **28**(10–12), 107–121 (1994)
31. Corless R.M., Fillion N.: Backward error analysis for perturbation methods. In: Fillion, N., Corless, R., Kotsireas, I. (eds.) Algorithms and Complexity in Mathematics, Epistemology, and Science. Fields Institute Communications, vol. 82. Springer, New York (2019). https://doi.org/10.1063/1.5019118
32. Hajdu, F.: Modelling of nonlinear active and passive suspension systems (in Hungarian). In: Volume of New National Excellence Program 2018/2019, Széchenyi István University, pp. 197–208 (2019)
33. Hajdu, F., Molnárka, Gy., Kuti, R.: One-at-a-time sensitivity study of a nonlinear fire truck suspension model. FME Trans. **48**(1), 90–95 (2020)
34. Zhu, Q., Ishitobi, M.: Chaotic vibration of a nonlinear full-vehicle model. Int. J. Solids Struct. **43**(3–4), 747–759 (2006)

35. Hajdu, F., Kuti, R.: Development of a simulation model for analyzing vibrations of a double cabin fire truck and their effects on firefighters. Int. Adv. Res. J. Sci. Eng. Technol. **6**(5), 74–79 (2019)
36. Maple Signal Processing. https://www.maplesoft.com/products/maple/features/Signal_Processing.aspx. Accessed 15 Oct 2020
37. International Standard 2631 (ISO 1974, 1985): Mechanical vibration and shock - Evaluation of human exposure to whole-body vibration

Merging Maple and GeoGebra Automated Reasoning Tools

Zoltán Kovács[1], Tomás Recio[2(✉)], and M. Pilar Vélez[2]

[1] The Private University College of Education of the Diocese of Linz, Salesianumweg 3, 4020 Linz, Austria
zoltan@geogebra.org

[2] Universidad Antonio de Nebrija, Calle Pirineos, 55, 28040 Madrid, Spain
{trecio,pvelez}@nebrija.es

Abstract. A branch of the Automated Deduction in Geometry (ADG) theory deals with the automatic proof and discovery of theses holding on a given collection of hypotheses. The mechanical proof and derivation of such statements, through computational complex algebraic geometry methods, will be exemplify in this paper through the performance of GeoGebra Automated Reasoning Tools. Then we will refer to some challenging issues that rise in this context, regarding the translation in algebraic terms of the given geometric facts, the verification or the finding of the sought properties, and the interpretation of the outcome. We will show how some of these involved issues could be be better approached through the collaboration of Maple packages for polynomial ideal manipulation, requiring, as well, diverse theoretical concepts recently introduced by the authors.

Keywords: Automated reasoning · Computational algebraic geometry · Elementary geometry · Maple · GeoGebra

1 GeoGebra's Automated Reasoning Tools

In the past years the authors have developed and integrated in the dynamic mathematics program GeoGebra [10] several tools and commands for automated proving and discovering in elementary geometry, as detailed in [19], see also [3]. Probably the most basic one is the *Relation* command, that automatically searches, from a given collection of possibilities: parallelism, equal size, concurrency, co-circularity, etc., if one of these relations[1] holds over some pair of geometric objects in a GeoGebra construction, specified by the user.

The command proceeds, first, making some numerical estimation (that was the only feature of the command in earlier times [16]) about the verification of

[1] See https://wiki.geogebra.org/en/Relation_Command for a full list.

The authors are partially supported by FEDER/Ministerio de Ciencia, Innovación y Universidades – Agencia Estatal de Investigación/MTM2017-88796-P (Symbolic Computation: new challenges in Algebra and Geometry and their applications).

R. M. Corless et al. (Eds.): MC 2020, CCIS 1414, pp. 252–267, 2021.
https://doi.org/10.1007/978-3-030-81698-8_17

the different properties and, when an answer sufficiently close to true is obtained for a particular query, a symbolic approach to deal with this property is turned on. Thus, the different steps of the geometric construction are internally and automatically translated into a collection of algebraic equations with variables representing the coordinates of the elements (free or constrained) in the construction. Likewise, the property under investigation is expressed as a polynomial equation, and, then, some automated reasoning algorithms [15,22] are executed to rigorously verify if the thesis (the property) holds for the given hypotheses (the geometric construction).

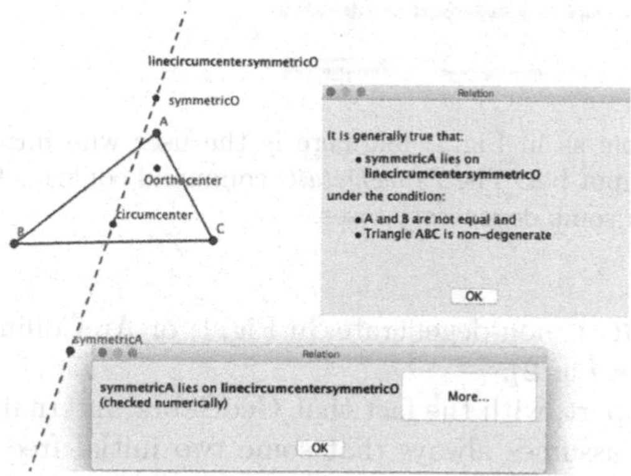

Fig. 1. An example of the *Relation* command executed in GeoGebra Classic 5. In the bottom, the numerical check. In the upper right, the symbolic answer provided by *Relation* after clicking the "More..." button.

Figure 1 shows the use of the *Relation* tool. An arbitrary triangle ABC is given and GeoGebra is asked about the existence of some relation holding between the symmetric point *symmetricA* of vertex A with respect to the midpoint of the opposite side, and the line *linecircumcentersymmetricO* defined by the circumcenter and the symmetric of the orthocenter O with respect to A. As shown in the figure, GeoGebra is able to automatically conjecture and, then, prove, the alignment of the three points *symmetricA*, circumcenter and *symmetricO*, as stated in Example 230 of Chou's benchmark [8].

In GeoGebra's automated reasoning tools, there are two other, quite natural, commands: *Prove* and *ProveDetails*. Both require the user to introduce, over a certain construction, a specific thesis. Then their output is, essentially, a true/false label, adding, in the former case, if using the *ProveDetails* command, details about instances that are considered degenerate and where the thesis does not hold. See Fig. 2.

Let us remark that, in general, it is quite involved to describe in detail the method implemented in GeoGebra to output the non-degeneracy requirements that appear in the *Relation* or in the *ProveDetails* window (e.g. A and B not

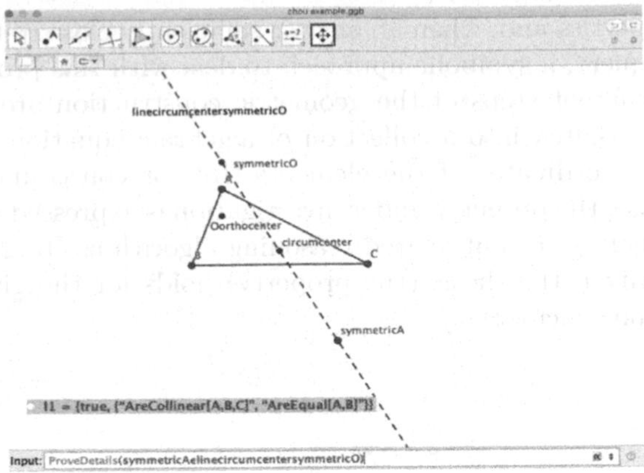

Fig. 2. Same example as in Fig. 1, but here is the user who introduces the conjectured thesis in the Input bar. The *ProveDetails* command confirms the truth of Chou's statement except for some degenerate cases.

equal, Triangle ABC is non-degenerate, in Fig. 1; or AreCollinear$\{A, B, C\}$ and AreEqual$\{A, B\}$) in Fig. 2).

It has to do, in part, with the fact that GeoGebra, internally and to simplify the computations, assumes always that some two initial free points of a given figure are $(0, 0)$ and $(0, 1)$ and, thus, it automatically outputs they are assumed to be different, even if it is not needed for the conclusion. It has to do, also, with the fact that the non-degeneracy conditions are algebraically related to the generators (and its factors!) of a certain ideal computed by GeoGebra internally (and that Maple computations help to understand, as described in the last two Sections of this chapter), and it might happen that the geometric translation of some factor or generator yields a property that is already consequence of another translated one, etc. It can also happen that there is not a sound geometric translation and GeoGebra then outputs just some {"..."}, etc. See https://wiki.geogebra.org/en/ProveDetails_Command.

Finally, the command *LocusEquation* is the last essential piece of the basic collection of GeoGebra's automated reasoning tools. Suppose we make a construction similar to the ones in the previous Figs. 1 and 2, but where we do not consider the symmetric of A with respect to the midpoint of the opposite side. Rather, we build a new, free point D and we consider the symmetric of A with respect to D. Obviously, this symmetric point does not lie, in general, in the line defined by the circumcenter and the symmetric of the orthocenter with respect to A. So we would like to discover a new theorem: where to place D so that this alignment takes place. And the answer is: place D on the red line (a necessary condition, maybe not sufficient), see Fig. 3 .

Indeed, in this case the necessary condition turns out to be also sufficient, as one can easily check with the *Prove* command, after adding the condition that D lies in the parallel to *linecircumcentersymmetricO* through the mid-point

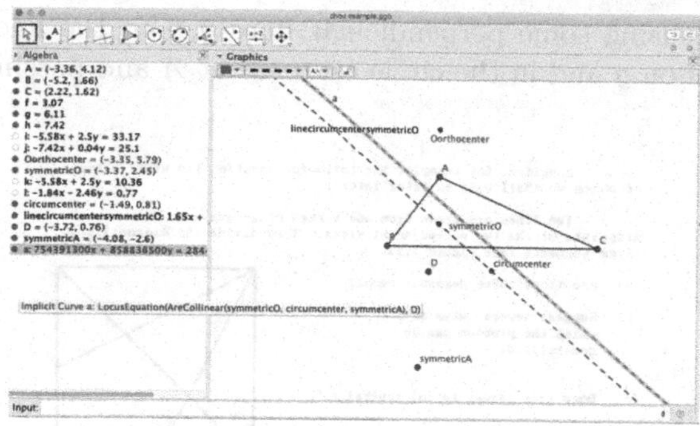

Fig. 3. Discovering a generalization of Chou's Example 230. The alignment holds, but *symmetricA* is now the symmetric with respect to any point on the (red) line, parallel to *linecircumcentersymmetricO* through the midpoint of side BC.

of side BC. It is, in many respects, a way to "discover" new theorems, keeping a given thesis but relaxing the hypotheses. It is quite surprising to apply this method to different, well known, theorems after dropping some—apparently—essential thesis, and see what happens.

Let us conclude this summary introduction to GeoGebra's automated reasoning tools by mentioning two recent novelties. One is the development of a specific fork of GeoGebra, that we have called GeoGebra Discovery[2], including the most recent versions of the automated reasoning commands we have described, plus the *Discover(P)* command. This command allows GeoGebra to formulate, almost without human intervention and following some heuristics, different geometric conjectures involving a concrete element (say, a point P, selected by the user) of a given geometric construction, and to confirm or deny them through the reasoning tools. It is available both on GeoGebra Classic 5, for Windows, Mac and Linux systems; and on GeoGebra Classic 6, ready for starting it in a browser and thus, adequate for tablets and smartphones.

The other is the *Geometer Automaton* (GA)[3], a module that allows GeoGebra to internally consider, following some combinatorial approach, and not requiring any human decision at all, the validity or failure of a collection of geometric conjectures involving all the elements of a given geometric construction. Consider, for instance, the following well known question posed in the usually called as the ICMI-Kuwait Report [12]: given a square $ABCD$, let E, F be the midpoints of the sides AB, BC, respectively. Let G, H be the intersection points of the diagonal AC and the lines DE and DF. Are the segments AG, GH, HC of equal length? See Fig. 4, reproducing a page from the report, and Fig. 5 where the construction of a square has been performed in GeoGebra starting from two

[2] https://github.com/kovzol/geogebra-discovery, http://autgeo.online/geogebra-discovery/.

[3] http://autgeo.online/ag/, https://github.com/kovzol/ag.

free points A, B, using some perpendicular lines g, h to the line $f = AB$ and selecting point D on g and in the circle centered at A and passing through B.

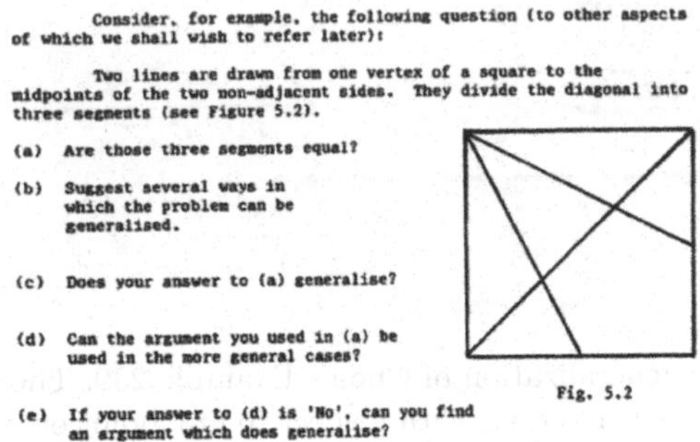

Consider, for example, the following question (to other aspects of which we shall wish to refer later):

Two lines are drawn from one vertex of a square to the midpoints of the two non-adjacent sides. They divide the diagonal into three segments (see Figure 5.2).

(a) Are those three segments equal?

(b) Suggest several ways in which the problem can be generalised.

(c) Does your answer to (a) generalise?

(d) Can the argument you used in (a) be used in the more general cases?

(e) If your answer to (d) is 'No', can you find an argument which does generalise?

Fig. 5.2

Fig. 4. A question from the ICMI-Kuwait Report.

Of course, we could address the posed question by using the *ProveDetails* command, asking for the equality of this pair of segments AG, GH and then, the same for GH, HC, see Fig. 5, where output $l1$ shows the truth of the first equality except when the square collapses ($A = B$) and $l2$ does the same for the second equality.

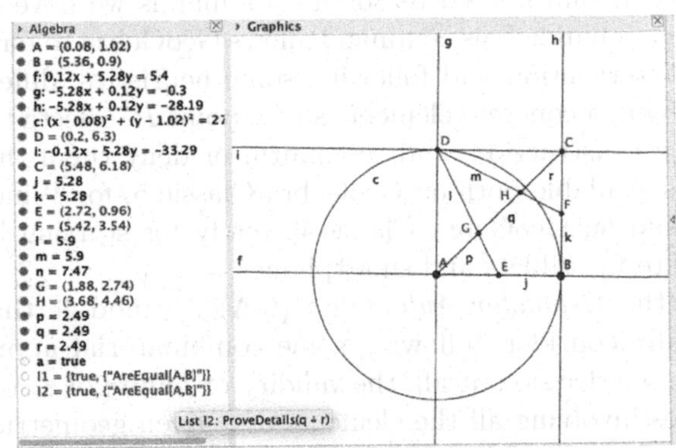

Fig. 5. Answering Kuwait Report question by using ProveDetails.

But here we want to show how the GA could deal with this situation in a different way. Indeed, formulating this precise question required, first, a human geometer observing that the segments AG, GH, HC could be of equal length, and then proving that this conjecture was correct. Now we would like to show how the GA plays the role of the human geometer, when receiving just a geometric diagram, see Fig. 6. The user just has to decide what kind of property the GA should consider, choosing among several options: *Collinearity of three points*, *Equality of distances between two points*, etc. (see bottom left in Fig. 6).

Then, after clicking the *Start Discovery* button, the obtained output (2 s computation time) is a collection of equalities between different points in the geometric diagram, as shown in Fig. 7. In particular we obtain the sought equalities and many more, some of them not so easy to visualize, like the equality $EH = FG$; some simply obvious, like $AB = AD$, as we are dealing with a square $ABCD$.

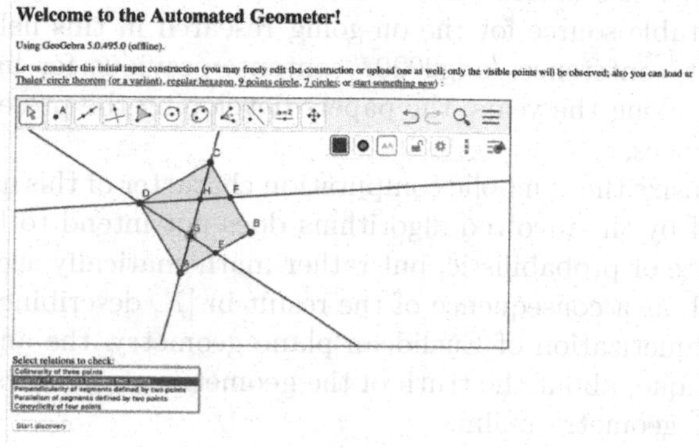

Fig. 6. The geometer automaton at work over a geometric figure.

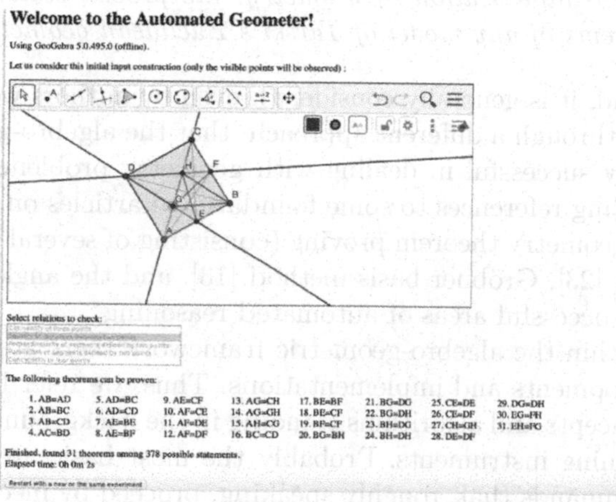

Fig. 7. Statements on equality of distances, automatically found on the given figure.

Details about these recent GeoGebra ART (Automated Reasoning Tools) improvements can be found in [4,5,21] and the references therein.

2 The Symbolic Computation Background: Basic Facts

Behind the quite successful performance of the above mentioned tools there are plenty of computational algebraic geometry algorithms, using the Giac [11] computer algebra system, embedded in GeoGebra [17].

Roughly speaking, the GeoGebra approach follows the well-known algebro-geometric path to automated reasoning in geometry described in the foundational work of [8], that also includes a detailed account of the basic concepts, precedents and performing examples. Moreover, since 1996, the proceedings of the biennial conference on Automated Deduction in Geometry (ADG) seem to be the most suitable source for the on-going research in this field, see https://www3.risc.jku.at/conferences/adg2020/?content=previous for links to the volumes collecting, along the years, the papers developing the presentations in the different conferences.

Let us emphasize the symbolic computation character of this approach, as the output provided by the involved algorithms does not intend to be just numerically approximate or probabilistic, but rather mathematically accurate and fully rigorous. Indeed, as a consequence of the result in [7], describing the formalization of the arithmetization of Euclidean plane geometry, the answers obtained using this technique, about the truth of the geometric statements, are also valid on the synthetic geometry realm:

> ...*The arithmetization of geometry paves the way for the use of algebraic automated deduction methods in synthetic geometry. Indeed, without a "back-translation" from algebra to geometry, algebraic methods only prove theorems about polynomials and not geometric statements. However, thanks to the arithmetization of geometry, the proven statements correspond to theorems of any model of Tarski's Euclidean geometry axioms.*

On the other hand, it is generally considered, even by authors working on automated reasoning through a different approach, that the algebro-geometric methods are extremely successful in dealing with geometry problems. For example, [2] states—including references to some foundational articles on this area—that "...Automated geometry theorem proving (consisting of several techniques such as Wu's methods [23], Gröbner basis method [13], and the angle method [9]) is one of the most successful areas of automated reasoning".

Of course, within the algebro-geometric framework there are many different theoretical developments and implementations. Thus, we refer to [18,20,22] for details on the concepts and algorithms standing in the background of GeoGebra's automated reasoning instruments. Probably the most basic are the *Prove* and *ProveDetails* commands that, roughly speaking, proceed by mechanically translating the hypotheses and thesis of a geometric statement $\{H \Rightarrow T\}$ into polynomial expressions in $K[X]$, where the variables $X = \{x_1, \ldots, x_n\}$ refer to the symbolic coordinates (after adopting a convenient coordinate system) involved in the algebraic description of the hypotheses $H = \{h_1 = 0, \ldots, h_r = 0\}$ and thesis $T = \{f = 0\}$, and K is the field of coefficients.

For example, if we are dealing with Pythagoras's Theorem, we are given $A = (a_1, a_2), B = (b_1, b_2), C = (c_1, c_2)$ so that AC, BC are perpendicular, that is $(c_1 - a_1)(c_1 - b_1) + (c_2 - a_2)(c_2 - b_2) = 0$. Moreover, we denote the length of the sides as h, i, j, thus verifying $h^2 = (b_1 - a_1)^2 + (b_2 - a_2)^2, i^2 = (c_1 - b_1)^2 + (c_2 - b_2)^2, j^2 = (c_1 - a_1)^2 + (c_2 - a_2)^2$. And the thesis is $h^2 = i^2 + j^2$. See Fig. 8.

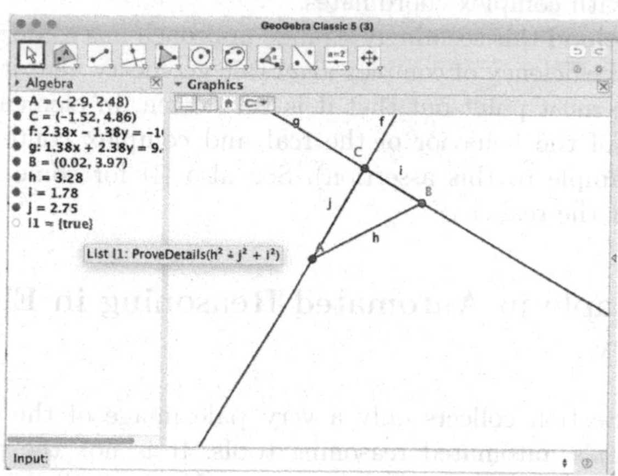

Fig. 8. Pythagoras's Theorem proved by GeoGebra.

Next, we associate the geometric instances verifying the hypotheses to the different solutions of a system of polynomial equations $V(H) = \{h_1 = 0, \ldots, h_r = 0\}$ (*hypotheses variety*), that in this case would be $\{(c_1 - a_1)(c_1 - b_1) + (c_2 - a_2)(c_2 - b_2)) = 0, h^2 - (b_1 - a_1)^2 \ (b_2 - a_2)^2 = 0, i^2 \ (c_1 - b_1)^2 - (c_2 - b_2)^2 = 0, j^2 - (c_1 - a_1)^2 - (c_2 - a_2)^2 = 0\}$. Analogously, the thesis is translated as the solution set of a polynomial $V(T) = \{f = h^2 - i^2 - j^2 = 0\}$ (*thesis variety*). Thus, we could say that Pythagoras's Theorem states that $V(H) \subseteq V(T)$, meaning that every instance in $V(H)$, i.e. a right triangle, verifies also the condition to be part of $V(T)$, i.e. that the square of the length of the hypotenuse is equal to the sum of the squares of the lengths of the other two sides.

Now, following [14], checking that $V(H) \subseteq V(T)$ is equivalent to proving that $V(H) \cap (K^n \setminus V(T)) = \emptyset$; and, using Hilbert's Nullstellensatz, assuming K is an algebraically closed field, this can be decided by showing that $1 \in \langle h_1, \ldots, h_r, f \cdot t - 1 \rangle$, a test that is available through the commands implemented in Maple's *PolynomialIdeals* package. The next lines show the corresponding Maple session:

```
> H:=<((c_1-a_1)*(c_1-b_1)+(c_2-a_2)*(c_2-b_2)),
h^2-(b_1-a_1)^2-(b_2-a_2)^2,i^2-(c_1-b_1)^2-(c_2-b_2)^2,
j^2-(c_1-a_1)^2-(c_2-a_2)^2>: T:=h^2-i^2-j^2:
1 in H + <T*t-1>;
                              true
```

Let us remark that whole algorithm relies on the consideration of the solutions of the hypotheses and thesis equations over an algebraically closed field (i.e. the complex numbers). That is, we are setting the validity or failure of the proposed statement for geometric instances over a field with such characteristics; if the statement is found true, it will be true for complex instances; if it is declared false, it could be true over the reals... but it would not be true for some instances with complex coordinates.

The reason behind this seemingly strange approach has to do with the higher development and efficiency of complex algebraic geometry algorithms versus the real case, yet we must point out that it is very often, in elementary geometry, the coincidence of the behavior of the real and complex contexts (see [6] for some counterexample to this assertion). See also [1] for some recent, specific developments for the real case.

3 Using Maple in Automated Reasoning in Elementary Geometry

The precedent Section collects only a very pale image of the complete story behind GeoGebra's automated reasoning tools. It is not the purpose of this paper to get into technical details concerning the theoretical background and algorithms implemented in GeoGebra (see, for instance [3,15]), but to exemplify how the cooperation of Maple can help understanding some challenging issues involved in the algebraic approach to the verification of some geometric property, contributing to the interpretation of the outcome.

Consider, for instance, Fig. 9, displaying the following construction: given a couple of points $A(a_1, a_2), B(b_1, b_2)$, we consider three points $D(d_1, d_2), F(f_1, f_2), G(g_1, g_2)$ on the line AB. Then we consider a new point $C(c_1, c_2)$ and points E, H, I that are, respectively the midpoints of the segments CD, CF, CG. Finally, we claim that E, H, I are aligned. Here the hypotheses H involves three equations, each one expressing that D, F, G are on the line AB, namely $H := \{(d_1 - a_1) \cdot (b_2 - a_2) - (d_2 - a_2) \cdot (b_1 - a_1) = 0, (f_1 - a_1) \cdot (b_2 - a_2) - (f_2 - a_2) \cdot (b_1 - a_1) = 0, (g_1 - a_1) \cdot (b_2 - a_2) - (g_2 - a_2) \cdot (b_1 - a_1) = 0\}$.

To formulate the thesis we consider the alignment of $E = ((d_1 + c_1)/2, (d_2 + c_2)/2), H = ((f_1 + c_1)/2, (f_2 + c_2)/2), I = ((g_1 + c_1)/2, (g_2 + c_2)/2))$, that is, the determinant of the matrix $\begin{pmatrix} 1 & (d_1+c_1)/2 & (d_2+c_2)/2 \\ 1 & (f_1+c_1)/2 & (f_2+c_2)/2 \\ 1 & (g_1+c_1)/2 & (g_2+c_2)/2 \end{pmatrix}$ yielding, after multiplying the output by 4, the thesis $T := d_1 f_2 - d_1 g_2 - d_2 f_1 + d_2 g_1 + f_1 g_2 - f_2 g_1 = 0$.

Now we perform the same test as in the precedent example, but this time the output is not the expected one:

```
> 1 in H + <T*t-1>;
```

false

When we arrive to this point, where our personal geometric intuition (the statement is certainly true!) seems to contradict the result of the computation (the

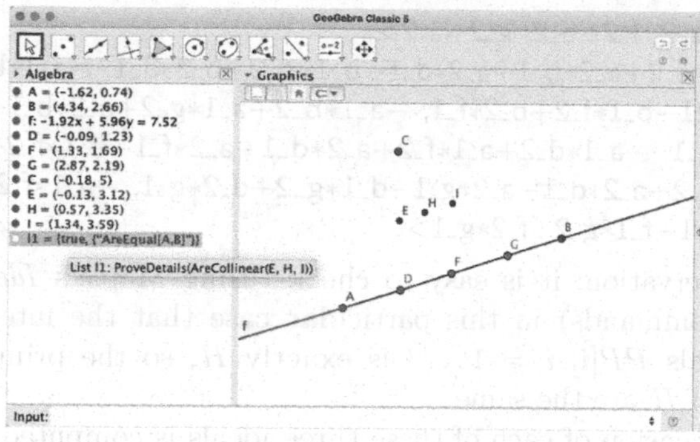

Fig. 9. Proving the collinearity of the mid-points of CD, CF, CG.

machine labels the statement as false!), it is natural to go back to the hypotheses set H, revisiting it in order to check that we have accurately collected in its description the geometric constraints of our given data. This is a task where the role of Maple is essential. To begin with, we would like to confirm that the set $V(H)$ has the expected dimension 9: two free coordinates for each of the initial points A, B, C and, then, since D, F, G are constrained to lie on the line AB, each of these points brings just a single free coordinate.

To verify this, we need, first, to declare H as an ideal living in $K[a_1, a_2, b_1, b_2, c_1, c_2, d_1, d_2, g_1, g_2, f_1, f_2]$:

```
> H:=<(d_1−a_1)∗(b_2−a_2)−(d_2−a_2)∗(b_1−a_1),(f_1−a_1)∗(b_2−a_2)−
−(f_2−a_2)∗(b_1−a_1), (g_1−a_1)∗(b_2−a_2)−(g_2−a_2)∗(b_1−a_1),
variables ={a_1,a_2,b_1,b_2,d_1,d_2,g_1,g_2, f_1 , f_2 ,c_1,c_2}>;
```

because c_1, c_2 are part of the initial construction, but these two variables are not subject to any constraint and thus they do not appear explicitly as part of the definition of H. And then we ask for the dimension of H:

```
> HilbertDimension(H);
                                10
```

getting, as unexpected answer, that it is ten!

Now, since the dimension of an ideal is the maximum of the dimension of its associated primary or prime ideals, we try to find the *PrimaryDecomposition*(H), to understand, checking the dimension on each primary component and trying to understand which one does not behave as expected. Unfortunately, after a long computation time without answer, we decide to abort the computation and switch to computing the *PrimeDecomposition* (over the field of coefficients involved in the description of H, namely, Q). This decomposition is easily computed by Maple, getting three prime ideals:

```
> PP:=PrimeDecomposition(H);
<b_1−a_1, b_2−a_2>,
```

```
<b_2−a_2, d_2−a_2, f_2−a_2, g_2−a_2>,
<−a_1*b_2+a_1*d_2+a_2*b_1−a_2*d_1−b_1*d_2+b_2*d_1, −a_1*b_2+a_1*f_2+
+a_2*b_1−a_2*f_1−b_1*f_2+b_2*f_1, −a_1*b_2+a_1*g_2+a_2*b_1−a_2*g_1−
−b_1*g_2+b_2*g_1, −a_1*d_2+a_1*f_2+a_2*d_1−a_2*f_1−d_1*f_2+d_2*f_1,
−a_1*d_2+a_1*g_2+a_2*d_1−a_2*g_1−d_1*g_2+d_2*g_1, −a_1*f_2+a_1*g_2+
+a_2*f_1−a_2*g_1−f_1*g_2+f_2*g_1>
```

A collateral observation: it is easy to check (using Maple's *IdealContainment* and *Intersect* commands) in this particular case that the intersection of the three prime ideals $PP[i], i = 1 \ldots 3$ is exactly H, so the prime and primary decomposition of H are the same.

Next the dimension of each of these three ideals is computed, obtaining

```
> for i from 1 to 3 do HilbertDimension(PP[i]) od;
```

$$10$$
$$8$$
$$9$$

so actually only the last component behaves as expected. Indeed we observe that the first one, $PP[1]$, has dimension 10 (e.g. free variables $\{a_1, a_2, c_1, c_2, d_1, d_2, f_1, f_2, g_1, g_2\}$), and corresponds to instances of our construction where $A = B$, a degenerate situation that we do not consider as being part of the "real" statement, yet it is a case that the equations of the hypotheses we have introduced do not dismiss.

The second component corresponds to instances where all the points A, B, D, F, G are on the line parallel to the X-axis through A, and thus this point contributes with two free coordinates, and each of the remaining B, D, F, G with only one: b_1, d_1, f_1, g_1. Finally C brings two more free variables, so $2+4+2 = 8$ in total.

```
> EliminationIdeal (PP[2],{a_1,a_2,b_1,c_1,c_2,d_1,f_1,g_1});
```

$$\langle 0 \rangle$$

We consider that such component includes some specific collection of "standard" instances, so we should test our thesis over them.

Finally, the last component $PP[3]$, has the expected dimension 9 and collection of free variables. Indeed

```
> EliminationIdeal (PP[3],{a_1,a_2,b_1,b_2, c_1,c_2,d_1,f_1,g_1});
```

$$\langle 0 \rangle$$

So, after these computations, we could conclude that the amended hypotheses ideal could rather be $HH := Intersect(PP[2], PP[3])$. And we can check that the rectified statement $HH \Rightarrow T$ is now correct, as expected:

```
> 1 in HH + <T*t−1>;
```

true

All these digressions are already captured by GeoGebra's automated reasoning tools, as shown in Fig. 9, *Algebra* column, displaying in the last entry that the collinearity of E, H, I is true except when $AreEqual[A, B]$ holds, that is, except for instances verifying $PP[1]$. Although GeoGebra's internal computations do not follow exactly the steps we have described here in this example, they highlight as well the same two crucial issues: the relevance of free variables and the relevance of the different irreducible components of the hypotheses variety $V(H)$.

To expand these essential ideas a little bit, in what follows we will consider K and L to be fields, with L an algebraically closed extension on K (for instance $L = \mathbb{C}$ and $K = \mathbb{Q}$), and an algebraically translated statement $\{H \Rightarrow T\}$. Let be $H = \langle h_1, \ldots, h_r \rangle$ and $T = \langle f \rangle$ the hypotheses and thesis ideals in the polynomial ring $K[X]$, where the variables $X = \{x_1, \ldots, x_n\}$ refer to the symbolic coordinates involved in the algebraic description of the hypotheses in K^n. And take the algebraic variety $V(H)$ (respectively, $V(T)$) in the affine space L^n defined over K.

Then consider (as a user choice, possibly following some geometric intuition) a subset $Y = \{x_1, \ldots, x_d\}$ of X being independent modulo H (i.e. $H \cap K[Y] = \langle 0 \rangle$), but not necessarily of maximum size, i.e. not necessarily $d = dimension(H)$. Now, the irreducible components of $V(H)$ where these variables Y do remain independent will be declared as *non-degenerate* components. With this terminology, the following definition involves a subtle modification of a similar one in [22] or [20], in that the cardinal of Y in our version below does not have to agree with the dimension of H.

Definition 1. *A statement $\{H \Rightarrow T\}$ is considered to be* generally true *if the thesis f vanishes on all non-degenerate K-components of the hypotheses variety $V(H)$. If the thesis holds over all the (degenerate or not) components the statement is labelled as* always true *or, simply,* true.

For example, it is easy to check, with Maple's command similar to the ones we have shown before, that Pythagoras's Theorem holds *always true* since the hypotheses variety has dimension 5, with independent variables given by the coordinates of A, B and one of the coordinates of C. It has two irreducible components, associated to different signs of i or of j, each are non-degenerate, of dimension 5, and the thesis is verified over both of them.

On the other hand, the collinearity of the three midpoints in Fig. 9 is *generally true*, with respect to the chosen independent variables $\{a_1, a_2, b_1, b_2, c_1, c_2, d_1, f_1, g_1\}$, but not *always true*, as the thesis does not hold over the first component $PP[1] = \langle b_1 - a_1, b_2 - a_2 \rangle$:

```
> 1 in <b_1-a_1, b_2-a_2>+<T*t-1>;
```

false

Of course, this computation is not strictly needed since it follows from the fact that we have already observed that 1 is not in $H + \langle T \cdot t - 1 \rangle$, but it is part of $(PP[2] \cap PP[3)] + \langle T \cdot t - 1 \rangle$. In general, we have the following characterization:

Proposition 1. *Let $\{H \Rightarrow T\}$ be a statement. Then it is* always true *if and only if $V(H) \subseteq V(T)$ if and only if $1 \in H + \langle T \cdot t - 1\rangle$ if and only if for all associated primes $PP[i], i = 1 \dots r$ of H, $V(PP[i]) \subseteq V(T)$ if and only if for all $i = 1 \dots r$, $1 \in PP[i] + \langle T \cdot t - 1\rangle$ if and only if for all $i = 1 \dots r$, $\{PP[i] \Rightarrow T\}$ is* always true.

Proof is immediate, bearing in mind Hilbert's Nullstellensatz for the algebraic/geometric interrelation.

On the other hand, we have the following modification of the first statement in Theorem 1 in [20], bearing in mind that in the proof of that particular item there it is not required that Y is of maximum size, with $d = dim(H)$.

Proposition 2. *Let $\{H \Rightarrow T\}$ be a statement and fix a set $Y = \{x_1, \dots, x_d\}$ of independent variables for the hypotheses ideal H (but d does not have to be equal to the dimension of H). Then, the statement is* generally true *if and only if $\langle h_1, \dots, h_r, f \cdot t - 1\rangle K[X,t] \cap K[Y] \neq \langle 0\rangle$.*

In fact, following the proof in [20] it is easy to conclude that the ideal $\langle h_1, \dots, h_r, f \cdot t - 1\rangle K[X,t] \cap K[Y]$ gives the algebraic description of the Zariski-closure for the projection, over the space K^d of $(x_1, \dots, x_d)$-coordinates, of the degenerate irreducible components of $V(H)$ where T is not always true.

As a toy example of this situation, we could think of $H = \langle x \cdot z\rangle K[x, y, z]$, with $\{y, z\}$ as free variables, with $\{x = 0\}$ as the non-degenerate component and with $\{z = 0\}$ as a degenerate component (since z is not free in this component). Let $T = \{x = 0\}$ be the thesis. Then $\{H \Rightarrow T\}$ is *generally true*, since T vanishes over the only non-degenerate component. In fact,

```
> EliminationIdeal(<x*z> + <x*t-1>,{y,z});
```

$$\langle z\rangle$$

and the output yields the equation $\{z = 0\}$ of the projection, over the $\{y, z\}$-plane, of the degenerate component $\{z = 0\}$. Notice that, if we select instead $\{x, y\}$ as free variables, then the same statement would not be *generally true*, since

```
> EliminationIdeal(<x*z> + <x*t-1>,{x,y});
```

$$\langle 0\rangle$$

because now $\{x = 0\}$ would be a degenerate component, yet it is the only component where the thesis holds true.

Again, using the above Proposition over the example of Fig. 9, we get

```
> EliminationIdeal(H + <T*t-1>,{a_1,a_2,b_1,b_2,c_1,c_2,d_1,g_1,f_1});
```

$$\langle b\_1 - a\_1, b\_2 - a\_2\rangle \quad (1)$$

confirming that the thesis vanishes identically over all non-degenerate components (considering $\{a_1, a_2, b_1, b_2, c_1, c_2, d_1, g_1, f_1\}$ as the set of variables that should be free in the non-degenerate components).

4 Conclusions

The above output of the *EliminationIdeal* command being not zero is, surely, a much simpler way—versus the previous approach, requiring the primary decomposition of the hypotheses ideal and the test over each component—to conclude that the involved statement (the collinearity of the three midpoints E, H, I) is true in all non-degenerate cases. It is, roughly speaking, the way the *ProveDetails* algorithm is implemented in GeoGebra, and depends on the internal choice of free variables, that GeoGebra selects following the construction steps performed by the user (introducing free or semi-free points, those constrained to lie on some one dimensional object by using the *Point on Object* command, etc.).

It is a simpler way, indeed, but one that does not provide any hint about the reasons behind the conclusion. It is in this respect that we believe that the contribution of Maple could be extremely clarifying to help humans to understand what is the detailed picture concerning the statement under consideration.

In particular, the precedent computations, yielding $\langle b_1 - a_1, b_2 - a_2\rangle$ as output of (1) explain why GeoGebra's affirmative answer to the *ProveDetails(AreCollinear(E,H,I))* requires discarding *AreEqual[A,B]*, as the equality between these two points is described, precisely, by the vanishing of all the polynomials in $\langle b_1 - a_1, b_2 - a_2\rangle$. In fact, let us remark that, if we are dealing with a *generally true* statement $\{H \Rightarrow T\}$, and if we consider any non-zero element $g \in \langle h_1, \ldots, h_r, f \cdot t - 1\rangle K[X, t] \cap K[Y]$, then adding $g \cdot z - 1$ as a new hypotheses yields a statement $\{H' = \langle h_1, \ldots, h_r, g \cdot z - 1\rangle \Rightarrow T\}$ that is *always true*. In fact, by construction, $g \in \langle h_1, \ldots, h_r, f \cdot t - 1\rangle$, so g is also in $\langle h_1, \ldots, h_r, g \cdot z - 1, f \cdot t - 1\rangle$, which implies this ideal is $\langle 1\rangle$.

Thus, in the Collinearity example above, the statement will be *always true* if we add the negation of any polynomial in $\langle b_1 - a_1, b_2 - a_2\rangle$, that is, if we are not in the case $A = B$. A similar consideration could explain the output shown in other statements addressed in Figs. 2 and 5.

We have exemplified in this paper a few, quite basic, issues concerning some surprising outcomes of GeoGebra's Automated Reasoning Tools, showing how they are better handled and interpreted with the cooperation of Maple's computations, trying to reproduce and deepen the involved algebraic geometry algorithms. We refer the interested reader to [20] for further, similar, situations, both concerning new surprising outputs in GeoGebra and the corresponding enhanced explanations using Maple.

References

1. Abar, C., Kovács, Z., Recio, T., Vajda, R.: Connecting Mathematica and GeoGebra to explore inequalities on planar geometric constructions, Brazilian Wolfram Technology Conference, Saõ Paulo, November 2019
2. Alvin, C., Gulwani, S., Majumdar, R., Mukhopadhyay, S.: Synthesis of geometry proof problems. In: Proceedings of the Twenty-Eighth Association for the Advancement of Artificial Intelligence Conference on Artificial Intelligence, pp. 245–252 (2014). https://www.microsoft.com/en-us/research/publication/synthesis-geometry-proof-problems/

3. Botana, F., et al.: Automated theorem proving in GeoGebra: current achievements. J. Autom. Reasoning **55**(1), 39–59 (2015). https://doi.org/10.1007/s10817-015-9326-4
4. Botana, F., Kovács, Z., Recio, T.: Towards an automated geometer. In: Fleuriot, J., Wang, D., Calmet, J. (eds.) AISC 2018. LNCS (LNAI), vol. 11110, pp. 215–220. Springer, Cham (2018). https://doi.org/10.1007/978-3-319-99957-9_15
5. Botana, F., Kovács, Z., Recio, T.: A mechanical geometer. Math. Comput. Sci. (2020)
6. Recio, T., Botana, F.: Where the truth lies (in automatic theorem proving in elementary geometry). In: Laganá, A., Gavrilova, M.L., Kumar, V., Mun, Y., Tan, C.J.K., Gervasi, O. (eds.) ICCSA 2004. LNCS, vol. 3044, pp. 761–770. Springer, Heidelberg (2004). https://doi.org/10.1007/978-3-540-24709-8_80
7. Boutry, P., Braun, G., Narboux, J.: Formalization of the arithmetization of Euclidean plane geometry and applications. J. Symbolic Comput. **90**, 149–168 (2019)
8. Chou, S.C.: Mechanical geometry theorem proving. D. Reidel Publishing Company, Dordrecht, Netherlands (1988)
9. Chou, S.C., Gao, X.S., Zhang, J.Z.: Machine proofs in geometry. Automated production of readable proofs for geometry theorems. In: Series on Applied Mathematics, vol. 6, World Scientific, Singapore (1994)
10. GeoGebra Homepage. http://www.geogebra.org. Accessed Dec 2020
11. Giac/Xcas Homepage. https://www-fourier.ujf-grenoble.fr/~parisse/giac.html. Accessed Dec 2020
12. Howson, G., Wilson, B. (eds.): ICMI Study Series: School Mathematics in the 1990's. Cambridge University Press, Cambridge, UK (1987)
13. Kapur, D.: Using Gröbner bases to reason about geometry problems. J. Symbolic Computat. **2**(4), 399–408 (1986)
14. Kapur, D.: A refutational approach to geometry theorem proving. Artif. Intell. **37**(1–3), 61–93 (1988)
15. Kovács, Z.: Computer based conjectures and proofs in teaching Euclidean geometry. Ph.D. Dissertation. Linz, Johannes Kepler University (2015)
16. Kovács, Z.: The relation tool in GeoGebra 5. In: Botana, F., Quaresma, P. (eds.) ADG 2014. LNCS (LNAI), vol. 9201, pp. 53–71. Springer, Cham (2015). https://doi.org/10.1007/978-3-319-21362-0_4
17. Kovács, Z., Parisse, B.: Giac and GeoGebra - improved Gröbner basis computations. In: Gutierrez, J., Schicho, J., Weimann, M. (eds.) Computer Algebra and Polynomials, LNCS, vol. 8942, pp. 126–138, Springer (2015). https://doi.org/10.1007/978-3-319-15081-9_7
18. Kovács, Z., Recio, T., Sólyom-Gecse, C.: Rewriting input expressions in complex algebraic geometry provers. Ann. Math. Artif. Intell. (1), 73–87 (2018). https://doi.org/10.1007/s10472-018-9590-1
19. Kovács, Z., Recio, T., Vélez, M.P.: Using automated reasoning tools in GeoGebra in the teaching and learning of proving in geometry. Int. J. Technol. Math. Educ. **25**(2), 33–50 (2018)
20. Kovács, Z., Recio, T., Vélez, M.P.: Detecting truth, just on parts. Revista Matemática Complutense **32**(2), 451–474 (2018). https://doi.org/10.1007/s13163-018-0286-1

21. Kovács, Z., Recio, T.: GeoGebra reasoning tools for humans and for automatons. In: Electronic Proceedings of the 25th Asian Technology Conference in Mathematics (ATCM 2020), 14–16 December 2020. Published by Mathematics and Technology, LLC (2020). http://atcm.mathandtech.org/EP2020/invited/21786.pdf (2020)
22. Recio, T., Vélez, M.P.: Automatic discovery of theorems in elementary geometry. J. Autom. Reasoning **23**, 63–82 (1999)
23. Wen-Tsün, W.: Basic principles of mechanical theorem proving in elementary geometries. J. Autom. Reasoning **2**(3), 221–252 (1986)

AGADE—A Maple Package for Computing Rational General Solutions of Parametrizable First-Order Algebraic ODEs

Johann J. Mitteramskogler(✉)

Research Institute for Symbolic Computation (RISC),
Johannes Kepler University Linz, Altenbergerstraße 69, 4040 Linz, Austria
johann.mitteramskogler@risc.jku.at

Abstract. In this paper the *Maple* package AGADE is introduced. This software package implements different methods for computing rational general solutions of first-order algebraic ODEs and planar rational systems. All implemented algorithms follow a special algebro-geometric solution approach. In the first stage, an affine algebraic set is associated to the differential equation(s) which is then parametrized with rational functions. Subsequently, a solution of the differential problem is obtained by finding a suitable reparametrization of these rational functions in the second stage. The basic usage of the software package is demonstrated at a number of examples. Finally, an outlook on future implementations is provided.

Keywords: Algebraic ordinary differential equation · Planar rational system · Rational general solution · Symbolic computation · Rational parametrization · Software package · Maple

1 Introduction

Computing symbolic solutions of differential equations or deciding whether a solution exists are notoriously difficult problems in the general (non-linear) case. For algebraic differential equations, i.e. differential equations where the independent variables and the dependent functions and their derivatives are related by a polynomial function, a general solution strategy is available if one restricts to a suitable class of solutions. The idea of this strategy, which is called the algebro-geometric method for solving algebraic differential equations [19], is a two stage process. First, an algebraic set is derived from the differential equations which is then parametrized with functions from the solution class. This is a purely algebraic problem. In the second stage, the parametrization is modified such that the differential aspect is satisfied as well. A solution of the differential equations is then extracted from this modified parametrization.

This research project was supported by the Austrian Science Fund (FWF) under grant no. P31327-N32.

R. M. Corless et al. (Eds.): MC 2020, CCIS 1414, pp. 268–287, 2021.
https://doi.org/10.1007/978-3-030-81698-8_18

This algebro-geometric approach is well understood for algebraic ordinary differential equations (AODEs) and rational solutions, cf. Winkler [19, Section 2]. Over a computable field of characteristic zero, such as the algebraic numbers, the algebro-geometric solution method has led to several algorithms which can be implemented on a computer. The aim of the software package `AGADE`[1] is to provide implementations of some of these methods for the widely used computer algebra system *Maple*. At present, the package implements four such algorithms: the methods for computing rational general solutions of first-order AODEs of Ngô and Winkler [11,12] and Vo et al. [18]; the special case of autonomous first-order AODEs from Feng and Gao [5,6]; and the algorithm for computing rational general solutions of planar rational systems of Ngô and Winkler [13]. Examples which show how these implementations can be used to find solutions are provided at the end of the paper.

The structure of this paper is as follows: Sect. 2 recalls basic concepts from algebraic geometry and provides a rigorous definition of the term *rational general solution*. Afterwards, the basic principles of the implemented algorithms are described in Sect. 3. Section 4 describes where the package `AGADE` can be obtained and how it can be used to find solutions. In addition, limitations and details on the implementation are discussed. Finally, an outlook on future implementations is given in Sect. 5.

2 Preliminaries

Denote by $\overline{\mathbb{Q}}$ the computable field of algebraic numbers and let x be an indeterminate. The polynomial ring $\overline{\mathbb{Q}}[x]$ and its quotient field $\overline{\mathbb{Q}}(x)$ can be seen as a differential ring and a differential field, respectively, by endowing both structures with the usual derivation $' := {}^{\mathrm{d}}\!/_{\mathrm{d}x}$. In either case, $\overline{\mathbb{Q}}$ is the field of constants.

Let y be a differential indeterminate representing an unknown differentiable function of x. A first-order *algebraic ordinary differential equation* (AODE) is an ordinary differential equation (ODE) of the form

$$A(x, y, y') = 0, \tag{1}$$

where A is a trivariate polynomial, say $A \in \overline{\mathbb{Q}}[x, u, v]$, of positive degree in v. The polynomial A will be of further importance in the sequel of this paper and is referred to as the *defining polynomial* of the AODE (1). Since the ultimate goal of the implemented algorithms is to compute (rational) general solutions, it is assumed w.l.o.g. that the defining polynomial of any AODE is irreducible.

Classically, by a general solution of a first-order ODE one understands a solution $\hat{y}$ of the differential equation which contains an arbitrary constant. In the case of AODEs a more rigorous definition is available in the language of differential algebra, cf. Ritt [14] or Kolchin [9] for details on the subsequent

[1] The acronym AGADE stands for <u>A</u>lgebro-<u>G</u>eometric methods for solving <u>A</u>lgebraic <u>D</u>ifferential <u>E</u>quations.

constructions. Denote by $\overline{\mathbb{Q}}(x)\{y\}$ the differential polynomial ring[2] over the differential field $(\overline{\mathbb{Q}}(x),')$. Let $P \in \overline{\mathbb{Q}}(x)\{y\}$ be an irreducible differential polynomial of order 1. The radical differential ideal generated by P, commonly denoted by $\{P\}$, can be decomposed into

$$\{P\} = (\{P\} : S_P) \cap \{P, S_P\},$$

where $S_P = {}^{\partial P}/_{\partial y'}$ is the separant of P. From this decomposition one can see that a root of P might be a root of the separant as well. These are the so-called singular solutions of the differential equation $P = 0$ and will not be of further interest in this paper. The component $\{P\} : S_P$ on the other hand does not contain the separant. Since P is irreducible, $\{P\} : S_P$ is a prime differential ideal and as such has a generic zero in some differential extension field [14, Chapter 2]. It is precisely the generic zero of this component that will be used to define the general solution of an AODE. In the literature, $\{P\} : S_P$ is usually referred to as the *general component* of P.

Given a first-order AODE $A(x, y, y') = 0$. The left-hand side can be considered as an element of the differential polynomial ring $\overline{\mathbb{Q}}(x)\{y\}$. Since $A \in \overline{\mathbb{Q}}[x, u, v]$ is assumed to be irreducible and of positive degree in v, $A(x, y, y')$ is an irreducible differential polynomial of order 1.

Definition 1. *Let $A(x, y, y') = 0$ be a first-order AODE and denote its left-hand side by $P = A(x, y, y')$. A* general solution *of $P = 0$ is a generic zero of the prime differential ideal $(\{P\} : {}^{\partial P}/_{\partial y'}) \subseteq \overline{\mathbb{Q}}(x)\{y\}$. If, in addition, a general solution $\hat{y}$ is of the form*

$$\hat{y} = \frac{a_m\, x^m + a_{m-1}\, x^{m-1} + \cdots + a_0}{x^n + b_{n-1}\, x^{n-1} + \cdots + b_0},$$

where a_i, b_j are constants in a differential extension field of $(\overline{\mathbb{Q}}(x),')$, then $\hat{y}$ is called a rational general solution.

In other words, a general solution of the first-order AODE $A(x, y, y') = 0$ is defined as a generic zero of the general component of $\{A(x, y, y')\} \subseteq \overline{\mathbb{Q}}(x)\{y\}$. If such a general solution is rational in the independent variable x, then it is called a rational general solution. An advantage of this definition is that there exists an algorithmic way to verify whether a solution is indeed a general solution via differential pseudo-remainder computations, cf. Ritt [14, Chapter 2, Sect. 13]. Note that if $\hat{y}$ is a rational general solution in the sense of Definition 1, then the coefficients a_i, b_j must contain a constant that is transcendental over $\overline{\mathbb{Q}}$. Consequently, $\hat{y}$ is also a general solution in the classical sense.

The implemented algorithms for computing rational general solutions of first-order AODEs derive a solution from a so-called proper rational parametrization

[2] Denote by $y^{(o)}$ the o-th derivative of y, then $\overline{\mathbb{Q}}(x)\{y\}$ is just the polynomial ring $\overline{\mathbb{Q}}(x)[y, y', y'', \ldots, y^{(o)}, \ldots]$, where the derivation $'$ extends to $\overline{\mathbb{Q}}(x)\{y\}$ by setting $(y^{(o)})' = y^{(o+1)}$ for all $o \in \mathbb{N}$.

of an associated affine algebraic variety. In order to define such parametrizations properly, it is necessary to recall certain basic terminology form algebraic geometry. Details can be found, for instance, in Shafarevich [17].

Let $\mathbb{F}$ denote an algebraically closed field of characteristic zero[3] and consider a non-constant polynomial $P \in \mathbb{F}[x_1, \ldots, x_n]$, where $n > 0$. The roots of P

$$\mathbf{Z}(P) := \{(a_1, \ldots, a_n) \in \mathbb{A}^n(\mathbb{F}) \mid P(a_1, \ldots, a_n) = 0\}$$

constitute a classical affine algebraic set, where $\mathbb{A}^n(\mathbb{F})$ denotes the n-dimensional affine space over $\mathbb{F}$. If P is irreducible, then $\mathbf{Z}(P)$ is irreducible as an algebraic set[4] and in this case $\mathbf{Z}(P)$ is called an *affine algebraic variety.*

Definition 2. *Let $\mathcal{V} \subseteq \mathbb{A}^n(\mathbb{F})$ be a d-dimensional affine algebraic variety. A rational map*

$$\mathcal{P}_{\mathcal{V}} : \mathbb{A}^d(\mathbb{F}) \dashrightarrow \mathcal{V},\ \mathbf{a} = (a_1, \ldots, a_d) \mapsto (\phi_1(\mathbf{a}), \ldots, \phi_n(\mathbf{a})),$$

where $\phi_i \in \mathbb{F}(t_1, \ldots, t_d)$ are rational functions in d variables over $\mathbb{F}$ is called a rational parametrization *of $\mathcal{V}$ if $\operatorname{im}(\mathcal{P})$ is Zariski-dense in $\mathcal{V}$. If, in addition, $\mathcal{P}$ is a birational equivalence, viz. has a rational inverse, then $\mathcal{P}$ is called a* proper rational parametrization *of $\mathcal{V}$.*

Typically, (proper) rational parametrizations will be denoted simply by a tuple of rational functions and we write $\mathcal{P}_{\mathcal{V}}(\mathbf{a}) = (\phi_1(\mathbf{a}), \ldots, \phi_n(\mathbf{a}))$. It is well-known that only irreducible affine algebraic sets can have a proper rational parametrization, hence it is sufficient to restrict to the case of varieties in the definition. If an affine algebraic variety $\mathcal{V}$ has a proper rational parametrization, i.e. is birationally equivalent to a full affine subspace, then $\mathcal{V}$ is called *rational* or a *rational variety.*

Only low-dimensional varieties will be of interest in this paper, in particular, irreducible curves and surfaces, i.e. affine algebraic varieties of dimension 1 and 2, respectively. For these types of varieties simple rationality criteria are known, hence the existence of a proper rational parametrization can be decided.

Theorem 1. *Let $\mathcal{C} \subseteq \mathbb{A}^2(\mathbb{F})$ and $\mathcal{S} \subseteq \mathbb{A}^3(\mathbb{F})$ be an irreducible curve and an irreducible surface, respectively. The curve $\mathcal{C}$ is rational if and only if its genus is 0. Furthermore, $\mathcal{S}$ is rational if and only if the arithmetic genus and the second plurigenus of $\mathcal{S}$ are both 0.*

A precise definition of these terms is beyond the scope of this section and can be found in standard literature on algebraic geometry such as Hartshorne [7]. Note that proper rational parametrizations of curves and surfaces can be computed. For parametrization methods of algebraic curves consult Sendra et al. [16] or van Hoeij [8]. Schicho [15] gave an algorithm for computing rational parametrizations of algebraic surfaces.

Remark 1. The *Maple* package `algcurves` offers methods for studying algebraic plane curves. In particular, the commands `algcurves:-genus` and `algcurves:-parametrization` can be used to compute the genus and a proper rational parametrization of an algebraic curve, respectively.

[3] In the sequel of this paper, the cases $\mathbb{F} = \overline{\mathbb{Q}}$ and $\mathbb{F} = \overline{\mathbb{Q}(x)}$ will be of importance.

[4] Viz. $\mathbf{Z}(P)$ is not the union of two strictly smaller affine algebraic sets.

3 Algorithms for Parametrizable AODEs

The aim of this section is to provide an overview of the implemented algorithms for computing rational general solutions of first-order AODEs. The common point of all methods is that they derive a solution from a proper rational parametrization of an associated affine algebraic set via a suitable reparametrization. In other words, a solution is found by substituting the variables of the parametrization with suitable rational functions.

3.1 Surface-Parametrizable AODEs

In Ngô and Winkler [11,12], an algorithm for computing rational general solutions of first-order AODEs based on surface parametrizations is described. The basic principle of their approach will be outlined briefly in this section. For the technical details, the reader is referred to the aforementioned references.

Let $A(x, y, y') = 0$ be a first-order AODE and $A \in \overline{\mathbb{Q}}[x, u, v]$ be the corresponding defining polynomial. The zero-locus of A

$$\mathcal{S}_A := \mathbf{Z}(A) = \{(a_1, a_2, a_3) \in \mathbb{A}^3(\overline{\mathbb{Q}}) \mid A(a_1, a_2, a_3) = 0\}$$

defines a surface in 3-dimensional affine space over $\overline{\mathbb{Q}}$. Since A is irreducible, $\mathcal{S}_A$ is irreducible as an algebraic set, i.e. an affine algebraic variety.

Definition 3. *The surface $\mathcal{S}_A$ obtained from a first-order AODE $A(x, y, y') = 0$ is referred to as the* associated surface *(of the AODE). Furthermore, if $\mathcal{S}_A$ has a proper rational parametrization, then $A(x, y, y') = 0$ is called* surface-parametrizable.

Assume that $A(x, y, y') = 0$ is surface-parametrizable and let

$$\mathcal{P}_{\mathcal{S}_A}(t_1, t_2) = (\phi_0(t_1, t_2), \phi_1(t_1, t_2), \phi_2(t_1, t_2)) \text{ with } \phi_0, \phi_1, \phi_2 \in \overline{\mathbb{Q}}(t_1, t_2) \quad (2)$$

be a proper rational parametrization of the associated surface $\mathcal{S}_A$. Given a rational (general) solution $\hat{y}$ of the AODE, such a solution generates a parametric curve[5] $\mathcal{C}_{\hat{y}}(x) = (x, \hat{y}(x), \hat{y}'(x))$ by interpreting the independent variable x as the parameter. Since $A(x, \hat{y}, \hat{y}') = 0$ the curve $\mathcal{C}_{\hat{y}}$ lies on the associated surface $\mathcal{S}_A$. The conditions when $\mathcal{P}_{\mathcal{S}_A}$ has a suitable reparametrization are then deduced from the preimage of $\mathcal{P}_{\mathcal{S}_A}$ under the parametric curve $\mathcal{C}_{\hat{y}}$. Recall that $\mathcal{P}_{\mathcal{S}_A}$ is a proper rational parametrization and therefore has a birational inverse.

Let $(\sigma(x), \tau(x)) = \mathcal{P}_{\mathcal{S}_A}^{-1}(x, \hat{y}(x), \hat{y}'(x))$, where $\mathcal{P}_{\mathcal{S}_A}^{-1}$ denotes the birational inverse of $\mathcal{P}_{\mathcal{S}_A}$. Application of $\mathcal{P}_{\mathcal{S}_A}$ on both sides yields $\mathcal{P}_{\mathcal{S}_A}(\sigma(x), \tau(x)) = (x, \hat{y}(x), \hat{y}'(x))$. From Eq. 2 the following conditions for $\sigma(x)$ and $\tau(x)$ are derived

$$\begin{cases} \phi_0(\sigma(x), \tau(x)) = x \\ \phi_2(\sigma(x), \tau(x)) = \dfrac{\mathrm{d}}{\mathrm{d}x}\phi_1(\sigma(x), \tau(x)). \end{cases} \quad (3)$$

[5] For rational general solutions this is actually a parametric family of rational curves.

The system (3) can be solved for $\sigma(x)$ and $\tau(x)$. A solution is of the form [11, Section 3]

$$\begin{cases} \sigma' = \dfrac{\phi_2(\sigma,\tau)\frac{\partial\phi_0(\sigma,\tau)}{\partial\tau} - \frac{\partial\phi_1(\sigma,\tau)}{\partial\tau}}{\frac{\partial\phi_0(\sigma,\tau)}{\partial\tau}\frac{\partial\phi_1(\sigma,\tau)}{\partial\sigma} - \frac{\partial\phi_0(\sigma,\tau)}{\partial\sigma}\frac{\partial\phi_1(\sigma,\tau)}{\partial\tau}} \\ \tau' = \dfrac{\phi_2(\sigma,\tau)\frac{\partial\phi_0(\sigma,\tau)}{\partial\sigma} - \frac{\partial\phi_1(\sigma,\tau)}{\partial\sigma}}{\frac{\partial\phi_0(\sigma,\tau)}{\partial\sigma}\frac{\partial\phi_1(\sigma,\tau)}{\partial\tau} - \frac{\partial\phi_0(\sigma,\tau)}{\partial\tau}\frac{\partial\phi_1(\sigma,\tau)}{\partial\sigma}}, \end{cases} \tag{4}$$

where the dependency on x is omitted for simplicity. Notice the special form of this system: Both equations are quasi-linear and the independent variable x does not appear explicitly, i.e. the system is autonomous. Furthermore, for rational general solutions of the system (4) the denominators do not vanish [12, p. 4].

Definition 4. *Let $A(x, y, y') = 0$ be a surface-parametrizable first-order AODE and $\mathcal{P}_{\mathcal{S}_A}(t_1, t_2) = (\phi_0(t_1, t_2), \phi_1(t_1, t_2), \phi_2(t_1, t_2))$ be a proper rational parametrization of the associated surface. The autonomous quasi-linear system of ODEs* (4) *is called the* associated planar system *(of the AODE) with respect to $\mathcal{P}_{\mathcal{S}_A}$.*

The following theorem can be found in Ngô and Winkler [12, Theorem 2.1].

Theorem 2. *There is a one-to-one correspondence between the rational general solutions of a surface-parametrizable first-order AODE $A(x, y, y') = 0$ and the rational general solutions[6] of the associated planar system (wrt. a proper rational parametrization of $\mathcal{S}_A$).*

In particular, if $(\hat{\sigma}(x), \hat{\tau}(x))$ is such a rational general solution of the associated planar system, then $\hat{y} = \phi_1(\hat{\sigma}(x+C), \hat{\tau}(x+C))$, where $C = x - \phi_0(\hat{\sigma}(x), \hat{\tau}(x))$, is a rational general solution of the original AODE [11, Theorem 3.15]. This leads to an algorithm for computing rational general solutions of surface-parametrizable AODEs which is summarized in Algorithm 1. Of course, this requires that computing the solutions of the associated planar system is actually algorithmic. This is indeed the case and the algorithm which performs that task is described briefly in Sect. 3.4.

Remark 2. If a first-order AODE has a rational general solution, then it is not necessarily surface-parametrizable. For example, the first-order AODE

$$x^2y'^2 - 2xyy' - y'^3 + y^2 - 2 = 0$$

has the rational general solution $\hat{y} = Cx + \sqrt{C^3 + 2}$, however, the arithmetic genus of the associated surface is non-zero.

3.2 Curve-Parametrizable AODEs

Recently, Vo et al. [18] presented another algorithm for computing rational general solutions of first-order AODEs based on curve parametrizations. As will be

[6] A precise definition of the term *rational general solution* for planar rational systems can be found in Ngô and Winkler [11, Definition 3.9].

Algorithm 1. Rational general solution surf.-param. first-order AODE (cf. Ngô and Winkler [11, Algorithm 1])

Input : First-order AODE $A(x, y, y') = 0$

Output: Rational general solution $\hat{y}$ or string message

if *the associated surface* $\mathcal{S}_A$ *is rational* **then**

  Compute a proper rational parametrization

$$\mathcal{P}_{\mathcal{S}_A}(t_1, t_2) = (\phi_0(t_1, t_2), \phi_1(t_1, t_2), \phi_2(t_1, t_2)), \text{ where } \phi_0, \phi_1, \phi_2 \in \overline{\mathbb{Q}}(t_1, t_2).$$

  Find a rational general solution $(\hat{\sigma}(x), \hat{\tau}(x))$ of

$$\begin{cases} \sigma' = \dfrac{\phi_2(\sigma,\tau)\frac{\partial \phi_0(\sigma,\tau)}{\partial \tau} - \frac{\partial \phi_1(\sigma,\tau)}{\partial \tau}}{\frac{\partial \phi_0(\sigma,\tau)}{\partial \tau}\frac{\partial \phi_1(\sigma,\tau)}{\partial \sigma} - \frac{\partial \phi_0(\sigma,\tau)}{\partial \sigma}\frac{\partial \phi_1(\sigma,\tau)}{\partial \tau}} \\ \tau' = \dfrac{\phi_2(\sigma,\tau)\frac{\partial \phi_0(\sigma,\tau)}{\partial \sigma} - \frac{\partial \phi_1(\sigma,\tau)}{\partial \sigma}}{\frac{\partial \phi_0(\sigma,\tau)}{\partial \sigma}\frac{\partial \phi_1(\sigma,\tau)}{\partial \tau} - \frac{\partial \phi_0(\sigma,\tau)}{\partial \tau}\frac{\partial \phi_1(\sigma,\tau)}{\partial \sigma}} . \end{cases}$$

  if *no such solution exists* **then**

    return *"AODE has no rational general solution"*

  else

    return $\hat{y} = \phi_1(\hat{\sigma}(x + C), \hat{\tau}(x + C))$, *where* $C = x - \phi_0(\hat{\sigma}(x), \hat{\tau}(x))$

else

  return *"AODE is not surface-parametrizable"*

seen, their approach is similar to the one from Sect. 3.1 and shall be presented at the same level of generality here.

Unlike before, the independent variable x will not be parametrized, but is adjoined to the coefficient field. In other words, given a first-order AODE $A(x, y, y') = 0$, the defining polynomial A is viewed as an element of the polynomial ring $\overline{\mathbb{Q}}(x)[u, v]$. Although this will not be required for the actual computations, the field $\overline{\mathbb{Q}}(x)$ must be, formally, algebraically closed to be consistent with the theory of rational parametrizations of Sect. 2. Thus, A is actually viewed as an element of $\overline{\mathbb{Q}(x)}[u, v]$. In this structure, the zero-locus of A defines the curve

$$\mathcal{C}_A := \mathbf{Z}(A) = \{(a_1, a_2) \in \mathbb{A}^2(\overline{\mathbb{Q}(x)}) \mid A(a_1, a_2) = 0\}$$

in 2-dimensional affine space over $\overline{\mathbb{Q}(x)}$. Notice that, although A is irreducible in $\overline{\mathbb{Q}}[x, u, v]$, the polynomial might factor in this new ring and $\mathcal{C}_A$ will not be an affine algebraic variety.[7] In this case $\mathcal{C}_A$ is not parametrizable as only varieties may possess a (proper) rational parametrization.

Definition 5. *The curve* $\mathcal{C}_A$ *obtained from the first-order AODE* $A(x, y, y') = 0$ *is referred to as the* associated curve *(of the AODE). Furthermore, if* $\mathcal{C}_A$ *has a proper rational parametrization, then* $A(x, y, y') = 0$ *is called* curve-parametrizable.

[7] For example, $A = v^2 - x$ is irreducible in $\overline{\mathbb{Q}}[x, u, v]$, but factors into $(v + \sqrt{x})(v - \sqrt{x})$ as an element of $\overline{\mathbb{Q}(x)}[u, v]$.

Consider a curve-parametrizable first-order AODE $A(x, y, y') = 0$ and let

$$\mathcal{P}_{\mathcal{C}_A}(t) = (\psi_1(t), \psi_2(t)) \text{ with } \psi_1, \psi_2 \in \overline{\mathbb{Q}(x)}(t) \tag{5}$$

be a proper rational parametrization of the associated curve $\mathcal{C}_A$. Further assume that ψ_1, ψ_2 actually can be chosen in $\overline{\mathbb{Q}}(x)(t)$.[8] This last assumption guarantees that a suitable rational reparametrization yields a rational solution. If some of the coefficients of the parametrization $\mathcal{P}_{\mathcal{C}_A}$ were actually in $\overline{\mathbb{Q}(x)} \setminus \overline{\mathbb{Q}}(x)$, then the reparametrization could yield an algebraic solution. Note that this additional requirement does not restrict the class of curve-parametrizable AODEs further and there exists an algorithm that produces parametrizations of this special form [18, Section 4 and Algorithm 1].

The conditions for a suitable reparametrization of $\mathcal{P}_{\mathcal{C}_A}$ are derived analogous to Sect. 3.1. Given a rational (general) solution $\hat{y}$, then $(\hat{y}, \hat{y}')$ defines a (family of) point(s) on the associated curve $\mathcal{C}_A$. Let $\omega = \mathcal{P}_{\mathcal{C}_A}^{-1}(\hat{y}, \hat{y}')$, where $\mathcal{P}_{\mathcal{C}_A}^{-1}$ is the birational inverse of $\mathcal{P}_{\mathcal{C}_A}$. Thus, $\mathcal{P}_{\mathcal{C}_A}(\omega) = (\hat{y}, \hat{y}')$ and by Equation (5)

$$\frac{\mathrm{d}}{\mathrm{d}x}\psi_1(\omega) = \psi_2(\omega). \tag{6}$$

A solution for ω of Equation (6) is of the form [18, Section 5]

$$\omega' = \frac{\psi_2(\omega) - \frac{\partial \psi_1(\omega)}{\partial x}}{\frac{\partial \psi_1(\omega)}{\partial \omega}}. \tag{7}$$

Equation (7) is a quasi-linear ODE. Unlike the associated planar system, however, it is not autonomous in general. Note that the denominator does not vanish for rational general solutions of Equation (7) [18, Lemma 5.2].

Definition 6. *Let $A(x, y, y') = 0$ be a curve-parametrizable first-order AODE and $\mathcal{P}_{\mathcal{C}_A}(t) = (\psi_1(t), \psi_2(t))$ be a proper rational parametrization of the associated curve such that $\psi_1, \psi_2 \in \overline{\mathbb{Q}}(x)(t)$. The quasi-linear ODE (7) is called the* associated quasi-linear equation *(of the AODE) with respect to $\mathcal{P}_{\mathcal{C}_A}$.*

This associated equation is constructed in such a way that rational general solutions are preserved. The following theorem can be found in Vo et al. [18, Theorem 5.3].

Theorem 3. *There is a one-to-one correspondence between the rational general solutions of a curve-parametrizable first-order AODE $A(x, y, y') = 0$ and the rational general solutions of the associated quasi-linear equation (wrt. a proper rational parametrization of $\mathcal{C}_A$) if the components of the parametrization are contained in $\overline{\mathbb{Q}}(x)(t)$.*

[8] In general, one would not expect to find a proper rational parametrization of $\mathcal{C}_A$—if such an object exists at all—whose components can be chosen in $\overline{\mathbb{Q}}(x)(t)$ [16, Corollary 5.9]. However, the coefficient field $\overline{\mathbb{Q}}(x)$ is quite special and it can be shown that a rational plane curve defined over $\overline{\mathbb{Q}}(x)$ does always have such a parametrization [18, Theorem 4.3].

In this case, $\hat{y} = \psi_1(\hat{\omega})$ is a rational general solution of the original AODE, where $\hat{\omega}$ is a rational general solution of the associated quasi-linear equation [18, Theorem 5.3]. The steps to compute a rational general solution of a curve-parametrizable first-order AODE from a solution of the associated quasi-linear equation are summarized in Algorithm 2. Note that quasi-linear ODEs are well-studied objects and ODE solvers of modern computer algebra systems should be able to find suitable solutions in most cases. An algorithm which exclusively produces rational general solutions of quasi-linear ODEs is described in Chen and Ma [3].

Remark 3. In fact, the associated quasi-linear equation must be of a special form in order to yield a suitable reparametrization. More elaborate, if Equation (7) is neither linear nor a Riccati equation, then there cannot exist a rational general solution [18, Theorem 5.4(ii)]. This is another termination condition which is not indicated in Algorithm 2 for the sake of brevity.

Remark 4. As with surface-parametrizable AODEs, not all first-order AODEs which possess a rational general solution are curve-parametrizable [18, Example following Definition 3.3]. However, if such an AODE has a rational general solution such that the arbitrary constant appears only rationally, i.e. $\hat{y} \in \overline{\mathbb{Q}}(x)(C)$, where C is a transcendental constant, then the AODE must be curve-parametrizable [18, Theorem 3.1]. Such solutions are called *strong rational general solutions* and for this solution class the algorithm presented in this section actually is a decision algorithm.

Algorithm 2. Rational general solution curve-param. first-order AODE (cf. Vo et al. [18, Algorithm 2])

Input : First-order AODE $A(x, y, y') = 0$
Output: Rational general solution $\hat{y}$ or string message

if *the associated curve C_A is rational* **then**
  Compute a proper rational parametrization

$$\mathcal{P}_{C_A}(t) = (\psi_1(t), \psi_2(t)) \text{ such that } \psi_1, \psi_2 \in \overline{\mathbb{Q}}(x)(t).$$

  Find a rational general solution $\hat{\omega}$ of

$$\omega' = \frac{\psi_2(\omega) - \frac{\partial \psi_1(\omega)}{\partial x}}{\frac{\partial \psi_1(\omega)}{\partial \omega}}.$$

  if *no such solution exists* **then**
    return *"AODE has no rational general solution"*
  else
    return $\hat{y} = \psi_1(\hat{\omega})$
else
  return *"AODE is not curve-parametrizable"*

3.3 Special Case for Autonomous AODEs

The algorithms described in Sect. 3.1 and Sect. 3.2 can, of course, be used to find rational general solutions of autonomous first-order AODEs as well. In this case the associated planar system and the associated quasi-linear equation simplify drastically. However, in the autonomous case even more can be said. As it turns out, the conditions for reparametrization reduce to simple pattern matching rather than solving an associated differential equation. The complete derivation of these results can be found in Feng and Gao [5,6].[9]

If a first-order AODE $A(x, y, y') = 0$ is autonomous, then the independent variable x does not appear explicitly in the equation and the defining polynomial A actually belongs to $\overline{\mathbb{Q}}[y, y']$. In this case

$$\mathcal{C}_A^a := \mathbf{Z}(A) = \{(a_1, a_2) \in \mathbb{A}^2(\overline{\mathbb{Q}}) \mid A(a_1, a_2) = 0\}$$

is just an irreducible plane curve over $\overline{\mathbb{Q}}$. To avoid confusion, the notation $A(y, y') = 0$ will be used to denote autonomous first-order AODEs.

Definition 7. *The curve $\mathcal{C}_A^a$ obtained from an autonomous first-order AODE $A(y, y') = 0$ is called the* associated autonomous curve *(of the AODE).*

Consider an autonomous first-order AODE $A(y, y') = 0$ such that $\mathcal{C}_A^a$ is rational and let

$$\mathcal{P}_{\mathcal{C}_A^a}(t) = (\chi_1(t), \chi_2(t)) \text{ with } \chi_1, \chi_2 \in \overline{\mathbb{Q}}(t)$$

be a proper rational parametrization. Further, let $\mu = \chi_2(t) / \frac{\partial \chi_1(t)}{\partial t}$ and notice that μ is well-defined, i.e. $\chi_1(t)$ cannot be a constant. One can show that $\mathcal{P}_{\mathcal{C}_A^a}$ has a suitable reparametrization if either $\mu = a$ or $\mu = a(t - b)^2$ for some $a, b \in \overline{\mathbb{Q}}$ with $a \neq 0$. In these cases

$$\hat{y} = \chi_1(a(x + C)) \text{ and } \hat{y} = \chi_1\left(b - \frac{1}{a(x + C)}\right)$$

are rational general solutions of the autonomous AODE, respectively, where C is a transcendental constant [5, Theorem 5]. In addition, if an autonomous first-order AODE has a rational general solution, then the associated autonomous curve must be rational [5, Section 3]. Unlike the general case of curve- and surface-parametrizable first-order AODEs, the rationality of the associated autonomous curve is a necessary condition for the existence of a rational general solution. All these considerations are summarized in Algorithm 3.

Remark 5. The efficiency of Algorithm 3 can be further improved. Based on known degree bounds for proper rational parametrizations of plane curves, one can derive additional necessary conditions that an autonomous first-order AODE

[9] Note that the authors focus on autonomous AODEs with coefficients in $\mathbb{Q}$. Nevertheless, the required theory for this section holds for AODEs with algebraic coefficients as well.

has a rational general solution. This reduces to simple verifications that the degrees of the defining polynomial in y and y' satisfy certain conditions, cf. [5, Theorem 3].[10]

Algorithm 3. Rational general solution autonomous first-order AODE (cf. Feng and Gao [5, Algorithm 1])

Input : Autonomous first-order AODE $A(y, y') = 0$
Output: Rational general solution $\hat{y}$ or string message

if *the associated autonomous curve* $\mathcal{C}_A^a$ *is rational* **then**
  Compute a proper rational parametrization

$$\mathcal{P}_{\mathcal{C}_A^a}(t) = (\chi_1(t), \chi_2(t)), \text{ where } \chi_1, \chi_2 \in \overline{\mathbb{Q}}(t)$$

  and let $\mu = \chi_2(t)/\frac{\partial \chi_1(t)}{\partial t}$.

  if $\mu = a \in \overline{\mathbb{Q}} \setminus \{0\}$ **then**
    return $\hat{y} = \chi_1(a(x + C))$
  else if $\mu = a(t - b)^2$ *for some* $a, b \in \overline{\mathbb{Q}}$, $a \neq 0$ **then**
    return $\hat{y} = \chi_1\left(b - \frac{1}{a(x+C)}\right)$
  else
    return *"AODE has no rational general solution"*
else
  return *"AODE has no rational general solution"*

3.4 Planar Rational Systems

An algorithm for computing rational general solutions of autonomous planar rational systems was proposed by Ngô and Winkler [13]. It is beyond the scope of this section to introduce the necessary objects and theory to describe a comprehensive derivation of their approach. Instead, only the basic objects required for the computations will be mentioned. The individual steps of the method are summarized in Algorithm 4.

Let σ, τ be differential indeterminates representing unknown differential functions of x. Consider the autonomous planar rational system

$$\begin{cases} \sigma' = \dfrac{M_\sigma}{N_\sigma} \\ \tau' = \dfrac{M_\tau}{N_\tau}, \end{cases} \tag{8}$$

where $M_\sigma, N_\sigma, M_\tau, N_\tau \in \overline{\mathbb{Q}}[\sigma, \tau]$ are such that $\gcd(M_\sigma, N_\sigma) = 1 = \gcd(M_\tau, N_\tau)$ and the denominators are non-zero. Solutions of such systems can be obtained by finding so-called (rational) first integrals.

[10] Additional improvements which replace the computationally expensive parametrization step by more efficient routines are described in Feng and Gao [6].

Definition 8. *A* rational first integral *of the autonomous planar rational system* (8) *is a non-constant rational function* $F \in \overline{\mathbb{Q}}(\sigma,\tau) \setminus \overline{\mathbb{Q}}$ *such that*

$$\frac{M_\sigma}{N_\sigma}\frac{\partial F}{\partial \sigma} + \frac{M_\tau}{N_\tau}\frac{\partial F}{\partial \tau} = 0.$$

Assume that the planar system (8) has a rational first integral $F = {}^{P}/_{Q}$, where $P, Q \in \overline{\mathbb{Q}}[\sigma,\tau]$ are coprime and let C be a transcendental constant. The irreducible factors of the polynomial $P - CQ$, viz. irreducible in $\overline{\mathbb{Q}(C)}[\sigma,\tau]$, determine a so-called *general invariant algebraic curve* of the system. If a proper rational parametrization of the curve determined by one such factor has a suitable reparametrization by a linear rational function, then one can construct a rational general solution of the planar rational system (8). The precise steps of this approach are described in Algorithm 4. Note that the irreducible factors of $P - CQ$ are conjugate over $\overline{\mathbb{Q}}(C)$ and the solvability of the system does not depend on which of these factors is parametrized [13, Remark 5.4 and Theorem 5.6]. Furthermore, the existence of a rational first integral of the planar rational system is a necessary condition that the system has a rational general solution [13, Theorem 5.6].

Algorithm 4. Rational general solution planar rational system (cf. Ngô and Winkler [13, Algorithm RATSOLVE])

Input : Autonomous planar rational system (8)
Output: Rational general solution $(\hat{\sigma},\hat{\tau})$ or string message

Compute a rational first integral $F = {}^{P}/_{Q} \in \overline{\mathbb{Q}}(\sigma,\tau) \setminus \overline{\mathbb{Q}}$ of the system (8), where $P, Q \in \overline{\mathbb{Q}}[\sigma,\tau]$ are coprime.
if *no such rational first integral exists* **then**
  return *"Planar rational system has no rational general solution"*
else
  Let C be a transcendental constant. Take any irreducible factor I of $P - CQ \in \overline{\mathbb{Q}(C)}[\sigma,\tau]$ and let $\mathcal{C}_I = \mathbf{Z}(I)$ be the plane curve defined by I.
  if $\mathcal{C}_I$ *is rational* **then**
    Compute a proper rational parametrization

$$\mathcal{P}_{\mathcal{C}_I}(t) = (\psi_1(t),\psi_2(t)), \text{ where } \psi_1,\psi_2 \in \overline{\mathbb{Q}(C)}(t).$$

    Find a linear rational function $\hat{\omega} \in \overline{\mathbb{Q}(C)}(x)$ that solves either[11]

$$\omega' = \frac{1}{\frac{\partial\psi_1(\omega)}{\partial\omega}}\frac{M_\sigma(\psi_1(\omega),\psi_2(\omega))}{N_\sigma(\psi_1(\omega),\psi_2(\omega))} \text{ or } \omega' = \frac{1}{\frac{\partial\psi_2(\omega)}{\partial\omega}}\frac{M_\tau(\psi_1(\omega),\psi_2(\omega))}{N_\tau(\psi_1(\omega),\psi_2(\omega))}.$$

    if *such a linear rational function exists* **then**
      return $(\hat{\sigma},\hat{\tau}) = (\psi_1(\hat{\omega}),\psi_2(\hat{\omega}))$
  return *"Planar rational system has no rational general solution"*

---

[11] Notice that it might happen that either ${}^{\partial\psi_1(\omega)}/_{\partial\omega} = 0$ or ${}^{\partial\psi_2(\omega)}/_{\partial\omega} = 0$. In this case, $\hat{\omega}$ should be a solution of the respective other differential equation. Since $\mathcal{P}_{\mathcal{C}_I}$ is a rational parametrization both partial derivatives cannot vanish simultaneously.

Remark 6. It can be shown that the rational first integrals of the planar rational system (8) coincide with those of the planar polynomial system [13, Section 5]

$$\begin{cases} \sigma' = M_\sigma N_\tau \\ \tau' = M_\tau N_\sigma. \end{cases}$$

Efficient algorithms for computing rational first integrals of planar polynomial systems can be found in Chèze [4] and Bostan et al. [1]. Notice that the computation of rational first integrals typically requires to pass a degree bound for these objects as additional input. Upper bounds for rational first integrals of planar polynomial systems are known in the generic situation [2], however, the general case is still open.

4 The Maple Package AGADE

The algorithms described in Sect. 3 have been implemented in the computer algebra system *Maple* and are available in the form of a package called `AGADE`[12] (Algebro-Geometric methods for solving Algebraic Differential Equations). This package is publicly available at the online repository https://github.com/JohannMitteramskogler/AGADE. In order to use the package, download the file `AGADE.mla` and make sure that it is found in the current library path by setting the value of the variable `libname`[13] accordingly. Afterwards, the package can be loaded with the usual `with(<PackageName>)` routine. In its current version, the following commands are available.

```
> with(AGADE);
  [RGSautonomousFOAODE, RGScurveParametrizableFOAODE,
      ↪ RGSplanarRationalSystem,
      ↪ RGSsurfaceParametrizableFOAODE]
```

Note that the abbreviations `RGS` and `FOAODE` stand for *rational general solution* and *first-order AODE*, respectively. A precise specification and current limitations of these commands are described in the remainder of this section. In addition, a couple of basic examples are given to demonstrate the usage of the software package. For spacial reasons it is not possible to provide larger examples or demonstrate the use of the optional arguments in this paper. For this, the interested reader may consult the demo file `AGADE_Demo.mw`, which can be found at the aforementioned online repository.

RGSautonomousFOAODE

This is an implementation of Algorithm 3 for computing rational general solutions of autonomous first-order AODEs. By default, the parametrization of the

[12] This package was developed and tested with *Maple 2020*.

[13] https://de.maplesoft.com/support/help/Maple/view.aspx?path=libname.

associated autonomous curve is computed with `algcurves:-parametrization`. Note that this routine implements the efficiency improvement mentioned in Remark 5 as well.

Calling sequence:

`RGSautonomousFOAODE(aode, depVar, indepVar, options)`

Parameters:

`aode`	- autonomous first-order AODE
`depVar`	- symbol; denotes the dependent variable
`indepVar`	- symbol; denotes the independent variable
`options`	- (optional) equations of the form `keyword=value`, where `keyword` can be `parametrization` or `extendedOutput`.

Options:

- The option `parametrization=[`χ_1, χ_2, t`]` can be used to supply a user-defined parametrization of the associated autonomous curve, where t is a symbol and $\chi_1, \chi_2 \in \overline{\mathbb{Q}}(t)$.
- If the option `extendedOutput=true` is passed, then the algorithm returns a record with additional information about the computation such as the used parametrization and the termination condition. The default value is `extendedOutput=false`.

Minimal example:

```
> aode1 := 27*y(x)^4 + diff(y(x),x)^3 + 27*diff(y(x),x)*y(
  ↪ x)^2 - 108*y(x)^3 + 4*diff(y(x),x)^2 - 54*diff(y(x),
  ↪ x)*y(x) + 166*y(x)^2 + 31*diff(y(x),x) - 116*y(x) +
  ↪ 31 = 0:

> AGADE:-RGSautonomousFOAODE(aode1, y, x);
       3                 2            2           3     2
    _C1  + (3 x + 1) _C1  + (3 x  + 2 x) _C1 + x  + x  + 1
   -------------------------------------------------------
                                     3
                           (x + _C1)
```

RGScurveParametrizableFOAODE

This is an implementation of Algorithm 2 for computing rational general solutions of first-order AODEs. By default, the parametrization of the associated curve is computed with `algcurves:-parametrization`. In case the resulting parametrization is not rational in the independent variable, the algorithm falls back to a less efficient implementation. Note that this routine will not attempt to find a solution of the associated quasi-linear equation if the latter is neither linear nor a Riccati equation, cf. Remark 3.

Rational general solutions of the associated quasi-linear equation are computed with dsolve using selected solving methods[14] which have been found to produce solutions of the desired form. However, dsolve cannot be used to decide existence of a rational general solution of the associated quasi-linear equation. Consequently, the implementation is unable to give a definite answer on the existence of a rational general solution of the original AODE if dsolve fails to find a solution.

Calling sequence:

`RGScurveParametrizableFOAODE(aode, depVar, indepVar, options)`

Parameters:

`aode`	- first-order AODE
`depVar`	- symbol; denotes the dependent variable
`indepVar`	- symbol; denotes the independent variable
`options`	- (optional) equations of the form `keyword=value`, where `keyword` can be `parametrization`, `solutionAssocODE` or `extendedOutput`.

Options:

- The option `parametrization=`$[\psi_1, \psi_2, t]$ can be used to supply a user-defined parametrization of the associated curve, where t is a symbol and $\psi_1, \psi_2 \in \overline{\mathbb{Q}}(x)(t)$.
- Use the option `solutionAssocODE=`$\hat{\omega}$ to pass a user-defined rational general solution of the associated quasi-linear equation, where $\hat{\omega}$ is a rational function in the independent variable.
- If the option `extendedOutput=true` is passed, then the algorithm returns a record with additional information about the computation such as the used parametrization, the associated quasi-linear equation and the termination condition. The default value is `extendedOutput=false`.

Minimal example:

```
> aode2 := diff(y(x),x)*x^6 + diff(y(x),x)^3*x^3 - 2*y(x)*
   ↪ x^5 - 3*y(x)*diff(y(x),x)^2*x^2 + 2*x^5 + 3*diff(y(x
   ↪ ),x)^2*x^2 + 3*y(x)^2*diff(y(x),x)*x - 6*diff(y(x),x
   ↪ )*y(x)*x - y(x)^3 + 3*diff(y(x),x)*x + 3*y(x)^2 - 3*
   ↪ y(x) + 1 = 0:

> AGADE:-RGScurveParametrizableFOAODE(aode2, y, x);
          2  2        3
       _C1   x  + _C1   + x
       -----------------
                 3
              _C1
```

[14] https://www.maplesoft.com/support/help/Maple/view.aspx?path=dsolve%2fdetails.

RGSplanarRationalSystem
This is an implementation of Algorithm 4 for computing rational general solutions of autonomous planar rational systems. Notice that this routine requires the external *Maple* package `RationalFirstIntegrals`[15] by Bostan et al. [1] for computing rational first integrals. This package provides the command `Generic-RationalFirstIntegral` which is used for the computation of such objects and requires a degree bound as input. Upper bounds for the degree of the rational first integrals of a planar rational system are not known in the general case, cf. Remark 6. Consequently, this routine cannot be used to decide the existence of a rational general solution. In the generic case, however, such a bound exists [2]. By default, the implemented method uses this generic bound.

The command `GenericRationalFirstIntegral` tends to be quite slow for moderate degree bounds. More efficient probabilistic/heuristic methods are provided by the package `RationalFirstIntegrals`, but these require some guidance by the user. In case the computation takes too long, one could try to find a rational first integral by one of these faster methods and then pass it to the algorithm via an optional argument. Another possibility is to specify a lower degree bound, again, via an optional argument.

Recall that in Step 5 of Algorithm 4 it might be necessary to factor a polynomial over the algebraic closure of a rational function field. Such a factorization is not yet available in *Maple*. The implemented method verifies certain trivial irreducibility criteria and then attempts to parametrize the polynomial in Step 5 using `algcurves:-parametrization`. If this attempt fails due to reducibility, then the algorithm cannot proceed to compute a solution.

Calling sequence:

```
RGSplanarRationalSystem(prs, depVars, indepVar, options)
```

Parameters:

`prs`	- list of two aut. quasi-linear ODEs; the planar rational system
`depVars`	- list of two symbols; denotes the dependent variables
`indepVar`	- symbol; denotes the independent variable
`options`	- (optional) equations of the form `keyword=value`, where `keyword` can be `generalIAC`, `degreeBoundRFI`, `parametrization` or `extended-Output`.

Options:

- Use the option `generalIAC=`$[I, C]$ to pass a user-defined (irreducible) general invariant algebraic curve, where C is a symbol and I is a polynomial over $\overline{\mathbb{Q}(C)}$ in the dependent variables.
- Use the option `degreeBoundRFI=`n to pass a user-defined degree bound of the rational first integrals, where n is a positive integer.

[15] This package is publicly available at http://www.unilim.fr/pages_perso/thomas.cluzeau/Packages/RFI/RationalFirstIntegrals.html.

- The option `parametrization=[`ψ_1, ψ_2, t`]` can be used to supply a user-defined parametrization of the general invariant algebraic curve, where t is a symbol and $\psi_1, \psi_2 \in \overline{\mathbb{Q}(C)}(t)$.
- If the option `extendedOutput=true` is passed, then the algorithm returns a record with additional information about the computation such as the computed invariant algebraic curve, the used parametrization and the termination condition. The default value is `extendedOutput=false`.

Minimal example:

```
> prs1 := [diff(u(x),x) = 1, diff(v(x),x) = (2*v(x) + 1)/(
    ↪ u(x) + 1)]:

> AGADE:-RGSplanarRationalSystem(prs1, [u, v], x);
                  2                  2
     [x, -_C1 x  - 2 _C1 x + 1/2 x  - _C1 + x]
```

RGSsurfaceParametrizableFOAODE

This is an implementation of Algorithm 1 for computing rational general solutions of first-order AODEs. Note that the associated planar system is solved by `RGSplanarRationalSystem`. To the best of the author's knowledge, there does not exist a complete implementation for computing proper rational parametrizations of algebraic surfaces in *Maple* at the moment. However, if the associated surface is a pencil of rational curves [10, Section 5.1], then a parametrization can be computed with `algcurves:-parametrization`. If this is not the case, a user-defined surface parametrization has to be passed via an optional argument. A discussion on surface parametrizations for AODEs of special geometric shape can be found in Ngô et al. [10, Section 5].

Calling sequence:

```
RGSsurfaceParametrizableFOAODE(aode, depVar, indepVar, options)
```

Parameters:

`aode`	- first-order AODE
`depVar`	- symbol; denotes the dependent variable
`indepVar`	- symbol; denotes the independent variable
`options`	- (optional) equations of the form `keyword=value`, where `keyword` can be `parametrization`, `solutionAssocODEs` or `extendedOutput`.

Options:

- The option `parametrization=[`$\phi_0, \phi_1, \phi_2, [t_1, t_2]$`]` can be used to supply a user-defined parametrization of the associated surface, where t_1, t_2 are symbols and $\phi_0, \phi_1, \phi_2, \in \overline{\mathbb{Q}}(t_1, t_2)$.

- Use the option `solutionAssocODEs=`$[\hat{\sigma}, \hat{\tau}]$ to pass a user-defined rational general solution of the associated planar system, where $\hat{\sigma}, \hat{\tau}$ are rational functions in the independent variable.
- If the option `extendedOutput=true` is passed, then the algorithm returns a record with additional information about the computation such as the used parametrization, the associated planar system and the termination condition. The default value is `extendedOutput=false`.

Minimal example:

```
> aode3  := x*diff(y(x),x)*y(x)^4 - 4*diff(y(x),x)*y(x)^3*x
  ↪  + y(x)^5 + 6*diff(y(x),x)*y(x)^2*x - 5*y(x)^4 -
  ↪ diff(y(x),x)^3 - 4*diff(y(x),x)*y(x)*x + 10*y(x)^3 +
  ↪  x*diff(y(x),x) - 10*y(x)^2 + 5*y(x) - 1 = 0:

> AGADE:-RGSsurfaceParametrizableFOAODE(aode3, y, x);
      3        2
   _C1  + _C1  x - 1
   -----------------
           2
       _C1  x - 1
```

5 Conclusion and Outlook

In this paper, four methods for computing rational general solutions have been presented: two for computing such solutions for first-order AODEs, one specialized method for autonomous first-order AODEs and, finally, one for autonomous planar rational systems. All these methods operate in an algebro-geometric manner, where the solution is obtained by modifying a proper rational parametrization of a derived algebraic set. These methods were implemented in the computer algebra system *Maple* and are available in the form of a software package named `AGADE`. The basic usage of the package was demonstrated with simple examples, with more advanced scenarios available in the online demo file.

Compared to *Maple's* `dsovle` command, an advantage of the implemented methods is that they provide a definite answer whether a (rational general) solution exists or not. Even when the implemented algorithms fail to give such an answer due to a technical limitation, enabling the extended output option usually gives enough information such that the user can easily deduce the answer by hand or provide the missing object to continue the computation. Furthermore, application of `dsovle` on first-order AODEs of higher degree often produces complicated integral solutions with large algebraic subexpressions. The output of the implemented methods is, by nature, much more manageable.

The complexity of the implemented methods is determined primarily by the parametrization step and, in the case of planar rational systems, the computation of a rational first integral. A complexity analysis for computing rational first integrals can be found in Bostan et al. [1]. The main parametrization method is based on the paper by van Hoeij [8]. However, the latter does not provide an explicit complexity estimation.

Finally, additional functionality for this package is planned. At the time of writing, an implementation for computing rational general solutions of systems of autonomous AODEs of algebro-geometric dimension 1 is under development. We hope that AGADE will be useful for further studies of symbolic solutions of AODEs.

References

1. Bostan, A., Chèze, G., Cluzeau, T., Weil, J.A.: Efficient algorithms for computing rational first integrals and Darboux polynomials of planar polynomial vector fields. Math. Comput. **85**, 1393–1425 (2016). https://doi.org/10.1090/mcom/3007
2. Carnicer, M.M.: The Poincaré problem in the nondicritical case. Ann. Math. **140**(2), 289–294 (1994). https://doi.org/10.2307/2118601
3. Chen, G., Ma, Y.: Algorithmic reduction and rational general solutions of first order algebraic differential equations. In: Differential Equations with Symbolic Computation, vol. 23, pp. 201–212. Birkhäuser Basel (2005). https://doi.org/10.1007/3-7643-7429-2_12
4. Chèze, G.: Computation of Darboux polynomials and rational first integrals with bounded degree in polynomial time. J. Complex. **27**(2), 246–262 (2011). https://doi.org/10.1016/j.jco.2010.10.004
5. Feng, R., Gao, X.S.: Rational general solutions of algebraic ordinary differential equations. In: Proceedings of the 2004 International Symposium on Symbolic and Algebraic Computation, pp. 155–162 (2004). https://doi.org/10.1145/1005285.1005309
6. Feng, R., Gao, X.S.: A polynomial time algorithm for finding rational general solutions of first order autonomous ODEs. J. Symbolic Comput. **41**(7), 739–762 (2006). https://doi.org/10.1016/j.jsc.2006.02.002
7. Hartshorne, R.: Algebraic Geometry. GTM, vol. 52. Springer, New York (1977). https://doi.org/10.1007/978-1-4757-3849-0
8. van Hoeij, M.: Rational parametrizations of algebraic curves using a canonical divisor. J. Symbolic Comput. **23**(2), 209–227 (1997). https://doi.org/10.1006/jsco.1996.0084
9. Kolchin, E.R.: Differential Algebra & Algebraic Groups. Academic Press, Cambridge (1973)
10. Ngô, L.X.C., Sendra, J.R., Winkler, F.: Classification of algebraic ODEs with respect to rational solvability. Comput. Algebraic Anal. Geom. Contemp. Math. **572**, 193–210 (2012). https://doi.org/10.1090/conm/572/11361
11. Ngô, L.X.C., Winkler, F.: Rational general solutions of first order non-autonomous parametrizable ODEs. J. Symbolic Comput. **45**(12), 1426–1441 (2010). https://doi.org/10.1016/j.jsc.2010.06.018
12. Ngô, L.X.C., Winkler, F.: Rational general solutions of parametrizable AODEs. Publicationes Mathematicae Debrecen **79**(3–4), 573–587 (2011). https://doi.org/10.5486/PMD.2011.5121
13. Ngô, L.X.C., Winkler, F.: Rational general solutions of planar rational systems of autonomous ODEs. J. Symbolic Comput. **46**(10), 1173–1186 (2011). https://doi.org/10.1016/j.jsc.2011.06.002
14. Ritt, J.F.: Differential Algebra. American Mathematical Society (1950)

15. Schicho, J.: Rational parametrization of surfaces. J. Symbolic Comput. **26**(1), 1–29 (1998). https://doi.org/10.1006/jsco.1997.0199
16. Sendra, J.R., Winkler, F., Pérez-Diaz, S.: Rational algebraic curves. In: Algorithms and Computation in Mathematics, Springer-Verlag, Berlin Heidelberg (2008). https://doi.org/10.1007/978-3-540-73725-4
17. Shafarevich, I.R.: Basic Algebraic Geometry 1. Springer-Verlag, Berlin Heidelberg, 3 edn. (2013). https://doi.org/10.1007/978-3-642-37956-7
18. Vo, N.T., Grasegger, G., Winkler, F.: Deciding the existence of rational general solutions for first-order algebraic ODEs. J. Symbolic Comput. **87**, 127–139 (2018). https://doi.org/10.1016/j.jsc.2017.06.003
19. Winkler, Franz: The algebro-geometric method for solving algebraic differential equations — a survey. J. Syst. Sci. Complex. **32**(1), 256–270 (2019). https://doi.org/10.1007/s11424-019-8348-0

Estimation of Travel Time for Additional Metrobus Route

V. Nieves-Cruz(✉) and P. E. Balderas-Cañas

National Autonomous University of Mexico, Mexico City, Mexico
patricia.balderas@ingenieria.unam.edu

Abstract. In this paper we present some results of a research with application case study format of a simulation model, as a tool that supports decision making. The simulation model is developed with Maple tools. The main objective was to estimate the travel time for the extension of the Metrobus line proposed, considering the time traveled between two adjacent stations, the waiting time at each station (intended for passengers to board and leave the units), and the waiting time at intersections with traffic lights. The simulation model takes as input parameters the probability functions that characterize the operating speed of the units on a current Metrobus line, the duration of each phase of the traffic light cycles, and the distance between stations and traffic lights, within the proposed route. With these parameters, random variables are generated to represent those durations and then, they are added to estimate the total duration of the trip. In the first part we introduce the context research. Then we give a brief literature review about Monte-Carlo simulation. After that, we explain the proposal and define the main variables of the simulation model. So that, in the next section we explain the structure of the model in a MAPLE 2016 worksheet, based on previously defined helper functions (algorithms). Finally, we discuss the results and compare them with the current travel time by buses, minibuses, and vans.

Keywords: Simulation model · Estimation of travel time · Metrobus route · Probability functions · Maple simulation

1 Introduction

Urban mobility is closely related to the daily movement that people need to carry out to achieve comprehensive development, involving the movement of people and goods from one place to another. Currently, most people use public transport systems as the main option for transportation, so the growth and urbanization of many regions demand accessible and efficient public transport systems that allow their comprehensive development. However, there are several factors that impede suitable operation of public transport units, harming urban mobility. Some of these factors are existing imbalance between user demand and available offer, lack of adequate infrastructure for the operation of public transport units, interaction of public transport units with private vehicles of independent travelers, which many times cause areas of vehicular congestion.

R. M. Corless et al. (Eds.): MC 2020, CCIS 1414, pp. 288–303, 2021.
https://doi.org/10.1007/978-3-030-81698-8_19

Because of the importance of urban mobility, the proper operation of transport systems and the consequences that their operation can generate, this paper addresses a proposal for connecting two transport systems in Mexico City: The Mexico City Subway System (MCSS) and the Metrobus System (MS), by means of the addition of two connecting routes of a current Metrobus line. This project was developed within the framework of a Mobility Project around the main campus of UNAM, University City (CU), whose objective was to prepare proposals to improve mobility in CU and its surroundings, within the framework of ecosystem services (Social Service 2020–12/81–166 program).

The model presented here is part of Nieves' project [1], a case study whose objective was to develop and analyze a proposal to connect the MCSS and MS systems, from different operational points of view and through several simulation models. This work corresponds to a naturalistic social application case [2] in CDMX, and was developed with a system approach [3, 4], where, in addition to identifying the elements that are involved (cars, public transport units, users, operational personnel, independent drivers, etc.), it is important the relationships among them and what are the behaviors emerging from their interaction. A valuable aspect of Nieves' work is the methodology adopted to prepare the proposal, starting with delimiting, understanding and describing the case study region, collecting and managing information for qualitative and quantitative analysis, as well as the application of analytical and simulation models as tools that support and validate decision-making. Each of its models has different objectives, for example, to estimate passenger demand on current routes that are carried out by buses or minibuses, or to estimate the average number of public transport units currently operating in the studied area, in periods of one hour.

In the model that is exposed here, the main objective was to estimate the travel duration of two proposed routes, to be made by Metrobus units. It is a Monte Carlo simulation model that can be easily implemented with various computational tools, whether it be specialized packages for the development of simulation models or programming languages (in general) which ones offer flexibility in the development of any type of model. We implemented the model with the Maple tools, unlike to Nieves, who used Excel sheets.

Although it is a simple model, one of our objectives is show to the reader a case of application in Mexico City, using real data exclusively collected for the development of the proposal and applying a holistic and naturalistic approach. Both the methodology and the model can be extended to other cases, not only in Mexico, but anywhere in the world. A strength of this research is the modeling and simulating with basis on data analysis of historical data, recorded on videos.

In the next section we expose some important concepts of simulation models. In the Sect. 3, we present the details of the proposal. In the Sects. 4 and 5 we describe the model and some characteristics of its implementation in Maple. In Sect. 6 we present some results from various model runs, then we compared with the duration of trips that are currently made by buses, minibuses, and vans. Finally, some conclusions are presented at the end in the Sect. 7.

2 Some Elements of the State of Art

Simulation is an Operations Research tool, whose development and use have grown widely in recent decades. The simulation process is based on graphical, physical, and computational models that represent in an abstract and simplified way situations or systems of the real world, or hypothetical in nature. These type of models allow us to analyze current situations taking into account from the modification of input parameters, the logical structure, or both, in order to make predictions and formulate control strategies on the original system [5]. For this reason, simulation models were a great tool to support decision making for this project.

Simulation models also offer high flexibility to represent systems of different types and allow us to propose improvements and solutions to problems of the real system, where mathematical model formulations, with complex analytical expressions impede finding solutions due to the complexity to solve them and even sometimes it is difficult or impossible to solve them [5]. In addition, they are widely used in situations where making changes and modifications to the real system is impossible, prohibited, not feasible, very expensive, long-term, unsafe and even illegal processes are involved [6]. Using simulation models, it is possible to easily perform actions on the system, avoiding damage or alterations in the real elements that participate. This defines simulation models as risk-free environments that could be produced by errors and bad decisions [7].

Simulation models can be static or dynamic [5]. In static models, over time is not a relevant element and they simply represent the system in a well-defined fixed time point. Conversely, in dynamic models, over time is a key element, since there is interest in knowing the state and behavior of the elements of the system (or the complete system), at different times, and many times it is necessary to analyze behaviors continuously, through time.

There are physical and computational simulation models. Physical models generally simulate processes and phenomena at real scales and allow elements of the real system to interact with the model directly. Many physical models constitute simulated environments that allow the training of staff to face different situations, for example, evacuations, fires, and floods. Meanwhile, computational models completely extract the elements of the system and do not allow a direct interaction between the real system and the modeled system. This work corresponds to the development of a computational simulation model, therefore, henceforth the *simulation model* concept is used to refer to a computational simulation model.

There are many characteristics to classify simulation models, for example, kind of application, implementation strategy, or kind of system being represented. Within the last classification are discrete and continuous systems. In discrete event systems there are well-defined states of the system and its elements change from one state to another instantaneously through jumps, at discrete points in time [5]. Meanwhile, in continuous systems, the state of its elements is constantly changing over time [5].

Regarding the implementation method, Robinson S. (2014), classifies the simulation models into four categories:

- *Discrete event simulation*
  This method simulates the interaction of entities in discrete systems. The main objective is to track in detail the movement of entities (people, objects, or information) through the system.
- *Monte Carlo Simulation*
- In this method, random variables are generated, and they interact with each other to generate results to support the analysis and prediction of the behavior of some variables of interest.
- *Dynamic System Simulation*
  It uses a continuous approach over time, and it refers to flows (of objects, people, vehicles, money, matter). These flows simultaneously integrate and disintegrate the blocks that make up the system. The size of these blocks keeps changing constantly depending on the flow that enters and leaves them.
- *Agent based simulation*
  With this simulation method, the objective is to model and analyze the community behavior of a population of agents, where each agent has characteristics and behaviors, but community behaviors emerge from the joint interaction.

Although each category is defined by different specific characteristics, frequently most situations can be modeled by more than one method. Robinson (2014), for example, mentions that there are many situations in which dynamic systems could be used instead of discrete event simulation and vice versa. When characteristics of two or more implementation methods are combined, hybrid simulation take place [8].

The key element in simulation models is the presence of random variables, whose behavior is generally defined by a probability distribution. For this reason, it is extremely important to have knowledge about it about the graphical form of probability distributions and their main areas of use and application.

With what has been mentioned so far, we can get a general idea of what simulation means. However, there are many technical questions that you can study independently in depth, for example: methodologies to collect and manage data and information, statistical procedures to adjust data and define the input parameters of a model, methods to generate random variables and simulation models implementation methods, computational tools to implement simulation models, analysis of the results of a simulation model, its interpretation and the formulation of conclusions from them.

By handling random variables in the simulation models, a set of results is obtained, and they represent the flexibility of the solution, a characteristic that analytical models lack. Given these circumstances, to provide an adequate and interpretable solution according to the model and situation being addressed, it is necessary to analyze sets of executions of the computational model, since each execution provides different results as a consequence of the generation and manipulation of random variables of the model. To reach a conclusion from a simulation model, some statistical methods are commonly used, such as confidence intervals and data fitting, corresponding to the output variable of interest. But one of the most common methods to analyze the outputs of a model and to define a stop condition in a simulation model, is to achieve a stable behavior of the variable of interest. Addressing an analysis of the results that leaves out of consideration the initial period of each replica of the simulation (or those replicates with few data), where the variable of interest shows large variations between replicates. This initial period is known as the warmup period [9]. There are various mathematical formulations that allow defining the warmup period, but one of the most used methods in practice is the graphical method, which allows us to easily identify when the variable of interest reaches a stable behavior.

A simple, but well-structured example of a simulation model corresponds to the work of White & Ingalls [6]. It is a discrete event model whose description is quite clear and detailed, so that it allows you a general understanding of the process of developing a simulation model.

3 Proposal of Metrobus Routes

The case study boarded in this paper is geographically located in Coyoacan, Mexico City. Two connecting routes for MS are proposed between an existing line of the MS and a terminal station of a line of the MCSS. This proposal was developed after a long period of observation and data collection within the region involved. Currently, the connection in this area, between the stations of both systems, is made by other transport systems which operate with Public Transport Units (PTU), like buses, minibuses, and vans, as well as taxis and private vehicles.

The proposal arises in response to the observation and analysis of the current situation, in which PTU routes generate congestion points due to the inadequate existing infrastructure on streets, continuous bus stops, the circulation of a large number of PTU, and the long duration of trips, aspects that harm urban mobility in the area. The details of this proposal and the analysis of its viability through simulation models can be consulted at the research work of Nieves [1].

Figure 1 illustrates the proposed connection between MS line (red color) and MCSS line (green color) in Mexico City. The proposed routes are from point A to point B and from point B to point C, where A and C are adjacent stations on a MS line, and B is a terminal station on an MCSS line. So, the proposed connection can be also thought as an extension of the MS line, since the proposed routes are for MS units, whose origin and destination points correspond to existing stations. Once the AB route connects the MS line to the MCSS station, the second route, BC, allows the connection in the opposite direction, in this way, after completing both routes, the MS units return to the original line (see Fig. 1).

In the design of this proposal, aspects such as the existing infrastructure of streets and avenues, traffic light signals and the direction of current vehicular flow were considered. The infrastructure and the existing traffic flow around station A prevent a direct connection from B to A, for this reason station C was considered.

It is worth mentioning that the two transport lines (and systems) involved have other connection points, but these are mainly in the north and center of Mexico City. In such a way this proposal offers one more point of connection between the lines and between both systems (MS and MCSS), but in the southern area, which could be quite beneficial

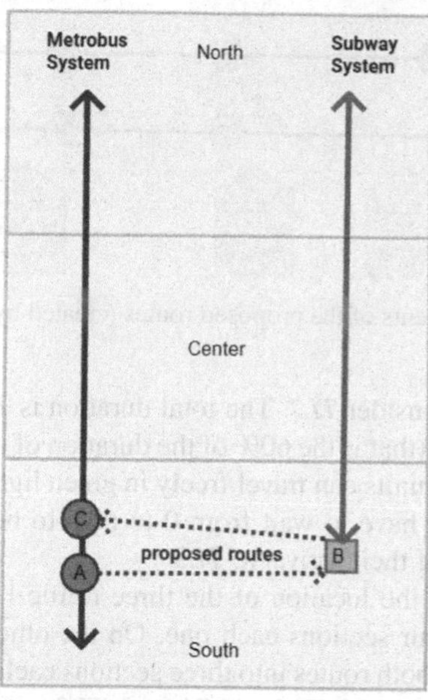

Fig. 1. Proposed connection routes in Mexico City (created by Nieves V.) (This and next figures, and graphs of this paper were presented in the Maple Conference 2020 which was done virtual, November 2 - 6, 2020. Available on https://www.maplesoft.com/mapleconference/Papers-and-Presentations.aspx.)

for workers, students and general users whose point of origin or destination is at the south of the city.

4 The Model to Document the Proposal

This section presents the structuring of the simulation model, including the definition of key variables used to develop the simulation model in Maple. First, look at the Fig. 1 that illustrates the surroundings of the proposed routes AB and BC.

As part of the proposal, two intermediate stations were defined on the routes: *D1* and *D2* (see Fig. 2), and three crossings with traffic light signals: *TL1*, *TL2* and *TL3* (see Fig. 2). Currently *TL1* and *TL2* exist, but *TL3* is part of the proposal.

For each traffic light, we define *TL* as the total duration of its traffic light cycle, which is composed of three phases: Green Light (*GL*), Yellow Light (*YL*), and Red Light (*RL*). The green light corresponds to the time that vehicles can continue moving, the yellow light is a caution alert that indicates to drivers that the red light is the next phase, and they must stop. The yellow light has a fixed duration of three seconds and the duration of the green light varies depending on the location of each traffic light. The total duration of one traffic light cycle is $TL = GL + YL + RL$, so if we know *TL* and *GL*, we can determine *RL* (because of *YL* is always 3 s). A very important assumption for this model is that only for the green light we allow the MS units to move, and for the yellow and red light we assume they stop.

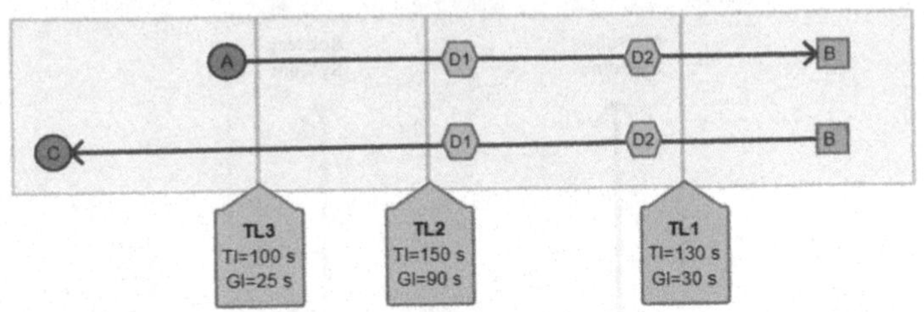

Fig. 2. Elements of the proposed routes (created by Nieves V.)

As an example, let's consider *TL2*. The total duration is $TL = 150$ s, and the green light duration is $GL = 90$ s (that is the 60% of the duration of the *TL2* traffic light cycle). Thus, 60% of the time MS units can travel freely in green light and, the remaining 40% of the time, MS units will have to wait from 0 to 60 s to be able to cross the cruise, according to the moment of their arrival to *TL2*.

From Fig. 2, note that the location of the three traffic light signals fragments the routes AB and BC into four sections each one. On the other hand, intermediate stop stations D1 and D2 divide both routes into three sections each one. For simplicity, given the closeness between TL1 and S2, TL2 and S1, and, TL3 and S1 the final fragmentation for both routes only considers the longest sections, as shown in Fig. 3. This simplification allows us to define the average speed for each section, as explained in the next section.

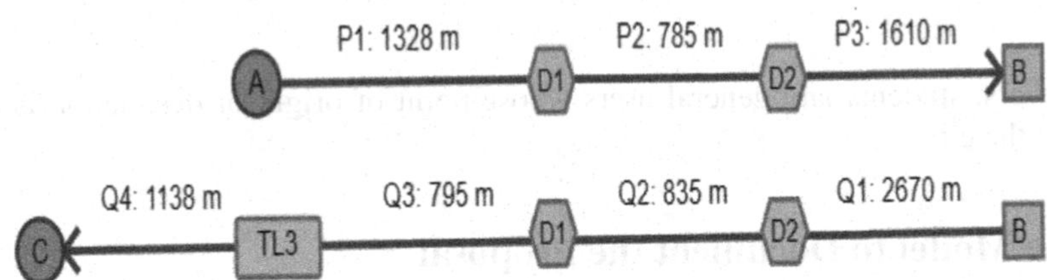

Fig. 3. Fragmentation of the proposed routes (created by Nieves V.)

4.1 Travel Time for Proposed Routes

Duration of each route considers three important variables. They are

T_T The *Travel Time* between two consecutive stations (or traffic light cruise)
T_S The *Stop Time* at bus stops
T_{TL} The *Traffic Lights Time* at cruise

So, total duration of a route is given by the following expression

$$T = T_T + T_S + T_{TL} \tag{1}$$

To estimate the duration of each of the three components, we were based on data obtained from observations made in February 2020 about the operation of line 1 of the MS units in Mexico City. The entire study for the development and analysis of the proposal is limited to the behavior of the system on workdays (Mon-day to Friday) and within specific schedule from 07:00 am to 10:00 am, in which there is high use of public transport in the study area. According to the results of the interviews carried out by Nieves [1], the operators of minibuses and combis ex-pressed that there is a strong relationship between this schedule and the beginning of work force and schooling activities in Mexico City. These activities are not presented in the same way on weekends, and that is exactly the reason why the data collection of the study was not carried out seven days a week.

The operating speed of the Metrobus units varies between each section defined by two adjacent stations on the line. When it comes to short length sections, Metrobus units cannot reach full throttle as they must anticipate stopping at the next station. In longer sections the average operating speed increases. That is, there is a direct relationship between the distance and the average operating speed of the Metrobus units, as long as in the section defined by two adjacent stations, there is no traffic light signal or intersection between roads that causes the stop of the units. This statement is reflected in the work of Nieves [1], as part of the description of the current operation of the Metrobus units.

Below is a brief description of each of the components of Eq. (1).

Travel Time (T_T). Duration of travel between two adjacent points was obtained from the current average operating speed.

The speed of the Metrobus units is not constant throughout each section (length between two adjacent stations), in fact, when leaving one station the units begin to accelerate, and before reaching the next station, they begin to slow down. A prospective study can undertake to analyze these behaviors in detail.

However, in this model, we consider the average speed per section, and we consider it as a random variable that presents different behaviors depending on the length of the route. We used three probability functions: Normal, Lognormal or Weibull[1].

The assigned probability function depends on the subsection with the greatest distance defined when is considering traffic light signals and stations (terminal or intermediate). In such a way that the average speed of operation of the Metrobus units on the smallest subsections corresponds to the Weibull distribution, those of medium distance were associated with the Normal distribution and those subsections of greater distance with the Lognormal distribution as indicated in Table 1.

This correspondence was obtained from the data adjustment of the speed variable, which was obtained by considering the distance of the routes (between adjacent stations and without intermediate traffic light signals), and the total time of the route (from when it begins its movement at one station until it stops at the next station). The distance of each section of the line was approximated with the distance function of Google Maps and the time of the journeys was measured by making tours on the units as a user, with the help of digital clocks. The details of the methodology to collect the data can be consulted in Nieves [1]. Regarding the adjustment of the data, goodness-of-fit tests were performed for

[1] Distribution functions characterizing the average operating speed of the MS units were defined from the fit of data, collected in the sections of the MS line defined by stations A, C and two adjacent stations, considering the operation in both directions of the line.

Table 1. Speed parameter per section on proposed routes (Created by Nieves V.)

Speed Key	Probability distribution	Parameters		Distance x for longest distance subsections (meter)
		location or shape	scale	
S1	Normal	11.010	1.1823	$701 \leq x \leq 1100$
S2	Lognormal	2.484	0.0853	$1101 \leq x$
S3	Weibull	9.8974	10.4064	$x \leq 700$

a set of continuous probability distributions. It was considered the representative function of the data as the one whose p-value is above the alpha significance level $\propto = 0.05$ to a greater extent (with respect to the set of probability distributions with the possibility of adequately representing the behavior of the data).This analysis was performed with the Minitab 18 statistical software tools [1]. The three probability distributions (Normal, Weibull and lognormal) are similar in that they all have the shape of "a mountain", which varies in thickness and height according to its parameters. The difference is that both the Weibull and the Lognormal allow slant to the left, while the normal is always symmetric.

So, to define the speed in each section of the proposed routes, first, we consider the distance of the longest subsection between each pair of adjacent intermediate stations of the routes, that is, the longest subsection that does not have intermediate traffic light signals. Then, assuming that the average operating speed in sections of similar length is similar, we define the average operating speed as a random variable that behaves in the same way as the average speed of the current sections of the Metrobus line.

Once calculated the average operating speed (*s*) of each section (from the distributions mentioned before), and considering the length (*l*), we obtained the travel time per section by the next expression:

$$T_T = \frac{length}{average\ operating\ speed} = \frac{l}{s} \tag{2}$$

Stop Time (T_S). To estimate this parameter, we collected information corresponding to the stop time of the MS units (in seconds) at 5 stations on the line, around A and C. The data showed an adequate fit to the Beta distribution (4.0878, 3.9075), with minimum value = 8.13 s. and maximum value = 13.45 s. Then the stop time (T_S) in both intermediate stations (*D1* and *D2*) is defined as a random variable that follows a beta distribution

$$T_S \sim Beta(4.0878, 3.9075, 8.13, 13.35) \tag{3}$$

Traffic Light Time at Cruise (T_{TL}). For each traffic light signal, the total duration of the traffic light cycle (*TL*) is considered, in such a way that in the fraction corresponding to the duration of green light (*GL*), a MS unit must wait 0 s to cross it, while the complement percentage in yellow (*YL*) or red light (*RL*), MS units must wait between 0 *TL-GL* seconds to cross it.

All the information related to the traffic lights was also collected by direct observation in the study area, within the schedule mentioning above, and with the help of digital watches and handmade paper notes.

5 Maple Algorithm for the Model

The implementation of the model with Maple tools involves the generation of random variables, defined by the Normal, Lognormal, Weibull and Beta probability functions. For this reason, the *Statistics* and *Random Tools* packages were used. On the other hand, *Array, Matrices* and *Graphics* functions were used to analyze and display the results of the model. Maple includes help documentation on all its libraries and functions, so users may consult this documentation from the help section, "*The Maple Help System*", or online way, from the official site [10].

One execution of the model consists of calculating the total travel time of a route from a starting point (A or B) to an end point (B or C), respectively, considering the Travel Time T_T (between stops), the Stop Time T_S (in intermediate stations), and the Traffic Light Time T_{TL} (in traffic light cruises). To calculate each of these components, the simulation model has three auxiliary functions **Section Travel Time Vector**, **Stop Time Vector** and **Traffic Light Vector**, which calculate the time of each component, depending on the input parameters and the specified route.

In the Maple algorithm, a route is defined as a vector expressed with an even number of input values. The odd inputs indicate the distance of the subsections of the route, while the even specify the probability distribution that represents the average operating speed according to the longest distance of each subsection (according to **Table 1**). For example, for route AB, this vector is **R1: =** [1328, ***S1***, 785, ***S1***, 1610, ***S2***]. Inputs in 1, 3 and 5 position represent the three distances of the subsections of the route AB (according to Fig. 3), and positions 2, 4 and 6 indicate the probability distribution of the mean operating speed for each subsection (as indicates Table 1). In the same way, for route BC, it is defined **R2: =** [2670, ***S2***, 835, ***S1***, 795, ***S3***, 1138, ***S2***].

Definition of routes does not specify intermediate stops or traffic light crossings since, in both R1 and R2, the two intermediate stations (*D1* and *D2*) and the three traffic light crossings (*TL1*, *TL2* and *TL3*) are considered. However, the model can be generalized to evaluate different routes that contain only some of these components, but not all.

The three auxiliary functions mentioned previously perform operations handling the seconds as a unit of time. Subsequently, through the **Route Evaluation** function, the results of these three functions are added together and the last result is provided in minutes, with the objective of offering to the user a perceptible result, easy to interpret.

The result of a single execution of the model is not enough to give conclusions about the random total travel time of the proposed routes. For this reason, to get an estimation, several sets of executions, called replicas, were analyzed. Each replica $\boldsymbol{r}$ consists of running the model $\boldsymbol{d}$ times. We considered the analysis of results through replicas, to focus our attention on the mean, as a new variable representative of each replica. In this sense, we look for the stability of the variable mean by choosing a suitable value of $\boldsymbol{d}$ and as the number of replicas increases.

Parameters $\boldsymbol{r}$ and $\boldsymbol{d}$ are provided by the user and can be easily modified. All results are stored in an array $\boldsymbol{M}_{\mathbf{r}\times\mathbf{d}}$, where each row corresponds to the results of one replica. In this case, 100 replicas were considered, each one with $\boldsymbol{d} = 100$ data.

Once results of independent executions have been calculated, statistics are obtained for each replica. This process of evaluating more than one execution is carried out through the outputs of the **Matrix Evaluation** function, which produces the matrix $\boldsymbol{M}$, and three vectors **Mmin**, **Mmax** and **Mmean**, which correspond to the sets of the minimum, maximum and average value of each replica.

After that, another function called **Model**, whose input parameters are $\boldsymbol{r}$, $\boldsymbol{d}$ and **Route** (*R1* or *R2*), uses all the auxiliary functions described above to provide the last results, considering $\boldsymbol{r}$ replicas, each one with $\boldsymbol{d}$ data. The results are shown graphically by representing the vectors **Mmin**, **Mmax** and **Mmean** and the 95% confidence interval of the mean of each replica. The confidence interval is also calculated as an auxiliary function of the model. Figure 4 shows the conceptual diagram of the complete simulation model created from the auxiliary functions, logical structures, and loops. As a complement, Fig. 5 illustrates the implementation structure of the model in Maple. This may be challenging for the curious reader interested in implementing a new version of the model. In addition, Fig. 6 shows the details of the implementation of the auxiliary function **Traffic Light Vector** (see Fig. 6 b), using the input parameter defined in a).

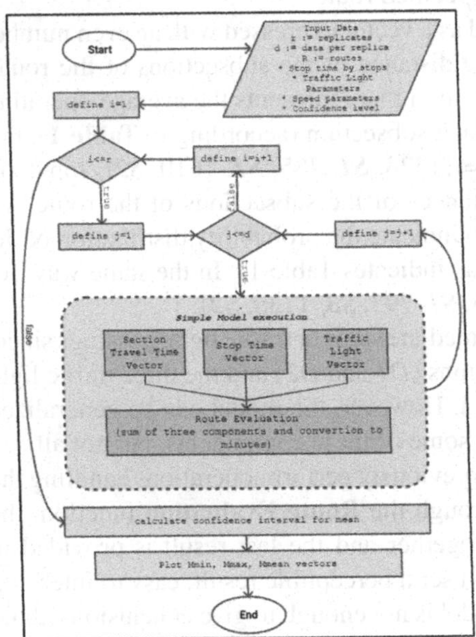

Fig. 4. Conceptual model for the simulation model (created by Nieves V.)

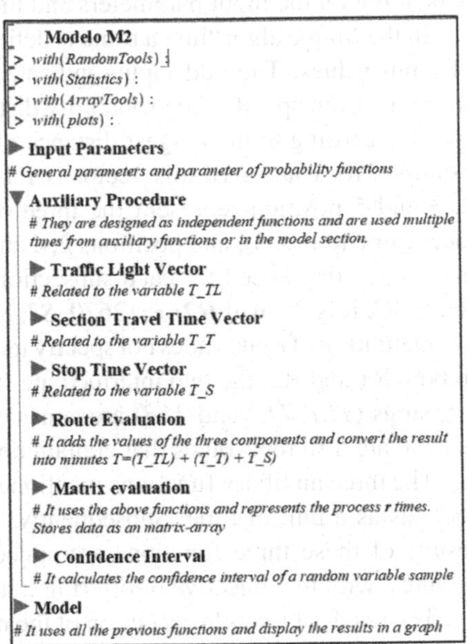

Fig. 5. Structure of the model in Maple. (created by Nieves V.)

As a representative result of the travel time of the routes, the global mean was considered, and the 95% confidence interval for population mean is calculated by taking all means of each replication.

a)

```
Input Parameters
> r := 100 : # Replicas
> d := 100 : # Data by replica
> ST := RandomVariable('Beta'(4.0878, 3.9075)) : #Stop Time by Stop
> a := 8.13 : b := 13.45 : # Inferior and superior limits for Beta Distribution
> alfa := 0.05 :
> # Traffic Light Parameters
> # [green light duration, red light duration] ;
  TL1 := [30, 100] :
  TL2 := [90, 60] :
  TL3 := [25, 75] :
```

b)

```
Traffic Light Vector
> TLV :=proc (d, TL) # d=number of data, TL=trafic light parameters
  v := Vector[row](d) :
  green := TL[1]; red := TL[2];
  for i from 1 by 1 to d do
  r := Generate('integer'('range'= 0 ..(green+red)));
  if r ≤ red then v[i] := r else v[i] := 0 end if;
  end do;
  #print(v);
  return v;
  end proc:
>
```

Fig. 6. Implementation of the **Traffic Light Vector** helper function in Maple (created by Nieves V.).

6 Results and Discussion

In the analysis of results, 100 replicates were considered, each one with 100 data. Due to the results of each replica are stored in a matrix $M_{100\times 100}$, Maple does not display them in the worksheet, however they can be exported and viewed with Microsoft Excel tools, if the user so wishes. The resulting graphs of the simulation model for R1 and R2 are Fig. 7 and Fig. 8, respectively.

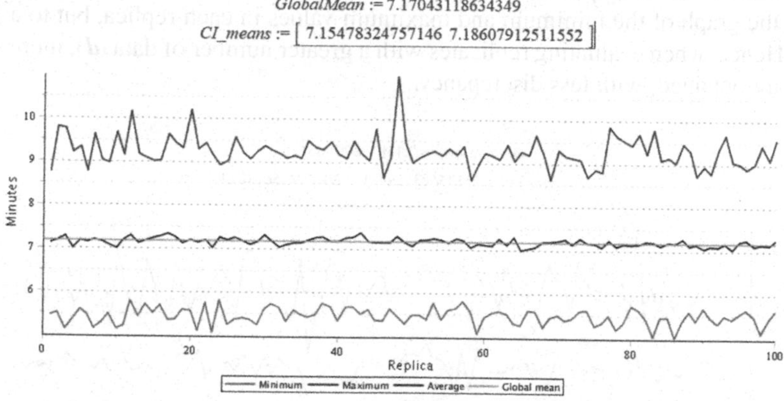

Fig. 7. Route R1 results, $r = 100$, $d = 100$ (generated with Maple software by Nieves V.)

The global mean for route R1 is approximately 7.17 min and for R2 it is 9.67 min. These model output parameters allow us to estimate the total duration of the two proposed routes, however this is a simulation model and manipulates random variables, so in each execution of the model different results can be obtained with small variations. For R1, 95% of the times, these variations are less than 0.016 min, according to the confidence interval, which is [7.154, 7.186] and whose range is approximately 0.032.

On the other hand, variation of the parameter d is reflected in the confidence interval; as the value of d increases, it becomes smaller. By repeatedly running the model and varying the amount d, as an experimental process, it is found that, for "small" values of d, the corresponding graph to the mean values of each replica presents greater oscillations around the global mean. To the contrary, for larger values of d, these oscillations are

$GlobalMean := 9.67425554235105$

$CI\_means := [\ 9.65742413104740\ \ 9.69108695365470\ ]$

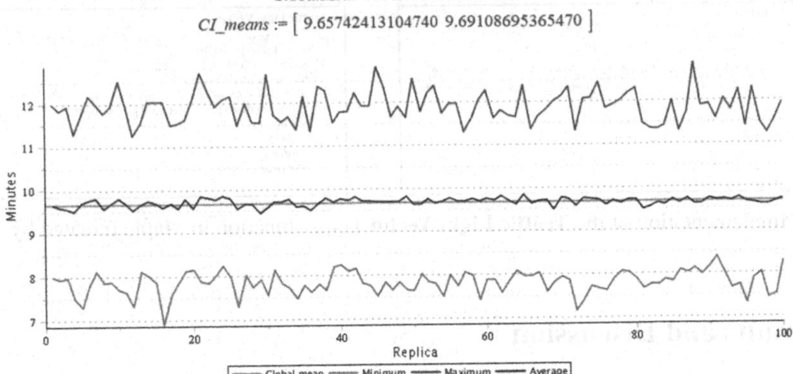

Fig. 8. Route R2 results, $\boldsymbol{r} = 100$, $\boldsymbol{d} = 100$ (generated with Maple software by Nieves V.)

minimal. For example, for $\boldsymbol{r}= 100$ and $\boldsymbol{d} = 20$, the results are presented in Fig. 9 (see black color line), where we can see greater oscillation in the average graph, compared to the Fig. 10, whose parameters are $\boldsymbol{r} = 100$ and $\boldsymbol{d} = 1000$. In the latter case, the average graph (black color line) appears almost as a straight line. Note that this behavior is also seen in the graph of the minimum and maximum values in each replica, but to a lesser extent. Hence, when evaluating replicates with a greater number of data (***d***), more stable results are obtained, with less discrepancy.

$GlobalMean := 7.16913177065758$

$CI\_means := [\ 7.13346488529701\ \ 7.20479865601814\ ]$

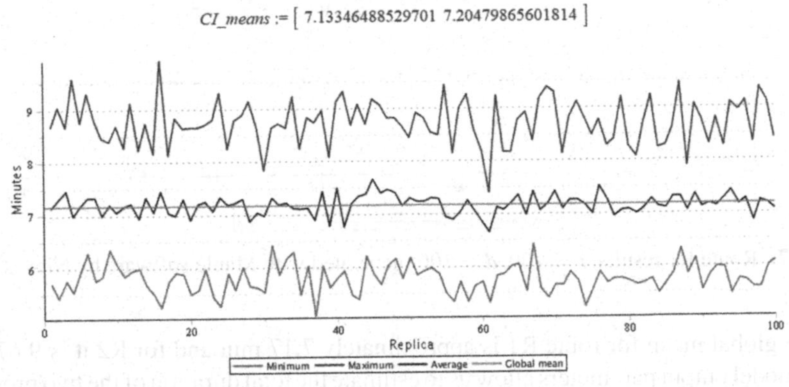

Fig. 9. Variation of the parameter ***d***, Route R1 results, $\boldsymbol{r} = 100$, $\boldsymbol{d}= \mathbf{20}$ (generated with Maple software by Nieves V).

But also, is necessary to mention that as the value of ***d*** increases, the execution of the model takes longer. This is a simple model where only one output variable is analyzed, and the execution time is almost imperceptible. But, in more complex models, where more variables are involved, this could make a significant difference in running time, which can take days, weeks, and even months. In such a way that the adequate definition of parameters (***d*** and ***r***) becomes restrictive in some cases.

GlobalMean := 7.16015058246546

CI_means := [7.15499526484062 7.16530590009029]

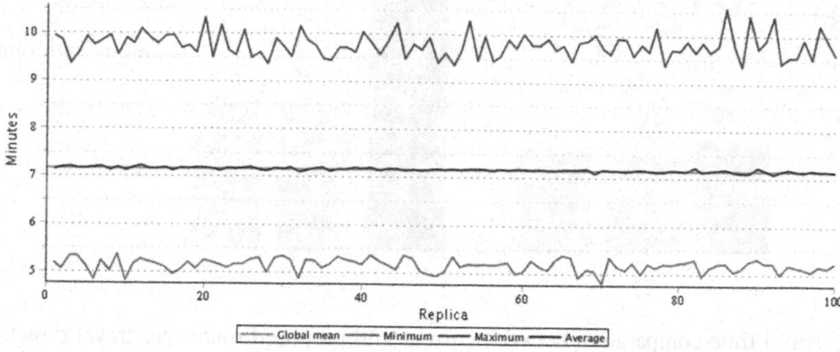

Fig. 10. Variation of the parameter ***d***, Route R1 results, $r = 100$, $d = 1000$ (generated with Maple software by Nieves V).

The results of this model are then compared with the duration of travels currently made by buses and vans[2] making the connection between the MS and MCSS systems. It should be noted that currently existing connections between these two systems only involve stations A and B, that is, there are routes AB and BA, but currently there is no connection between station B and C. In a generic way, we named Travel T1 to the routes from the MS line to the MCSS, and Travel T2 to the routes in the opposite direction, from MCSS to MS, as shows Fig. 11.

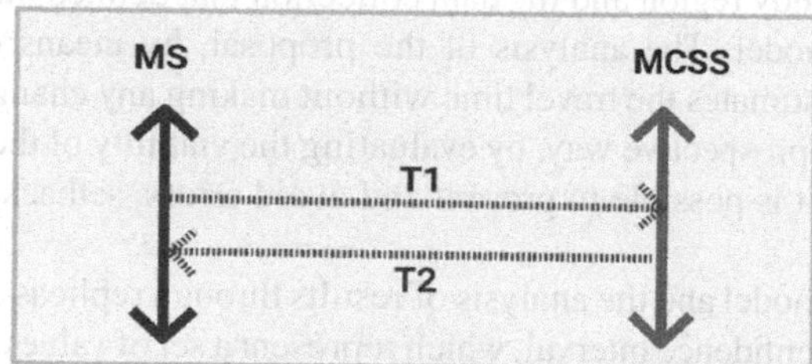

Fig. 11. Generic travels directions between MS y MCSS, (created by Nieves V.)

In Fig. 12 we compare the travel time of the current and proposed travels T1 and T2. It indicates that in T1, the travel by MS units is shortest (7.17 min), compared to the current bus and van travel times (7.88 and 75 min respectively). On the other hand, regarding the T2 routes, the travel time of the proposed route (9.67 min) is similar to vans (9.62 min), and is approximately 36.1% (5.48 min) less than the current travels made by buses and minibuses, even when the proposed route is longer for considering station C.

2 The travel time of current trips made by buses, minibuses and vans was estimated from time measurements taken during running travels on weekdays, between 7:00 A.M. and 10:00 A.M. The information in **Fig. 12** corresponds to the average value of the data collected in each case.

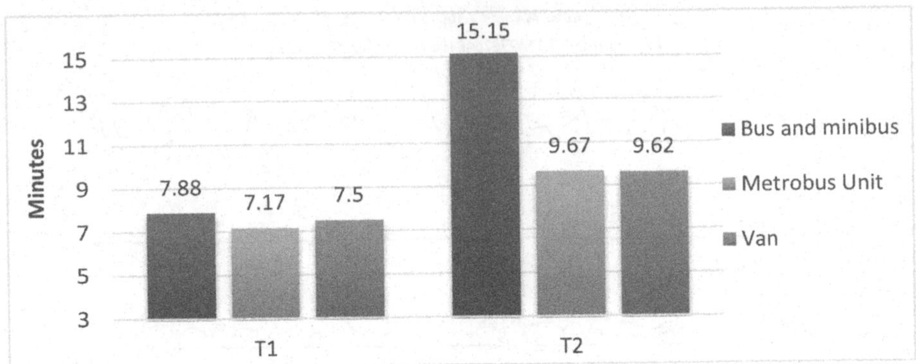

Fig. 12. Travel time comparation between current and proposed routes per travel direction and type of transport unit, in minutes (created by Nieves V.)

7 Conclusive Notes

The proposed routes offer an alternative connection between two transport systems in Mexico City and are favorable in terms of travel time. The development, feasibility and analysis of this proposal are studied in more detail and from different perspectives in a broader investigation of Nieves [1].

The present work constitutes an example of the application case study of simulation as a tool to estimate the travel time of routes proposed by Metrobus units, as an extension of a currently existing line. Something valuable about this work is the methodology for the observation of the study region and the data collection that defined the input parameters of the simulation model. The analysis of the proposal, by means of a Monte Carlo simulation model, estimates the travel time without making any changes to the routes of the real system. In a prospective way, by evaluating the viability of the proposal through simulation models, it is possible to prevent and avoid errors, setbacks, and expenses in the real system.

The simulation model and the analysis of results through replicas allowed to obtain a mean value and its confidence interval, which represent a set of values and not an only fix value as result. In such a way that the management and the adequate interpretation of the results turn into effective tools for decision making to connect two public transportation systems.

On the other hand, Maple is a useful programming language to develop Monte Carlo simulation models because of it has packages and functions that help to structure these models, for example, statistics packages allow to generate, manipulate and analyze sets of random variables. In addition, it has graphical tools which facilitate the interpretation of results. Likewise, the environment of this programming language has visual and structural qualities that facilitate the implementation of algorithms and models, not only for simulation models, but also of a general nature.

References

1. Nieves, V.: Modelos de simulación para documentar una propuesta cuyo objetivo es reducir el congestionamiento vehicular. Un caso de estudio en Coyoacan, CDMX. National Autonomous University of Mexico, Mexico City (2020)
2. Guba, E.G., Lincoln, Y.S.: Naturalistic Inquiry, p. 466. Sage Publications, Beverly Hills (1985)
3. Chen, G.K.C.: What is the system aproach. Interfaces **6**, 32–37 (1975)
4. Ackoff, R.L.: Systems thinking and thinking systems. Syst. Dyn. Rev. **10**, 175–188 (1994)
5. Rubinstein, R.Y., Kroese, D.P.: Simulation and the Monte Carlo Method. Wiley, Hoboken (2017)
6. White, K.P., Ingalls, R.G.: The basics of simulation. In: Proceeding of the 2017 Winter Simulation Conference, pp. 505–519 (2017)
7. Grigoryev, I.: Anylogic 7 in three days. A quick course in simulation modeling (2015)
8. Brailsford, S.C., Eldabi, T., Kunc, M., Mustafee, N., Osorio, A.F.: Hibrid simulation modelling in operational reseach: a state of the art review. Eur. J. Oper. Res. 721–737 (2019)
9. Currie, C.S.M., Cheng, R.C.: A practical introduction to analysis of simulation output data. In: de Proceeding of the 2016 Winter Simulation Conference (2016)
10. Maplesoft (2020). Maplesoft.com. [En línea]. https://www.maplesoft.com/support/help/. Último acceso: 2020
11. Law, A.M.A.: Simulation Modeling and Análisis, 5th edn. McGraw Hill Education, New York (2015)
12. Maplesoft: Statistics and Data Analysis (2020). (En línea) https://www.maplesoft.com/products/maple/features/statistics.aspx
13. Maplesoft: Programming (2020). (En línea). https://de.maplesoft.com/products/maple/features/programming.aspx.
14. EOD: Encuesta Origen Destino, en hogares de la Zona Metropolitana del Valle de México. Presentación de resultados por INEGI. Instituto Nacional de Estadística y Geografía, Mexico City (2017)

Understanding Math Concepts in Music and Vice-Versa

Gabriel Picioroaga(✉)

Department of Mathematical Sciences, University of South Dakota,
414 E. Clark Street, Vermillion, SD 57069, USA
Gabriel.Picioroaga@usd.edu

Abstract. The aim of this paper is pedagogical in nature and two-folded. It is primarily intended for instructors who teach undergrad math courses (such as Linear Algebra, Calculus or Fourier Analysis), or courses in basic Music Theory. In the modern classrooms of today, educational software is ubiquitous. Sometimes it is a daunting task to find, use and take advantage of applications that allow for an efficient instructional activity. In this work we show how to use the AudioTools, SignalProcessing and Curve Fitting packages in Maple to illustrate and explain at a deeper level the mathematics needed to implement and play various musical concepts such as: tones, overtones, chords, and more complex paradigms from Music Theory, the counterpoint, and Euler's space of (horizontal) fifths and (vertical) major thirds. Our Maple code and proposed student projects are elementary, and allow for the math and music concepts to be transparent and easy to grasp in the application context. A few lectures based on these ideas, in an independent or hybrid course setting, may bring important education benefits to a large spectra of undergraduate students taking math/music courses, regardless of their major.

Keywords: Signal processing · Sound wave · (over)tone · Chords · Maple · Pitch · Musical scale

1 Introduction

Here is a question that can be studied from a multitude of fields of science: What is sound? For the purposes of this paper we will start with the short answer, from the math/physics perspective: sound is a wave created by differences in air pressure. Although not a straightforward explanation, sound is perceived (encoded/decoded) because ears and brains are equipped with "hardware" and "software" tools able to detect and interpret/decode these vibrations. Organisms have evolved quite a complex mechanism to perform these jobs. Explaining it (how/why it works) requires learning anatomy, neuroscience, physics, mathematics, and probably more. For example it is not clear how/why brains transforms the auditive information into a complex cognitive experience (e.g. see [13]).

R. M. Corless et al. (Eds.): MC 2020, CCIS 1414, pp. 304–318, 2021.
https://doi.org/10.1007/978-3-030-81698-8_20

In this work we will bound ourselves to the realm of mathematics and computers science to process sound as digital signals. We will relate to very specific sounds, namely those bits that occur in music. The implementation of the basic musical pitches is quite easy due to the periodic nature of such waves. Also, multiple connections with mathematics can be obtained, starting from an elementary level (algebra, trigonometry) then going up to more abstract or technical levels (group theory, Fourier analysis, and signal processing to mention a few).

Starting with a fundamental frequency f_0 (e.g. conventionally the nowadays 440 Hz for note A on the piano) all musical pitches are encoded in Maple as sine waves using the nice logarithmic formula between pitches and frequencies (in musical terms, we have implemented the "temperate tuning"). Hence all pitch classes are associated with the numbers $0 = \text{C}$, $1 = \text{C}\#$, $2 = \text{D}$, ..., $11 = \text{B}$, $12 = \text{C}$ an octave higher, and so on. The procedure $Note(i, t, x)$ encodes the pitch i, played for t seconds, at amplitude x. A melody then becomes a sequence (array) of notes that can be played either straightforward and/or saved as a .wav file.

The paper is structured as follows: In the first section we present an elementary mathematical background needed to implement sound waves as functions of a single variable. The next section serves a similar purpose from the music theory point of view. The terms collected should be sufficient to a reader with more math backround than music to continue further to next sections. In the third section we present the relationship between frequencies and tuning along with the historical examples of temperate, Pythagorean, and just tuning. The section also contains justifications for Pythagorean and just tunings which due to the simple algebraic manipulations can be implemented even in a highschool algebra course. The fourth section contains Maple code used for implementing the 12-scale musical notes. From here onward it is almost straightforward to go on to implement chords code, that is combinations of notes played at the same time. In the last section we touch upon the simpler species of the counterpoint technique, and some mathematics behind. This subject is vast and besides a simple example to obtain counterpoint-like sound with Maple, we mentioned a few mathematical connections and bibliography needed to further study this topic. Throughout the paper we propose Maple projects that could be interesting to either math or music majors, and can be used to better absorb mathematical facts and concepts. Of course, most of the projects proposed are about writing Maple "apps" that manipulate sound/music. The amount of software available to the professional or amateur musician is staggering, however such software behaves like a "black box" from the user point of view, and for good reason. The simpler Maple projects are intended to make clear to the student or educator, some of the operations that such "black boxes" are based on.

The intersection between music and science as a topic is vast, and references abound. We kept the bibliography section light so that it is available to either students, musicians, and mathematicians. Most of the references included should be accessible to STEM undergrads, except maybe [9] and [14] where a graduate level preparation is advised.

2 Basic Terms/Math

One can start the mathematical representation of sound waves with the solutions of a simple differential equation,

$$my''(t) + ky(t) = 0$$

where the unknown function $y(t)$ models the (free, undamped) vibrations in time of a spring of mass m, and spring constant $k > 0$. As taught in any undergraduate Differential Equations course, the general solution is

$$y(t) = c_1 \sin(2\pi f_0 t) + c_2 \cos(2\pi f_0 t)$$

where $f_0 := \frac{1}{2\pi} \cdot \sqrt{\frac{k}{m}}$ is called the "natural" or "fundamental" frequency. Using trigonometry one can rewrite the general solution in the form $A \sin(2\pi f_0 t + \varphi)$, where φ is called the angle of phase-shift and A the amplitude. One can of course rewrite the solution $y(t)$ in cosine form, and use the cosine "wave" when implementing musical tones without affecting the sound result. Either way, one can think of these sine and cosine stemming from the spring equation above, as fundamental "bits" of sound, which model the vibrations produced by sound traveling through the air. We have thus arrived at a simple model of a sound wave, or sound function. Due to J. Fourier's research into heat conduction at the start of the XIX-century, and his representation of periodic functions as sums of sine and cosine waves, a more general sound function can be assumed to have the following form (Fig. 1):

$$\begin{aligned} F(t) = A_0 &+ A_1 \sin(2\pi f_0 t + \varphi_1) + A_2 \sin(2\pi 2 f_0 t + \varphi_2) + \\ &+ A_3 \sin(2\pi 3 f_0 t + \varphi_3) + A_4 \sin(2\pi 4 f_0 t + \varphi_4) + \\ &+ A_5 \sin(2\pi 5 f_0 t + \varphi_5) + \ldots \end{aligned}$$

where A_n are amplitudes, f_0 is the fundamental (frequency), nf_0 are partials, and φ_n are phases. The summation above is of course a "series", and modern analysis has specific tools to deal with types of convergence, recovery of the

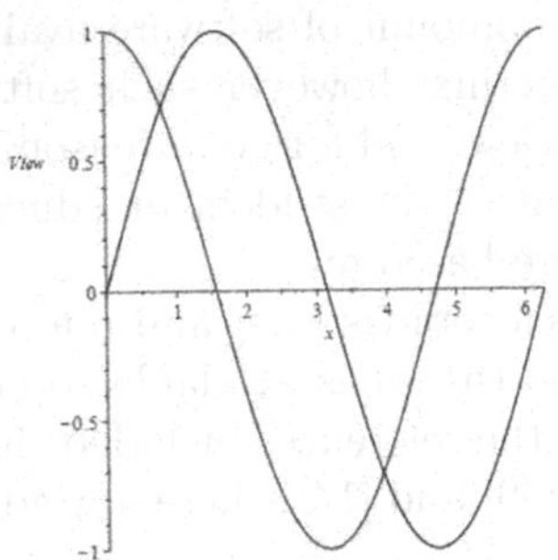

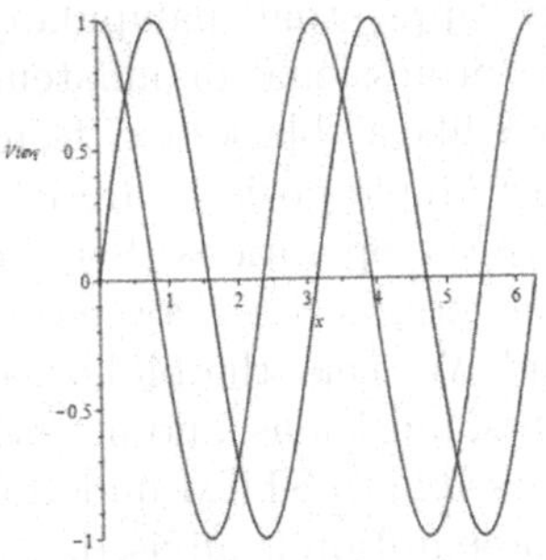

Fig. 1. sin(t), cos(t), sin($2t$), and cos($2t$)

coefficients A_n etc. For the purpose of implementing the function digitally and be able to experiment its "sound", only a few terms are needed (finite summation) and a little care in choosing the coefficients. Nonetheless, for more realistic sound experiments and situations, acoustic models are constructed in multiple dimensions. The equations giving rise to such sounds can be quite complicated (partial) differential equations (see [3] and references therein). For a realistic model, think of the sound wave as tucked inside a surface, with each partial inside its own envelope. Such considerations are beyond the elementary exposition and purpose of this paper. The bibliography on the subject is vast, but the interested reader may consult the books [2] and [15]. Here we will consider 1-dimensional (time variable) waves only. However, the envelope function will be implemented into the sound, which simply means multiplying the sine wave(s) by it. In music such a function bears the acronym ADSR (attack, decay, sustain, release). Examples (see Figs. 2 and 3) are built in Maple using a few points in the plane and the (unique) polynomial curve passing through those points. For the second envelope we used the Spline command. We warn the reader that some issues may be encountered when creating audio files by means of spline functions. It seems that spline functions do not accept the evalhf command which is needed to sample the audio.

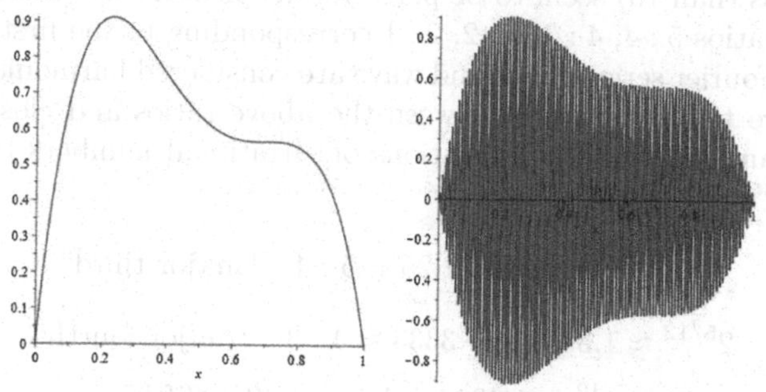

Fig. 2. ADSR polynomial enveloped sound wave

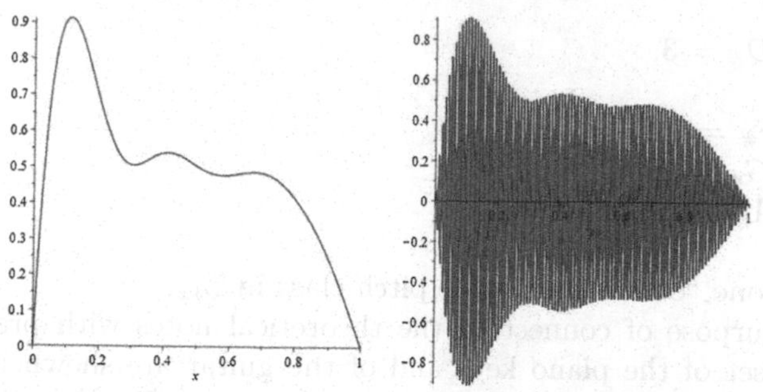

Fig. 3. ADSR spline enveloped sound wave

Maple Project 2.1. *Create more ADSR envelopes and check the differences of the sound this produced. Experiment with Maple CurveFitting package and solving the system of linear equations based on sample points on a polynomial ADSR curve. Notice that equidistant points on the x-axis give rise to a Vandermonde coefficient matrix, which is "famously ill-conditioned" [7] and may create issues when solving such linear systems.*

3 Basic Terms/Music

What is music? Music of course is a complex phenomenon which can be studied and created from multiple corners of science, culture and civilization. In this section we describe a few basic concepts for the benefit of a reader who has no previous knowledge of music theory but is equipped with minimal mathematical background. We will not need music notation and will not mention other important basic concepts such as rhythm, meter, time signature etc. We claim that after reading this article the interested reader should have sufficient knowledge to cover these concepts by herself/himself from elementary texts such as [12]. Coming back to the question above, from the laymen perspective here is an oversimplified attempt at a definition: music can be thought of as a succession of "nice" sound bites. Frequency ratios close to rationals p/q with "smallish" p and q (e.g. less than 10) seem to be perceived as pleasant by (most) beings. For example the ratios $5:4$, $4:3$, $3:2$, $2:1$ corresponding to the first five terms in the previous Fourier series of a sound wave are considered harmonious. In reality it is quite rare to discriminate between the above ratios and close approximations. For example, the (approximations of) irrational numbers below are also perceived as "nice" by most humans:

$$2^{4/12} \approx 1.2599 \approx 1.25 = 5:4 \quad \text{"major third"}$$

$$2^{5/12} \approx 1.3348 \approx 1.3333 \approx 4:3 \quad \text{"major fourth"}$$

$$2^{7/12} \approx 1.4983 \approx 1.5 = 3:2 \quad \text{"fifth"}$$

Notes (tones and semitones) are identified with numbers as follows: $C = 0$ (Do),
$C_{\#} = 1$,
$D = 2$ (Re), $D_{\#} = 3$,
$E = 4$ (Mi),
$F = 5$ (Fa), $F_{\#} = 6$,
$G = 7$ (Sol), $G_{\#} = 8$,
$A = 9$ (La), $A_{\#} = 10$,
$B = 11$ (Ti),
$C = 12$ (Do) one "octave" up, same pitch class in $\mathbb{Z}_{12}$.

For the purpose of connecting the theoretical notes with practical instruments, a subset of the piano keys and of the guitar are shown in the figures, together with numbering on consecutive 12-scale lists (Figs. 5 and 6).

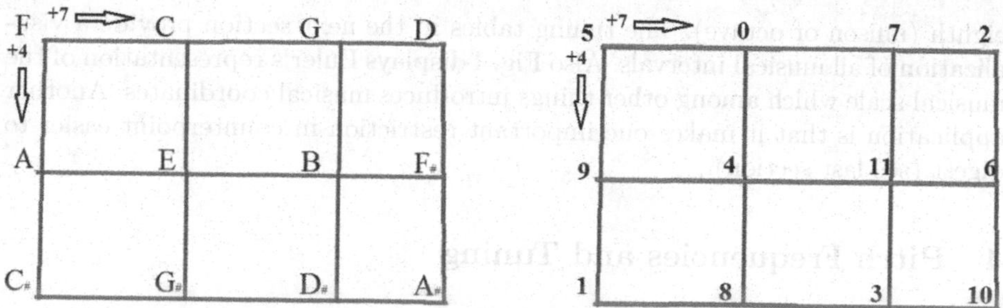

Fig. 4. Euler's space of fifths and major thirds

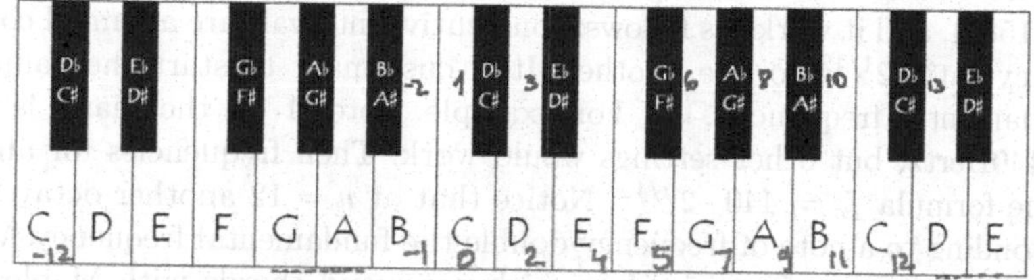

Fig. 5. Subset of piano keys

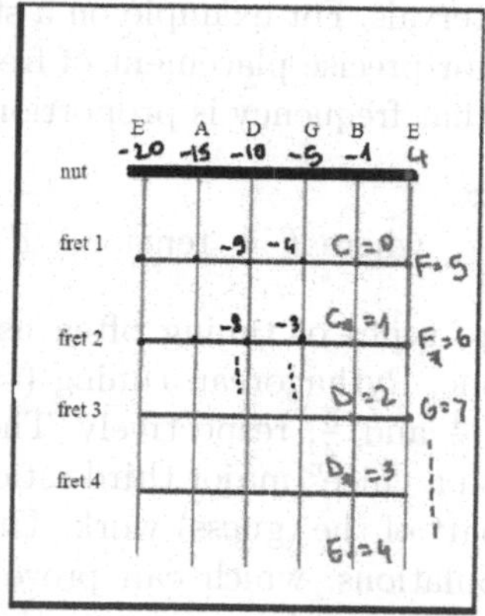

Fig. 6. Guitar strings and notes

Working with equivalence classes of the tones (pitches) modulo 12, we can define "musical intervals" terms. For example, a major third represents the interval from $C = 0$ to $E = 4$. The namesake comes from counting on the 8-notes scale, avoiding the sharp notes. Any interval of length 4 in $\mathbb{Z}_{12}$ is a major third: $F = 5$ to $A = 9$, $A = 9$ to $C_{\#} = 1 = 13 \bmod 12$ etc. Fifth (perfect fifth): Interval from $C = 0$ to $G = 7$ or any interval of length 7 in $\mathbb{Z}_{12}$. The complete list includes: minor/major second, minor third, perfect fourth, tritone (augmented fourth or diminished fifth), minor/major sixth, minor/major seventh, perfect eighth (unison or octave). The tuning tables in the next section provide a visualization of all musical intervals. Also Fig. 4 displays Euler's representation of the musical scale which among other things introduces musical coordinates. Another application is that it makes one important restriction in counterpoint easier to digest (see last section).

4 Pitch Frequencies and Tuning

There are many types of tuning in music depending on a variety of factors. One type that established itself as classic throughout the ages is the so-called temperate tuning. Its discovery is attributed to Chinese mathematician Zhu Zaiyu (1584), and it works as follows: consecutive intervals are assumed constant frequency ratio $2^{1/12}$ to one another. It is customary to start the game with a "fundamental frequency", f_0. For example, note A on the piano is set at $f_0 := 440$Hertz, but other settings would work. Then frequencies for all notes obey the formula $f_n = 440 \cdot 2^{n/12}$. Notice that at $n = 12$ another octave starts corresponding to a note of frequency double the fundamental frequency. We will implement this type of tuning along with notes and chords with Maple in the next section. We will not look at the technique of tuning an instrument. For this one needs to identify the physical place on a component of the instrument where to obtain musical intervals. For example on a string instrument, physical characteristics are needed for precise placement of frets or fingers. This is done according to the principle that frequency is proportional to the inverse of string length L:

$$f = \frac{1}{2L}\sqrt{\frac{T}{d}} \quad \text{where } T = \text{ tension}, \quad d = \text{ density}$$

Next we describe two more types of tuning often used in the development of western music. The first one, Pythagorean tuning (≈500 BC) displays "nice" fourth and fifth intervals: $\frac{4}{3}$ and $\frac{3}{2}$, respectively. The second one, just tuning (Middle Age) improves with a "nice" major third ratio $\frac{5}{4}$. We display all interval ratios for both and show part of the (guess) work. These values pop up because of simple algebraic manipulations, which can prove illuminating to any high school student. Before doing the work we mention the somehow notorious place of the tritone (e.g. from C to $F_{\#}$) whose ratio is not that "nice" given the large numerator and denominator (in comparison to the other ratios in the table): $\frac{729}{512}$ (Pythagorean) and $\frac{45}{32}$ (just).

In **Pythagorean tuning** the frequency ratios are of type $2^p 3^q$ where $p, q \in \mathbb{Z}$.

n=Note	Interval name $[0, n]$	Ratio f_n/f_0
0=C	unison or perfect eighth	1
1=C$_\#$	minor second	$\frac{256}{243}$
2=D	major second	$\frac{9}{8}$
3=D$_\#$	minor third	$\frac{32}{27}$
4=E	major third	$\frac{81}{64}$
5=F	fourth	$\frac{4}{3}$
6=F$_\#$	tritone	$\frac{729}{512}$
7=G	fifth	$\frac{3}{2}$
8= G$_\#$	minor sixth	$\frac{128}{81}$
9=A	major sixth	$\frac{27}{16}$
10= A$_\#$	minor seventh	$\frac{16}{9}$
11= B	major seventh	$\frac{243}{128}$

To explain the values in the table we start with the fourth: from $C = 0$ to $F = 5$ make the correspondence $r^5 \mapsto \frac{4}{3}$.

$r^{2 \cdot 5} = r^{10} \mapsto \left(\frac{4}{3}\right)^2$, so $10 = \mathrm{A}_\#$ has frequency ratio $\frac{16}{9}$.
$r^{3 \cdot 5} = r^{15} \mapsto \left(\frac{4}{3}\right)^3$, an octave up, so 15 mod 12 $= \mathrm{D}_\#$ has frequency ratio $\frac{1}{2} \cdot \left(\frac{4}{3}\right)^3 = \frac{32}{27}$.
$r^{4 \cdot 5} = r^{20} \mapsto \left(\frac{16}{9}\right)^2$, so 20 mod 12 $= \mathrm{G}_\#$ has frequency ratio $\frac{1}{2} \cdot \left(\frac{16}{9}\right)^2 = \frac{128}{81}$.

Here we stop this correspondence as one more iterate would yield ratios $\frac{p}{q}$ with too large p or q in comparison to what is achieved in the next step: take the fifth (interval): from $C = 0$ to $G = 7$ make the correspondence $s^7 \mapsto \frac{3}{2}$.

Proceed as above with consecutive powers of s^7: $s^{14} \mapsto \frac{9}{4}$, one octave up: 14 mod 12 $=$ D, ratio $\frac{1}{2} \cdot \frac{9}{4} = \frac{9}{8}$, and continue in this fashion until the tables fills out.

In **just tuning** the frequency ratios are of type $2^p 3^q 5^l$ where $p, q, l \in \mathbb{Z}$.

n=Note	Interval name $[0, n]$	Ratio f_n/f_0
0=C	unison or perfect eighth	1
1=C$_\#$	minor second	$\frac{16}{15}$
2=D	major second	$\frac{9}{8}$
3=D$_\#$	minor third	$\frac{6}{5}$
4=E	major third	$\frac{5}{4}$
5=F	fourth	$\frac{4}{3}$
6=F$_\#$	tritone	$\frac{45}{32}$
7=G	fifth	$\frac{3}{2}$
8= G$_\#$	minor sixth	$\frac{8}{5}$
9=A	major sixth	$\frac{5}{3}$
10= A$_\#$	minor seventh	$\frac{16}{9}$
11= B	major seventh	$\frac{15}{8}$

In this case the guesswork is murkier than Pythagorean: there are multiple choices to attempt the nicest possible ratios. Throughout its evolution many authors came up with different values, see e.g. [2] for a variety of tunings. We will proceed similarly to Pythagorean, using correspondences. First, keep $r^5 \mapsto \frac{4}{3}$ and $s^7 \mapsto \frac{3}{2}$, and the major second's ratio $\frac{9}{8}$. Then add a new correspondence $t^4 \mapsto \frac{5}{4}$. In combination with r^5 and s^7 we obtain the major sixth: $t^4 r^5 \mapsto \frac{5}{3}$. Also the minor second: $t^{-4} r^5 \mapsto \frac{16}{15}$ (notice another option could be $t^8 r^5 \mapsto \frac{25}{24}$).

For the tritone : $t^4 s^{14} \mapsto \frac{1}{2} \cdot \frac{45}{16} = \frac{45}{32}$. However the tritone can be given a nicer treatment as follows: $t^8 r^{10} \mapsto \frac{1}{2} \cdot \frac{25 \cdot 16}{16 \cdot 9} = \frac{25}{18}$. The idea is to continue these correspondences until all values are filled, and produce ratios with the smallest possible numerators and denominators.

5 Creating Notes in Maple

We return to temperate tuning and shift the frequency formula f_n by 9. Thus the note C corresponds to $n = 0$.

$$w(n) = 440 \cdot 2^{(n-9)/12}$$

The Maple code below creates a sound wave at frequency f, and amplitude $am \in [-1, 1]$ such that the sine curve sampled at 44100 values per second in $[0, t]$.

```
with(AudioTools):
with(SignalProcessing):
Tone := proc(f,t,am)
    local x, final;
 final:= Create( (x) -> evalhf(am*sin(x/44100*4*Pi*f)), duration=t);
 return final:
end proc:
```

Remark 5.1. *The code above represents the bare minimum needed to create a "pure" tone. We will modify (improve) the code in two ways: by multiplying with an envelope function, and by frequency modulation, that is the sampled sine curve of the form* $\sin(ft+\sin(kft))$, *where k is a constant. This idea was used by John Chowning in 1967 at Stanford and implemented in the first synthesizers. We note in passing that the command Modulate available in Maple is completely different. Its description in the Help section is as follows: "consists of multiplying each sample in the audArray by the corresponding sample in the maskArray, and writing the result to the output. Notice that this operation is commutative; the data and mask can be interchanged and will still give the same result." This Modulate tool for sound is similar with Mask for images.*

To create notes all we have to do now is to compose the Tone procedure with the shifted $w(n)$ defined previously:

```
with(AudioTools):
with(SignalProcessing):
```

```
note:= proc(n::integer, t, am)
local q;
q:= Tone(w(n), t, am):
return q;
end proc:
```

Next we put together the C-major scale notes from C to a C an octave higher and back. For this we used the straightforward "Extend" Maple command which concatenates arrays. The audio result may sound a bit harsh at the place where the notes are stitched; however when we "smooth out" the notes using the above techniques (enveloping and modulating) the result is more pleasant. For comparison we plot the spectrograms of three variations of the C-major scale: with respect to pure and enveloped, modulated tones in temperate tuning (first two spectrograms), then enveloped, modulated tones in Pythagorean tuning (third spectrogram) (Figs. 7, 8 and 9).

```
with(AudioTools):
with(ArrayTools):
with(SignalProcessing):
CMscale := Extend(note(0,.5,1), note(2,.5,1), note(4,.5,1), note(5,.5,1),
 note(7,.5, 1), note(9,.5,1), note(11,.5,1), note(12,1,1),
 note(11,.5,1), note(9,.5,1), note(7,.5,1), note(5,.5,1),
 note(4,.5,1), note(2,.5,1), note(0,.75,1),
inplace = false)
```

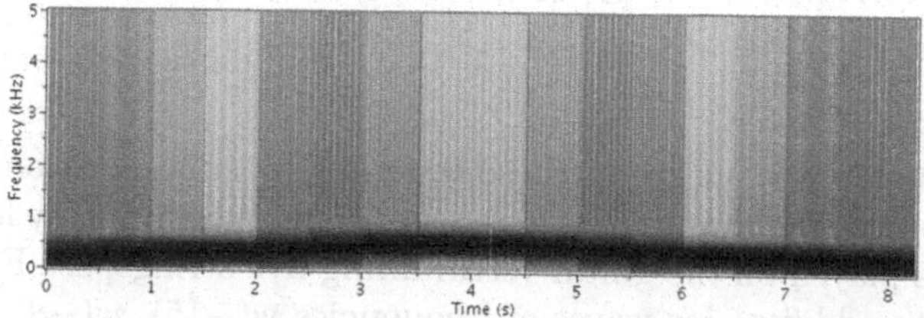

Fig. 7. C major scale with pure tones, temperate

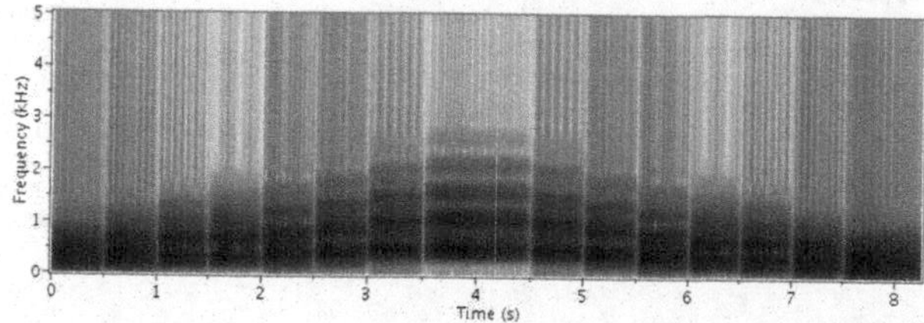

Fig. 8. C major scale enveloped and modulated, temperate

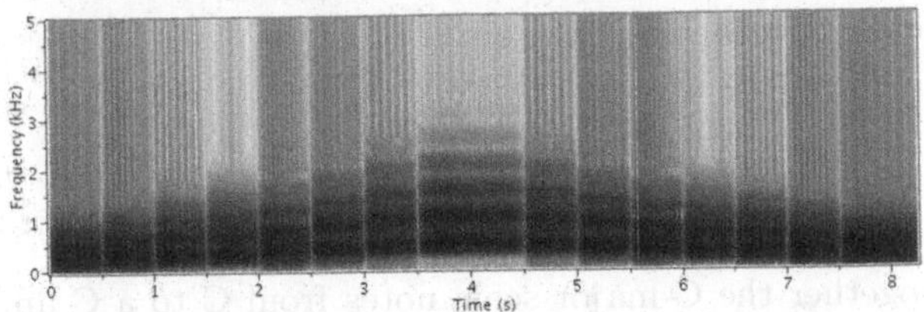

Fig. 9. C major scale enveloped and modulated, Pythagorean

Maple Project 5.2. *Experiment with overtones: to note(k) add sine waves of frequency integer multiples of $w(k)$. Add/avoid some, and investigate suitable ADSR envelopes in order to obtain instrument-like sounds.*

Remark 5.3. *Let us note that it is cumbersome to fully implement Pythagorean or just tunings. Passing to another note or octave is not recurrent as in temperate tuning. Below we show the Maple code for the Pythagorean C major scale, one octave in length.*

```
with(AudioTools):
with(ArrayTools):
with(SignalProcessing):
CMscalePyth := Extend(Tone((1/27)*(440*16),.5,1), Tone(9*(440*16)/(27*8),.5,1),
Tone(81*(440*16)/(27*64),.5,1), Tone((1/27)*(440*16)*(4/3),.5,1),
Tone((1/27)*(440*16)*(3/2),.5,1), Tone(440,.5,1), Tone((1/8)*(440*9),.5,.9),
Tone(16*(2*440)*(1/27),1,1), Tone((1/8)*(440*9),.5,.9), Tone(440,.5,1),
Tone((1/27)*(440*16)*(3/2),.5,1), Tone((1/27)*(440*16)*(4/3),.5,1),
Tone(81*(440*16)/(27*64),.5,1), Tone(9*(440*16)/(27*8),.5,1),
Tone((1/27)*(440*16),.75,1),
inplace = false)
```

Having built notes we can group these together by superposition (addition) in order to implement the musical concept "chord". For example, by looking at how the A-chord is played on the guitar (strumming five strings, see Fig. 10), we see that we need to add five sine waves of frequencies $w(-15)$, $w(-8)$, $w(-3)$, $w(1)$, and $w(4)$, respectively. After multiplying by an envelope function and composing with modulation we obtain a complex sound of chord A, suitable to combine with other chords.

```
with(AudioTools):
with(ArrayTools):
with(SignalProcessing):
chA :=(t->Create( x->(evalhf(1/(5)* env(x/(t*44100))*sin(x/44100*2*Pi*w(-15))+
1/(5)* env(x/(t*44100))*sin(x/44100*2*Pi*w(-8))+
1/(5)* env(x/(t*44100))*sin(x/44100*2*Pi*w(-3))+
1/(5)*env(x/(t*44100))* sin(x/44100*2*Pi*w(1))+
1/(5)* env(x/(t*44100))*sin(x/44100*2*Pi*w(4)+sin(x/44100*2*Pi*w(4))))), duration=t))
```

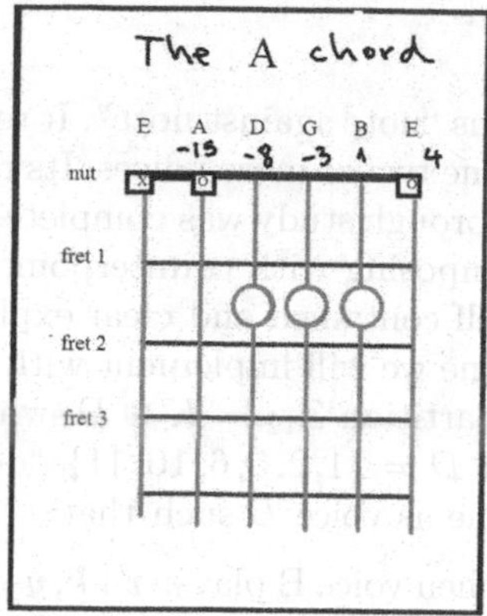

Fig. 10. Chord A on the guitar

Maple Project 5.4. *Write code for all chords. One may find lists of chords in any "....for beginners" book. For example we have used [4] which contains all thorough guitar chords descriptions. This project can be developed further with a deeper mathematical flavor, by studying and observing distances of the strings frequencies in a chord composition and their connections to distances in Cayley graphs of the group* $\mathbb{Z}_{12}$ *(see next section on counterpoint).*

A melody example that uses the chords A, D, and E is displayed below, together with its spectrogram. The chords are played for a duration t, e.g. $chA(0.5)$ means chord A is played half a second (Fig. 11).

```
with(AudioTools):
with(ArrayTools):
with(SignalProcessing):
Melody:= Extend(chA(1), chA(1), chD(1.5), chD(.5),
chA(1),chA(1), chE(2), chA(1), chA(1), chE(1.5),
chD(1), chA(1), chE(2));
Play(Melody);
```

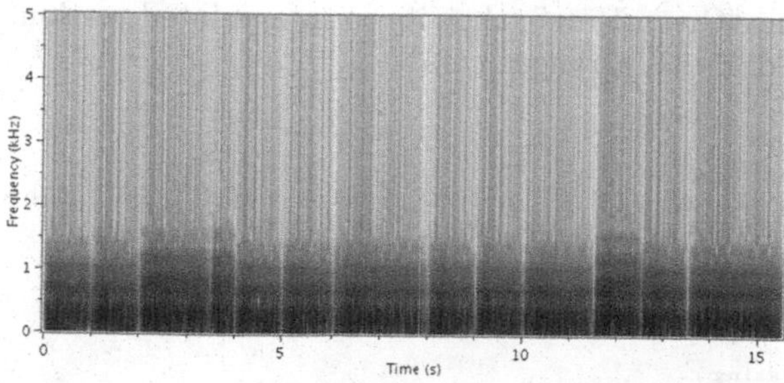

Fig. 11. Melody spectrogram

6 Counterpoint

Counterpoint in music means "note against note". It represents a sum of composing techniques which combine two or more voices. Its roots can be traced back to the 9th century. The first thorough study was completed by Johann J. Fux [6] who spelled out the rules of composing with counterpoint. There are five species of counterpoint, see [12] for self contained and clear explanations. The first species of the counterpoint is the one we will implement with Maple. We follow in parts [10]. Define the following partition $\mathbb{Z}_{12} = K \cup D$, where $K = \{0, 3, 4, 7, 8, 9\}$ is called "cantus firmus" and $D = \{1, 2, 5, 6, 10, 11\}$ "decantus". Assume voice A plays notes at the same time as voice B such that:

• If voice A plays: $x, y, z...$ then voice B plays: $x+k, y+l, z+p...$ where $k, l, p \in K$.
• There are restrictions: "parallel" fifths are forbidden, i.e. consecutive distances k, l in the sequence above can't be both equal to 7mod12. One can read this rule on Euler's coordinates, see Fig. 4. This restriction can be analyzed mathematically ([10]) from a group theory vantage point, which we will explain a bit further down. The full list of exceptions however may have to do more with the musical esthetic that Fux embraced. For example the distances 3, 4 (minor/major third) and 8, 9 (minor/major sixth) "are fine but no more than three in a row", see e.g. [12].

In [9] and [10] the peculiar choice of the partition K, D is explained beautifully through actions of symmetries T of the discrete torus $\mathbb{Z}_3 \times \mathbb{Z}_4$ such that $T^2 = id$, $T(D) = K$ and T preserves the (Cayley graph's) distance. These considerations stem from this Abstract Algebra exercise: Prove that there exists precisely five such transformations T (Hint: $T(x) = (ax + b)\text{mod } 12$, then ask $T^2 = id$). [10] claims each of these five symmetries correspond to a counterpoint species.

Let us note here that we can work with the original group $\mathbb{Z}_{12}$, with the Cayley graph generated by group elements 3 and 4. Then the set K from which the counterpoint distances are chosen (except restrictions) consists of zero length, unit length $3, 4, -3 = 9, -4 = 8$ paths, and the two-unit path $7 = 3 + 4$. This raises the question why is 5 (path of length 2 on the graph) out? This is an interesting subject because throughout the centuries many composers considered the fourth as consonant, i.e. 5 should belong to K.

The code below lists the first notes of the lower guitar strings needed to play the popular song "Greensleeve", then its counterpointed version. We took some liberty on note duration arrangements.

```
with(AudioTools):
with(ArrayTools):
with(SignalProcessing):
audio1:= Extend( note(-3,.5,1), note(0,.75,1), note(2,.5,1), note(4,.75,1),
note(5,.35,1), note(4,.35,1), note(2,1,1), note(-1,.5,1),
note(-5,1,1), note( -3,.5,1),..., inplace = false)
```

```
with(AudioTools):
with(ArrayTools):
with(SignalProcessing):
audio2:=Extend( note(-15,.5,1), note(-4, .75,1), note(-1, .5, 1), note(1,.75,1),
note(1,.35, 1), note(-4,.35,1), note(-1,1,1), note(-4,.5,1),
note(-5,1,1), note(-7,.5,1),... , inplace = false)
```

To obtain the final counterpointed melody in Maple, one can mix the two (or more!) audio files as weighted sum.

$$counterpoint = \alpha \cdot audio1 + \beta \cdot audio2, \quad \alpha + \beta = 1$$

Remark 6.1. *One can experiment with Mask command to combine the files, i.e. audio1 $\times$ audio 2. This is the command Modulate. It results in shifting one audio's frequencies by the other's, which corresponds to Fourier transform of a product, thus in shifting (convolution) of the frequencies.*

Maple Project 6.2. *Build a "composition" Maple application:*
Write $Notes(n, t, am)$ in a window editor, then compile to produce audio files. Add sound effects such as reverberate and distortion. Write code for "automated" compositions (e.g. Wolfram tones uses cellular automata). For advanced students, study and explain the theory behind "Vocoder" technology [5], i.e. Tempo/Pitch change without affecting the listening experience.

Acknowledgements. The author thanks his nephew Philip Eitner who was available to discuss and explain guitar playing, during schools lock-down in Spring 2020 due to the Covid-2019 pandemic. Many thanks go to Paul Lombardi from the Music Department at USD who generated a great deal of interest in the subject overlapping mathematics and music. The author is grateful to the referees whose valuable insights and corrections helped improve the paper's exposition.

References

1. Backus, J.: The Acoustical Foundations of Music, 2nd edn. Norton, New York (1977)
2. Berg, R.E., Stork, D.G.: The Physics of Sound, 2nd edn. Prentice Hall, Hoboken (1995)
3. Brown, J.W., Churchill, R.V.: Fourier Series and Boundary Value Problems, 7th edn. McGraw-Hill, New York (2008)
4. Fleming, T.: Guitar for Beginners. Metro Books (2014)
5. Flanagan, J.L., Golden, R.M.: Phase vocoder. Bell Labs Tech. J. **45**(9), 1493–1509 (1966)
6. Fux, J.: Gradus ad Parnassum (1725). Norton, The Study of Counterpoint. Translated and edited by Alfred Mann (1971)
7. Williams, G.: Linear Algebra with Applications, 9th edn. Jones & Bartlett Learning, Burlington (2019)
8. Maple software (2018)
9. Mazzola, G.: Geometrie der Tone. Birkhauser, Basel (1990)
10. Mazzola, G., et al.: Basic Music Technology. An Introduction. Computational Music Science, Springer, Cham (2018). https://doi.org/10.1007/978-3-030-00982-3
11. Rigden, J.: Physics and the Sound of Music, 2nd edn. Wiley, New York (1985)
12. Rush, T.W.: Music Theory for Musicians and Normal People. https://tobyrush.com

13. Sacks, O.: Musicophilia. Tales of Music and the Brain. Vintage Books, New York (2008)
14. Tymoczko, D.: The geometry of musical chords, supporting online material. Science **313**, 72 (2006)
15. White, H.E., White, D.H.: Physics and Music: The Science of Musical Sound, Dover Books on Physics (1980)
16. http://tones.wolfram.com

Numerical Solution for Radial Distortion Rectification in Optical Systems

Obed I. Rios-Orellana[1(✉)], Rigoberto Juarez-Salazar[2], and Victor H. Diaz-Ramirez[1]

[1] Instituto Politécnico Nacional-CITEDI, Av. Instituto Politécnico Nacional 1310, 22435 Nueva Tijuana, Tijuana, B.C., Mexico
orios@citedi.mx, vdiazr@ipn.mx

[2] CONACYT–Instituto Politécnico Nacional-CITEDI, Av. Instituto Politécnico Nacional 1310, 22435 Nueva Tijuana, Tijuana, B.C., Mexico
rjuarezsa@conacyt.mx

Abstract. Accurate homography estimation is a crucial step for many computer vision applications. Nevertheless, nonlinear optical camera imaging effects can introduce radial distortion, making unfeasible the pinhole model for homography estimation. In this paper, an algorithm to rectify radially distorted images using the Maple software is proposed. First, the effect of radial distortion is modeled and analyzed. Next, an inverse distortion model is developed. The proposed algorithm allows us to estimate both the homography matrix and the distortion parameters by processing images of a calibration target using the Gauss-Newton approach. Successful estimation of homographies and distortion parameters to correct real-world images is reported.

Keywords: Radial lens distortion · Homography estimation · Distortion parameters estimation

1 Introduction

Imperfection in optical lens construction introduces distortions such as chromatic aberration, astigmatism, and geometrical distortions such as radial and tangential distortion [1,4,5]. Moreover, computer vision applications usually assume the theoretical linear pinhole camera model. Nevertheless, when non-linear distortions (such as radial distortion) are present, the pinhole camera model can not be directly applied. Particularly, radial distortion is a relevant issue in computer vision [11,13] because of its impact in tasks such as 3D imaging techniques and pattern recognition [8,10,16].

Warped images must be rectified through a model that maps between rectified and distorted images. Radial distortion can be modeled as a process that

This work was supported by the *Consejo Nacional de Ciencia y Tecnología* (CONACYT) by the projects *Cátedras*-880 and A1-S-28112. Authors thank the support of *Instituto Politécnico Nacional* by the project SIP-20210845.

R. M. Corless et al. (Eds.): MC 2020, CCIS 1414, pp. 319–333, 2021.
https://doi.org/10.1007/978-3-030-81698-8_21

involves a geometric mapping $T : \mathbb{R}^2 \mapsto \mathbb{R}^2$. That is, radial distortion deviates ideal image coordinates into a new modified coordinates image pixels as shown Fig. 1. A distorted image can be restored by applying the inverse radial warp process. Within state-of-the-art, radial distortion is usually modeled with polynomials [2,6,7,13,15]. This approach does not require to consider complex functions (such as logistical or trigonometric).

In practice, the polynomials do not provide a general method to invert the radial distortion effect [1,10,14]. For even and odd polynomial warp formulations, their inverse can be found by approximation [2,3,6]. Additionally, there are other existing methods that model radial distortion using a single parameter (known as *division model*) which can directly be inverted [3,7,9]. In this work, we propose an inverse process and inverse warp function to counteract the radial distortion effects of warped images. Additionally we propose an algorithm based on the Gauss-Newton approach to find a numerical solution for radial distortion parameters and a proper homography.

This work is organized as follows. In Sect. 2 we consider the direct warp transformation process to model the radial distortion effect with its radial distortion parameters. Also, we consider the transformation of the inverse warp process to model the inverse distortion parameters and homography to counteract radial distortion effects. In Sect. 3 we formulate an algorithm to find a numerical solution for radial distortion parameters and a suitable homography. In Sect. 4 we present experimental results using a Maple implementation of the proposed algorithm. In Sect. 5 a discussion about the findings is given. Section 6 presents the conclusions. Furthermore, Appendix A provides a brief introduction regarding Homogeneous coordinates with Maple implementations. Finally, the appendices B and C provide the rest of the Maple implementations for the proposed algorithm for radial distortion correction of images.

2 Direct and Inverse Warp Process Transformation

Consider direct warp process transformation as shown in Fig. 1, where $\boldsymbol{\mu} = \mathcal{H}^{-1}[G\mathcal{H}[\boldsymbol{\rho}]]$ is an ideal image point, $\mathcal{H}[\cdot]$ is the homogeneous transformation operator[1], G is an homography matrix, and $\boldsymbol{\rho}$ are the ideal or undistorted reference points. Then the deviated image points $\tilde{\boldsymbol{\mu}}$, obtained by applying a warp function, are given by

$$\tilde{\boldsymbol{\mu}} = r_d(\boldsymbol{\mu}, \boldsymbol{\lambda})\boldsymbol{\mu} = \mathrm{warp}(\boldsymbol{\mu}, \boldsymbol{\lambda}), \tag{1}$$

where r_d is a warp radial distortion function and $\boldsymbol{\lambda}$ is a distortion parameter vector. Now consider $\boldsymbol{\mu} = r_d(\tilde{\boldsymbol{\mu}}, \boldsymbol{\lambda}^{-1})\tilde{\boldsymbol{\mu}} = \mathrm{warp}(\tilde{\boldsymbol{\mu}}, \boldsymbol{\lambda}^{-1})$. We propose the inverse warp functions to counteract warped effects as follows:

[1] More detailed information about homogeneous coordinates can be found in Appendix A.

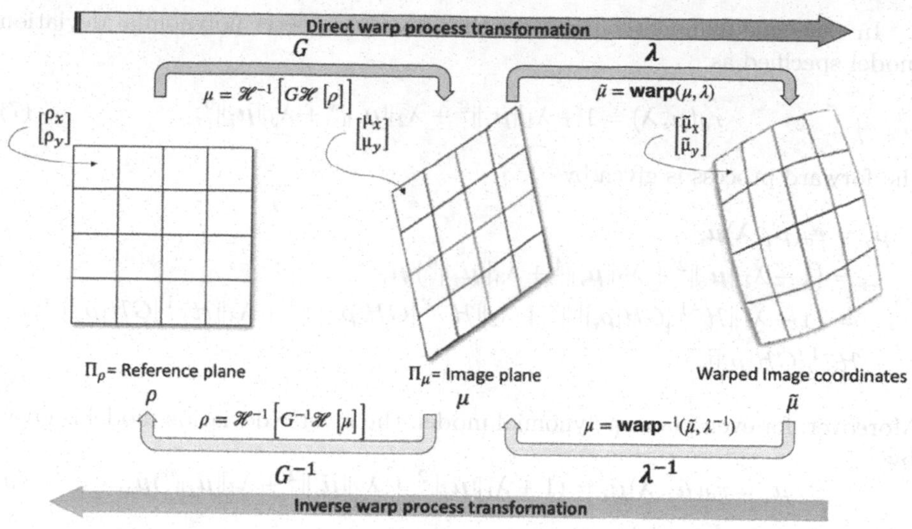

Fig. 1. Direct and inverse warp coordinates process in radially distorted images.

$$\begin{aligned}\boldsymbol{\rho} &= \mathcal{H}^{-1}\left[G^{-1}\mathcal{H}[\boldsymbol{\mu}]\right] \\ &= \mathcal{H}^{-1}\left[G^{-1}\mathcal{H}[\text{warp}(\tilde{\boldsymbol{\mu}}, \boldsymbol{\lambda}^{-1})]\right] \\ &= \mathcal{H}^{-1}\left[G^{-1}\mathcal{H}[r_d(\tilde{\boldsymbol{\mu}}, \boldsymbol{\lambda}^{-1})\tilde{\boldsymbol{\mu}}]\right].\end{aligned} \tag{2}$$

We suggest, along with the proposed inverse process, the use of an inverse-based Fitzgibbon's division model and an even polynomial model with three distortion parameters specified as follows. Consider the Fitzgibbon's division based model deviation function given by

$$r_d(\boldsymbol{\mu}_i, \lambda) = 1 + \lambda\|\boldsymbol{\mu}_i\|^2. \tag{3}$$

The forward process can be written as

$$\begin{aligned}\tilde{\boldsymbol{\mu}}_i &= r_d(\boldsymbol{\mu}_i, \lambda)\boldsymbol{\mu}_i \\ &= \left(1 + \lambda\|\boldsymbol{\mu}_i\|^2\right)\boldsymbol{\mu}_i \\ &= \left(1 + \lambda\|\mathcal{H}^{-1}\left[G\mathcal{H}[\boldsymbol{\rho}_i]\right]\|^2\right)\mathcal{H}^{-1}\left[G\mathcal{H}[\boldsymbol{\rho}_i]\right].\end{aligned} \tag{4}$$

Now, for the inverse model, consider the inverse radial distortion deviation model given by

$$\boldsymbol{\mu}_i = r_d(\tilde{\boldsymbol{\mu}}_i, \lambda)\tilde{\boldsymbol{\mu}}_i = (1 + \lambda\|\tilde{\boldsymbol{\mu}}_i\|^2)\tilde{\boldsymbol{\mu}}_i. \tag{5}$$

Next the inverse process given by

$$\begin{aligned}\boldsymbol{\rho}_i &= \mathcal{H}^{-1}\left[G^{-1}\mathcal{H}[\boldsymbol{\mu}_i]\right] \\ &= \mathcal{H}^{-1}\left[G^{-1}\mathcal{H}[(1 + \lambda\|\tilde{\boldsymbol{\mu}}_i\|^2)\tilde{\boldsymbol{\mu}}_i]\right].\end{aligned} \tag{6}$$

In the same manner, consider for the three parameter polynomial deviation model specified as

$$r_d(\boldsymbol{\mu}, \boldsymbol{\lambda}) = 1 + \lambda_1 \|\boldsymbol{\mu}_i\|^2 + \lambda_2 \|\boldsymbol{\mu}_i\|^4 + \lambda_3 \|\boldsymbol{\mu}_i\|^6, \tag{7}$$

the forward process is given by

$$\begin{aligned}
\tilde{\boldsymbol{\mu}}_i &= r_d(\boldsymbol{\mu}_i, \boldsymbol{\lambda})\boldsymbol{\mu}_i \\
&= \left(1 + \lambda_1 \|\boldsymbol{\mu}_i\|^2 + \lambda_2 \|\boldsymbol{\mu}_i\|^4 + \lambda_3 \|\boldsymbol{\mu}_i\|^6\right) \boldsymbol{\mu}_i \\
&= \left(1 + \lambda_1 \|\mathcal{H}^{-1}\left[G\mathcal{H}[\boldsymbol{\rho}_i]\right]\|^2 + \lambda_2 \|\mathcal{H}^{-1}\left[G\mathcal{H}[\boldsymbol{\rho}_i]\right]\|^4 + \lambda_3 \|\mathcal{H}^{-1}\left[G\mathcal{H}[\boldsymbol{\rho}_i]\right]\|^6\right) \\
&\quad \mathcal{H}^{-1}\left[G\mathcal{H}[\boldsymbol{\rho}_i]\right].
\end{aligned} \tag{8}$$

Moreover, for even power polynomial model, the inverse deviation model is given by

$$\boldsymbol{\mu}_i = r_d(\tilde{\boldsymbol{\mu}}_i, \boldsymbol{\lambda})\tilde{\boldsymbol{\mu}}_i = (1 + \lambda_1 \|\tilde{\boldsymbol{\mu}}_i\|^2 + \lambda_2 \|\tilde{\boldsymbol{\mu}}_i\|^2 + \lambda_4 \|\tilde{\boldsymbol{\mu}}_i\|^6)\tilde{\boldsymbol{\mu}}_i, \tag{9}$$

and the inverse process is

$$\begin{aligned}
\boldsymbol{\rho}_i &= \mathcal{H}^{-1}\left[G^{-1}\mathcal{H}[\boldsymbol{\mu}_i]\right] \\
&= \mathcal{H}^{-1}\left[G^{-1}\mathcal{H}[(1 + \lambda_1 \|\tilde{\boldsymbol{\mu}}_i\|^2 + \lambda_2 \|\tilde{\boldsymbol{\mu}}_i\|^4 + \lambda_3 \|\tilde{\boldsymbol{\mu}}_i\|^6)\tilde{\boldsymbol{\mu}}_i]\right].
\end{aligned} \tag{10}$$

Note that, with the proposed the inverse processes, we can solve numerically the distortion parameters and the homography using the Gauss-Newton approach to counteract the forward process warp effects.

3 Numerical Solution Algorithm for Radial Distortion Parameters with Gauss-Newton Approach

In this section, we integrate the proposed inverse radial distortion warp functions into the Gauss-Newton approach. This section shows the general formulation steps to estimate of the radial distortion parameters. For simplicity of notation we consider the inverse homography G^{-1} as

$$K = G^{-1} = \begin{bmatrix} k_1 & k_2 & k_3 \\ k_4 & k_5 & k_6 \\ k_7 & k_8 & k_9 \end{bmatrix} = \begin{bmatrix} \bar{\boldsymbol{k}}_1^T \\ \bar{\boldsymbol{k}}_2^T \\ \bar{\boldsymbol{k}}_3^T \end{bmatrix}, \tag{11}$$

with $k_9 = 1$. First, let a general deviation model be given by

$$\boldsymbol{\mu}_i = r_d(\tilde{\boldsymbol{\mu}}, \boldsymbol{\lambda})\tilde{\boldsymbol{\mu}}, \tag{12}$$

where Eq. (12) can be used with the derived deviations models from Eq. (5) and Eq. (9). Therefore considering the inverse process from Eq. (2) we can obtain

Algorithm 1: Distortion parameter estimation algorithm.

```
Input:
μ̃ /* Distorted image points                                  */
ρ /* Exact reference points                                   */
K0 /* Initial guess homography matrix                         */
λa /* Initial guess distortion parameter λa                   */
λb /* Initial guess distortion parameter λb                   */
λc /* Initial guess distortion parameter λc                   */
N /* Iteration max number                                     */
ε /* Stop condition tolerance                                 */
Output:
KΘ /* Estimated inverse homography                            */
λ1 /* Estimated distortion parameter λ1                       */
λ2 /* Estimated distortion parameter λ2                       */
λ3 /* Estimated distortion parameter λ3                       */
begin
    Θ0 ← Vec3,3(K0);
    Θ0 ← [Θ0,k with k = 1...8, λa, λb, λc];
    y ← [ρx,i ρy,i]^T;
    for i ← 1 to N do
        ŷ ← f(Θ0[1..8], Θ[9], Θ[10], Θ[11], μ̃);
        JΘ ← J(Θ0[1..8], Θ[9], Θ[10], Θ[11], μ̃);
        Θ = Θ0 + (JΘ^T JΘ)^-1 JΘ^T (y − ŷ);
        /* Verify stop condition                              */
        if ||Θ − Θ0|| < ε then
            KΘ = Vec3,3^-1([Θ 1]^T);
            λ1 = Θ[9];
            λ2 = Θ[10];
            λ3 = Θ[11];
            return KΘ,λ1,λ2,λ3;
        else
            Θ0 = Θ;
        end
    end
    return Error: Can't find solution for N iterations.
end
```

the following expression

$$
\begin{aligned}
\boldsymbol{\rho}_i &= \mathcal{H}^{-1}[K\mathcal{H}[\boldsymbol{\mu}_i]] \\
&= \mathcal{H}^{-1}[K\mathcal{H}[r_d(\tilde{\boldsymbol{\mu}}_i, \boldsymbol{\lambda})\tilde{\boldsymbol{\mu}}_i]] \\
&= \mathcal{H}^{-1}\left[\begin{bmatrix} \bar{\boldsymbol{k}}_1^T \\ \bar{\boldsymbol{k}}_2^T \\ \bar{\boldsymbol{k}}_3^T \end{bmatrix} \mathcal{H}[r_d(\tilde{\boldsymbol{\mu}}_i, \boldsymbol{\lambda})\tilde{\boldsymbol{\mu}}_i]\right] \\
&= \frac{1}{\bar{\boldsymbol{k}}_3^T \mathcal{H}[r_d(\tilde{\boldsymbol{\mu}}_i, \boldsymbol{\lambda})\tilde{\boldsymbol{\mu}}_i]} \begin{bmatrix} \bar{\boldsymbol{k}}_1^T \mathcal{H}[r_d(\tilde{\boldsymbol{\mu}}_i, \boldsymbol{\lambda})\tilde{\boldsymbol{\mu}}_i] \\ \bar{\boldsymbol{k}}_2^T \mathcal{H}[r_d(\tilde{\boldsymbol{\mu}}_i, \boldsymbol{\lambda})\tilde{\boldsymbol{\mu}}_i] \end{bmatrix}
\end{aligned}, \tag{13}
$$

or, equivalently

$$
\begin{bmatrix} \rho_{x,i} \\ \rho_{y,i} \end{bmatrix} = \begin{bmatrix} \frac{[k_1,k_2]\boldsymbol{\mu}_i + k_3}{[k_7,k_8]\boldsymbol{\mu}_i + 1} \\ \frac{[k_4,k_5]\boldsymbol{\mu}_i + k_6}{[k_7,k_8]\boldsymbol{\mu}_i + 1} \end{bmatrix} = \begin{bmatrix} \frac{\tilde{\mu}_{x,i} r_d(\tilde{\boldsymbol{\mu}}_i^T, \boldsymbol{\lambda}) k_1 + \tilde{\mu}_{y,i} r_d(\tilde{\boldsymbol{\mu}}_i^T, \boldsymbol{\lambda}) k_2 + k_3}{\tilde{\mu}_{x,i} r_d(\tilde{\boldsymbol{\mu}}_i^T, \boldsymbol{\lambda}) k_6 + \tilde{\mu}_{y,i} r_d(\tilde{\boldsymbol{\mu}}_i^T, \boldsymbol{\lambda}) k_7 + 1} \\ \frac{\tilde{\mu}_{x,i} r_d(\tilde{\boldsymbol{\mu}}_i^T, \boldsymbol{\lambda}) k_4 + \tilde{\mu}_{y,i} r_d(\tilde{\boldsymbol{\mu}}_i^T, \boldsymbol{\lambda}) k_5 + k_6}{\tilde{\mu}_{x,i} r_d(\tilde{\boldsymbol{\mu}}_i^T, \boldsymbol{\lambda}) k_6 + \tilde{\mu}_{y,i} r_d(\tilde{\boldsymbol{\mu}}_i^T, \boldsymbol{\lambda}) k_7 + 1} \end{bmatrix}. \tag{14}
$$

From Eq. (14) we specify the general objective function as

$$f(K,\boldsymbol{\lambda},\boldsymbol{\mu}) = \begin{bmatrix} \frac{[k_1,k_2]\boldsymbol{\mu}_1+k_3}{[k_7,k_8]\boldsymbol{\mu}_1+1} \\ \vdots \\ \frac{[k_1,k_2]\boldsymbol{\mu}_i+k_3}{[k_7,k_8]\boldsymbol{\mu}_i+1} \\ \frac{[k_4,k_5]\boldsymbol{\mu}_1+k_6}{[k_7,k_8]\boldsymbol{\mu}_1+1} \\ \vdots \\ \frac{[k_4,k_5]\boldsymbol{\mu}_i+k_6}{[k_7,k_8]\boldsymbol{\mu}_i+1} \end{bmatrix}, \tag{15}$$

where $\boldsymbol{\mu}_i = r_d(\tilde{\boldsymbol{\mu}},\boldsymbol{\lambda})\tilde{\boldsymbol{\mu}}$. From Eq. (14) the Jacobian for Gauss-Newton approach is given by

$$J(K,\boldsymbol{\lambda},\boldsymbol{\mu}) = \begin{bmatrix} \frac{\partial f(K,\boldsymbol{\lambda},\boldsymbol{\mu}_1)}{\partial k_1} & \frac{\partial f(K,\boldsymbol{\lambda},\boldsymbol{\mu}_1)}{\partial k_2} & \cdots & \frac{\partial f(K,\boldsymbol{\lambda},\boldsymbol{\mu}_1)}{\partial k_8} & \frac{\partial f(K,\boldsymbol{\lambda},\boldsymbol{\mu}_1)}{\partial \lambda_1} & \frac{\partial f(K,\boldsymbol{\lambda},\boldsymbol{\mu}_1)}{\partial \lambda_2} & \cdots & \frac{\partial f(K,\boldsymbol{\lambda},\boldsymbol{\mu}_1)}{\partial \lambda_n} \\ \vdots & \vdots & & \vdots & \vdots & \vdots & & \vdots \\ \frac{\partial f(K,\boldsymbol{\lambda},\boldsymbol{\mu}_i)}{\partial k_1} & \frac{\partial f(K,\boldsymbol{\lambda},\boldsymbol{\mu}_i)}{\partial k_2} & \cdots & \frac{\partial f(K,\boldsymbol{\lambda},\boldsymbol{\mu}_i)}{\partial k_8} & \frac{\partial f(K,\boldsymbol{\lambda},\boldsymbol{\mu}_i)}{\partial \lambda_1} & \frac{\partial f(K,\boldsymbol{\lambda},\boldsymbol{\mu}_i)}{\partial \lambda_2} & \cdots & \frac{\partial f(K,\boldsymbol{\lambda},\boldsymbol{\mu}_i)}{\partial \lambda_n} \end{bmatrix}. \tag{16}$$

Finally, the output variables $\boldsymbol{y}$ and $\hat{\boldsymbol{y}}$ are defined as

$$\begin{aligned} \boldsymbol{y} &= \left[\rho_{x,1},\cdots,\rho_{x,i},\rho_{y,1},\cdots,\rho_{y,i}\right]^T, \\ \hat{\boldsymbol{y}} &= f(K,\boldsymbol{\lambda},\boldsymbol{\mu}))\,|_{K=K_0,\lambda=\lambda_0,\mu=\tilde{\mu}_0}, \end{aligned} \tag{17}$$

where $K_0, \boldsymbol{\lambda}_0, \tilde{\boldsymbol{\mu}}_0$ are the initial guess for the Gauss-Newton approach. The Algorithm 1, summarizes the steps presented in this section. In addition, the Maple implementation can be found in Appendix C.

Table 1. Initial homographies G_0 guess for each input image.

Image No.	g_1	g_2	g_3	g_4	g_5	g_6	g_7	g_8
1	0.4728	0.0335	−0.0060	−0.0161	0.3903	−0.0229	−0.0192	−0.0805
2	0.4752	0.0390	0.0325	−0.0115	0.3964	−0.0238	0.0664	−0.0802
3	0.5286	0.0453	0.0330	−0.0149	0.4407	−0.0259	0.0866	−0.1050
4	0.5261	0.0368	−0.0212	−0.0219	0.4366	−0.0257	0.0599	−0.1035
5	0.5108	0.0501	0.0748	−0.0096	0.4439	−0.0245	0.0643	−0.1087
6	0.4863	0.0380	0.0015	−0.0160	0.4025	−0.0223	−0.0700	−0.0918
7	0.4935	0.0366	0.0104	−0.0156	0.4085	−0.0228	0.0928	−0.0888
8	0.4462	0.0358	0.0264	−0.0105	0.3722	−0.0217	−0.0018	−0.0687
9	0.4678	0.0285	−0.0394	−0.0221	0.3930	−0.0225	0.0620	−0.0782
10	0.4334	0.0382	0.0485	−0.0063	0.3715	−0.0215	−0.0774	−0.0727

4 Results

We evaluated the proposed inverse process for radial distortion image rectification using the algorithm presented in Algorithm 1 for one and three distortion parameters (deviations models from Eq. (5) and Eq. (9) respectively). As input for the algorithm for one and three distortion parameters, we used ten calibration checkboard images with a slightly radial distortion, which were captured using a Fuji DSLR Finepix s5800 camera, as shown in Fig. 2(a)–(e). A checkerboard calibration target with squares size of 2.2 cm is employed. This target provides 54 corner points $\boldsymbol{\rho}_0$ on the checkerboard. Then, from the captured images, the Harris corner detector is used to extract the 54 corner points $\tilde{\boldsymbol{\mu}}_0$ on the image for each checkerboard image. The image corner points $\tilde{\boldsymbol{\mu}}_0$ are taken as the initial distortion image points for the algorithm.

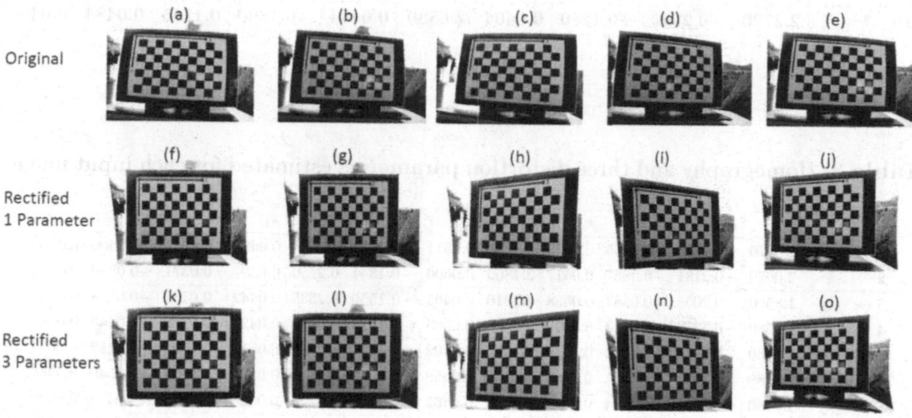

Fig. 2. Radial distortion rectification with one and three parameters using the proposed method. (a)–(e) The first five of ten original radially distorted images. Rectified images applying radial distortion correction using (f)–(j) one distortion parameter, and (k)–(o) three distortion parameters.

The initial guess parameters for one and three distortion parameters where set as $\lambda = 0.5$, and $\boldsymbol{\lambda} = [0.35, -0.25, 0.15]^T$ respectively, a tolerance $\varepsilon = 1 \times 10^{-6}$, and a maximum number of iterations $N = 25$. Also, the initial guess homographies used for the experiment are reported in Table 1. The Algorithm 1 was applied to estimate the homography and distortion parameters for each input image. Two cases are considered, first, when a single distortion parameter is estimated, and, second, when three distortion parameters are estimated. For the first case, the homography, distortion parameter, and reprojection error[2] ρ_ε

[2] *Reprojection error* is an error measure defined as the Euclidean distance between the estimated reference points (obtained using the estimated homography and distortion parameters) and the exact coordinates of the reference points.

are presented in Table 2. Similarly, the Table 3 presents the estimation results for the second case.

Table 2. Homography and one distortion parameter estimated for each input image.

Image No.	k_1	k_2	k_3	k_4	k_5	k_6	k_7	k_8	λ	ρ_ε
1	2.0750	−0.1772	0.0065	0.0874	2.5260	0.0511	0.0471	0.2014	0.0499	0.0464
2	2.0650	−0.2179	−0.0666	0.0517	2.4780	0.0499	−0.1341	0.2150	0.0499	0.0466
3	1.8480	−0.2054	−0.0587	0.0537	2.2200	0.0461	−0.1559	0.2535	0.0496	0.0541
4	1.8570	−0.1471	0.0442	0.0878	2.2580	0.0504	−0.1029	0.2449	0.0497	0.0534
5	1.9170	−0.2541	−0.1481	0.0341	2.1970	0.0408	−0.1210	0.2575	0.0488	0.0538
6	2.0160	−0.1943	−0.0160	0.0864	2.4500	0.0457	0.1500	0.2130	0.0495	0.0490
7	1.9860	−0.1822	−0.0144	0.0666	2.4090	0.0462	−0.1797	0.2328	0.0499	0.0485
8	2.2050	−0.2249	−0.0652	0.0620	2.6550	0.0495	0.0081	0.1832	0.0497	0.0431
9	2.0970	−0.1347	0.0898	0.1119	2.5230	0.0540	−0.1218	0.2073	0.0495	0.0452
10	2.2720	−0.2592	−0.1280	0.0464	2.6880	0.0484	0.1800	0.1765	0.0484	0.0430

Table 3. Homography and three distortion parameters estimated for each input image.

Image No.	k_1	k_2	k_3	k_4	k_5	k_6	k_7	k_8	λ_1	λ_2	λ_3	ρ_ε
1	2.0770	−0.1773	0.0065	0.0875	2.5280	0.0511	0.0471	0.2014	0.0402	0.0365	−0.0400	0.0459
2	2.0670	−0.2181	−0.0667	0.0517	2.4800	0.0500	−0.1341	0.2150	0.0397	0.0371	−0.0396	0.0461
3	1.8500	−0.2055	−0.0587	0.0538	2.2210	0.0461	−0.1559	0.2535	0.0442	0.0132	−0.0088	0.0537
4	1.8580	−0.1472	0.0443	0.0879	2.2590	0.0505	−0.1029	0.2449	0.0421	0.0200	−0.0148	0.0529
5	1.9160	−0.2541	−0.1482	0.0340	2.1970	0.0408	−0.1210	0.2575	0.0511	−0.0122	0.0157	0.0537
6	2.0180	−0.1945	−0.0161	0.0865	2.4530	0.0458	0.1500	0.2130	0.0361	0.0502	−0.0540	0.0486
7	1.9880	−0.1824	−0.0144	0.0667	2.4110	0.0462	−0.1797	0.2328	0.0363	0.0462	−0.0458	0.0480
8	2.2060	−0.2250	−0.0652	0.0620	2.6570	0.0495	0.0081	0.1832	0.0399	0.0424	−0.0528	0.0427
9	2.1000	−0.1348	0.0901	0.1120	2.5250	0.0541	−0.1218	0.2073	0.0357	0.0489	−0.0492	0.0445
10	2.2740	−0.2595	−0.1284	0.0464	2.6900	0.0484	0.1800	0.1765	0.0346	0.0590	−0.0698	0.0422

The estimated parameters were used for rectification of radial distortion of the input images. For this, the inverse transformation process by the warp functions given by Eqs. (5) and (9) are applied. Then, a bilinear interpolation was used to obtain the final representation of the rectified image. The Figs. 2(f-j) show the result of the described distortion correction process using the parameters in Table 2. Similarly, Figs. 2(k)–(o) show the corresponding corrected images using the parameters in Table 3.

5 Discussion

Radial distortion rectification problems may be summarized as finding the polynomial distortion coefficients of the model for the inverse process transformation. For multiple distortion parameters, there is no direct deviate reverse solution, unlike a single distortion parameter. To counteract the natural radial distortion effect, we proposed an inverse process to rectify images using the Gauss-Newton approach to find a numerical solution for the homography and distortion parameters. We formulated the inverse process transformation algorithm to adapt smoothly distinct warp functions. The results show a proper rectification of the input images using the proposed algorithm in two cases (estimating one and three distortion parameters) as shown in Fig. 2. It is worth mentioning that similar homographies and reprojection errors were obtained as shown in Tables 2 and 3.

The presented experiments suggests that models with higher degrees may not be necessary. Using only a second-order deviation model should be sufficient to rectify a radial distorted image using the proposed algorithm. Some caveats regarding the algorithm's use are the initial guess parameters needed because of the Gauss-Netwon approach's nature. The proposed initial settings were selected using a uniform-random range initial guess distortion parameters. Also, a final refinement may be needed because the algorithm does not contemplate noise environments.

Since the proposed algorithm is tightly related to pinhole image formation, it may be used as a pre-calibration stage delivering satisfactory results. Nevertheless, other images coming from another kind of image formation nature, the algorithm might not work as expected. For instance, Fig. 3 shows the image captured with a fisheye lens, and the rectified image using our proposed algorithm. Note that the image is not entirely rectified.

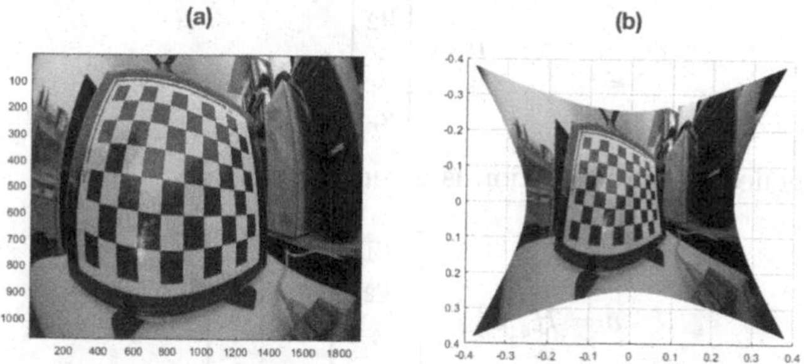

Fig. 3. (a) Image acquired with an Arducam OV5647 fisheye camera lens. (b) Image corrected for radial distortion using the proposed method. Note that there is a remaining radial distortion.

In summary, the proposed algorithm for radial distortion image rectification can deal with distinct warp image models. With the Gauss-Newton approach, the algorithm can find a numerical solution for radial distortion parameters and the homography. The algorithm can be applied in pre-calibration stages, and some caveats may need to be considered, such as noise environments and the image formation nature to rectify images successfully.

6 Conclusion

In this work, we analyzed the forward and inverse warp process, particularly for radial distorted images. We proposed an inverse transformation process for two deviation models with one and three distortion parameters, respectively. Then we formulated an algorithm to achieve a numerical solution for the distortion parameters and their homography. We provided Maple implementation for the proposed algorithm, and we experimentally tested it with real images. We showed that images from pinhole formation nature along with radial distortion are rectified successfully using the Gauss-Netwon approach to find a numerical solution for radial distortion parameters and a suitable homography. On the other hand, we tested the algorithm with a fisheye formation nature, and we found that the resulting image is not entirely rectified. As future work, this approach may be used as a baseline to contemplate distinct image formation natures such as fisheye, panoramic, and catadioptric cameras.

A Homogeneous coordinates

In this section we introduce the homogeneous coordinates and the direct linear transform (DLT) method for homography estimation. We use the homogeneous coordinates operator $\mathcal{H}[.]$ as in [12] to represent an n-dimensional point in homogeneous coordinates. To clarify, let a Cartesian's coordinate point $\boldsymbol{u} \in \mathbb{R}^n$ specified as

$$\boldsymbol{u} = \begin{bmatrix} u_1 \\ u_2 \\ \vdots \\ u_n \end{bmatrix}, \tag{18}$$

whose homogeneous representation is given by

$$\boldsymbol{v} = \mathcal{H}_s[\boldsymbol{u}] = \begin{bmatrix} v_1 \\ v_2 \\ \vdots \\ v_{n+1} \end{bmatrix} = \begin{bmatrix} u_1 \\ u_2 \\ \vdots \\ s \end{bmatrix}, \tag{19}$$

such that $v \in \mathbb{R}^{n+1}$. Inversely to obtain the Cartesian's coordinates from homogeneous coordinates representation we use the inverse homogeneous operator defined as

$$\boldsymbol{u} = \mathcal{H}_s^{-1}[\boldsymbol{v}] = \frac{s}{\mathcal{S}[\boldsymbol{v}]}\mathcal{H}_0^{-1}[\boldsymbol{v}], \tag{20}$$

where $\mathcal{S}$ is the scale operator which returns the last element of $\boldsymbol{v}$ and the null inverse homogeneous operator specified as $\mathcal{H}_0^{-1}[\boldsymbol{v}] = [v_1, v_2, \ldots, v_n]^T$.

A.1 Correspondence Between Planes

Homography estimation is critical in many vision-based applications such as camera calibration [16], perspective correction [12], among others [8]. Projective transformations are linear operations that relate points in homogeneous coordinate space. Homographies are a special kind of projective transformations that define direct and inverse mapping relationships between planes in homogeneous coordinates such as points, lines, and other objects. In this work, we use the point paradigm for correspondences between two planes (Π_μ, Π_ρ) which are denoted as $\mu_i \leftrightarrow \rho_i$ with $\mu_i \in \Pi_\mu$ and $\rho_i \in \Pi_\rho$. Equally important, consider direct and inverse mappings

$$\rho_i = \mathcal{H}^{-1}\left[G\mathcal{H}[\mu_i]\right], \tag{21}$$

and

$$\mu_i = \mathcal{H}^{-1}\left[G^{-1}\mathcal{H}[\rho_i]\right], \tag{22}$$

we may state the problem to estimate the homography that satisfies the equation for the direct mapping and inverse mapping given a set of correspondence points. To estimate a homography given a pair correspondences $\mu_i \leftrightarrow \rho_i$ we may minimize the algebraic distance

$$\mathcal{H}[\mu_i] \times \mathcal{H}[\rho_i] = \mathbf{0}_3, \tag{23}$$

From Eq. (23), using the matrix form of cross product $[\mathcal{H}[\mu_i]]_\times \mathcal{H}[\rho_i] = \mathbf{0}_3$ we may write

$$\begin{bmatrix} 0 & -1 & \mu_{y,i} \\ 1 & 0 & -\mu_{x,i} \\ -\mu_{y,i} & \mu_{x,i} & 0 \end{bmatrix} \begin{bmatrix} \bar{g}_1{}^T\mathcal{H}[\rho_i] \\ \bar{g}_2{}^T\mathcal{H}[\rho_i] \\ \bar{g}_3{}^T\mathcal{H}[\rho_i] \end{bmatrix} = \mathbf{0}_3, \tag{24}$$

where $\bar{g}_i{}^T$ are the matrix rows of G, and rearranging terms from Eq. (24) we obtain the measurements matrix M and the homography vector $\boldsymbol{g}$ specified as follows

$$\underbrace{\begin{bmatrix} \mathbf{0}^T & -\mathcal{H}[\rho_i]^T & \mu_{y,i}\mathcal{H}[\rho_i]^T \\ \mathcal{H}[\rho_i]^T & \mathbf{0}^T & -\mu_{x,i}\mathcal{H}[\rho_i]^T \\ -\mu_{y,i}\mathcal{H}[\rho_i]^T & \mu_{x,i}\mathcal{H}[\rho_i]^T & \mathbf{0}^T \end{bmatrix}}_{M} \underbrace{\begin{bmatrix} \bar{g}_1{}^T \\ \bar{g}_2{}^T \\ \bar{g}_3{}^T \end{bmatrix}}_{g} = \mathbf{0}_3. \tag{25}$$

Now from Eq. (25) the third row is a linear combination of the first two rows of the measurement matrix and promptly we may simplify it as

$$\underbrace{\begin{bmatrix} \mathbf{0}^T & -\mathcal{H}[\rho_i]^T & \mu_{y,i}\mathcal{H}[\rho_i]^T \\ \mathcal{H}[\rho_i]^T & \mathbf{0}^T & -\mu_{x,i}\mathcal{H}[\rho_i]^T \end{bmatrix}}_{M} \underbrace{\begin{bmatrix} \bar{g}_1{}^T \\ \bar{g}_2{}^T \\ \bar{g}_3{}^T \end{bmatrix}}_{g} = \mathbf{0}_3. \tag{26}$$

The Eq. (26) is the principle of a well-known method for homography estimation called the Direct Linear Transformation (DLT). As we may observe, the DLT requires at least four correspondences points to estimate the homography matrix

G, but we may supply more than four point correspondences and rewrite Eq. (26) as

$$\begin{bmatrix} \mathbf{0}^T & -\mathcal{H}[\rho_1]^T & \mu_{y,1}\mathcal{H}[\rho_1]^T \\ \mathbf{0}^T & -\mathcal{H}[\rho_2]^T & \mu_{y,2}\mathcal{H}[\rho_2]^T \\ \vdots & \vdots & \vdots \\ \mathbf{0}^T & -\mathcal{H}[\rho_i]^T & \mu_{y,i}\mathcal{H}[\rho_i]^T \\ \mathcal{H}[\rho_1]^T & \mathbf{0}^T & -\mu_{x,1}\mathcal{H}[\rho_1]^T \\ \mathcal{H}[\rho_2]^T & \mathbf{0}^T & -\mu_{x,2}\mathcal{H}[\rho_2]^T \\ \vdots & \vdots & \vdots \\ \mathcal{H}[\rho_i]^T & \mathbf{0}^T & -\mu_{x,i}\mathcal{H}[\rho_i]^T \end{bmatrix} \begin{bmatrix} \bar{g_1}^T \\ \bar{g_2}^T \\ \bar{g_3}^T \end{bmatrix} = \mathbf{0}_{2n}, \tag{27}$$

which we may be solved through the singular value decomposition method (SVD) [16]. A detailed Maple implementation for homogeneous coordinates transformation and homography estimation by the DLT method can be found in Appendix A.2.

A.2 Homogeneous Coordinates and DLT Maple Listings

For this Maple listings section, it is necessary to load the following libraries using

```
with(LinearAlgebra);
with(ArrayTools);
```

Listing 1.1. Homogeneous coordinates operator.

```
H := proc(V, scale:=1)
 map(v -> <v, scale>, V);
end proc;
```

Listing 1.2. Inverse homogeneous coordinates operator.

```
HInvPoint := proc(w, scale:=1)
 local n, s;
 n := Size(w, 1);
 s := w[n][1];
 convert(scale*w[1 .. n - 1]/s, Vector[column]);
end proc;

HInv := proc(W, s:=1)
 map(x -> HInvPoint(x, s), W);
end proc;
```

Listing 1.3. DLT method for homography estimation.

```
DLT2D := proc(Pmu, Prho, scale := 1)
 local Hmu, Hrho, M, zeros, u, s, vt, G;
 zeros = Vector(3);
 Hrho = H(Prho, scale)
 M := <seq(Matrix([
  <zeros, -Hrho[i], Pmu[i](2).Hrho[i]>,
  <Hrho[i], zeros, -(Pmu[i](1).Hrho[i])>])^%T,
  i=1..Size(Prho,2))>;
 u, s, vt := SingularValues(M, output = ['U','S','Vt']);
 vt := vt^%T;
 G := Reshape(vt[1.., Size(vt, 2)], [3, 3])^%T;
 G/G(-1);
end proc;
```

B Support Functions Maple Listings

Listing 1.4. Simple Euclidean distance.

```
EuclidNorm := p -> sqrt(add(map(x -> x^2, p)))
```

Listing 1.5. Symbolic gradient computing.

```
Gradient := proc(f, variables)
 local v;
 <seq(diff(f, v), v in variables)>;
end proc;
```

Listing 1.6. Symbolic Jacobian computing.

```
Jacobian := proc(vfunc, variables)
 local v;
 <seq(Gradient(v, variables)^%T, v in vfunc)>;
end proc;
```

Listing 1.7. Grid point generation function.

```
Grid2DGenerator := proc(u__seq, v__seq)
 local p, i, j;
 p := [];
 for i in u__seq do
  for j in v__seq do
    p := [op(p), <i, j>];
  end do;
 end do;
end proc;
```

C Numerical Solution for Radial Distortion Maple Listings

Listing 1.8. Inverse radial distortion process (single parameter).

```
LInv1 := proc(G0, Smu, lambda)
  local n, ld, Snu, i;
  n := Size(Smu, 2);
  ld := map(nu -> 1 + lambda*EuclidNorm(nu)^2, Smu);
  Snu:= [seq(ld[i]*Smu[i], i = 1 .. n)];
  return Hinv(map(nu -> (MatrixInverse(G0)) . nu, H(Snu)));
end proc;
```

Listing 1.9. Inverse radial distortion process (three parameters).

```
LInv2 := proc(G0, Smu, lambda)
  local n, ld, Snu, i;
  n := Size(Smu, 2);
  ld := map(nu -> 1 + ((lambda[1]) . (EuclidNorm(nu)^2)) + ((lambda[2])
     . (EuclidNorm(nu)^4)) + ((lambda[3]) . (EuclidNorm(nu)^6)), Smu);
  Snu := [seq(ld[i]*Smu[i], i = 1..n)];
  return HInv(map(nu -> (MatrixInverse(G0)) . nu, H(Snu)));
end proc;
```

Listing 1.10. Estimation of homography and distortion parameters.

```
RectifyRadialDistortion3 := proc(G0, mud, rho0, lambda0 := <0.35, -0.25,
    0.15>, N:=25, epsilon:=0.1*10^(-5), output := false)
  local iteration, n0, L2, J, J0, f, Theta0, y, M, yt, rhot, Theta,
      varepsilon, K0, K, k, lambdaa, lambdab, lambdac, lambdan, lambdav,
      i;
```

```
n0 := Size(mud, 2);
lambdaa := lambda0[1];
lambdab := lambda0[2];
lambdac := lambda0[3];
lambdav := <lambda1, lambda2, lambda3>;
#
k[9] := 1;
K := Reshape(Matrix(<k[i] $ (i=1..9)>), [3, 3])^%T;
# Forward transformation function
L2 := (mu, lambda) -> 1 + ((lambda[1]) . (EuclidNorm(mu)^2)) + ((
    lambda[2]) . (EuclidNorm(mu)^4)) + ((lambda[3]) . (EuclidNorm(mu)
    ^6));
# Objective function definition
f := (K, Smu, lambda, n) -> <seq((K[1]) . <Smu[i](1)*L2(Smu[i], lambda
    ), Smu[i](2)*L2(Smu[i], lambda), 1>/((K[3]) . <Smu[i](1)*L2(Smu[i
    ], lambda), Smu[i](2)*L2(Smu[i], lambda), 1>), i = 1..n), seq((K
    [2]) . <Smu[i](1)*L2(Smu[i], lambda), Smu[i](2)*L2(Smu[i], lambda)
    , 1>/((K[3]) . <Smu[i](1)*L2(Smu[i], lambda), Smu[i](2)*L2(Smu[i],
    lambda), 1>), i = 1..n)>;
#Jacobian function definition
J := (K, S, lambda, n) -> Jacobian(f(K, S, lambda, n), [k[i] $ (i=1
    ..8), lambda[1], lambda[2], lambda[3]]);
K0 := MatrixInverse(G0);
Theta0 := convert(Matrix([Reshape(K0^%T, [1,9])[.., 1..8], lambdaa,
    lambdab, lambdac]), Vector[column]);
y := convert(<Vector(rho0(1)), Vector(rho0(2))>, Vector);
if evalb(output) then
  printf("%s\t %s\t %s\t\n","varepsilon","iteration","approx");
  printf("————————————————————————————————————————\n");
end if;
# Minimize objective function iterative stage
for iteration to N do
 rhot := Linv2(MatrixInverse(K0), mud, <lambdaa, lambdab, lambdac>);
 yt := convert(<Vector(rhot(1)), Vector(rhot(2))>, Vector);
 J0 := eval(J(K, mud, lambdav, n0), [lambda1 = lambdaa, lambda2 =
     lambdab, lambda3 = lambdac, seq(k[i] = convert(Reshape(K0^%T, [1,
     9]), Vector)[1..8][i], i = 1..8)]);
 M := (MatrixInverse((J0^%T) . J0, method = pseudo)) . (J0^%T);
 Theta := Theta0 + (M . (y - yt));
 varepsilon := Norm(Theta - Theta0, 2, conjugate = false);
 if varepsilon < epsilon then
  if evalb(output) then
   printf("%g\t %2d \t <%2g, %2g, %2g>\n", varepsilon, iteration,
       lambdaa, lambdab, lambdac);
  end if;
  break;
 end if;
 #
 Theta0 := Theta;
 K0 := Reshape(<Theta[1..8], 1>, [3,3])^%T;
 lambdan := <Theta[9], Theta[10], Theta[11]>;
 lambdaa := lambdan[1];
 lambdab := lambdan[2];
 lambdac := lambdan[3];
 # Enable algorithm output status if needed
 if evalb(output) then
  printf("%g\t %2d \t <%2g, %2g, %2g>\n", varepsilon, iteration,
      lambdaa, lambdab, lambdac);
 end if;
end do;
return [Reshape(<Theta[1..8], 1>,[3,3])^%T, <lambdaa, lambdab, lambdac
    >];
end proc;
```

References

1. Benligiray, B., Topal, C.: Lens distortion rectification using triangulation based interpolation. In: Bebis, G., et al. (eds.) ISVC 2015. LNCS, vol. 9475, pp. 35–44. Springer, Cham (2015). https://doi.org/10.1007/978-3-319-27863-6_4
2. Brown, D.: Decentering distortion of lenses. Photogrammetric Engineering and Remote Sensing (1966)
3. Bukhari, F., Dailey, M.N.: Automatic radial distortion estimation from a single image. J. Math. Imaging Vis. **45**(1), 31–45 (2013)
4. Cao, V.-T., Park, Y.-Y., Shin, J.-H., Lee, J.-H., Cho, H.-M.: A simple method for correcting lens distortion in low-cost camera using geometric invariability. In: Huang, D.-S., Zhang, X., Reyes García, C.A., Zhang, L. (eds.) ICIC 2010. LNCS (LNAI), vol. 6216, pp. 325–333. Springer, Heidelberg (2010). https://doi.org/10.1007/978-3-642-14932-0_41
5. Corke, P.: Robotics, Vision and Control: Fundamental Algorithms in MATLAB® Second, Completely Revised, vol. 118. Springer (2017)
6. Devernay, F., Faugeras, O.: Straight lines have to be straight. Mach. Vis. Appl. **13**(1), 14–24 (2001)
7. Fitzgibbon, A.W.: Simultaneous linear estimation of multiple view geometry and lens distortion. In: Proceedings of the 2001 IEEE Computer Society Conference on Computer Vision and Pattern Recognition. CVPR 2001, vol. 1, pp. I-I. IEEE (2001)
8. Geng, J.: Structured-light 3D surface imaging: a tutorial. Adv. Optics Photonics **3**(2), 128–160 (2011)
9. Hughes, C., Denny, P., Jones, E., Glavin, M.: Accuracy of fish-eye lens models. Appl. Opt. **49**(17), 3338–3347 (2010). https://doi.org/10.1364/AO.49.003338
10. Hughes, C., Glavin, M., Jones, E., Denny, P.: Review of geometric distortion compensation in fish-eye cameras. In: IET Irish Signals and Systems Conference 2008. IET (2008)
11. Hwang, K., Kang, M.G.: Correction of lens distortion using point correspondence. In: Proceedings of IEEE. IEEE Region 10 Conference. TENCON 1999. Multimedia Technology for Asia-Pacific Information Infrastructure'(Cat. No. 99CH37030), vol. 1, pp. 690–693. IEEE (1999)
12. Juarez-Salazar, R., Diaz-Ramirez, V.H.: Operator-based homogeneous coordinates: application in camera document scanning. Optic. Eng. **56**(7), 070801 (2017). https://doi.org/10.1117/1.OE.56.7.070801
13. Kasturi, R., Jain, R.: Computer Vision: Principles, vol. 1. IEEE Computer Society (1991)
14. Mallon, J., Whelan, P.F.: Precise radial un-distortion of images. In: Proceedings of the 17th International Conference on Pattern Recognition, 2004. ICPR 2004, vol. 1, pp. 18–21. IEEE (2004)
15. Tsai, R.: A versatile camera calibration technique for high-accuracy 3D machine vision metrology using off-the-shelf tv cameras and lenses. IEEE J. Robot. Autom. **3**(4), 323–344 (1987)
16. Zhang, Z.: Flexible camera calibration by viewing a plane from unknown orientations. In: Proceedings of the Seventh IEEE International Conference on Computer Vision. vol. 1, pp. 666–673. IEEE (1999)

A Simplified Introduction to Virus Propagation Using *Maple*'s *Turtle Graphics* Package Suitable for Children

Eugenio Roanes-Lozano[1](✉) and Eugenio Roanes-Macías[2]

[1] Instituto de Matemática Interdisciplinar & Departamento de Didáctica de las Ciencias Experimentales, Sociales y Matemáticas, Facultad de Educación, Universidad Complutense de Madrid, Madrid, Spain
eroanes@ucm.es

[2] Departamento de Álgebra, Geometría y Topología, Universidad Complutense de Madrid, Madrid, Spain
roanes@ucm.es

Abstract. In March 2020 the Spanish authorities ordered a nation-wide home confinement in an effort to avoid the spread of COVID-19 pandemic. This paper takes the current COVID-19 pandemic as motivation for a simple growth model designed for explaining virus propagation to children and was initially prepared in *Scracth 3* for the son of the first author. The mathematical model used is that of fractal growth trees, which are graphically rendered in order to provide a strong visual message of the nature of exponential growth. The rendering is done in *Maple's Turtle Graphics* package. This work is situated within a history of *Turtle Geometry*, starting with its beginnings in the classic *Logo* programming language, and describing how it fits within the current landscape of powerful software tools. The implementation within *Maple* is described, with relevant vignettes of code included. The complete *Maple* version of the tale is available from *MaplePrimes*.

Keywords: Turtle geometry · Fractals · Virus propagation · Modelling · Social conscience · Visualization · Childrens' mathematics education

MSC 2010: 28A80 · 92D30 · 97M10 · 97P70 · 97R60

1 Introduction

In March 2020 the Spanish authorities ordered a nation-wide home confinement in an effort to avoid the spread of COVID-19 (a pandemic of a coronavirus disease first identified at the end of 2019). The first author, father of a 13 year old son, decided to prepare a very simplified justification (for children), that was initially implemented in *Scratch 3* computer language. It relates virus propagation [1] to fractal trees and it is based on the *Turtle Geometry* computer graphics [2] (see Sect. 1.1 below). It shows how one infected cat can spread a contagious illness in a cat colony (the average number of cats infected by each ill cat can be easily changed). This justification was later written

R. M. Corless et al. (Eds.): MC 2020, CCIS 1414, pp. 334–349, 2021.
https://doi.org/10.1007/978-3-030-81698-8_22

as a tale and recorded in a 5 min video (in Spanish) that are available from the *Instituto de Matemática Interdisciplinar (IMI)* of the *Universidad Complutense de Madrid* web page [3] and an improved version of the tale (in English) was presented at *ESCO 2020* conference on June 2020 [4] and published in [5].

Note that the authors have experience in designing and developing applications for teaching and decision making in medicine (with CAS), like the early contributions [6, 7], and pharmacokinetics with CAS [8]. Moreover, they developed in the 90's a *Maple*[1] [9–14] implementation of the *Turtle Geometry* [15] (that was incorporated to the *Maple Share Library*).

Now, an improved new version of the tale (in English), which underlying code is written in *Maple 2020* (using an updated version of the *Maple* implementation of *Turtle Geometry* mentioned above, that takes advantage of using exact arithmetic for storing the geometric coordinates of the points), is available from *MaplePrimes* [16]. This advantage is important when dealing with geometric designs that somehow concatenate geometric objects, as is the case of fractals (something already mentioned in [17]). The possibilities of *Maple's Turtle Geometry* package and this particular application are analysed.

We would like to explicitly remark that the model of exponential growth presented has just an educational purpose: it tries to communicate the nature of exponential growth, and the need to reduce the spread of the virus by limiting contact between people (or cats). Note that there are far more sophisticated and accurate models for research purposes (that take into account many factors that occur in practice). As a further reading, [18] presents interesting models of real epidemics using fictional creatures.

1.1 Turtle Geometry and *Logo* Programming Language

Logo [19, 20] was a very powerful programming language mainly remembered because of its peculiar *turtle geometry* (also known as *turtle graphics*) [2] and its educational use with children: working with geometric concepts and learning programming. The *Turtle Geometry* applies constructionist ideas [21].

The turtle is a graphic cursor for drawing, which basic movements (forward, back, turn right, turn left) are not related to the usual Cartesian coordinates but to the turtle's position and heading at each moment. In fact the turtle is frequently introduced to children as a little animal living and moving on the screen, what is far more intuitive than Cartesian coordinates. The main advantages with respect to working with Cartesian coordinates are:

- trigonometric calculations are performed internally (it is oriented to be used by children),
- repeating a certain design elsewhere only requires to allocate the turtle in the new position and new heading and to apply the same list of commands.

In the very beginning, the *turtle* was a mechanical device moving on the floor or on a table (the monitors only displayed text characters).

[1] All product names, trademarks and registered trademarks are property of their respective owners.

Despite its extraordinary success in the '80s (Fig. 1), *Logo* is now sparsely used and considered outdated, except for some specific applications such as robotics or fractals [22–24] and other special curves [25].

We believe there are two main reasons for such a decay:

- *Logo* is a very friendly programming language, easy to use and really easy for beginners. Nevertheless, there isn't nowadays an agreement on whether introducing children to basic computer programming is a must or not [26]. And for introducing programming to children there are more modern options like the well-known *Scratch 3* [27] and *Snap!* [28] (both of them including *turtle graphics*).
- The range, availability and spread of mathematical software have increased enormously. Nowadays, Dynamic Geometry Systems (DGS), and, more specifically, *GeoGebra*, have possibly occupied the place left by *Logo* language in teaching geometry with the aid of technology (although they have very different approaches: roughly speaking, DGS are based on ruler and compass constructions performed with the mouse, meanwhile *Logo* is based on the use of *Turtle Geometry* and traditional programming). Other areas of mathematics teaching, like algebra and calculus, are covered by computer algebra systems (CAS). Apart from the very powerful leaders *Maple* [9–14] and *Mathematica* [29], there are many other CAS, of general purpose (*Maxima*, *Reduce*, *Axiom*, *SageMath*, *Xcas*,...) as well as specific purpose (*CoCoA*, *Singular*,...).

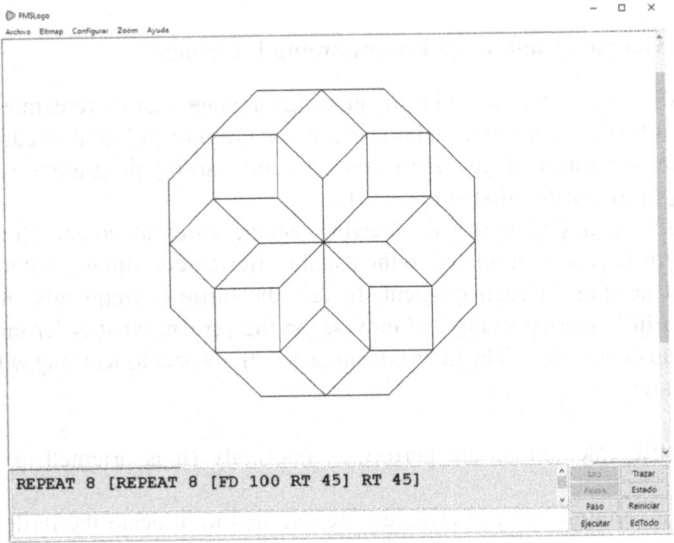

Fig. 1. Eight regular octagons, all sharing a vertex and each one turned 45° with respect to the previous one, drawn using a modern version of *Logo*.

There are many implementations of *Logo* available nowadays. An updated comprehensive list of *Logo* implementations and related software, including more than 250 references to different dialects, can be found in [30]. We could underline *FMSLogo* [31], StarLogo [32, 33] and *NetLogo* [34].

Some *Logo* implementations include 3D extensions. Moreover, there are many impressive extensions of turtle graphics, for instance to spherical [35], elliptic [36] and hyperbolic geometries [37].

Some *Logo* implementations include multiple turtles. Nevertheless, the intensive use of multiple turtles, although existent [22, 38, 39], is scarce. When multiple turtles are available the user can activate a certain turtle and the subsequent commands will be obeyed by this turtle, until another turtle is activated. Examples of *Logo* implementations including multiple turtles are *FMSLogo, StarLogo* and *NetLogo.*

1.2 Other Implementations of *Turtle Geometry*

Turtle graphics-like packages or implementations could or can be found in many programming languages. Possibly the closer descendant of *Logo* language is *Scratch*, that includes a simplified version of *turtle graphics* [27, 40]. Regarding *Scratch*, we could mention that:

- it is based on the use of intuitive *graphical programming blocks*,
- the classic turtle has been substituted by a cat,
- the costumes of the sprites (graphic cursor) can be easily changed,
- it is not difficult to develop animations,
- there are no blocks corresponding to *Logo*'s `BACK` (*turtle*) and `RETURN` (for procedures) commands,
- it has limited recursive capabilities.

The later drawbacks are corrected in the not so well-known computer language *Snap!*, that can be considered a descendent or evolution of *Scratch*, with wider programming capabilities (for instance, regarding recursion, it has a `report` block, similar to *Logo*'s `RETURN` command).

Other languages including implementations of *turtle geometry* are:

- *PythonTurtle* [41], a learning environment for *Python*, inspired by Logo,
- *Haskell* (a functional language) [42], where simplified versions of *turtle graphics have* been implemented [43, 44],
- *Java*, where different implementations of turtle graphics like *Java TurtleGraphics* [45], *Jurtle* and *Pencil Code Online* [46] (the latter implemented in *CoffeeScript*) are available.

Some pieces of mathematical software like the CAS *Xcas* include implementations more or less standard of turtle graphics.

There are also modern attempts to put together different computer approaches, for instance the reasoning power of *ProLog* with the drawing capabilities of turtle graphics (the *turtles* considered as agents) [47].

Let us finally remark that the *One Laptop per Child* project includes turtle activities [48]. And, for instance, *Berkeley Logo* (*UCB Logo*)} [49] is available for the *XO*.

The authors have developed implementations of turtle graphics in:

- *Turbo-Pascal* and *Turbo-C* languages [50] (at that time *Turbo-Prolog* [51] included a reduced implementation of *turtle graphics*),
- *Maple*, that was incorporated to the *Maple Share Library* [15, 17],
- *Derive* [52].

Other authors developed similar implementations for the CAS *Reduce* [53] and the *TI-92* calculator [54, 55].

Let us also mention that a peculiar implementation of *turtle graphics* has been incorporated to the DGS *GeoGebra* [56].

Finally, there are implementations for smartphones like *JTurtleLib* [57].

2 Revisiting the 1994 *Maple* Implementation of *Turtle Geometry*

The idea was simple. The *turtle* can only draw segments. These segments are stored in the global variable `dib` (a list). Each forward or backward movement adds a segment to list `dib` (if the pen is down). The coordinates and heading of the turtle, the pen colour it is using and whether the pen is down or up are stored in global variables.

Just to give the flavour of the implementation, we include afterwards the main procedures of the implementation.

2.1 Main Procedures of the Implementation

Below the code can be found. Only procedure `FullScreen()` had to be updated in order to work in *Maple 2020*. Some other minor changes (like substituting the `fi` by `end if`; the `od` by `end do`; the `end` by `end proc`; and `CURVES` by `line`, from `plottools` package) have been included in order to modernize the implementation without changing it too much (maintaining the flavour of the original code).

Procedure `ClearScreen()` clears the screen and resets the *turtle*.

```
ClearScreen:=proc()
  global XCor, YCor, Heading, PenColor,
         posicion_lapiz, dib;
   XCor := 0; YCor := 0;
   Heading := 0;
   PenColor := COLOUR(RGB,0,0,0);
   posicion_lapiz := 1;
   dib:={};
   NULL;
 end proc: #ClearScreen
```

`Forwd(n)` moves the *turtle* `n` steps forward in its present direction.

```
Forwd:=proc(distancia:algebraic)
  global XCor, YCor, dib;
  local antigua_absc_tort, antigua_ord_tort,
        angulo_radianes;
   angulo_radianes := Pi * (90-Heading) / 180;
   antigua_absc_tort := XCor;
   antigua_ord_tort := YCor;
   XCor := XCor + distancia * cos(angulo_radianes);
   YCor := YCor + distancia * sin(angulo_radianes);
   if posicion_lapiz = 1 then
         dib:={op(dib),line([evalf(antigua_absc_tort),
                             evalf(antigua_ord_tort)],
                            [evalf(XCor),
                             evalf(YCor)],
                             color=PenColor)}
    end if;
   NULL;
 end proc: #Forwd
```

`TurnRight(a)` turns the *turtle* a degrees clockwise its present position. `TurnLeft(a)` does the same counterclockwise.

```
TurnRight:=proc(angulo:algebraic)
  global Heading;
   Heading := Modu(Heading + angulo);
   NULL;
 end proc: #TurnRight
```

At any point we can ask *Maple* to plot what the turtle has drawn by typing `FullScreen()` (it doesn't automatically draw the updated situation after each new command).

```
FullScreen:=proc()
  display(dib,axes=none,scaling=constrained);
end proc: #FullScreen
```

2.2 List of Commands Implemented in *Maple*

The following list will be clear for any acquainted user of *turtle geometry*. There are some pure *Turtle Geometry* commands:

```
Forwd(n),
Back(n),
Home(),
ClearScreen(),
TurnRight(angle),
TurnLeft(angle),
SetHeading(angle),
SetHeadingTowards(x1,x2),
PenUp(),
PenDown(),
SetPenColor(color)
```

some Cartesian coordinates-related *turtle geometry* commands:

```
SetPosition(x1,x2),
SetX(x1),
SetY(x2)
```

and the auxiliary procedure (detailed in Sect. 2.1): `FullScreen()`.

2.3 Available Information About the Status of the *Turtle*

There are global variables that can be accessed by the user and return information about the status of the *turtle* (they are usual *turtle geometry* commands):

```
XCor,
YCor,
Heading,
PenColor.
```

2.4 Examples of the Use of the *Turtle Geometry Maple* Package

Many geometric designs are very easy to draw using *turtle graphics*. For instance the star of Fig. 2 can be produced in *Maple* just typing:

```
> ClearScreen();
> TurnRight(90);
> for i to 9 do Forwd(100);
>                TurnRight(160)
>  end do;
> FullScreen();
```

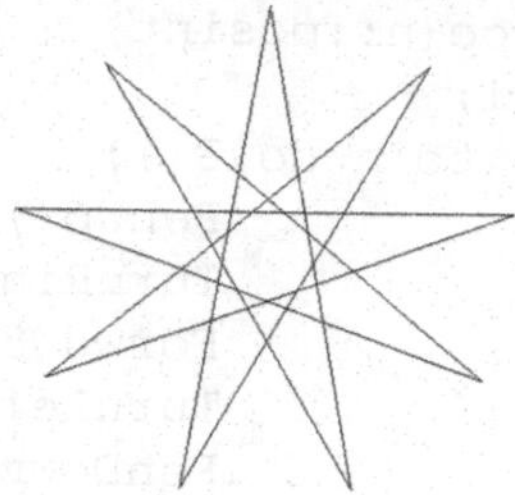

Fig. 2. A star drawn using *Maple's Turtle Graphics* package.

(after loading the package: `read('C:/.../Maple/turtle2021.mpl'):`).

A design can be stored in a procedure. For instance, when executed, procedure `T()` generates the drawing of Fig. 3.

```
> T:=proc()
>    Forwd(50);
>    TurnRight(90);
>    Forwd(20);
>    Back(40);
>    Forwd(20);
>    TurnLeft(90);
>    Back(50);
>  end proc:
> ClearScreen();
> T();
> FullScreen();
```

Fig. 3. An uppercase T letter drawn using *Maple's Turtle Graphics* package.

Turtle geometry is very well suited for drawing periodic designs [58–61]. The drawing of Fig. 3 has a vertical symmetry axis. If replicated horizontally, we can easily obtain a *FM1* frieze. It is very easy to implement a procedure that replicates it n times:

```
> FM1:=proc(n::posint)
>  local i;
>   for i to n do T();
>                 PenUp();
>                 TurnRight(90);
>                 Forwd(50);
>                 TurnLeft(90):
>                 PenDown();
>    end do;
>  end proc:
```

and now `FM1(6);` generates the plot of Fig. 4.

Fig. 4. An *FM1* frieze drawn using *Maple's Turtle Graphics* package.

As said above, once a design is constructed, allocating it in another place or direction is trivial. For instance,

```
> ClearScreen();
> TurnLeft(30);
> FM1(6);
> FullScreen();
```

turns the design of Fig. 4 an angle of 30° counterclockwise.

From a *FM1* frieze it is easy to obtain a *pm* plane crystallographic group replicating the frieze vertically. Procedure `pm(n,m)` generates a *pm* pattern with m rows and n columns:

```
> pm:=proc(n::posint,m::posint)
>  local j;
>   for j to m do FM1(n);
>                 PenUp();
>                 TurnLeft(90);
>                 Forwd(n*50);
>                 TurnRight(90):
>                 Back(65);
>                 PenDown();
>    end do;
>  end proc:
```

For instance, `pm(6,3);` generates the drawing of Fig. 5.

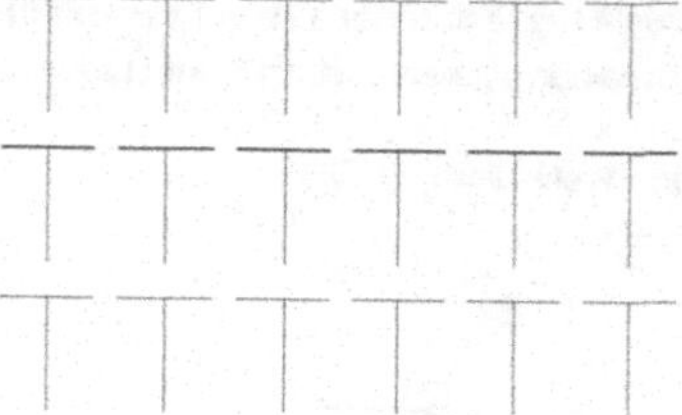

Fig. 5. A *pm* plane crystallographic group drawn using *Maple's Turtle Geometry* package

3 The Origin of this Work

The 13 year old son of the first author was very disappointed when a lockdown was ordered in Spain (at the beginning of the pandemic). It wasn't easy to explain to him why he should stay at home (without meeting his friends and relatives). Therefore we decided to develop a visual, simplified explanation of virus propagation (in the form of a tale and a video), taking into account the audience (children and young people). The tale and video have an elementary but clear mathematical background:

- they use fractals trees
  - the number of branches at each level of the tree can be related to the average number of animals infected by each ill animal,
  - the depth of the tree can be related to the time passed till the animals stop meeting.
- they are made visual by using the *Turtle Geometry*,
- they insist on the social conscience.

It was initially implemented in the computer language *Scratch 3*. *Scratch's* graphic cursor (a cat) was used to simulate the animals infected:

4 The Original Tale (*Scratch 3* Version)

As said above, the original version of the tale and a 5 min video (both in Spanish) are available from the *Instituto de Matemática Interdisciplinar (IMI)* of the *Universidad Complutense de Madrid* web page [3] and an improved version of the tale (in English) was presented at *ESCO 2020* conference [4]. A screenshot of part of an intermediate page of the original tale in Spanish can be found in Fig. 6.

The complete *Scratch 3* improved version of the tale (in English) can be found in [5].

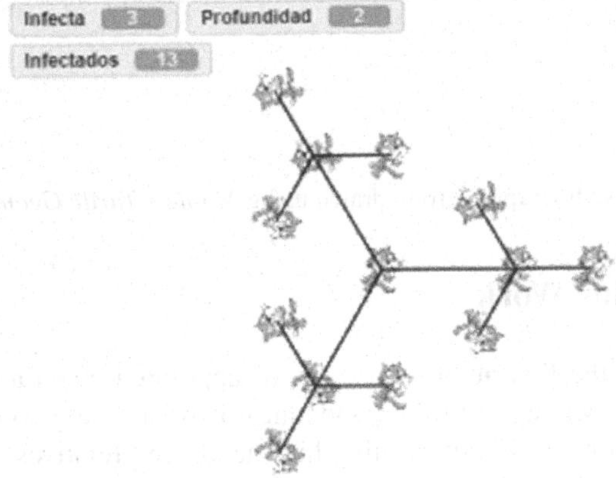
Si cada uno de los gatos dela figura anterior a su vez tiene contacto con a sus amigos y familiares pasamos a la situación siguiente, con 13 infectados:
Infecta 3
Profundidad 2
Infectados 13

Fig. 6. Part of an intermediate page of the original tale (in Spanish), that uses *Scratch 3*.

5 The *Maple* Version of the Tale

Surprisingly, the plots in *Scratch* have a low resolution (worse than modern *Logo* dialects). Therefore, we decided to try to port the implementation of the fractals illustrating the virus propagation in the tale to *Maple*. As said above, we had already underlined the advantages of the *Maple* implementation of [17] for plotting fractals.

The code turned out to be simple to translate. Two procedures are required for the general case, and the input for the main procedure (`arbolnb`) are the average number of animals infected by each ill one, the depth of the fractal considered and the length of the first branch. The animals are represented by red squares. The complete code can be found below.

```
> arbolnb:=proc(infects,depth,length)
>  global times, infectados;
>   times:=depth;
>   ClearScreen();
>   infectados:=0;
>   auxnb(1,length,infects);
>   print(infectados);
>   FullScreen();
>  end proc:
> auxnb:=proc(n,l,ve)
>  local i;
>  global times, infectados;
>   if n < times + 2 then
>      if n=1 then PenUp()
>             else PenDown()
>       end if;
>      Forwd(l);
>      RedDiamond(XCor,YCor);
>      infectados := infectados + 1;
>      TurnLeft(180 - 180 / ve);
>      for i to ve do auxnb(n + 1,l / 2,ve);
>                     TurnRight(360 / ve);
>       end do;
>      TurnLeft(360 / ve + 180 - 180 / ve);
>      if n=1 then PenUp()
>             else PenDown()
>       end if;
>      Forwd(-l)
>    end if;
>  end:
```

For instance, `arbolnb(5,2,300)` produces the plot of Fig. 7 and `arbolnb(3,5,300)` produces the plot of Fig. 8. In the later one, many animals are represented. The way *Turtle Graphics* are implemented in *Maple* (internally working in exact arithmetic and approximating only for the final plot) implies a lower speed but a perfect quality (what is especially important if a scalable graphics format is used for saving the plot).

As said above, a *Maple* version of the tale can be found in *MaplePrimes* [16].

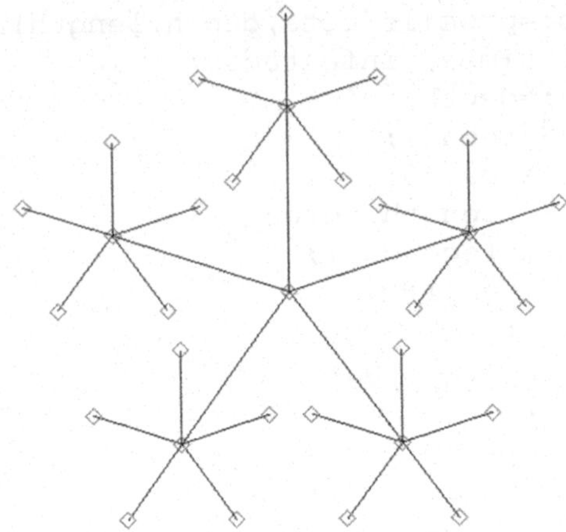

Fig. 7. One of the drawings of the tale produced with *Maple*.

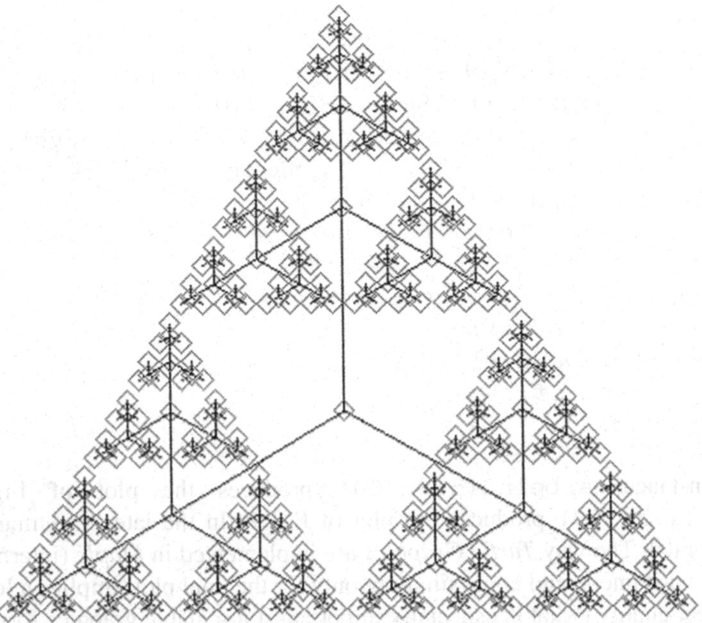

Fig. 8. A dense drawing of the tale (also produced with *Maple*).

6 Conclusions

Turtle Geometry is almost forgotten nowadays. Nevertheless, it is a very convenient approach to graphic generation in a lot of cases, as well as for teaching certain geometric topics and programming.

More concretely, our old *Maple* implementation (1994) has been easily updated to work in *Maple 2020* and has shown its possibilities for producing high quality plots for the tale about virus propagation previously written by the authors.

The new *Turtle Graphics* file (`turtle2021.mpl`) and a worksheet with the examples of Sect. 5 (`Maple_Turtle_Virus_6.mw`) are available from the first author's web page, at [62] and [63], respectively.

Acknowledgments. Partially funded by the research project PGC2018–096509-B-100 (Government of Spain).

The authors would sincerely like to thank the anonymous reviewers of this article for their most valuable comments, which have greatly contributed to the improvement of this article.

References

1. Faisst, S.: Propagation of viruses | Animal. In: Grano, A., Webster, R.G. (eds.) Encyclopedia of Virology, 2nd edn., pp. 1408–1413. Academic Press, San Diego (1999)
2. Abbelson, H., diSessa, A.: Turtle Geometry The Computer as a Medium for Exploring Mathematics. MIT Press, Cambridge (1981)
3. Instituto de Matemática Interdisciplinar (IMI) Other activities 2020. "Por qué quedarse en casa es bueno para evitar la propagación de un virus? Una explicación simplificada para jóvenes usando fractales y Scratch. Eugenio Roanes Lozano. Video y texto divulgativo. https://ucm.es/imi/other-activities-2020
4. ESCO 2020. https://www.esco2020.femhub.com/
5. Roanes-Lozano, E., Solano-Macías, C.: Using fractals and Turtle Geometry to visually explain the spread of a virus to kids: a STEM multitarget activity. Math. Comput. Sci. (2021). https://doi.org/10.1007/s11786-021-00500-9
6. Laita, L.M., Roanes-Lozano, E., Maojo, V., Roanes-Macias, E., de Ledesma, L., Laita, L.: An expert system for managing medical appropriateness criteria based on computer algebra techniques. Comp. Math. Appl. **42**(12), 1505–1522 (2001). https://doi.org/10.1016/S0898-1221(01)00258-9
7. Pérez-Carretero, C., Laita, L.M., Roanes-Lozano, E., Lázaro, L., González-Cajal, J., Laita, L.: A logic and computer algebra-based expert system for diagnosis of anorexia. Math. Comput. Simul. **58**(3), 183–202 (2002). https://doi.org/10.1016/S0378-4754(01)00370-6
8. Roanes-Lozano, E., González-Bermejo, A., Roanes-Macías, E., Cabezas, J.: An application of computer algebra to pharmacokinetics: the Bateman equation. SIAM Rev. **48**(1), 133–146 (2016). https://doi.org/10.1137/050634074
9. https://www.maplesoft.com/products/Maple/
10. Bernardin, L., et al.: Maple Programming Guide. Maplesoft, Waterloo Maple Inc., Waterloo, Canada (2020). https://www.maplesoft.com/documentation_center/maple2020/ProgrammingGuide.pdf
11. Maplesoft: Maple User Manual. Maplesoft, Waterloo Maple Inc., Waterloo, Canada (2020). https://www.maplesoft.com/documentation_center/maple2020/UserManual.pdf
12. Corless, R.: Essential Maple An Introduction for Scientific Programmers. Springer, New York (1995). https://doi.org/10.1007/978-1-4757-3985-5
13. Heck, A.: Introduction to Maple. Springer, New York (2003). https://doi.org/10.1007/978-1-4613-0023-6
14. Roanes-Macías, E., Roanes-Lozano, E.: Cálculos Matemáticos por Ordenador con Maple V.5. Editorial Rubiños-1890, Madrid (1999)

15. Roanes-Lozano, E., Roanes-Macías, E.: An Implementation of "Turtle Graphics" in Maple V. MapleTech Special Issue, 82–85 (1994)
16. MaplePrimes. Why Staying at Home is Good to Avoid the Spread of a Virus? A tale of fractals, cats and virus. https://www.mapleprimes.com/posts/212674-Why-Staying-At-Home-Is-Good-To-Avoid
17. Roanes-Lozano, E., Roanes-Macías, E.: "Turtle Graphics" in Maple V. In: Lopez, R.J. (ed.) Maple V: Mathematics and Its Applications, pp. 3–12. Birkhäuser, Boston-Basel-Berlin (1994)
18. Smith, R.: Mathematical Modelling of Zombies. University of Ottawa Press (2014)
19. Abelson, H.: Logo for the Apple II. BYTE/McGraw-Hill, Peterborough (1980)
20. Wikipedia: Logo (programming language) https://en.wikipedia.org/wiki/Logo_(programming_language)#:~:text=Logo%20is%20an%20educational%20programming,logos%2C%20meaning%20word%20or%20thought
21. Papert, S.: Mindstorms: Children, Computers and Powerful Ideas. Basic Books, New York (1980)
22. Give'on, Y.S.: Teaching recursive programming using parallel multi-turtle graphics. Comput. Educ. **16/3**, 267–280 (1991)
23. Goldman, R., Schaefer, S., Ju, T.: Turtle geometry in computer graphics and computer–aided design. Comp. Aid. Des. **36**(14), 1471–1482 (2004)
24. Ju, T., Schaefer, S., Goldman, R.: Recursive turtle programs and iterated affine transformations. Comput. Graph. **28**(6), 991–1004 (2004)
25. Trott, M.: Wolfram Demonstrations Project. Turtle-Graphics. http://demonstrations.wolfram.com/TurtleGraphics/
26. Shein, E.: Should everybody learn to code? Commun. ACM **57**(2), 16–18 (2014)
27. https://scratch.mit.edu/
28. https://snap.berkeley.edu/
29. https://www.wolfram.com/mathematica/index.html.es?footer=lang
30. Logo Tree. https://pavel.it.fmi.uni-sofia.bg/logotree/
31. FMSLogo: An Educational Programming Environment. http://fmslogo.sourceforge.net/
32. Resnick, M.: New paradigms for computing, new paradigms for thinking. In: diSessa, A., et al. (eds.) Computers and Exploratory Learning, pp. 31–43. NATO ASI Series, no. 146. Springer, Heidelberg (1995). https://doi.org/10.1007/978-3-642-57799-4_3
33. Introduction to StarLogo. https://education.mit.edu/project/starlogo-tng/
34. Wilensky, U.: NetLogo 5.0.5 User Manual. http://ccl.northwestern.edu/netlogo/docs/NetLogo%20User%20Manual.pdf
35. Cabezas, J., Hernández Encinas, L.: Geometría esférica en Logo. Gac. Mat. **1**, 13–24 (1988)
36. Sims-Coomber, H., Martin, R.R.: An implementation of LOGO for elliptic geometry. Comput. Graph. **18**(4), 543–552 (1994)
37. Sims-Coomber, H., Martin, R.R.: A non-Euclidean implementation of LOGO. Comput. Graph. **15**(1), 117–130 (1991)
38. Neuwirth, E.: Turtle Ballet: Simulating Parallel Turtles in a Nonparallel LOGO Version. In Futschek, G. (ed.) European Logo conference Eurologo 2001, a turtle odyssey, pp. 263–270. Österreichische Computer Gesellschaft (2001)
39. Resnick, M.: Turtles, Termites, and Traffic Jams Explorations in Massively Parallel Microworlds. The MIT Press, Cambridge (1997)
40. Roanes-Lozano, E.: Geometría de la Tortuga con Scratch 2.0 y Enseñanza de Matemática Elemental. https://webs.ucm.es/info/secdealg/ApuntesLogo/INF_MATN_Scratch18-19_v11.pdf
41. Rachum, R.: PythonTurtle. http://pythonturtle.org/
42. The Haskell Programming Language. http://www.haskell.org/haskellwiki/Haskell

43. Boiten, E.: Turtle Graphics: Exercises in Haskell. Technical Report No. 11-04, University of Kent, Canterbury (2004)
44. Graphics.X11.Turtle. http://hackage.haskell.org/package/xturtle-0.1.5/docs/Graphics-X11-Turtle.html
45. Haas, G.M.: BFOIT. Introduction to Computer Programming. Java TurtleGraphics. http://guyhaas.com/bfoit/itp/JavaTurtleGraphics.html
46. Pencil Code Online Guide. http://guide.pencilcode.net/
47. Sancho, F.: NetProLogo. http://www.cs.us.es/~fsancho/?e=23
48. Wikipedia. LOGO. http://wiki.laptop.org/go/LOGO
49. Harvey, B.: Berkeley Logo (UCBLogo). http://www.cs.berkeley.edu/~bh/logo.html
50. Roanes-Lozano, E., Roanes-Macías, E.: Nuevas Tecnologías en Geometría. Complutense, Madrid (1994)
51. Anonymous. Turbo PROLOG the natural language of artificial intelligence. Borland Int. Inc., Scotts Valley, CA (1986)
52. Lechner, J., Roanes-Lozano, E., Roanes-Macías, E., Wiesenbauer, J.: An Implementation of "Turtle Graphics" in Derive 3. Bull. DERIVE User Group **25**, 15–22 (1997)
53. C. Cotter. Turtle Graphics Interface for REDUCE Version 3. https://www.semanticscholar.org/paper/Turtle-Graphics-Interface-for-REDUCE-Version-3-Cotter/4be30e3d124eea67dec1dd70e640ab91aaa9fbbb
54. Kutzler, B, Stoutemyer, D.R.: Great TI-92 Programs (Vol. 1). bk teachware, Hagenberg, Austria (1997)
55. ticalc org project. TI-92 Turtle Graphics v1.0. http://www.ticalc.org/archives/files/fileinfo/13/1376.html
56. https://www.geogebra.org/m/RSaep6ne#material/reSARTjy
57. JTurtleLib. Java Turtle Graphics for Android. http://www.aplu.ch/home/apluhomex.jsp?site=123
58. Garbayo, M., Roanes-Lozano, E.: Implementación de un paquete de dibujo de rosetones (Grupos de Leonardo). Bol. Soc. "Puig Adam" **37**, 87–96 (1994)
59. Garbayo, M., Roanes-Lozano, E.: Implementación de un paquete de dibujo de frisos. Bol. Soc. "Puig Adam" **40**, 39–53 (1995)
60. Garbayo, M., Roanes-Lozano, E.: Implementación de un paquete de dibujo de grupos cristalográficos planos. Bol. Soc. "Puig Adam" **43**, 71–77 (1996)
61. Garbayo, M., Roanes-Lozano, E.: Tort–decó: a "turtle geometry"–based package for drawing periodic designs. Math. Comp. Mod. **33**, 321–340 (2001)
62. https://webs.ucm.es/info/secdealg/gato/turtle2021.mpl
63. https://webs.ucm.es/info/secdealg/gato/Maple_Turtle_Virus_6.mw

A Maple Toolchain for Rigid Body Dynamics of Serial, Hybrid and Parallel Robots

Moritz Schappler(✉), Tim-David Job, and Tobias Ortmaier

Institute of Mechatronic Systems, Leibniz University Hannover, Garbsen, Germany
{moritz.schappler,tim-david.job,tobias.ortmaier}@imes.uni-hannover.de

Abstract. A new Maple toolchain for generating rigid body dynamics in symbolic form for robot manipulators is presented. The peculiarity compared to existing tools lies in the framework of Bash scripts controlling the full workflow of the toolchain with a high degree of automation. The optimized Matlab code generated by Maple is automatically converted to function files with proper documentation and input assertions. This renders manual post-processing of the results unnecessarily. The focus of the paper is on the implemented unit-testing framework according to the method of test-driven development. By providing the test framework together with the generated code in a stand-alone version, a good test coverage and a good software quality can be achieved. The results of the open source project provide a basis for dynamics simulations for robot dimensional synthesis or in model-based control of robot manipulators in research or in industrial context. The general software approach can be applied to other fields where theoretical models are derived with Maple.

Keywords: Rigid body dynamics · Robotics · Symbolic code · Toolchain · Test-driven development · Maple computer algebra system

1 Introduction and State of the Art

Using a symbolic rather than numeric implementation of dynamics models for robots is highly beneficial regarding the computational efficiency [9]. Some aspects of the models can only be obtained in a useful way via symbolic derivation, such as the identification model [12]. Using models in simulations for a comparison of different robots requires an automatic, general and efficient approach.

To be able to find the robot that is suited best for a given task, first a set of robot kinematics has to be created, which is the outcome of the *structural synthesis*. The structural synthesis of serial robots can be performed using screw theory [17], Denavit-Hartenberg parameters [23] or variants thereof, such as the traveling coordinate system method [10]. The synthesis of serial chains is mostly discussed in literature in the context of parallel robots, which contain several serial kinematic leg chains connected to a moving platform. Leg chains are generated with the virtual-chain approach and screw theory [15] or using the theory of linear transformations and evolutionary morphology [11]. Following [19, p. 25]

R. M. Corless et al. (Eds.): MC 2020, CCIS 1414, pp. 350–364, 2021.
https://doi.org/10.1007/978-3-030-81698-8_23

(for parallel robots), for the assessment of the robot's performance, the dimensions of the design parameters (e.g. lengths of the links) are as important as the kinematic structure itself (i.e. number, type and alignment of joints). This leads to the requirement of a *combined robot synthesis*, as proposed in [16] for parallel robots, to determine which robot structure is suited best for a given task. A *dimensional synthesis*, i.e. an optimization of the robot's dimensions, has to be performed for all possible structures. First investigations on the combined synthesis for serial robots [23] have shown that the approach is feasible in principle. However, the practicability of such an extensive optimization of hundreds of robots with several optimization parameters each and highly nonlinear models strongly depends on the implementation. The robot models and the objective and constraints function within the optimization problem need an efficient implementation to be able to generate substantiated results.

A simulation of the *robot dynamics model* (i.e. the relation of force and motion) has to be evaluated in each iteration of the aforementioned optimization. The rigid body dynamics for robots itself is a mathematical problem that can be considered solved in that context for serial [12], hybrid [6,9,13,24,26] and parallel robots [1,3,5,19]. For serial kinematic chains the NEWTON-EULER algorithm is mainly used in software dedicated for robot dynamics [13,14,24]. Hybrid robots, i.e. serial robots with additional closed kinematic loops, are mainly modeled based on the serial chain dynamics with additional variational principles to take closed loops into account [24] (D'ALEMBERT, JOURDAIN). For parallel robots, different definitions of the system coordinates are possible based on these principles of energy equivalence [1,3,5].

There exist a variety of software toolchains for modeling dynamics equations for robot manipulators. A probably non-exhaustive list contains the symbolic tools Robotran [6,9,24], SYMORO [13], openSYMORO [14], MapleSim [31], Neweul-M^2 [18], the Peter Corke Matlab toolbox [4] and some open source toolboxes from single research projects, such as FloBaRoID [2], SymPyBotics (or SageRobotics [27]) and the dVRK Dynamic Model Identification Package [30]. Several numeric tools are available for simulating the inverse and forward dynamics of general multibody systems, which includes the robot manipulators in this work. Prominent examples are MSC Adams [20], Matlab Simscape Multibody (SimMechanics) [29], the Rigid Body Dynamics Library [8], based on Featherstone's theory [7], and Drake [28]. Some symbolic programs also provide the possibility for a numeric simulation of the systems, such as MapleSim [31].

These toolchains do not directly meet the requirements for creating a model database required for the combined synthesis. Extensions regarding batch-processing, unit-testing and post-processing are required, since most tools require user interaction, which is not feasible for hundreds of robots. A key method for ensuring software quality is systematic testing using unit testing frameworks, which is central to test-driven development. This is often disregarded in software for scientific projects [32]. Available toolchains presumably all give correct results, but it is not always possible to completely verify this by the end user. Open source tools, such as [2,14,27], typically come explicitly without warranty raising the need for additional validation of the results. Misinterpretation of interfaces of not well documented, but still error-free, software modules may introduce errors in the further use within a bigger project. Creating a robot database with the claim to include every unique robot structure will generate all possible test cases and will raise all existing software bugs. Not using a proper testing environment therefore can put unnecessary risks on projects relying on the results of the software tools.

To encounter these issues for the case of robot dynamics for the proposed application, a new toolchain[1] was developed. It is based on Maple as symbolic engine and Bash scripts for an automation of the model generation process. This provides the flexibility to test the implementation of different algorithms, e.g. an efficient formulation for parallel robots [1] or an unconventional method for hybrid robot dynamics [25]. The contributions of this paper are

- a comparison of existing tools for the symbolic form of robot dynamics,
- elaborations on performing unit testing for parts of theoretical models at the example of robot dynamics,
- details on the implementation of the new toolchain, which may be used to structure similar programs in other fields,
- the application of the toolchain to a robot model database.[2]

The remainder of the paper is structured as follows. An overview of existing programs is given Sect. 2. Theoretical fundamentals of robot dynamics are summarized in Sect. 3 with a focus on how to perform unit testing. The structure of the toolchain is presented in Sect. 4. The robot database as application example is introduced in Sect. 5 and Sect. 6 concludes the paper.

2 Comparison of Existing Toolboxes for Robot Dynamics

As sketched in the previous section, several tools already exist for generating the rigid body dynamics of robots in symbolic form. An extensive comparison of the tools is given in Table 1. Some older software packages, e.g. referenced in [27] are left out of the comparison due to their presumed deprecation. Commercial software, such as Robotran and MapleSim, is available at a mature stage of development. Since OpenSymoro is publicly available, the necessity escapes to use Symoro+ with similar features. A variety of open source projects for robot manipulators is implemented in Python using the sympy library as computer algebra system (CAS) which helps avoiding licensing costs for software like Maple, Mathematica, MapleSim and Matlab. The drawback of open source tools is the dependency on single researchers supporting them, as can be seen by the status "unmaintained" of SymPyBotics or the GitHub list of issues of OpenSymoro.

Robotran, MapleSim and Neweul-M^2 allow the derivation of multibody dynamics of general mechanisms, which includes both tree-like and closed-loop systems and therefore all types of common robot manipulators. The Python tools

[1] Available under free license at https://github.com/SchapplM/robsynth-modelgen.

[2] The database is available at https://github.com/SchapplM/robsynth-serroblib for serial robots, ...-serhybroblib for hybrid and ...-parroblib for parallel robots.

mainly focus on robotic applications, e.g. only serial robots [27] or additionally robot kinematics with closed loops [14,30]. Some have very specific focus, such as humanoid robots [2] or dynamics model identification [2,30]. Parallel robots (PKM, parallel kinematic machines) require a specific modeling approach (see Sect. 3.3), which is not available in the open source tools, but can be obtained by MapleSim or Robotran. If a general closed-loop robot model not in minimal (platform) coordinates is used instead, this leads to a less efficient implementation, since coordinates of platform coupling joints remain in the equations.

Some multibody tools have graphical user interfaces that reduce the need of expert knowledge. For the batch creation of a robot database, this strength can become a weakness, if it is not possible to automatically generate the dynamics equations from a standardized description of the robot. A new toolchain was developed, partly to avoid dependencies on dedicated commercial tools (while allowing the dependency on a commercial CAS), partly due to the fact that most open source tools were not accessible at the begin of the work in 2015. Since an institutional license was available, the core tools of the proposed toolchain are Maple for the symbolics engine and Matlab for the model evaluation and simulations. This design decision distinguishes the proposed toolchain from the Python toolboxes which have no commercial dependencies.

Table 1. Comparison of different tools for symbolic robot dynamics (legend below)

Name	Ref.	Area (1)	License (2)	CAS (3)	IM (4)	FlB (5)	Year	UI (6)
Robotran	[6]	Multibody	Comm. (7)	MBS (8)	Yes	Yes	1990	GUI/CMD
Robotica	[22]	OL Rob.	OSS	Mathematica	No	No	1994	CMD+Vis.
Symoro+	[13]	OL/CL Rob.	Comm.	Mathematica	Yes	No	1997	GUI
MapleSim	[31]	Multibody	Comm.	Maple	?	?	2000	GUI
Neweul-M^2	[18]	Multibody	Pr. (9)	Matlab	No	Yes	2007	GUI/CMD
ParaDyn	[5]	OL/PKM	Pr. (10)	Maple	Yes	No	2009	CMD
RVC toolbox	[4]	OL Rob	OSS	Matlab	No	No	2012	CMD+Vis.
OpenSymoro	[14]	OL/CL Rob.	OSS	Python	Yes	Yes	2014	CMD+Vis.
SymPyBotics	[27]	OL Rob.	OSS	Python	Yes	No	2014	CMD+Vis.
FloBaRoID	[2]	OL Rob.	OSS	Python	Yes	Yes	2016	CMD+Vis.
dVRK DMI	[30]	OL/CL Rob.	OSS	Python	Yes	No	2019	CMD
Proposed		OL/CL/PKM	OSS	Maple	Yes	Yes	2019	CMD

Legend for Table 1 (referenced by round brackets in table headings and rows):
1: Area of application: multibody: general m.b. dynamics; OL Rob.: open loop robots; CL Rob.: closed-loop robots; PKM: parallel kinematic machines (parallel robots).
2: OSS: Open source software; comm.: commercial software; pr.: proprietary tool.
3: Additional license required for Mathematica, Matlab Symbolics Toolbox or Maple.
4: Identification model of the inverse dynamics (linear in the dynamics parameters).
5: Floating base model for the inverse dynamics (non-fixed base link with six DoF).
6: UI: User interface; GUI: graphical; CMD: command line; Vis.: visualisation of the results (but no visual interface for input).
7: Free for teaching and academic research.
8: Dedicated CAS for multibody systems (see [24]), accessed via web-based service.
9: Access to the software provided for project partners from industry and academia.
10: The tool was used for several projects from 2009 to 2017 at the author's institute.

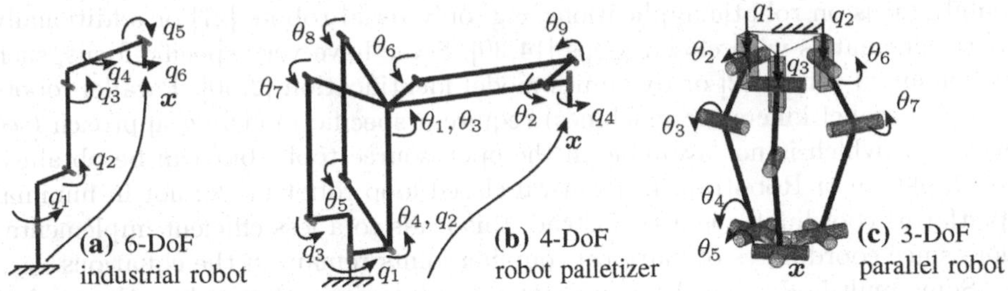

Fig. 1. Examples of different types of robots with annotation of coordinates $\boldsymbol{q}$, $\boldsymbol{\theta}$, $\boldsymbol{x}$. Cylinders and cuboids mark revolute and prismatic joints.

3 Robot Dynamics and Unit Testing Framework

The following section contains a high-level summary of the kinematic and dynamics models of the three types of robots implemented in the proposed toolchain. The modeling usually starts from a kinematic sketch of the robot, which may originate from an existing CAD model. Detailed derivations and explanations can be obtained from standard textbooks, such as [3,12,19,24], and the research papers that are referenced. The theory is the basis for the unit testing framework and therefore every part of the model (structured into numbered properties) is followed by an elaboration on how to perform a unit test on it. It is assumed that the model is derived symbolically but tests are performed numerically, since the output of the tool are functions which implement single terms of the model. A symbolic check for equality of two expressions is often not feasible, especially if the derivation is by different approaches. The structure of this section enables to follow the transfer from theory to test cases and allows an adaption of the approach to theoretical models from other fields. The theoretical framework is restricted to rigid body dynamics for three different types of robots, which each require a specific approach to derive efficient models. The three types of robots are sketched in Fig. 1. Serial robots (Fig. 1,a) are discussed in Sect. 3.1, serial-hybrid robots (Fig. 1,b) in Sect. 3.2 and parallel robots (Fig. 1,c) in Sect. 3.3.

3.1 Serial-Link Robots

Fundamentals for serial robots are taken from the standard textbook [12]. The derivation of the theory is structured according to the basic dynamics principles of Newton-Euler and Lagrange, which start with the relations of position and velocity (kinematics), over the definition of energy to forces (dynamics).

The following list of examples is not exhaustive for serial robots, but gives an impression on how to prove the validity of the results for all steps of the derivation of the kinematic and dynamics equations. The theoretical properties and test cases of the models are partly summarized in Fig. 2. For a good test coverage, each property block should have a dashed connection to a test case block.

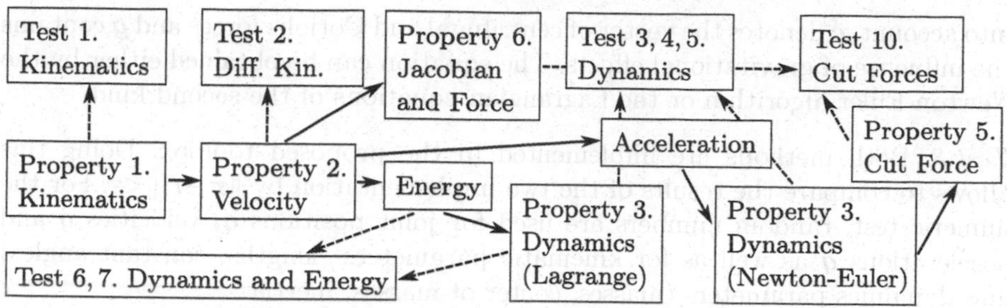

Fig. 2. Overview of the properties and tests for serial robots

Property 1. The end effector pose $\boldsymbol{x}$ can be calculated via $\boldsymbol{x}=\boldsymbol{f}(\boldsymbol{q})$ for a given vector of n joint coordinates $\boldsymbol{q}$. The *forward kinematics* is implemented using homogenous transformation matrices and the modified Denavit-Hartenberg parameters from [12] for a minimal-parameter representation of the joint transformations. The pose $\boldsymbol{x}$ is without loss of generality a vector of position and three Euler angles expressing orientation.

Test 1. The function $\boldsymbol{f}(\boldsymbol{q})$ can be validated graphically with plots, such as Fig. 1,a. A CAD model can facilitate this, if available, but this is not mandatory.

Property 2. The *velocity* of the end effector follows the linear relation $\boldsymbol{v}=\boldsymbol{J}_{\mathrm{g}}(\boldsymbol{q})\dot{\boldsymbol{q}}$ and $\dot{\boldsymbol{x}}=\boldsymbol{J}(\boldsymbol{q})\dot{\boldsymbol{q}}$, which represents the *differential kinematics*. The velocity $\boldsymbol{v}$ contains linear velocity and angular velocity, while $\dot{\boldsymbol{x}}$ contains the time derivative of the representation of orientation instead of the angular velocity. The geometric Jacobian matrix $\boldsymbol{J}_{\mathrm{g}}(\boldsymbol{q})$ can be derived by a geometric formula or by performing the partial derivative $\boldsymbol{J}_{\mathrm{g}}=\partial\boldsymbol{v}(\boldsymbol{q},\dot{\boldsymbol{q}})/\partial\dot{\boldsymbol{q}}$ based on the kinematics of velocities of the rigid bodies in the robot. The latter is beneficial for hybrid robots.

Test 2. The implementation of the Jacobian can be tested as follows: Let $\boldsymbol{q}_1$ and $\Delta\boldsymbol{q}$ be arbitrary vectors in $\mathbb{R}^n$ with $\|\Delta\boldsymbol{q}\|\ll 1$ and $\boldsymbol{q}_2=\boldsymbol{q}_1+\Delta\boldsymbol{q}$. The pose difference $\Delta\boldsymbol{x}=\boldsymbol{x}_2-\boldsymbol{x}_1=\boldsymbol{f}(\boldsymbol{q}_2)-\boldsymbol{f}(\boldsymbol{q}_1)$ can also be obtained by using the Jacobian with $\Delta\boldsymbol{x}'=\boldsymbol{J}(\boldsymbol{q}_1)\Delta\boldsymbol{q}$. The dash only denotes the second implementation. If the implementation of the Jacobian is correct, we have $\|\Delta\boldsymbol{x}-\Delta\boldsymbol{x}'\|<\varepsilon$. This relation is trivial from a mathematical point of view, since the test only uses differential calculus from the derivation of $\boldsymbol{J}$ in Property 2. However, this has to be explicitely implemented as a test to ensure the correctness of the implementation of $\boldsymbol{J}(\boldsymbol{q})$. Throughout this paper the threshold $\varepsilon\approx 10^{-9}$ is used to check for numeric equality. This accounts for linearization error within differential relations and rounding errors of floating point numbers in the numerous operations.

Property 3. The *inverse dynamics* equation $\boldsymbol{\tau}=\boldsymbol{M}(\boldsymbol{q})\ddot{\boldsymbol{q}}+\boldsymbol{c}(\boldsymbol{q},\dot{\boldsymbol{q}})+\boldsymbol{g}(\boldsymbol{q})$ for the rigid body robot model in joint coordinates of the robot gives torques $\boldsymbol{\tau}$ of the joints required to perform the motion $\boldsymbol{q}$, $\dot{\boldsymbol{q}}$, $\ddot{\boldsymbol{q}}$. The general equation describes both prismatic and revolute joints. The inertia matrix $\boldsymbol{M}$ takes inertial couplings into account, $\boldsymbol{c}$ denotes the vector of centrifugal and Coriolis forces and $\boldsymbol{g}$ contains the influence of gravitational effects. The equation can be obtained either by the Newton-Euler algorithm or the Lagrangian equations of the second kind.

Test 3. Both methods are implemented in the proposed toolbox. Doing this allows to compare the results of the two implementation by $\|\boldsymbol{\tau}-\boldsymbol{\tau}'\|<\varepsilon$. For the numeric test, random numbers are used for joint positions $\boldsymbol{q}$, velocities $\dot{\boldsymbol{q}}$ and accelerations $\ddot{\boldsymbol{q}}$ as well as for kinematic parameters (lengths, constant angles) and dynamics parameters (masses, center of masses, inertia).

Test 4. Several *properties of the dynamics equations* can be exploited to perform further tests on the implementations of the single terms. A Coriolis matrix $\boldsymbol{C}$ can be obtained from the mass matrix $\boldsymbol{M}$ using Christoffel symbols. It follows the relation $\boldsymbol{c}'(\boldsymbol{q},\dot{\boldsymbol{q}}) = \boldsymbol{C}(\boldsymbol{q},\dot{\boldsymbol{q}})\dot{\boldsymbol{q}}$. This can be exploited by the test $\|\boldsymbol{c}-\boldsymbol{c}'\|<\varepsilon$.

Test 5. The term $\dot{\boldsymbol{M}}(\boldsymbol{q},\dot{\boldsymbol{q}})-2\boldsymbol{C}(\boldsymbol{q},\dot{\boldsymbol{q}})$ has to be a skew matrix.

Test 6. The *kinetic energy* can be derived symbolically using mechanics principles as $E_{\mathrm{kin}}(\boldsymbol{q},\dot{\boldsymbol{q}})$. The mass matrix (in generalized coordinates $\boldsymbol{q}$) is connected with the kinetic energy via $E'_{\mathrm{kin}} = \dot{\boldsymbol{q}}^{\mathrm{T}}\boldsymbol{M}(\boldsymbol{q})\dot{\boldsymbol{q}}$. Comparing the two implementations leads to the test inequality $\|E_{\mathrm{kin}}-E'_{\mathrm{kin}}\|<\varepsilon$.

Test 7. The gravitational model can be tested with a *forward dynamics* simulation. An ODE simulation is performed for $\ddot{\boldsymbol{q}} = -\boldsymbol{M}^{-1}(\boldsymbol{q})[\boldsymbol{c}(\boldsymbol{q},\dot{\boldsymbol{q}})+\boldsymbol{g}(\boldsymbol{q})]$ using the Runge-Kutta numerical integration `ode45`. This gives a time series $\boldsymbol{q}(t)$, $\dot{\boldsymbol{q}}(t)$ and $\ddot{\boldsymbol{q}}(t)$. The *sum of energies* over this trajectory is calculated using the potential energy $E_{\mathrm{pot}}(\boldsymbol{q})$, which has to be implemented within the Lagrange approach. The test now checks, if the sum of energies $E_{\mathrm{total}} = E_{\mathrm{kin}}+E_{\mathrm{pot}}$ stays constant, by using the inequality $\|E_{\mathrm{total}}(t=0)-E_{\mathrm{total}}(t=t_{\mathrm{end}})\|<\varepsilon$.

Property 4. The dynamics equations can be formulated in different *sets of parameters*: barycentric parameters $\boldsymbol{p}_{\mathrm{B}}$ (mass, center of mass, inertia), inertial parameters $\boldsymbol{p}_{\mathrm{I}}$ (mass, first and second moments of mass) and a minimal parameter vector $\boldsymbol{p}_{\mathrm{M}}$ which is a linear combination of the inertial parameters, regrouping parameters with the same effect on the dynamics. The latter implementation is very efficient and essential for the identification of the parameters of a real robot. The former approaches are more intuitive. The inertial parameters only occur in a linear relation in the dynamics equations, allowing to write $\boldsymbol{\tau}' = \boldsymbol{\Phi}_{\mathrm{I}}(\boldsymbol{q},\dot{\boldsymbol{q}},\ddot{\boldsymbol{q}})\boldsymbol{p}_{\mathrm{I}}$ and $\boldsymbol{\tau}'' = \boldsymbol{\Phi}_{\mathrm{M}}(\boldsymbol{q},\dot{\boldsymbol{q}},\ddot{\boldsymbol{q}})\boldsymbol{p}_{\mathrm{M}}$. This *identification model of the dynamics* requires a specific approach to the derivation of velocity and energy.

Test 8. To compare the different implementations regarding sets of parameters, a consistent set of parameters $\boldsymbol{p}_{\mathrm{B}}$, $\boldsymbol{p}_{\mathrm{I}}$ and $\boldsymbol{p}_{\mathrm{M}}$ is created using the parallel axis theorem. Then the tests $\|\boldsymbol{\tau}-\boldsymbol{\tau}'\|<\varepsilon$ and $\|\boldsymbol{\tau}-\boldsymbol{\tau}''\|<\varepsilon$ are performed.

Test 9. It is tested numerically if $\boldsymbol{\Phi}_{\mathrm{M}}\boldsymbol{p}_{\mathrm{M}}$ is a minimal form of $\boldsymbol{\Phi}_{\mathrm{I}}\boldsymbol{p}_{\mathrm{I}}$. Via QR decomposition it is checked that the information matrix for a virtual identification problem with random virtual trajectory samples $\boldsymbol{q}_1,\dot{\boldsymbol{q}}_1,\ddot{\boldsymbol{q}}_1,\boldsymbol{q}_2,\ldots$ has $\mathrm{rank}([\boldsymbol{\Phi}_{\mathrm{I}}^{\mathrm{T}}(\boldsymbol{q}_1,\dot{\boldsymbol{q}}_1,\ddot{\boldsymbol{q}}_1),\boldsymbol{\Phi}_{\mathrm{I}}^{\mathrm{T}}(\boldsymbol{q}_2,\dot{\boldsymbol{q}}_2,\ddot{\boldsymbol{q}}_2),\ldots]^{\mathrm{T}}) = \dim(\boldsymbol{p}_{\mathrm{M}})$.

Property 5. The *internal cut forces* from the rigid body dynamics are calculated with $\boldsymbol{w}^{\mathrm{T}} = [\boldsymbol{w}_0^{\mathrm{T}}, \boldsymbol{w}_1^{\mathrm{T}}, ..., \boldsymbol{w}_n^{\mathrm{T}}]^{\mathrm{T}}$, where $\boldsymbol{w}_i$ contains the stacked cut force and cut moment for rigid body i (cut at the corresponding joint) from the Newton-Euler approach.

Property 6. The geometric approach for the $6{\times}n$ Jacobian matrix $\boldsymbol{J}_{\mathrm{g}}$ can be extended to obtain a $6 \times 6(n{+}1)$ cut force Jacobian $\boldsymbol{J}_{\mathrm{g,cut}}$. The internal cut forces from an external wrench (force and moment) can be obtained using this matrix with $\boldsymbol{w}' = \boldsymbol{J}_{\mathrm{g,cut}}^{\mathrm{T}}(\boldsymbol{q})[\boldsymbol{f}_{\mathrm{ext}}^{\mathrm{T}}, \boldsymbol{m}_{\mathrm{ext}}^{\mathrm{T}}]^{\mathrm{T}}$.

Test 10. Further, it has to be ensured that the implementations of kinematics and dynamics match each other. For the test, we set $\ddot{\boldsymbol{q}} = \dot{\boldsymbol{q}} = \boldsymbol{0}$ and set only one mass of the robot to be non-zero. Here, $\boldsymbol{f}_{\mathrm{ext}} = \boldsymbol{f}_{\mathrm{grav}}$ is the force resulting from the test mass gravity and $\boldsymbol{m}_{\mathrm{ext}} = \boldsymbol{0}$. Both expressions for the cut force have to be identical, which can be tested numerically with $\|\boldsymbol{w}{-}\boldsymbol{w}'\|{<}\varepsilon$.

3.2 Serial-Hybrid Robots

Serial-hybrid robots consist of a serial main structure connecting the base and the end effector with additional closed kinematic loops, as depicted in Fig. 1,b. The closed loops are used to constrain degrees of freedom or to shift the position of motors within the structure [21]. The kinematics of closed loops require a different approach than of open loops. The joint coordinates are separated into the generalized (active joint) coordinates $\boldsymbol{q}$ and passive joints coordinates $\boldsymbol{\theta}$. The theory can be viewed in detail in the textbooks [12,24] and e.g. in [6,9,13,21,26].

Property 7. The default approach uses *loop equations in the implicit formulation* $\boldsymbol{h}(\boldsymbol{q}, \boldsymbol{\theta}) = \boldsymbol{0}$. For simple mechanisms, such as the planar parallelograms in Fig. 1,b, an *inverse geometric model* can be formulated explicitely as $\boldsymbol{\theta} = \boldsymbol{\theta}(\boldsymbol{q})$. The kinematic model of serial-hybrid robots is set up with the extended version of the modified DH parameters [3,12] taking the branching in the kinematic tree structure into account. Additionally to the open loop model, the loop closing conditions are modeled as symbolic equations for $\boldsymbol{h}(\boldsymbol{q}, \boldsymbol{\theta})$ and $\boldsymbol{\theta}(\boldsymbol{q})$ by hand.

Test 11. While creating the model, visual plausibility is checked with a kinematic sketch as in Fig. 1,b.

Test 12. After this, the test $\|\boldsymbol{h}(\boldsymbol{q}, \boldsymbol{\theta})\|{<}\varepsilon$ is performed. The passive joint coordinates $\boldsymbol{\theta}(\boldsymbol{q})$ are obtained symbolically or – if not possible – numerically using the Newton-Raphson algorithm on $\boldsymbol{h} = \boldsymbol{0}$. The kinematic parameters and test configurations for $\boldsymbol{q}$ can not be chosen randomly as in the serial robot case, but have to be chosen as plausible values by visual inspection or from CAD data.

Property 8. The *kinematic constraints* can be formulated in the *differential form* $\dot{\boldsymbol{\theta}} = \boldsymbol{J}_{\theta}\dot{\boldsymbol{q}}$. The constraints Jacobian $\boldsymbol{J}_{\theta}$ can be obtained from the implicit form $\boldsymbol{h}$ of the constraints as $\boldsymbol{J}_{\theta} = -(\partial\boldsymbol{h}/\partial\boldsymbol{\theta})^{-1}(\partial\boldsymbol{h}/\partial\boldsymbol{q})$. Using the elimination approach [25], $\boldsymbol{\theta}(\boldsymbol{q})$ is available in symbolic form and the differential relation $\boldsymbol{J}_{\theta}' = \partial\boldsymbol{\theta}/\partial\boldsymbol{q}$ [21] can be obtained.

Test 13. The two implementations (implicit and explicit form) are tested with the identity of $\boldsymbol{J}_\theta$ and $\boldsymbol{J}'_\theta$ up to rounding errors ε.

Property 9. Creating the robot model requires the definition of an open-loop tree structure with the coordinates $\boldsymbol{q}_{\mathrm{OL}}$, which contains the coordinates $\boldsymbol{q}$ and $\boldsymbol{\theta}$. For this model, all tests and definitions from Sect. 3.1 can be used.

Test 14. Velocities of the rigid bodies of the robot are now generated by both models based on Property 2. For the elimination approach $\boldsymbol{v} = \boldsymbol{J}_{\mathrm{g}}(\boldsymbol{q})\dot{\boldsymbol{q}}$ and for the open-loop structure (implicit approach) $\boldsymbol{v}' = \boldsymbol{J}_{\mathrm{g,OL}}(\boldsymbol{q}_{\mathrm{OL}})\dot{\boldsymbol{q}}_{\mathrm{OL}}$ is used. The entities $\boldsymbol{q}$ and $\boldsymbol{q}_{\mathrm{OL}}$ as well as $\dot{\boldsymbol{q}}$ and $\dot{\boldsymbol{q}}_{\mathrm{OL}}$ are chosen consistently with random numbers like in Test 12. Using this within the test $\|\boldsymbol{v}-\boldsymbol{v}'\|<\varepsilon$ proves the validity of the implementation of the velocities within the algorithm. The velocity and Jacobians can be set up for any rigid body of the mechanism, not limited to the end effector link.

The same approach can be performed for the accelerations.

Property 10. The *dynamics equations* are again deduced by two different approaches to allow testing the results. Using the elimination approach [25], the passive joints $\boldsymbol{\theta}$ are completely eliminated from the symbolic equations already at the kinematics stage. The Lagrangian equations of the second kind are used to deduce the dynamics $\boldsymbol{\tau}(\boldsymbol{q}, \dot{\boldsymbol{q}}, \ddot{\boldsymbol{q}})$ in the closed-loop robots minimal coordinates $\boldsymbol{q}$. The projection approach leads to $\boldsymbol{\tau}' = \boldsymbol{\tau}_q + \boldsymbol{J}_\theta \boldsymbol{\tau}_\theta$, where $\boldsymbol{\tau}_q$ and $\boldsymbol{\tau}_\theta$ are the components of the open-loop dynamics $\boldsymbol{\tau}_{\mathrm{OL}}(\boldsymbol{q}_{\mathrm{OL}}, \dot{\boldsymbol{q}}_{\mathrm{OL}}, \ddot{\boldsymbol{q}}_{\mathrm{OL}})$ corresponding to the entries of $\boldsymbol{q}$ and $\boldsymbol{\theta}$ in $\boldsymbol{q}_{\mathrm{OL}}$.

Test 15. The implementations are again tested numerically via $\|\boldsymbol{\tau}-\boldsymbol{\tau}'\|<\varepsilon$. Random values for $\boldsymbol{q}, \dot{\boldsymbol{q}}, \ddot{q}$ and consistent values for $\boldsymbol{\theta}, \dot{\boldsymbol{\theta}}, \ddot{\boldsymbol{\theta}}$ are selected.

Some other tests on the dynamics from Sect. 3.2, such as the test of energy consistency, are also applied.

3.3 Parallel Robots

Parallel robots, as given in Fig. 1,c have a similar modeling approach as hybrid robots since they also contain closed kinematic loops. A detailed overview on the dynamics is given in [3]. Usually the platform coordinates $\boldsymbol{x}$ are chosen as minimal coordinates of the system [19]. The relation between platform velocity $\boldsymbol{v}$ and the time derivative $\dot{\boldsymbol{x}}$ of the platform coordinates has to be regarded in the algorithm [1,19] and is considered in the implementation, but is omitted here for the sake of brevity and only entities related to $\dot{\boldsymbol{x}}$ are presented.

Property 11. For the symbolic derivation of the dynamics the two-step projection approach from [1] with a claim on high efficiency is used and gives the dynamics $\boldsymbol{\tau}_x(\boldsymbol{x}, \dot{\boldsymbol{x}}, \ddot{\boldsymbol{x}}, \boldsymbol{q}_{\mathrm{OL}})$ and the inverse Jacobian matrix $\boldsymbol{J}^{-1}(\boldsymbol{x}, \boldsymbol{q}_{\mathrm{OL}})$ with $\dot{\boldsymbol{q}} = \boldsymbol{J}^{-1}\dot{\boldsymbol{x}}$. The dynamics $\boldsymbol{\tau}_x$ in platform coordinates can be projected into the active joint coordinates with $\boldsymbol{\tau} = \boldsymbol{J}^{\mathrm{T}}\boldsymbol{\tau}_x$. This represents the actuator force necessary to achieve the robot motion given by $\boldsymbol{x}$, $\dot{\boldsymbol{x}}$ and $\ddot{\boldsymbol{x}}$ – a value necessary for simulation and control.

Property 12. This implementation is verified by a second approach, which is taken from [5] and presents a more general approach than standard algorithms [19]. It mainly corresponds to a general form of the state-of-the-art approach for the dynamics of closed kinematic loops from [6] with focus on using the platform coordinates $\boldsymbol{x}$. The *implicit definition of the constraints* $\boldsymbol{h}(\boldsymbol{x}, \boldsymbol{q}_{\mathrm{OL}}) = \boldsymbol{0}$ is only partially implemented symbolically due to the high computational demand in the general case. Again, $\boldsymbol{q}_{\mathrm{OL}}$ includes the active joint coordinates $\boldsymbol{q}$ and the passive joints coordinates $\boldsymbol{\theta}$ of the kinematic leg structure (without the platform).

Property 13. The *differential formulation* $\boldsymbol{h}_{\boldsymbol{x}}\dot{\boldsymbol{x}}+\boldsymbol{h}_{\boldsymbol{q}_{\mathrm{OL}}}\dot{\boldsymbol{q}}_{\mathrm{OL}} = \boldsymbol{0}$ with $\boldsymbol{h}_{\boldsymbol{x}} = \partial\boldsymbol{h}/\partial\boldsymbol{x}$ and $\boldsymbol{h}_{\boldsymbol{q}_{\mathrm{OL}}} = \partial\boldsymbol{h}/\partial\boldsymbol{q}_{\mathrm{OL}}$ can be obtained from the constraints equations. This leads to the *inverse Jacobian matrix* $\boldsymbol{J}_{\mathrm{OL}}^{-1} = -\boldsymbol{h}_{\boldsymbol{q}_{\mathrm{OL}}}^{-1}\boldsymbol{h}_{\boldsymbol{x}}$ for the full joint vector, relating $\dot{\boldsymbol{q}}_{\mathrm{OL}} = \boldsymbol{J}_{\mathrm{OL}}^{-1}\dot{\boldsymbol{x}}$. Selecting only the rows corresponding to the active joint coordinates gives the inverse Jacobian matrix $\boldsymbol{J}'^{-1}$, where the dash only marks the second implementation in demarcation of the first one.

Test 16. The gradient matrices $\boldsymbol{h}_{\boldsymbol{x}}$ and $\boldsymbol{h}_{\boldsymbol{q}_{\mathrm{OL}}}$ are tested against the constraints formulation $\boldsymbol{h}$ by defining $\Delta\boldsymbol{x}$ and $\Delta\boldsymbol{q}_{\mathrm{OL}}$ with $\|\Delta\boldsymbol{x}\|\ll 1$ and $\|\Delta\boldsymbol{q}_{\mathrm{OL}}\|\ll 1$. Let $\boldsymbol{x}_1$ and $\boldsymbol{q}_{\mathrm{OL},1}$ be arbitrary random numbers with $\boldsymbol{h}_1 = \boldsymbol{h}(\boldsymbol{x}_1, \boldsymbol{q}_{\mathrm{OL},1}) \neq \boldsymbol{0}$. The values $\boldsymbol{x}_2 = \boldsymbol{x}_1+\Delta\boldsymbol{x}$, $\boldsymbol{q}_{\mathrm{OL},2} = \boldsymbol{q}_{\mathrm{OL},1}+\Delta\boldsymbol{q}_{\mathrm{OL}}$ and $\boldsymbol{h}_2 = \boldsymbol{h}(\boldsymbol{x}_2, \boldsymbol{q}_{\mathrm{OL},2})$ are calculated. As a second step, $\boldsymbol{h}_2' = \boldsymbol{h}_1+\boldsymbol{h}_{\boldsymbol{x}}\Delta\boldsymbol{x}+\boldsymbol{h}_{\boldsymbol{q}_{\mathrm{OL}}}\Delta\boldsymbol{q}_{\mathrm{OL}}$ is calculated and the two implementations are tested with $\|\boldsymbol{h}_2-\boldsymbol{h}_2'\|<\varepsilon$. This of course is (again) mathematically trivial, but necessary, as elaborated upon in Test 2. Due to the complexity of the terms the implementation is otherwise prone to errors.

Test 17. Both implementations $\boldsymbol{J}^{-1}$ and $\boldsymbol{J}'^{-1}$ from Properties 11 and 13 are tested for equality up to rounding errors of ε within the numerical computation.

Similar tests can be defined for the second time derivative of the constraints equation, which is used to determine the acceleration relations.

Property 14. The second implementation of the *dynamics of parallel robots* [5] is determined numerically. The approach is very similar to the case of hybrid robots using the constraints Jacobians [6, 24]. Expressed in platform frame, it results $\boldsymbol{\tau}_{\boldsymbol{x}}' = \boldsymbol{\tau}_{\mathrm{P}}(\boldsymbol{x}, \dot{\boldsymbol{x}}, \ddot{\boldsymbol{x}})+\boldsymbol{J}_{\mathrm{OL}}^{-\mathrm{T}}(\boldsymbol{q}_{\mathrm{OL}}, \boldsymbol{x})\boldsymbol{\tau}_{\mathrm{OL}}(\boldsymbol{q}_{\mathrm{OL}}, \dot{\boldsymbol{q}}_{\mathrm{OL}}, \ddot{\boldsymbol{q}}_{\mathrm{OL}})$. The dependencies on $\boldsymbol{q}_{\mathrm{OL}}$ and $\boldsymbol{x}$ and their time derivatives are added for clarity. These quantities have to fulfill the constraints equations $\boldsymbol{h} = \boldsymbol{0}$, $\dot{\boldsymbol{h}} = \boldsymbol{0}$ and $\ddot{\boldsymbol{h}} = \boldsymbol{0}$. The dynamics of the platform as a rigid body in Cartesian space is considered with the term $\boldsymbol{\tau}_{\mathrm{P}}$ and $\boldsymbol{\tau}_{\mathrm{OL}}$ contains the open-loop dynamics of the single leg chains that are deduced with the methods presented in Sect. 3.1.

Test 18. Both implementations are then tested using the inequality $\|\boldsymbol{\tau}_{\boldsymbol{x}}-\boldsymbol{\tau}_{\boldsymbol{x}}'\|<\varepsilon$.

Other tests for the dynamics, such as energy consistency by time-integration of the forward dynamics, can be performed as presented in Sect. 3.1.

4 Description of the Proposed Toolchain

The fundamentals of kinematics and dynamics of Sect. 3 are implemented in a toolchain to obtain the robot models for serial, hybrid and parallel robots following the requirements introduced at the end of Sect. 1. The program is structured within several Maple worksheets which each contain only a limited set of fundamental equations, corresponding to one of the numbered properties in Sect. 3. This allows a convenient debugging with the graphical user interface of Maple. All intermediate symbolic expressions are exchanged between worksheets via data files which are saved in one worksheet and read in the next. Therefore each worksheet can run independently, once previous parts are generated. This structural decision can be justified also by the experiences with another toolbox [5], which was implemented solely based on Maple procedures. This made debugging and extending the tool an impossible task regarding 280 interleaved procedures.

The proposed toolchain has three workflows corresponding to the robot type, where the *serial robot* case is the most central one. The workflow for *parallel robots* is modular. It first generates the corresponding serial kinematic leg chain which is then used by the approach of Property 11 in Sect. 3.3. The use of the Lagrange equations of the second kind allows a modular reuse of the worksheets for serial robots also for *hybrid robots* using the elimination approach. The second implementation for hybrid robots of Property 10 in Sect. 3.2 is implemented in a modular way similar to parallel robots using the workflow of the open-loop tree structure first and then applying the worksheets for the implicit constraints.

The overall workflow is summarized in Fig. 3. As *step 1*, serial robots are described with an input definition file using DH parameters from [12] and parallel robots by an additional definition file referring to the leg chain and alignment of the base joint. For hybrid robots, a separate manually created worksheet for the constraints of Sect. 3.2 is necessary. *Step 2* comprises (automatically) running all Maple worksheets. To enable batch and partially parallel processing, all worksheets (`.mw` files) are saved separately in a text format (with `.mpl` extension), which can be run by the terminal application of Maple. Every worksheet exports the symbolic expressions of the model equations as optimized code in Matlab syntax using the Maple `CodeGeneration` package. Following basic principles for software quality [32], in *step 3* the automatically generated optimized Matlab code for all symbolic expressions is post-processed to reach a certain standard. A Bash script creates a function file with a header comment with short description of the function and its inputs and outputs, assertions to prevent unexpected user input, compiler information, statistics of the code generation and finally the optimized code itself. After all function files are (automatically) generated, the unit test framework is run in *step 4*. If all tests are passed, the results can be used for their designated purpose in *step 5*.

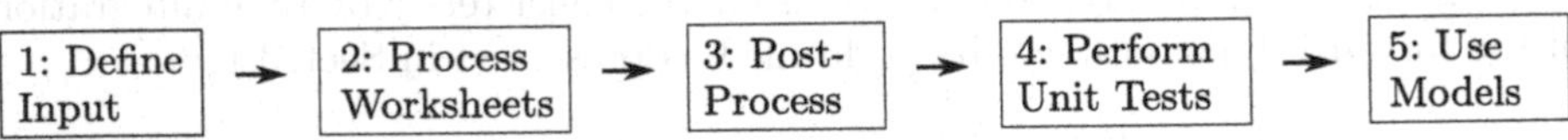

Fig. 3. Overview of the overall workflow of the toolchain

5 Application to a Model Database Framework

As introduced in Sect. 1, one of the purposes of the toolchain is to create an algorithm that is able to determine the best robot for a given task. In a first step of structural synthesis all possible structures have to be identified and their description stored systematically. Then all robot models have to be generated in symbolic form. Finally, the batch-optimization of all robots is performed and the best robot is selected. All serial kinematic chains are generated using the approach from [23], which is similar to the evolutionary morphology from [11]. Leg chains for parallel robots are created with the same approach with modifications on the requirements from parallel robot structural synthesis [11, 15]. This leads to a database of the kinematic descriptions, i.e. Denavit-Hartenberg parameters for serial robots [12]. For parallel robots lists of possible leg chains and alignments of the base and platform coupling joints are stored. The databases are saved in text-based csv-tables to facilitate software version control using Git.

The models in the serial robot database are created as Matlab functions by batch-processing the robot definition files with the proposed toolchain. This stored input and output data, referenced in the footnotes on p. 3 of the paper, can be regarded as a case study for the validity of the toolchain. At the current stage, 616 unique kinematic chains (with 3 to 6 joints) are stored, which by elimination of isomorphisms represent all possible structures [23]. The database contains approximately 36 thousand Matlab files with 6 million lines of automatically generated code. The size of the complete database is around 300 MB and therefore still feasible. Generating all robot models takes about 5 days of CPU time on a standard desktop computer. Due to partially parallel execution, all model files for one serial robot can be generated within one hour. By also generating the whole database with the tools SymPyBotics [27] and OpenSymoro [14], the inverse dynamics of Property 3 is validated against another reference. The number of operations of these different implementations are counted in the generated code and are compared in Fig. 4. The proposed Maple toolchain (with Newton-Euler) has a similar efficiency as the Python-based references. Using Lagrange is less efficient, as expected from literature [12].

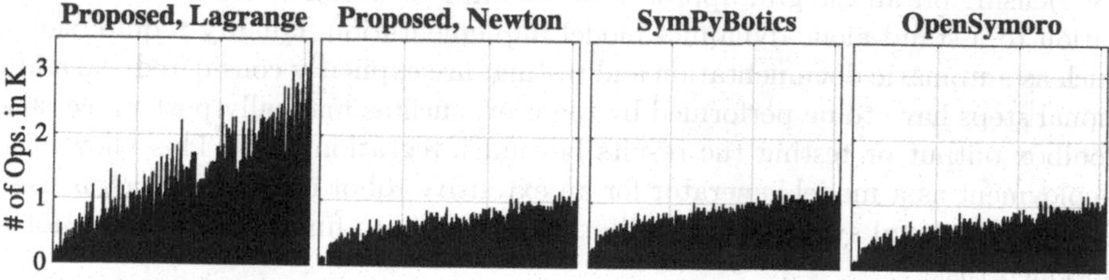

Fig. 4. Comparison of the number of operations for inverse dynamics of the proposed toolbox with two methods (Lagrange, Newton-Euler) and two reference toolboxes (SymPyBotics, OpenSymoro) over all 616 serial kinematics on the horizontal axis. Sorted by increasing total number of joints and then by number of revolute joints.

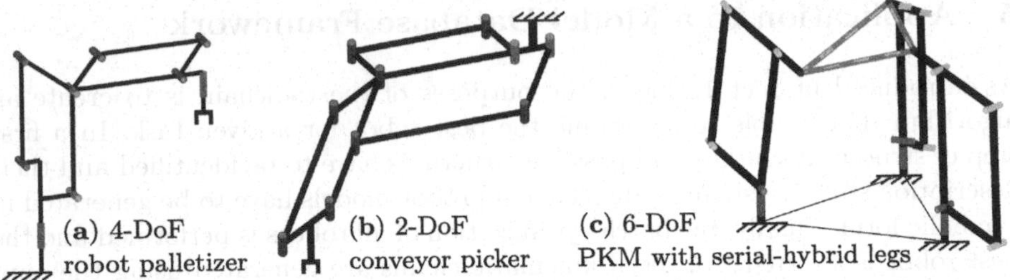

Fig. 5. Examples of hybrid robots implemented in the program framework.

Code generation for parallel robots shows that the selected symbolic approach [1] of Property 11 in Sect. 3.3 is very efficient for simple kinematic structures with a low number of kinematic parameters (resulting from a low number of mechanical joints). For general kinematic leg chains with revolute joints instead of universal and spherical joints, the necessary computation time can reach several days and is not feasible any more. The case of parallel robots with platform coupling joints that are not spherical is not included in the symbolic approach [1]. In summary, only symbolic code for 91 parallel robot models was included in the database, which in total consists of a few thousand symmetric parallel robots with 3 to 6 platform degrees of freedom (DoF). To perform dynamics simulations for parallel robots therefore the approach [5] of Property 12 in an extended formulation is used, allowing also robots with non-spherical coupling joints.

Only some examples of industrial robots with closed loops are implemented manually, such as robot palletizers (see Fig. 1,b or Fig. 5,a) and the 2-DoF pick-and-place machine for conveyor belts from Fig. 5,b. An automated systematic synthesis of serial-hybrid robots or parallel robots based on serial-hybrid leg chains as (the manually created example) in Fig. 5,c is not implemented yet.

6 Conclusion

The presented new toolchain for robot dynamics stands out against existing tools by focusing on an integral approach of a complete workflow from a robot definition to a stand-alone dynamics model implementation. Quality requirements, such as automatic documentation and testing, are explicitly considered. No additional steps have to be performed by the user, such as manually post-processing toolbox output or testing the results beyond integration tests. This allows the deployment as a model generator for an extensive robot database which is used for a dimensional synthesis over all existing robots to find the best robot for a specified task.

Acknowledgements. This work was developed over five years using funding from the German Research Foundation (DFG, grant 341489206), the Federal Ministry of Education and Research of Germany (BMBF, grant 16SV6175) and the European Union's Horizon 2020 research and innovation programme (grant 688857).

References

1. Abdellatif, H., Heimann, B.: Computational efficient inverse dynamics of 6-DOF fully parallel manipulators by using the Lagrangian formalism. Mech. Mach. Theory **44**(1), 192–207 (2009). https://doi.org/10.1016/j.mechmachtheory.2008.02.003
2. Bethge, S., Malzahn, J., Tsagarakis, N., Caldwell, D.: FloBaRoID — a software package for the identification of robot dynamics parameters. In: Ferraresi, Carlo, Quaglia, Giuseppe (eds.) RAAD 2017. MMS, vol. 49, pp. 156–165. Springer, Cham (2018). https://doi.org/10.1007/978-3-319-61276-8_18. https://github.com/kjyv/FloBaRoID
3. Briot, S., Khalil, W.: Dynamics of Parallel Robots. MMS, vol. 35. Springer, Cham (2015). https://doi.org/10.1007/978-3-319-19788-3
4. Corke, P.: Robotics, vision and control. Springer Tracts in Advanced Robotics (2011). https://doi.org/10.1007/978-3-642-20144-8. The CodeGenerator extension was added by Jörn Malzahn, available at https://github.com/petercorke/robotics-toolbox-matlab/tree/master/@CodeGenerator; a Python version of the toolbox is under development at https://github.com/petercorke/robotics-toolbox-python
5. Do Thanh, T., Kotlarski, J., Heimann, B., Ortmaier, T.: On the inverse dynamics problem of general parallel robots. In: IEEE International Conference on Mechatronics, pp. 1–6 (2009). https://doi.org/10.1109/ICMECH.2009.4957202
6. Docquier, N., Poncelet, A., Fisette, P.: ROBOTRAN: a powerful symbolic gnerator of multibody models. Mech. Sci. **4**(1), 199–219 (2013). https://doi.org/10.5194/ms-4-199-2013. https://www.robotran.be/
7. Featherstone, R.: Rigid Body Dynamics Algorithms. MMS, Springer, Boston (2008). https://doi.org/10.1007/978-1-4899-7560-7
8. Felis, M.L.: RBDL: an efficient rigid-body dynamics library using recursive algorithms. Auton. Robots **41**(2), 495–511 (2016). https://doi.org/10.1007/s10514-016-9574-0. project homepage: https://rbdl.github.io
9. Fisette, P., Postiau, T., Sass, L., Samin, J.C.: Fully symbolic generation of complex multibody models. Mech. Struct. Mach. **30**(1), 31–82 (2002). https://doi.org/10.1081/SME-120001477
10. Gogu, G.: Families of 6R orthogonal robotic manipulators with only isolated and pseudo-isolated singularities. Mech. Mach. Theory **37**(11), 1347–1375 (2002). https://doi.org/10.1016/S0094-114X(02)00048-4
11. Gogu, G.: Structural Synthesis of Parallel Robots, Part 1: Methodology, vol. 866. Springer (2008). https://doi.org/10.1007/978-1-4020-5710-6
12. Khalil, W., Dombre, E.: Modeling, identification and control of robots. Hermes Penton Sci. (2002). https://doi.org/10.1016/B978-1-903996-66-9.X5000-3
13. Khalil, W., Creusot, D.: Symoro+: a system for the symbolic modelling of robots. Robotica **15**(2), 153–161 (1997). https://doi.org/10.1017/S0263574797000180
14. Khalil, W., Vijayalingam, A., Khomutenko, B., Mukhanov, I., Lemoine, P., Ecorchard, G.: OpenSYMORO: an open-source software package for symbolic modelling of robots. In: IEEE/ASME International Conference on Advanced Intelligent Mechatronics, Besançon, France, pp. 1206–1211 (2014). https://doi.org/10.1109/AIM.2014.6878246. https://github.com/symoro/symoro
15. Kong, X., Gosselin, C.M.: Type Synthesis of Parallel Mechanisms. Springer, Heidelberg (2007). https://doi.org/10.1007/978-3-540-71990-8
16. Krefft, M.: Aufgabenangepasste Optimierung von Parallelstrukturen für Maschinen in der Produktionstechnik. Ph.D. thesis, Technische Universität Braunschweig, Germany (2006). ISBN 3802786890, Vulkan-Verlag GmbH

17. Kuo, C.H., Dai, J.S.: Structural synthesis of serial robotic manipulators subject to specific motion constraints. In: Proceedings of the ASME 2010 International Design Engineering Technical Conferences and Computers and Information in Engineering Conference, Montreal (2010). https://doi.org/10.1115/DETC2010-28947
18. Kurz, T., Eberhard, P., Henninger, C., Schiehlen, W.: From Neweul to Neweul-M^2: Symbolical equations of motion for multibody system analysis and synthesis. Multibody Syst. Dyn. **24**(1), 25–41 (2010). https://doi.org/10.1007/s11044-010-9187-x. https://www.itm.uni-stuttgart.de/software/neweul-m/
19. Merlet, J.P.: Parallel robots, Solid Mechanics and its Applications, 2nd edn., vol. 128. Springer (2006). https://doi.org/10.1007/1-4020-4133-0
20. MSC Software Corporation: MSC Adams, website. https://www.mscsoftware.com/product/adams. Accessed 2 Dec 2020
21. Nakamura, Y., Ghodoussi, M.: Dynamics computation of closed-link robot mechanisms with nonredundant and redundant actuators. IEEE Trans. Robot. Autom. (1989). https://doi.org/10.1109/70.34765
22. Nethery, J.F., Spong, M.W.: Robotica: a Mathematica package for robot analysis. IEEE Robot. Automat. Mag. **1**(1), 13–20 (1994). https://doi.org/10.1109/100.296449. github.com/RoboticSwarmControl/robotica
23. Ramirez, D.A.: Automatic generation of task-specific serial mechanisms using combined structural and dimensional synthesis. Ph.D. thesis, Gottfried Wilhelm Leibniz Universität Hannover, Germany (2018). https://doi.org/10.15488/4571
24. Samin, J.C., Fisette, P.: Symbolic Modeling of Multibody Systems. Solid Mechanics and its Applications, vol. 112. Springer, Dordrecht (2003). https://doi.org/10.1007/978-94-017-0287-4
25. Schappler, M., Lilge, T., Haddadin, S.: Kinematics and dynamics model via explicit direct and trigonometric elimination of kinematic constraints. In: Proceedings of the 15th IFToMM World Congress (2019). https://doi.org/10.1007/978-3-030-20131-9_311
26. Shi, P., McPhee, J.: Dynamics of flexible multibody systems using virtual work and linear graph theory. Multibody Syst. Dyn. **4**(4), 355–381 (2000). https://doi.org/10.1023/A:1009841017268
27. Sousa, C.D., Cortesão, R.: SageRobotics: open source framework for symbolic computation of robot models. In: Proceedings of 27th Annual ACM Symposium Applied Computing, pp. 262–267 (2012). https://doi.org/10.1145/2245276.2245329. Project homepage (successor): https://github.com/cdsousa/SymPyBotics
28. Tedrake, R.: the Drake Development Team: Drake: Model-based design and verification for robotics (2019). Project homepage: https://drake.mit.edu
29. The MathWorks Inc: Matlab Simscape Multibody. https://mathworks.com/products/simmechanics.html. Accessed 2 Dec 2020
30. Wang, Y., Gondokaryono, R., Munawar, A., Fischer, G.S.: A convex optimization-based dynamic model identification package for the da Vinci research kit. IEEE Robot. Autom. Lett. **4**, 3657–3664 (2019). https://doi.org/10.1109/LRA.2019.2927947. Project homepage: https://github.com/WPI-AIM/dvrk_dynamics_identification
31. Waterloo Maple Inc.: MapleSim, website. https://maplesoft.com/products/maplesim/. Accessed 2 Dec 2020
32. Wilson, G., et al.: Best practices for scientific computing. PLoS Biol. **12**(1), e1001745 (2014). https://doi.org/10.1371/journal.pbio.1001745

Rational Trigonometry Using Maple

Thomas Schramm(✉)

HafenCity University Hamburg, Hamburg, Germany
thomas.schramm@hcu-hamburg.de

Abstract. In 2005, Norman Wildberger presented a concept for a geometry without transcendental functions in his book *Divine Proportions: Rational Geometry for Universal Geometry*. Inspired by ancient Babylonian and Greek mathematics, he introduced spreads and quadrances instead of angles and lengths to describe triangles and more. With this concept, most tasks and proofs of Euclidean geometry can easily be carried out without sine and cosine functions and without introducing a differential calculus. Using Maple, we introduce the concept and definitions in this paper and then compare some basic calculations to the way they are normally solved. This concept has a clear didactic advantage and shows some parallels to the way surveyors carry out their calculations, avoiding transcendental functions wherever they can.

Keywords: Rational trigonometry · Linear algebra · Geometry

1 Norman J. Wildberger's Dream

Inspired by ancient Greek and Babylonian mathematics [4, 5, 8] N. J. Wildberger introduced concepts for trigonometry, algebra [11] and calculus [12] using only rational numbers. For engineering math, this method makes no difference in accuracy but it is easier to understand and avoids logical problems of infinite objects. Trigonometry can be fully developed avoiding transcendent functions like sine, cosine, and tangent.

The two main ideas following Euclid are:

- Square areas are more powerful than distances.
- Angles as linear scale length on the unit circle is without integral calculus a problematic concept.

Solution: Quadrances as squares over distances instead of distances and spreads instead of angles as the squared ratio of the side opposite the spread to the hypotenuse in a rectangular triangle.

Result: All canonical problems in/with triangles can be solved with simple algebra with rational numbers and at most square roots.

These ideas can easily be extended to standard 3D or hyperbolic geometry (and e.g. be applied in special relativity) [6, Chapter 22, 13].

Wildberger laid down this new concept of geometry in his book *Divine Proportions: Rational Trigonometry to Universal Geometry* in 2005 [6]. We follow his vector approach

R. M. Corless et al. (Eds.): MC 2020, CCIS 1414, pp. 365–375, 2021.
https://doi.org/10.1007/978-3-030-81698-8_24

as in his *Introduction to Hyperbolic Geometry* 39 [13]. All calculations as shown are carried out in Maple 2020. Maple inputs begin with " >", outputs are shown in blue. A complete Maple worksheet can be found in the Application Center of Maplesoft [16].

2 Re-definitions (Vector Version)

We chose the vector version of the formulation of rational trigonometry, since the proofs of the theorems only follow from the calculations with vectors with abstract components, which can be easily performed with Maple.

At first we introduce two vectors to define the necessary mathematical objects

$> \vec{u} := \langle x_1, y_1 \rangle : \vec{v} := \langle x_2, y_2 \rangle :$

2.1 Quadrance

The quadrance is the squared area over a distance. (Compare the discussion of this definition in [6, p4].) For vectors: the sum of the squared coordinates. We define for Maple

$> Q := \vec{w} \rightarrow \|\vec{w}\|_2^2$

e.g.

$> Q(\vec{u})$

$$|x_1|^2 + |y_1|^2$$

2.2 Spread

In a rectangular triangle, the spread is the squared ratio of the opposite to the hypotenuse. (Note: The spread is essentially the square of the sine of the angle between the two vectors, but our purpose is to avoid transcendent functions and stay with rational numbers.)

For vectors:

$> s := (\vec{u}, \vec{v}) \rightarrow 1 - \frac{(\vec{u}.\vec{v})^2}{Q(\vec{u}) \cdot Q(\vec{v})} :$

$> s(\vec{u}, \vec{v}) : \% = simplify(\%, symbolic)$

$$1 - \frac{(x_1 x_2 + y_1 y_2)^2}{(|x_1|^2 + |y_1|^2)(|x_2|^2 + |y_2|^2)} = \frac{(x_1 y_2 - x_2 y_1)^2}{(x_1^2 + y_1^2)(x_2^2 + y_2^2)}$$

Example:

Spread between two intersecting lines or two vectors.

$> u := \langle 1,\ 3 \rangle : v := \langle 4,\ 2 \rangle :$

$> u_v := \frac{u.v}{\|v\|_2^2} \cdot v : h := u - u_v :$

$> s(u, v) = \frac{Q(h)}{Q(u)};$

$$\frac{1}{2} = \frac{1}{2}$$

Note that for two intersecting lines all four spreads in the point of intersection are identical. Compare the application below (Fig. 1).

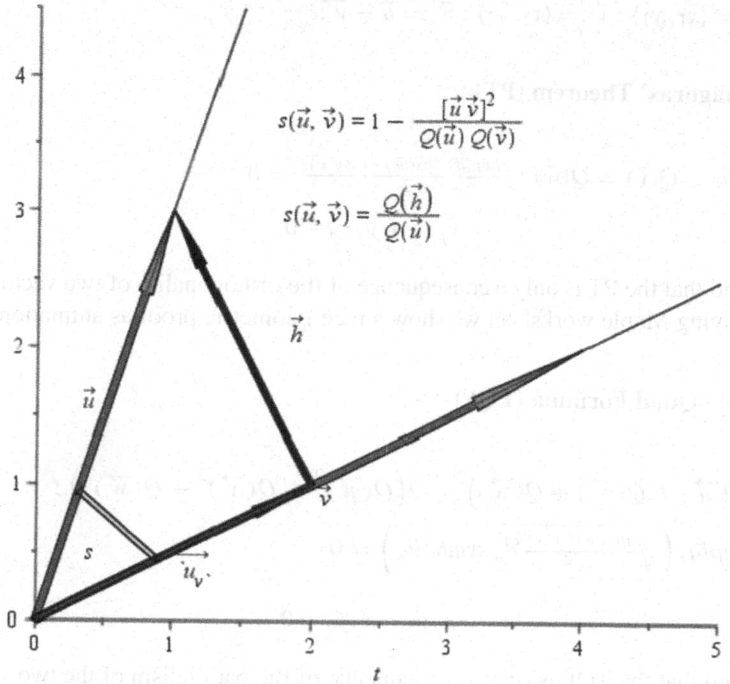

Fig. 1. Two vectors or intersecting lines and the construction of the spread s.

For the practical measurement of spreads in a seminar, we designed a setsquare with a spread measuring scale and printed it out on foil (Fig. 2).

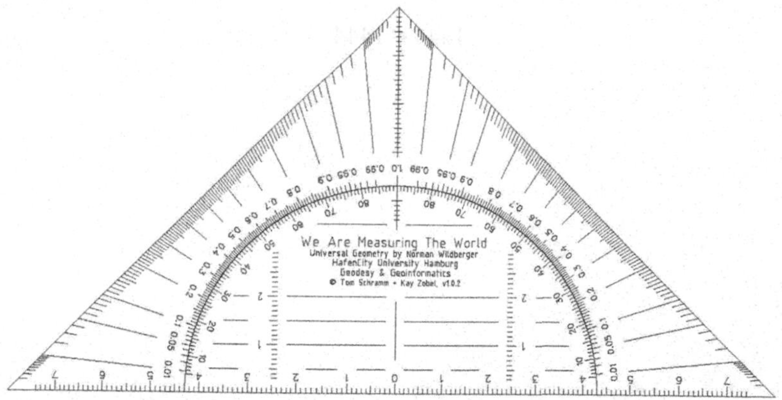

Fig. 2. A protractor spread ruler or set square with a nonlinear spread scale.

3 Five Laws for Rational Trigonometry

First, we introduce three vectors

$> \vec{u} := \langle x_1, y_1 \rangle : \vec{v} := \langle x_2, y_2 \rangle : \vec{w} := \vec{u} - \vec{v}:$

3.1 Pythagoras' Theorem (PT)

$\vec{u} \perp \vec{v} \Leftrightarrow$

$> Q(\vec{u}) + Q(\vec{v}) = Q(\vec{w}) : \frac{simplify(lhs(\%)-rhs(\%))}{2} = 0$

$$x_1 x_2 + y_1 y_2 = 0$$

We find that the PT is only a consequence of the orthogonality of two vectors. In the accompanying Maple worksheet we show a nice geometric proof as animation.

3.2 Triple Quad Formula (TQF)

$\vec{u} \| \vec{v} \Leftrightarrow$

$> (Q(\vec{u}) + Q(\vec{v}) + Q(\vec{w}))^2 = 2\left(Q(\vec{u})^2 + Q(\vec{v})^2 + Q(\vec{w})^2\right) :$

$> simplify\left(\sqrt{\frac{lhs(\%)-rhs(\%)}{4}}, symbolic\right) = 0$

$$x_1 y_2 - x_2 y_1 = 0$$

We find that the TQF is only a consequence of the parallelism of the two vectors. If we extend the two vectors by adding a third zero–coordinate, we see that the TQF states simply that the cross product is also zero.

Example: Quadrances for two parallel vectors (Fig. 3).

$> u := \langle 5,\ 0 \rangle : v := \langle 3, 0 \rangle : w := u - v :$

$> (Q(u) + Q(v) + Q(w))^2 = 2(Q(u)^2 + Q(v)^2 + Q(w)^2)$

$$1444 = 1444$$

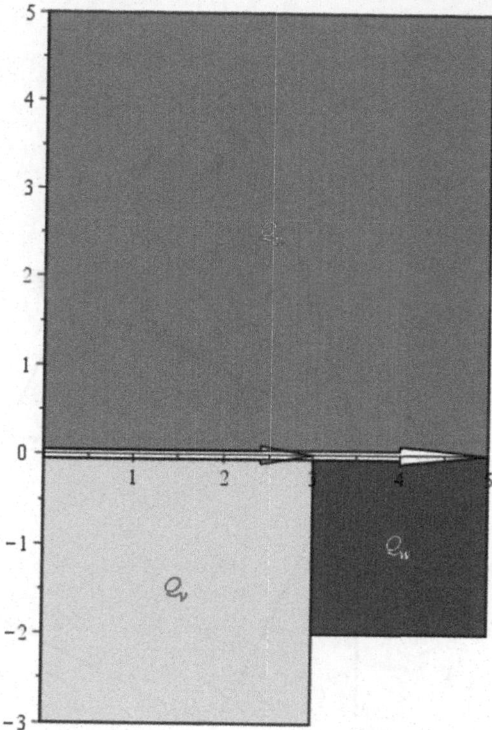

Fig. 3. The figure shows the two example vectors, their difference and the related quadrances.

3.3 Cross Law (CL)

The CL replaces in some sense the cosine law.

$> (Q(\vec{u}) + Q(\vec{v}) - Q(\vec{w}))^2 = 4 \cdot Q(\vec{u}) \cdot Q(\vec{v}) \cdot (1 - s(\vec{u}, \vec{v})) : simplify(\%)$

$$4\,(x_1 x_2 + y_1 y_2)^2 = 4\,(x_1 x_2 + y_1 y_2)^2$$

Example: A triangle defined by two vectors (Fig. 4).

$> u := \langle 1, 3 \rangle : v := \langle 4, 2 \rangle : w := u - v :$

$> (Q(u) + Q(v) - Q(w))^2 = 4 \cdot Q(u) \cdot Q(v) \cdot (1 - s(u, v))$

$$400 = 400$$

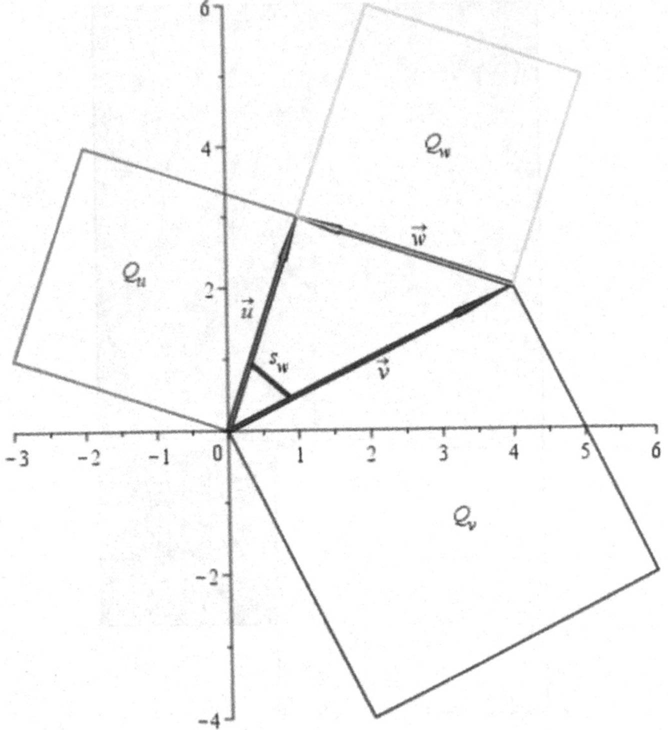

Fig. 4. Two example-vectors, their spread, their difference vector and the related quadrances.

3.4 Spread Law (SL)

The SL replaces in some sense the sine law.

$$\frac{s_u}{Q_u} = \frac{s_v}{Q_v} = \frac{s_w}{Q_w}$$

$> \frac{s(\vec{v},\vec{w})}{Q(\vec{u})} - \frac{s(\vec{w},\vec{u})}{Q(\vec{v})} = 0,\ \frac{s(\vec{v},\vec{w})}{Q(\vec{u})} - \frac{s(\vec{u},\vec{v})}{Q(\vec{w})} = 0,\ \frac{s(\vec{w},\vec{u})}{Q(\vec{v})} - \frac{s(\vec{u},\vec{v})}{Q(\vec{w})} = 0:$

$simplify([\%_1, \%_2, \%_3], symbolic)$

$$[0 = 0, 0 = 0, 0 = 0]$$

Example: A triangle defined by two vectors (Fig. 5).

$> u := \langle 1,\ 3\rangle : v := \langle 4,\ 2\rangle : w := u - v :$

$> \frac{s(u,v)}{Q(w)}, \frac{s(v,w)}{Q(u)}, \frac{s(u,w)}{Q(v)}$

$$\frac{1}{20}, \frac{1}{20}, \frac{1}{20}$$

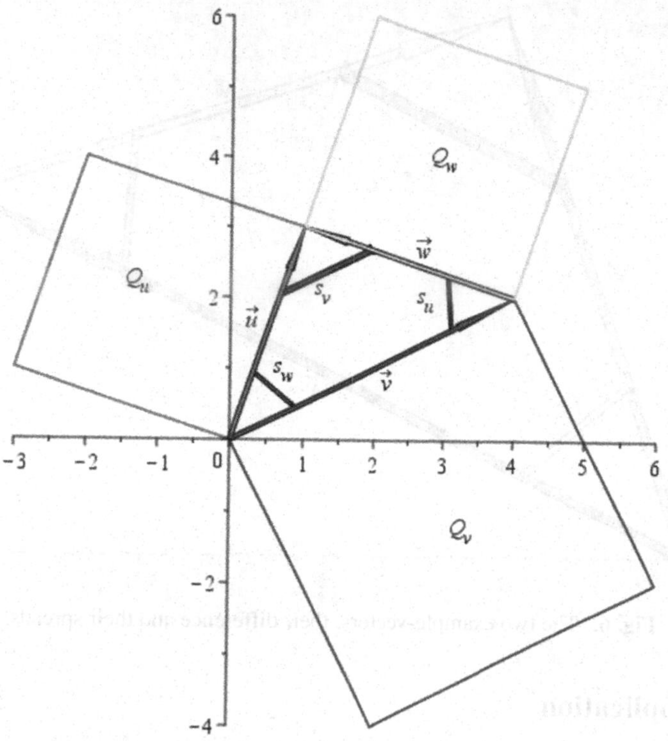

Fig. 5. Two example-vectors, their difference, their related spreads and related quadrances.

3.5 Triple Spread Formula (TSF)

The TSF replaces in some sense the sum of angles rule of triangles.

$(s_u + s_v + s_w)^2 = 2 \cdot (s_u^2 + s_v^2 + s_w^2) + 4 \cdot s_u \cdot s_v \cdot s_w$

$> (s(\vec{v}, \vec{w}) + s(\vec{u}, \vec{w}) + s(\vec{u}, \vec{v}))^2 - 2 \cdot (s^2(\vec{v}, \vec{w}) + s^2(\vec{u}, \vec{w}) + s^2(\vec{u}, \vec{v})) - 4 \cdot s(\vec{v}, \vec{w}) \cdot s(\vec{u}, \vec{w}) \cdot s(\vec{u}, \vec{v}) = 0 :$ *simplify*(%, *symbolic*)

$$0 = 0$$

Example: A triangle defined by two vectors (Fig. 6).

$> u := \langle 1,\ 3 \rangle : v := \langle 4,\ 2 \rangle : w := u - v :$

$> (s(v, w) + s(u, w) + s(u, v))^2 - 2 \cdot (s^2(v, w) + s^2(u, w) + s^2(u, v)) - 4 \cdot s(v, w) \cdot s(u, w) \cdot s(u, v) = 0$

$$0 = 0$$

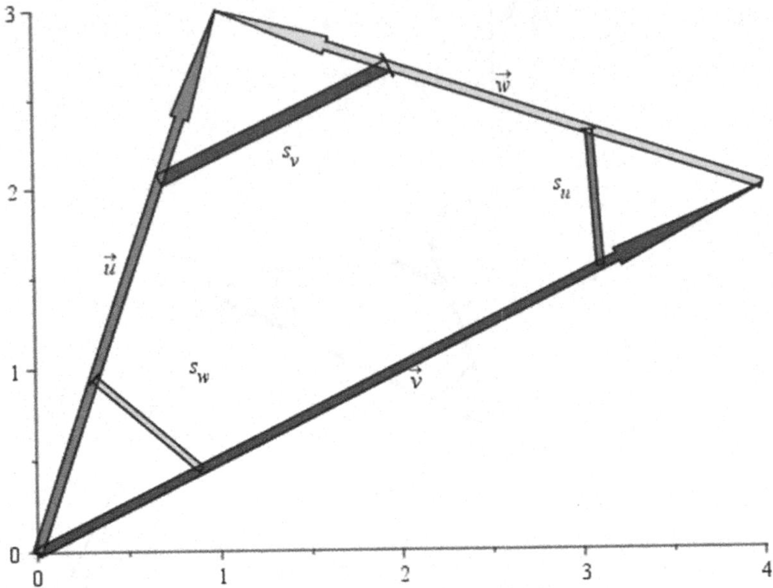

Fig. 6. The two example-vectors, their difference and their spreads.

4 An Application

A simplified determination of the height of a lighthouse. The spreads s_1, s_2 and the quadrance Q_3 are measured and we look for the quadrance H or the related height h (Fig. 7).

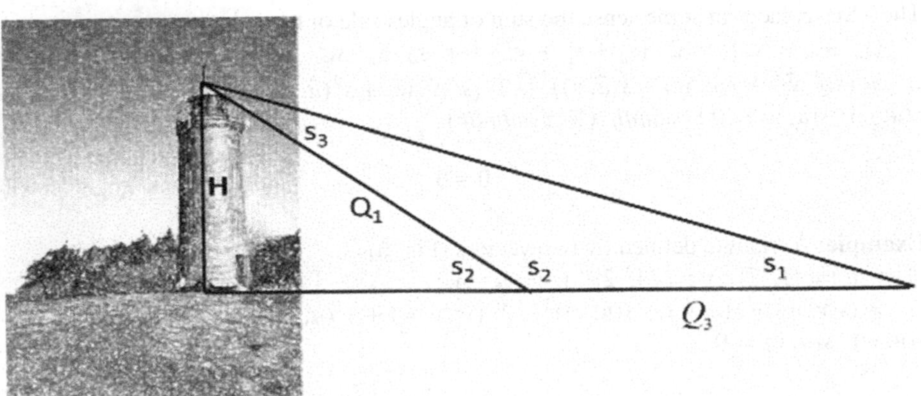

Fig. 7. A surveyor's simplified method for determining the height of a tower. Here the famous lighthouse of St. Peter-Böhl in Germany [15].

In this example the spreads s_1 and s_2 are measured from two standpoints at equal heights with quadrance Q_3. For the old fashioned, transcendental calculations, the spreads are converted into angles and quadrances into distances. (In practice, of course, the other way around until we have the appropriate devices to measure.)

4.1 The Rational Way

The calculation is easily done in Maple using the Triple Spread Formula for the measured values:

$> s_1 := 0.10 : \; s_2 := 0.35 : \; Q_3 := 540.10$
$> solve(\{(s_1 + s_2 + s_3)^2 = 2 \cdot (s_1^2 + s_2^2 + s_3^2) + 4 \cdot s_1 \cdot s_2 \cdot s_3\}); assign(\%_1)$

$$\{s_3 = 0.09381823957\}, \{s_3 = 0.6661817604\}$$

and the Spread Law twice.

$> solve\left(\left\{\frac{s_3}{Q_3} = \frac{s_1}{Q_1}\right\}\right); assign(\%);$

$$\{Q_1 = 575.6876299\}$$

$> solve\left(\left\{\frac{s_2}{H} = \frac{1}{Q_1}\right\}\right); assign(\%);$

$$\{H = 201.4906705\}$$

$> h := evalf\left(\sqrt{H}\right)$

$$h := 14.22777860$$

We find roughly the real height of the well-known lighthouse.

4.2 The Transcendent Way

The converted values ($s_i \to \alpha_i$, $Q_i \to q_i$) are

$> \alpha_1 := 0.322 : \; \alpha_2 := 2.509 : \; q_3 := 23.240 :$

From the sum of angles law it follows

$> solve(\{\alpha_1 + \alpha_2 + \alpha_3 = Pi\}); assign(\%);$

$$\{\alpha_3 = 0.3105926536\}$$

And from the sine law

$> solve\left(\left\{\frac{q_1}{\sin(\alpha_1)} = \frac{q_3}{\sin(\alpha_3)}\right\}\right); assign(\%);$

$$\{q_1 = 24.06439633\}$$

And finally

$> h := q_1 \cdot \sin(\alpha_2)$

$$h := 14.22777860$$

A similar result. The slight difference comes from the cancellation of decimals.

$> s_1 \approx sin^2(\alpha_1), s_2 \approx sin^2(\alpha_2), s_3 \approx sin^2(\alpha_3)$

$$0.10 \approx 0.1001497172, 0.35 \approx 0.3495620308,$$
$$0.09381823957 \approx 0.09340541006$$

5 Conclusion

We introduced Wildberger's consistent and systematic approach to trigonometry by using only rational numbers.

The concept reminds of a concept of surveying-mathematics to avoid the computations of trigonometric functions wherever possible.

It is easier to understand and to apply and as accurate as normal engineering mathematics.

It could be useful as didactic and methodological concept in schools and even universities before and independent of the introduction of differential calculus.

Acknowledgement. I would like to thank the referees for helpful comments and Lars Schmeink for carefully reading the manuscript.

References

1. Fuhrmann, T., Navratil, G.: Ausgleichungsrechnung mit Gröbnerbasen, zfv (Zeitschrift für Geodäsie, Geoinformation und Landmanagement), 138. Jg. **6**, 399–404 (2013)
2. Henle, M.: Divine proportions: rational trigonometry to universal geometry by N. J. Wildberger. Rev. Am. Math. Monthly **114**(10), 933–937 (2007). https://www.jstor.org/stable/27642383, October 2020
3. Kosheleva, O.: Rational trigonometry: computational viewpoint. Geombinatorics **1**(1), 18–25 (2008)
4. Mansfield, D., Wildberger, N.J.: Written in stone: the world's first trigonometry revealed in an ancient Babylonian tablet. In: Pitici, M. (ed.) The Best Writing on Mathematics 2018. Princeton University Press (2019)
5. Spiegel Online, Mathematiker wollen Rätsel babylonischer Tontafel gelöst haben, October 2020. https://www.spiegel.de/wissenschaft/mensch/babylonische-tontafel-plimpton-322-mathematiker-liefern-neue-erklaerung-a-1164527.html
6. Wildberger, N.J.: Divine Proportions: Rational Trigonometry to Universal Geometry, Wild Egg Pty Ltd., (2005). https://www.researchgate.net/publication/266738365_Divine_Proportions_Rational_Trigonometry_to_Universal_geometry, October 2020
7. Wildberger, N.J.: A rational approach to trigonometry, October 2020. https://doi.org/10.1080/10724117.2007.11974738. https://www.researchgate.net/publication/242364807_A_Rational_Approach_to_Trigonometry
8. Wildberger, N.J.: The ancient Greeks present: rational Trigonometry. arXiv:0806.3481v1 [math.MG] (2008)
9. Wildberger, N.J.: An introduction to rational trigonometry and chromogeometry (2015). https://doi.org/10.13140/RG.2.1.1280.6488. https://www.researchgate.net/publication/284572258_An_Introduction_to_Rational_Trigonometry_and_Chromogeometry, October 2020

10. Wildberger, N.J.: Real numbers: a critique and way forward (2015). https://doi.org/10.13140/RG.2.1.3673.8406, October 2020. https://www.researchgate.net/publication/280387376_Real_numbers_A_critique_and_way_forward
11. Wildberger, N.J.: The fundamental dream of algebra, October 2020. https://njwildberger.com/2015/11/19/the-fundamental-dream-of-algebra/
12. Wildberger, N.J.: Invitation to a more logical, solid and careful analysis | Algebraic Calculus One, October 2020. https://youtu.be/rTw6XbmO8Nc
13. Wildberger, N.J.: Rational trigonometry: an overview | Universal Hyperbolic Geometry 39 | https://youtu.be/dVk3CpjHR4Y, October 2020. For an introduction compare: Universal Hyperbolic Geometry 0: Introduction. https://youtu.be/N23vXA-ai5M
14. Schramm, T.: Divine Proportions -- Norman Wildbergers andere - rationale Trigonometrie. In: Proceedings 15 Workshop Mathematik in ingenieurwissenschaftlichen Studiengängen. Rostock-Warnemünde 2019, Heft 02/19, Wismarer Frege-Reihe, Hochschule Wismar (2019)
15. Image: Lighthouse St.-Peter-Böhl, Unukorno - Own work, CC BY 3.0. https://commons.wikimedia.org/w/index.php?curid=32442857
16. Schramm, T.: Rational Trigonometry, Maplesoft Application Center. https://www.maplesoft.com/applications/view.aspx?SID=154670

Power Series Representations of Hypergeometric Type Functions

Bertrand Teguia Tabuguia(✉) and Wolfram Koepf

Kassel University, Heinrich-Plett-Str. 40., 34132 Kassel, Germany
{bteguia,koepf}@mathematik.uni-kassel.de

Abstract. In 1992, Koepf proposed a symbolic approach to compute power series. This algorithm was extended for a larger family of expressions thanks to Petkovsek's and van Hoeij's algorithms (1993 and 1998) which compute hypergeometric term solutions of any given holonomic recurrence equation (RE). Mark van Hoeij's algorithm whose outputs are bases is available in Maple through the command `LREtools[hypergeomsols]`, and Koepf's algorithm through `convert` and the built-in module `FormalPowerSeries`. `LREtools[hypergeomsols]` is internally used by `convert/FormalPowerSeries`.

However, using van Hoeij's algorithm one cannot compute m-fold hypergeometric term solutions of holonomic REs, for integers $m > 1$. Given a field K of characteristic zero, a term $a(n)$ is said to be m-fold hypergeometric if the term ratio $a(n+m)/a(n)$ is rational over K. Note that the hypergeometric term case corresponds to $m = 1$. If one adds for example an odd hypergeometric function, like $\arcsin(z)$, and an even hypergeometric function, like $\cos(z)$ (which both are two-fold hypergeometric), then van Hoeij's algorithm cannot find those by solving the resulting recurrence equation. Due to this limitation, the computation of many power series is missed by Maple, in particular, linear combinations of power series having m-fold hypergeometric term coefficients are generally not detected.

We overcome these issues by using a new algorithm called **mfoldHyper**, proposed in the first author's Ph.D. thesis to compute bases of the subspace of m-fold hypergeometric term solutions of holonomic REs. It turns out that **mfoldHyper** linearizes the computation of hypergeometric type power series, i.e. every linear combination of hypergeometric type power series is detected. This paper describes our Maple implementation of an algorithm that conclusively extends Maple's capabilities regarding the computation of hypergeometric type power series.

Keywords: Hypergeometric type power series · m-fold hypergeometric term · Holonomic recurrence equation

1 Introduction

By connection to the generalized hypergeometric series, the term "hypergeometric type" had been introduced in [5] to denote expressions whose power series coefficients

---

Supported by DAAD Erasmus Plus.

R. M. Corless et al. (Eds.): MC 2020, CCIS 1414, pp. 376–393, 2021.
https://doi.org/10.1007/978-3-030-81698-8_25

lead to a two-term recurrence relation with polynomial coefficients. The type of such series is defined as the positive difference of the indeterminate sequence indices in the equation. However, despite the fact that we keep this terminology, we are considering a much larger family of expressions, mostly linear combinations of holonomic meromorphic functions. Indeed, the original definition does not consider arbitrary holonomic recurrence equations[1], and therefore neglects the possibility to have finitely many different types for the same power series. Specifically, we have the following definition.

Definition 1 (Hypergeometric type power series). *Let $\mathbb{K}$ be a field of characteristic zero. For an expansion around $z_0 \in \mathbb{K}$, a series $s(z)$ is said to be of hypergeometric type if it can be written as*

$$s(z) := T(z) + \sum_{j=1}^{J} s_j(z),\; s_j = \sum_{n=n_{j,0}}^{\infty} a_{j,n}(z-z_0)^{n/p_j} \tag{1}$$

where n is the summation variable, $T(z) \in \mathbb{K}[z, 1/z, \ln(z)]$, $n_{j,0} \in \mathbb{Z}$, $J, p_j \in \mathbb{N}$, and $a_{j,n}$ is such that there exists a positive integer m_j so that $a_{j,n+m_j}/a_{j,n} \in \mathbb{K}(n)$.

Thus a hypergeometric type power series is a linear combination of Laurent-Puiseux series whose coefficients are m-fold hypergeometric terms[2]. A hypergeometric function is a function that can be expanded as a hypergeometric type power series. T is called the Laurent polynomial part of the expansion, and the p_j's are its Puiseux numbers.

The presence of $\ln(z)$ in a hypergeometric type expansion is justified by the solution of the underlying holonomic differential equation (see [4]). The definition in [5] reduces to the case $T = 0$ and $J \leqslant m$, where m is the unique type ($m_1 = \cdots = m_J = m$) encountered in Definition 1.

Algorithmic attempts were proposed [1] to determine the Puiseux numbers in expression (1), but this was limited and could not be taken into account in the general case. This is why commonly used approaches to Puiseux number calculation are based on heuristic tests on the input function. But this generally only works in the case where one deals with a single Puiseux number (all the p_j's are identical in (1)). Based on a symbolic approach, our algorithm offers a clear procedure to determine all different Puiseux numbers involved in a power series expansion. Note that the scope here is more general than what could be done using the Frobenius method (see [7]) which rather forces Puiseux numbers to appear additively ($z^{n+r}, r \in \mathbb{Q}$) in the power of the indeterminate.

Likewise, instead of checking expressions, our algorithm also proposes a symbolic approach to determine the Laurent polynomial part in (1).

---

[1] We recall that a recurrence equation is said to be holonomic if it is linear and homogeneous having polynomial coefficients.

[2] An m-fold hypergeometric term is implicitly defined in Definition 1 by the property of hypergeometric type series coefficients. This is admitted in the remaining part of the paper.

Remark 1.

- An m-fold hypergeometric term a_n encodes m linearly independent hypergeometric terms. These can be enumerated by the m different representations of the corresponding ratios as

$$\frac{a_{n+m+l}}{a_{n+l}} \in \mathbb{K}(n),\ 0 \leqslant l \leqslant m-1. \tag{2}$$

- Without loss of generality, we assume $z_0 = 0$ since the non-zero case easily reduces to it. This is also implemented.

A generic representation of hypergeometric type functions for which we compute power series is

$$f(z) = T_0(z) + \sum_{i=1}^{I} T_i(z) \cdot f_i(z), \tag{3}$$

where $T_0(z), T_i(z) \in \mathbb{K}[z, 1/z, \ln(z)]$, and the f_i's are of hypergeometric type. The linear combination of hypergeometric type power series found by our algorithm is what stops the recursive aspect[3] of representation (3). Therefore having an expression $f(z)$ of the form (3) does not guarantee that its power series representation has I hypergeometric type power series, and $T_0(z)$ is not necessarily equal to $T(z)$ in (1). For a better view of this fact, let us first recall the steps (with ramifications in the last step) of the method described in [5]. Given an expression $f(z)$, the algorithm proceeds as follows.

(Step 1) Compute a holonomic differential equation satisfied by $f(z)$.
(Step 2) Convert the obtained differential equation into a recurrence equation for the power series coefficients of $f(z)$.
(Step 3) Find all m-fold hypergeometric term solutions of the resulting recurrence equation, and use some initial values to find a linear combination corresponding to the power series of $f(z)$.

Although we have implementations that can be used to improve the efficiency (the function `HolonomicDE` in our Maple package `FPS` is generally faster than Maple's `DEtools[FindODE]` for expressions of the form (3)) for (Step 1), this article puts emphasis on (Step 3), given that (Step 2) is straightforward.

Let $f(z)$ be as in (3). One may think of using Koepf's algorithm on every summand of $f(z)$, but this will not always work as some f_i in (3) could lead to a recurrence equation with more than two terms in (Step 2). In such a situation, the current Maple implementation internally uses van Hoeij's algorithm. The issue with the latter algorithm is the fact that it only looks for hypergeometric term solutions ($m = 1$) which is just a particular case of what should be considered; also, it may find hypergeometric terms that are equivalent to all the needed m-fold hypergeometric terms but by using unnecessary extension fields. Usually when this happens, the current Maple implementation fails to find a linear combination for the power series representation sought.

We give some details explaining why Maple fails to find the power series of $z^2\sin(z) + z^4 \ln\left(1+z+z^2+z^3\right)$.

[3] Recursive because hypergeometric type functions are used in (3).

Example 1.

```
> f:=z^2*sin(z)+z^4*ln(1+z+z^2+z^3):
> convert(f,FormalPowerSeries,z,n)
```

$$z^2 \sin(z) + z^4 \ln\left(z^3 + z^2 + z + 1\right)$$

Hence no representation is found. Sometimes this could happen if the holonomic differential equation to be used has an order larger than 4, but this is not the case since a holonomic recurrence equation can be found without such a specification.

Example 2. The resulting differential equation is converted into the following recurrence equation:

```
> RE:=FormalPowerSeries[SimpleRE](f,z,a(n))
```

$$\begin{aligned} RE := {} & 12\,(n-5)\,(n-4)\,(n-3)\,(n-2)\,a(n) + 3\,(n-21)^2\,a(n-17) \\ & + (n-20)\,(11\,n-221)\,a(n-16) + (n-19)\left(3\,n^3 - 174\,n^2 + 3364\,n - 21685\right) \\ & a(n-15) + (n-18)\left(11\,n^3 - 606\,n^2 + 11080\,n - 67263\right)a(n-14) \\ & + (n-17)\left(19\,n^3 - 1006\,n^2 + 17597\,n - 101774\right)a(n-13) \\ & + (n-16)\left(23\,n^3 - 1190\,n^2 + 20137\,n - 111758\right)a(n-12) \\ & + 2\,(n-15)\left(11\,n^3 - 550\,n^2 + 8904\,n - 46849\right)a(n-11) \\ & + 2\,(n-14)\left(7\,n^3 - 334\,n^2 + 5120\,n - 25155\right)a(n-10) \\ & + (n-13)\left(54\,n^3 - 1900\,n^2 + 22141\,n - 84715\right)a(n-9) \\ & + (n-12)\left(170\,n^3 - 5316\,n^2 + 54793\,n - 185801\right)a(n-8) \\ & + (n-11)\left(359\,n^3 - 10198\,n^2 + 95556\,n - 295473\right)a(n-7) \\ & + (n-10)\left(567\,n^3 - 14774\,n^2 + 127704\,n - 366363\right)a(n-6) \\ & + (n-9)\left(703\,n^3 - 16454\,n^2 + 128059\,n - 331272\right)a(n-5) \\ & + (n-8)\left(643\,n^3 - 13694\,n^2 + 96543\,n - 225156\right)a(n-4) \\ & + 8\,(n-5)\,(n-7)\left(58\,n^2 - 816\,n + 2811\right)a(n-3) \\ & + 16\,(n-5)\,(n-6)\,(16\,n-123)\,(n-4)\,a(n-2) \\ & + 24\,(n-3)\,(n-4)\,(n-5)\,(3\,n-19)\,a(n-1) = 0 \end{aligned} \tag{4}$$

Since (4) has more than two terms, `LREtools[hypergeomsols]` is internally used.

Example 3.

```
> LREtools[hypergeomsols](RE,a(n),{},output=basis)
```

$$\left[\frac{(-1)^n}{n-4}, \frac{(-i)^n}{n-4}, \frac{i^n}{n-4}, \frac{i^n}{\Gamma(n-1)}, \frac{(-i)^n}{\Gamma(n-1)}\right]$$

It is not difficult to prove that the above basis of hypergeometric terms can be used to represent the power series sought, but this is missed by `convert/FormalPowerSeries`. Moreover, using this built-in command on individual summands does not give much improvement on the result because the same issue occurs for $z^4 \cdot \ln(1+z+z^2+z^3)$ whose recurrence equation also has more than two terms. Example 7 shows by using our implementation that 'simpler' (no extension field used) coefficients exist over the rationals.

It may happen, moreover, that the power series coefficients of two distinct hypergeometric type functions reduce to a single m-fold hypergeometric term computed in (Step 3). This is another reason why calling `convert/FormalPowerSeries` for individual summands does not always give the best possible representation.

Example 4. Applying `convert/FormalPowerSeries` to each summand of $\sin(z)^3 - \cos(z)^3$ yields

```
> F:=sin(z)^3-cos(z)^3:
> map(f->convert(f,FormalPowerSeries,z,n),F)
```

$$\sum_{n=0}^{\infty}\left(-1/4\,\frac{3^n\sin(1/2\,n\pi)}{n!}+3/4\,\frac{\sin(1/2\,n\pi)}{n!}\right)z^n + \sum_{n=0}^{\infty}\frac{\left(-1/8\,(3\,i)^n-1/8\,(-3\,i)^n-3/8\,i^n-3/8\,(-i)^n\right)z^n}{n!} \tag{5}$$

which is much more simplified avoiding algebraic extensions using our implementation as follows:

```
> FPS[FPS](F,z,n)
```

$$\left(\sum_{n=0}^{\infty}-\frac{(-1)^n\,\left(9^n+3\right)\,z^{2\,n}}{4\,(2\,n)!}\right)+\left(\sum_{n=0}^{\infty}\frac{3\,(-1)^n\,\left(3^{2\,n+2}-1\right)\,z^{2\,n+3}}{4\,(2\,n+3)!}\right)$$

Note that the latter output is not the same as what is obtained using the internal command directly.

```
> convert(F,FormalPowerSeries,z,n)
```

$$\sum_{n=0}^{\infty}\frac{\left((-3/8+3/8\,i)\,(-i)^n-(3/8+3/8\,i)\,i^n-(1/8+i/8)\,(-3\,i)^n-(1/8-i/8)\,(3\,i)^n\right)z^n}{n!}$$

Many more examples of this kind can be provided.

As one can see, our algorithm is implemented in our Maple package `FPS` under the name `FPS`, presented as the main function of the package[4].

The main ingredient of our approach is algorithm **mfoldHyper** from [8, Chapter 7] that we implemented with the same name. This algorithm computes a basis of the subspace of all m-fold hypergeometric term solutions of any given holonomic recurrence equation.

[4] `FPS` contains some other results that will not be discussed in this paper.

Example 5. We come back to the power series sought in Example 1. A basis of m-fold hypergeometric term solutions of (4) can be represented as

```
> FPS[mfoldHyper](RE,a(n))
```

$$\left[\left[1, \left\{\frac{(-1)^n}{27\,(n-4)}\right\}\right], \left[2, \left\{\frac{(-1)^n}{9\,(n-2)}, \frac{(2\,n-1)\,n\,(-1)^n}{72\,(2\,n)!}\right\}\right]\right]$$

Note that each 2-fold solution above corresponds to two hypergeometric terms. By default, the algorithm computes terms corresponding to $l = 0$ in (2). Once we know that 2-fold hypergeometric term solutions exist, we can call the algorithm again to get the other representations.

Example 6.

```
> FPS[mfoldHyper](RE,a(n),ml=[2,1])
```

$$\left\{\frac{(-1)^n}{27\,(2\,n-3)}, \frac{n\,(-1)^n}{(2\,n)!}\right\}$$

Finally with all these m-fold hypergeometric terms we look for a linear combination using appropriate initial values and get the representation

Example 7.

```
> FPS[FPS](f,z,n)
```

$$\sum_{n=0}^{\infty} \frac{(-1)^n\,z^{n+5}}{n+1} + \sum_{n=0}^{\infty} \frac{(-1)^n\,z^{2\,n+6}}{n+1} + \sum_{n=0}^{\infty} \frac{(-1)^n\,z^{2\,n+3}}{(2\,n+1)!}$$

Observe that some shifts may be applied to the coefficients according to the starting point obtained from the Laurent polynomial part of the series. This is always used even when the corresponding Laurent polynomial part is zero, because it leads to appropriate starting points.

Note that (Step 3) can be divided into two important sub-steps. Indeed, finding a linear combination after obtaining m-fold hypergeometric terms requires a certain number of initial values and evaluations that can be determined from the obtained basis of m-fold hypergeometric terms. If a precise matching between evaluations and initial values is not correctly made then the representation sought might be missed. Therefore the algorithm (see [10], [8]) behind our method could work as a decision procedure to decide whether a given holonomic meromorphic function is of hypergeometric type.

On the other hand, it is proved in [8, Theorem 7.2, 7.3] that the *exp-like* and the *rational function* series types considered in [5] are both of hypergeometric type. Therefore it is not necessary to split our development into these particular cases.

In the following sections we give an overview of algorithm **mfoldHyper** and some details about the steps of our algorithm. Many examples where the current Maple `convert/FormalPowerSeries` misses results will be presented.

2 An Overview of Algorithm mfoldHyper

m-fold hypergeometric terms have sometimes been referred to as m-hypergeometric sequences in [6], m-interlacings of hypergeometric sequences (see the conclusion of [11]) that are also considered as a particular case of Liouvillian sequences in [2]. We use the phrase m-fold hypergeometric term from the most recent paper about this notion in [3]. However, none of the approaches described in these previous works corresponds to the method used by **mfoldHyper**. Indeed, most of the effort on finding m-fold hypergeometric term solutions of holonomic recurrence equations has been focused on extending Petkovšek's or van Hoeij's algorithm. Specifically, the aim is usually to compute right factors of the form $\tau^m - r(n)$, where τ denotes the shift operator, and $r(n)$ is a rational function over a field of characteristic zero, of the given recurrence operator, and adapt the steps of Petkovšek's or van Hoeij's algorithm to the m-fold case. As mentioned in [6], these approaches usually increase the complexity dramatically, which may explain the lack of implementations. **mfoldHyper** uses a completely different strategy, the algorithm in [9], Petkovšek's or van Hoeij's algorithm, can be used as a black box. The method results from a study of holonomic recurrence equations. In the sequel, we give the theorem upon which **mfoldHyper** is based and present its steps towards computing m-fold hypergeometric term solutions of holonomic recurrence equations.

Definition 2 **(m-fold holonomic recurrence equation).** *Let m be a positive integer. A holonomic recurrence equation is said to be m-fold holonomic if it has at least two non-zero terms, and the difference between every pair of indices in the equation is a multiple of m.*

Example 8.

- Hypergeometric type power series considered in [5] lead to an m-fold holonomic recurrence equation with two terms.
- One can always write an m-fold holonomic recurrence equation as

$$P_{md}(n)a_{n+md} + P_{m(d-1)}a_{n+m(d-1)} + \cdots + P_0(n)a_n = 0 \tag{6}$$

An important point to notice is that representation (6) is just a particular notation. We may have many different representations of m-fold holonomic recurrence equations in the equation of study, and these have to be considered separately. The following definition is used to identify these differences.

Definition 3 **(m-fold distinct holonomic recurrence equations).** *Let m be a positive integer. Two m-fold holonomic recurrence equations are said to be m-fold distinct, if the difference between any index taken from one and another taken from the second is not a multiple of m.*

Example 9.

$$\begin{aligned} RE_1 &: P_{1,3} \cdot a_{n+7} + P_{1,2} \cdot a_{n+4} + P_{1,1} \cdot a_{n+1} = 0, \\ RE_2 &: P_{2,4} \cdot a_{n+11} + P_{2,3} \cdot a_{n+8} + P_{2,2} \cdot a_{n+5} + P_{2,1} \cdot a_{n+2} = 0. \end{aligned} \tag{7}$$

RE_1 and RE_2 are 3-fold holonomic distinct.

We can now state the fundamental theorem behind algorithm **mfoldHyper** (see [8, Theorem 7.1]).

Theorem 1 (Structure of holonomic recurrence equations having m-fold hypergeometric term solutions). *Let $m \in \mathbb{N}$, $\mathbb{K}$ a field of characteristic zero, and h_n be an m-fold hypergeometric term which is not u-fold hypergeometric over $\mathbb{K}$ for all positive integers $u < m$. Then h_n is a solution of a given holonomic recurrence equation, if that equation can be written as a linear combination of m-fold holonomic recurrence equations; such that h_n is solution of each of the m-fold distinct holonomic recurrence equations of that linear combination.*

Remark 2. In Theorem 1 the aim of the assumption that the recurrence equation should be written as a linear combination of m-fold holonomic recurrences is to eliminate those with only one (non-zero) term which are not taken into account by Definition 2.

Combined with the fact that the given recurrence equation order plays the role of a bound for the value of m (see [2, 8]), Theorem 1 leads to the following main steps to determine a basis of the subspace of all m-fold hypergeometric term solutions.

Algorithm 1. Compute a basis of m-fold hypergeometric term solutions of a given holonomic recurrence equation (RE)

- Set $m = 1$.
- Repeat
  1. If the given RE is a linear combination of m-fold holonomic REs then go to item 2. Otherwise go to item 4.
  2. Compute bases of m-fold hypergeometric term solutions of each m-fold distinct holonomic RE in the linear combination found in item 1. These latter are computed after applying the substitution that transforms m-fold holonomic REs to 1-fold holonomic REs and allows computations of m-fold hypergeometric terms as hypergeometric term. Petkovšek's or van Hoeij's algorithm can then be used. However, we recommend the algorithm in [9] for the purpose of computing power series.
  3. Collect all m-fold hypergeometric terms that are linearly dependent to an element of each basis of m-fold hypergeometric term solutions computed in item 2.
  4. Increment m and go back to item 1.
- Until $m = d$.
- Return the collected m-fold hypergeometric terms.

Example 10. Consider the recurrence equation

$$(2+n)\cdot(4+n)\cdot(6+n)\cdot a_{n+6} - 2\cdot(2+n)\cdot(4+n)\cdot a_{n+4} + 4\cdot(2+n)\cdot a_{n+2} - 8\cdot a_n = 0. \tag{8}$$

- For $m = 2$, we find that (8) is 2-fold, we then apply the substitution

$$\begin{cases} 2\cdot k = n \\ s_k = a_{2\cdot k} \end{cases}, \tag{9}$$

that transforms (8) into the recurrence equation

$$(2+2\cdot k)\cdot(4+2\cdot k)\cdot(6+2\cdot k)\cdot s_{k+3}-2\cdot(2+2\cdot k)\cdot(4+2\cdot k)\cdot s_{k+2} \\ +4\cdot(2+2\cdot k)\cdot s_{k+1}-8\cdot s_k=0. \quad (10)$$

We solve (10) using the algorithm in [9] and substitute the initial variable back to get the following basis of 2-fold hypergeometric terms solutions

$$\left\{\frac{1}{n!}\right\}. \quad (11)$$

- For $m=4$, we find a combination of two 4-fold holonomic REs, namely,

$$-2\cdot(2+n)\cdot(4+n)\cdot a_{n+4}-8\cdot a_n=0,$$

and

$$(2+n)\cdot(4+n)\cdot(6+n)\cdot a_{n+6}+4\cdot(2+n)\cdot a_{n+2}=0.$$

These lead to the same basis of 4-fold hypergeometric term solutions, which is

$$\left\{\frac{(-1)^n}{(2\cdot n)!}\right\}. \quad (12)$$

- No more linear combination is found. Therefore the final output is

$$\left[\left[2,\left\{\frac{1}{n!}\right\}\right],\left[4,\left\{\frac{(-1)^n}{(2\cdot n)!}\right\}\right]\right]. \quad (13)$$

Our Maple package has an implementation of Algorithm 1 under the name `mfoldHyper`.

Example 11. (8) is obtained by computing the recurrence equation for the power series of $\exp(z^2)+\cos(z^2)$. Let us recover the solution as in (13).

```
> RE:=FPS[FindRE](exp(z^2)+cos(z^2),z,a(n)):
```

`FindRE` is our variant of `FormalPowerSeries[SimpleRE]`.

```
> FPS[mfoldHyper](RE,a(n))
```

$$\left[\left[2,\left\{(n!)^{-1}\right\}\right],\left[4,\left\{\frac{(-1)^n}{(2\,n)!}\right\}\right]\right]$$

Of course, such a solution cannot be detected by van Hoeij's algorithm.

```
> LREtools[hypergeomsols](RE,a(n),{},output=basis)
```

$$0$$

3 Computing Hypergeometric Type Power Series

Having presented how **mfoldHyper** works, we can now give some details about how our procedure builds a hypergeometric type power series from a given expression. Remember, as mentioned in the introduction, we consider a much larger family of expressions than what is described in [5] or currently internally used by Maple. The decision property of the algorithm in [10] could not be reached by previous approaches since m-fold hypergeometric terms were barely accessible. However, we highlight a possible gap between the algorithm and its implementation since limitations can be encountered due to unavailability of computer algebra tools to deal with hypergeometric terms over larger algebraic extension fields; an example will be presented. Nevertheless, as our Maxima implementation (see [8])[5], our Maple implementation demonstrates an important improvement that covers a very large family of hypergeometric type functions given as in (3), which can also be used to show equivalences between them.

Once a holonomic recurrence equation satisfied by the power series coefficients of a given expression is computed, determining the following items is the essential focus of our procedure.

- Puiseux numbers.
- Laurent polynomial part and starting points.
- A basis (in its complete form[6]) of m-fold hypergeometric term solutions of the obtained holonomic recurrence equation.
- A linear combination of hypergeometric type power series.

We use our Maple implementation to describe these steps for some interesting examples.

Consider $f(z) = \arctan(z) + \ln(1 + z^2) + \exp(z^3)$, we want to find the power series representation of f around $z_0 = 0$. This first example is used to give some details about the way we get the complete basis of m-fold hypergeometric terms and how to find the needed linear combination of hypergeometric type power series. The recurrence equation found is the following.

Example 12.

```
> f:=arctan(z)+ln(1+z^2)+exp(z^3):
> RE:=FPS[FindRE](f,z,a(n))
```

[5] Currently being discussed for integration into the system.

[6] Complete form means that all representations of m-fold hypergeometric terms are given. These are m linearly independent terms.

$$\begin{aligned} RE := {} & 6\,(n-1)\,n\,(n-3)\,a\,(n) - 18\,(n-9)^2\,a\,(n-9) - 9\,(n-8)\,(n-7) \\ & a\,(n-8) - 36\,(n-8)\,(n-7)\,a\,(n-7) + 6\,(n-18)\,(n-6)^2\,a\,(n-6) \\ & + 3\,(n-5)\,(n^2-25\,n+102)\,a\,(n-5) \\ & + 3\,(n-4)\,(4\,n^2-71\,n+267)\,a\,(n-4) + 6\,(n-3)\,(2\,n^2-24\,n+57)\,a\,(n-3) \\ & + 2\,(n-2)\,(5\,n^2-51\,n+118)\,a\,(n-2) + (n-1)\,(11\,n^2-93\,n+166)\,a\,(n-1) \\ & + 2\,(n-1)\,(n-2)\,(n+1)\,a\,(n+1) + 2\,(n+2)\,(n+1)\,(n-1)\,a\,(n+2) = 0. \end{aligned} \tag{14}$$

Using `mfoldHyper`, we find the following basis of hypergeometric terms (incomplete form).

Example 13.

```
> FPS[mfoldHyper](RE,a(n))
```

$$\left[\left[2, \left\{1/2\,\frac{(-1)^n}{n}\right\}\right], \left[3, \left\{(n!)^{-1}\right\}\right]\right]$$

This reveals that we have one more 2-fold hypergeometric term, and two more 3-fold hypergeometric terms. They can be computed using `mfoldHyper` as follows.

Example 14.

```
> FPS[mfoldHyper](RE,a(n),ml=[2,1])
```

$$\left\{\frac{(-1)^n}{2\,n+1}\right\}$$

```
> foldl(`union`,{},
> seq(FPS[mfoldHyper](RE,a(n),ml=[3,i]),i=1..2))
```

$$\left\{\frac{1}{(3\,n+1)\,(1/3)_n}, \frac{1}{(3\,n+2)\,(2/3)_n}\right\}$$

Hence we obtain the basis of m-fold hypergeometric term solutions in its complete form. We emphasize on repeated use of `mfoldHyper` because thanks to [9], it represents all its outputs in appropriate normal forms. We recall that $(1/3)_n$ denotes the Pochhammer symbol or rising factorial.

We have 2-fold and 3-fold hypergeometric terms, therefore we expect series expansions with the following powers

$$z^{2n},\ z^{2n+1},\ z^{3n},\ z^{3n+1},\ z^{3n+2}. \tag{15}$$

We need to know the number of evaluations to make with the obtained m-fold hypergeometric terms, and the number of initial coefficients of the Taylor expansion of $f(z)$ that should be used. This way we will get a linear system of 5 unknowns representing the coefficients of the linear combination sought. Note that integer roots of the recurrence equation leading coefficient are automatically taken into account when computing the coefficients thanks to the appropriate integer shifts applied in [9]. We establish (see [8, Chapter 8]) that the number of initial coefficients to be used from the Taylor expansion of $f(z)$ can be taken as

$$\left(\sum_{m\in\{m_1,\ldots,m_\mu\}} l_m - 1\right)\cdot \operatorname{lcm}(m_1,\ldots,\cdot m_\mu) + m_\mu - 1 \tag{16}$$

where the m_i, $i = 1,\ldots,\mu$, $\mu \in \mathbb{N}$ are the types involved in the hypergeometric type power series; l_{m_i} is the number of coefficient of type m_i; and m_μ is the maximum of these types. This number corresponds to the number of linear equations which might be reduced sometimes, but in general taking a lower number of equations may result in missing of the representation sought. Applied to our example one gets $(2+1-1)\cdot(2\cdot 3)+(3-1)=14$. We finally obtain the following power series representation.

Example 15.

```
>   FPS[FPS](f,z,n)
```

$$\sum_{n=0}^{\infty}\frac{(-1)^n z^{2n+2}}{n+1}+\sum_{n=0}^{\infty}\frac{(-1)^n z^{2n+1}}{2n+1}+\sum_{n=0}^{\infty}\frac{z^{3n}}{n!}$$

Remark 3.

- Sometimes the linear system has many solutions leading to different (but equivalent) representations of the same power series. We have observed that in certain cases our Maxima implementation yields a different representation than our Maple implementation. This could be explained by the way both CASes represent solutions of linear systems. Our Maple implementation uses `Solve[Linear]` and sometimes `solve`.
- Although the number of initial coefficients can be computed using limit computations, in our implementation we rather use Maple's `series` command. Note, however, that this command does not always give expansions of the required orders due to internal cancellations, therefore it is always important to check the degree of the obtained Taylor polynomial. That is one difference we encountered between Maple and Maxima which mostly handles Taylor polynomials as desired.

Our next example is $f(z) := -1+(1+z^2)\cdot\exp(z)+\operatorname{arcsech}(\sqrt{z})$. In the previous case, we could neither expect shifted starting points nor Puiseux numbers. With this new example one may expect the hypergeometric type part of the expansion starting summations at 1 (or with z^1 instead of z^0) since a constant term appears in f. One also observes that a possible Puiseux number is 2 as $\sqrt{z}$ appears. However, our procedure does not make any checking on its inputs, everything is deduced from the holonomic recurrence equations which encode all this information.

Example 16. The computed recurrence equation is

```
> f:=(1+z^2)*exp(z)+arcsech(sqrt(z))-1:
> FPS[FindRE](f,z,a(n))
```

$$-9(n-1)n(2n+1)a(n)-2(2n-7)(n-4)a(n-4)+(n-3)\left(4n^2-52n+103\right)a(n-3)+(n-2)\left(14n^2-67n+49\right)a(n-2)-(n-1)\left(14n^2-112n+95\right)a(n-1)+2(5n-12)(n+1)^2a(n+1)+4(n+1)(n+2)^2a(n+2)=0. \quad (17)$$

Observe that $7/2$ (or $1/2$ after normalizing the equation) is a root of the trailing polynomial coefficient of (17). What we have established is that for hypergeometric type power series, the least common multiple of the leading and trailing polynomial coefficient root denominators should be taken as the Puiseux number of the representation sought. Thus by computing the power series of $f(z^p)$, where p denotes that Puiseux number, and replacing z by $z^{1/p}$ in the final representation will automatically generate all the Puiseux numbers of the representation. Hence next we compute a holonomic recurrence equation for $f(z^2)$.

Example 17.

```
> FPS[FindRE](subs(z=z^2,f),z,a(n))
```

$$-4(n-10)(n-11)a(n-11)+2(n-9)\left(n^2-32n+190\right)a(n-9)+(n-7)\left(7n^2-109n+362\right)a(n-7)-(n-5)\left(7n^2-154n+589\right)a(n-5)-9(n-5)(n-2)(n-3)a(n-3)+(5n-39)(n-1)^2a(n-1)+2(n-1)(n+1)^2a(n+1)=0 \quad (18)$$

Remark that all the rational roots of the leading and trailing polynomial coefficient are now integers. This is even more advantageous since it allows `mfoldHyper` to get nicer formulas for m-fold hypergeometric terms.

For computing starting points, by developing a procedure to find finite sequence (coefficients of the Laurent polynomial part) solutions of holonomic recurrence equations, we established that the algorithm behind the following Maple code could generally give the starting point and the Laurent polynomial part of a hypergeometric type power series.

```
LPolyPart := proc(CRE,f,z,n)
        local d,M,N,P0;
        description "Compute the Laurent polynomial part"
        "of a hypergeometric type power series"
        "from normalized polynomial coefficients";
        d:=numelems(CRE)-1;
        M:=foldl(`union`,{},isolve(subs(n=n-d,CRE[-1])));
        M:=map(rhs,M);
        N:=foldl(`union`,{},isolve(CRE[1]));
        N:=map(rhs,N);
```

```
        if numelems(N)<1 then
                PO:=0;
                N:=min(M)
        else
                N:=max(N)+1;
                PO:=convert(series(f,z=0,N),polynom)
        end if;
        return PO, N
end proc:
```

Normalized polynomial coefficients means that the coefficients are collected with the recurrence equation written with a_n (index n) as trailing term. For the present example we find the following Laurent polynomial part and starting point

$$\ln(2) - 1/2\ln(z),\ \ 1. \tag{19}$$

We mention that mis-consideration of starting points may lead to wrong power series representations. This could explain why in some examples the built-in Maple approach gives an incorrect representation for $\exp(z) + \ln(1+z)$. Some other similar examples can be found.

Finally using the other steps described in the first example, we get the representation below.

Example 18.

```
>   FPS[FPS](f,z,n)
```

$$\ln(2) - 1/2\ln(z) + \sum_{n=0}^{\infty} -\frac{\left(-(n+1)!\,n^2 + 2^{-2n-2}(2n+1)! - (n+1)!\,n - (n+1)!\right) z^{n+1}}{\left((n+1)!\right)^2}$$

Our implementation groups coefficients with same z-powers together.

Remark 4. As we mentioned earlier, rational functions are all of hypergeometric type. However some of their power series representations need extension fields where computations cannot easily be handled. The main issue with such cases relies on the linear system to be solved to find the corresponding linear combination of hypergeometric type power series.

Example 19. Consider for example $f(z) = 1/(1+z+z^4)$. We find the recurrence equation

```
>   f:=1/(1+z+z^4):
>   RE:=FPS[FindRE](f,z,a(n))
```

$$RE := (n+1)\,a(n) + (n+1)\,a(n-3) + (n+1)\,a(n+1) = 0,$$

and the following hypergeometric term ($m = 1$ only) solutions

```
>   FPS[mfoldHyper](RE,a(n),C)
```

$$\Big[\big[1, \Big\{ \left(RootOf\left(\_Z^4 + \_Z^3 + 1, index = 1\right)\right)^n, \left(RootOf\left(\_Z^4 + \_Z^3 + 1, index = 2\right)\right)^n, \left(RootOf\left(\_Z^4 + \_Z^3 + 1, index = 3\right)\right)^n, \left(RootOf\left(\_Z^4 + \_Z^3 + 1, index = 4\right)\right)^n \Big\}\big]\Big]. \quad (20)$$

The argument `C` in `mfoldHyper` is used to allow computations over extension fields. This happens in the algorithm whenever no solution exists over the rationals. When looking for a linear combination by solving the underlying linear system, we get the following error message.

Error, (in evala/Normal/preproc0) numeric exception: division by zero

This is a particular issue that we should try to overcome while finalizing our implementation.

Let us now present more examples describing our algorithm.

Example 20.

```
> FPS[FPS](sin(2*arcsin(z))+cos(3*arccos(z)),z,n)
```

$$4\,z^3 - 3\,z + \left(\sum_{n=0}^{\infty} -\frac{2\,(2\,n)!\,4^{-n}\,z^{2\,n+1}}{(2\,n-1)\,n!^2}\right)$$

```
> FPS[FPS](cosh(z)+z*cos(z)+sin(z^3),z,n)
```

$$\sum_{n=0}^{\infty} \frac{z^{2\,n}}{(2\,n)!} + \sum_{n=0}^{\infty} \frac{(-1)^n\,z^{2\,n+1}}{(2\,n)!} + \sum_{n=0}^{\infty} \frac{(-1)^n\,z^{6\,n+3}}{(2\,n+1)!}$$

```
> FPS[FPS](exp(z)+hypergeom([a, b], [c],
> z^2),z,n,fpstype=SpecialFunctions)
```

$$\sum_{n=0}^{\infty} \frac{z^n}{n!} + \sum_{n=0}^{\infty} \frac{(b)_n (a)_n z^{2\,n}}{n!\,(c)_n}$$

For special functions like the generalized hypergeometric function, the approach used to compute holonomic differential equations must slightly be modified. That is why in this previous example the option `fpstype=SpecialFunctions` is used.

Example 21.

```
> FPS[FPS](1/((p-z^2)*(q-z^3)),z,n)
```

$$\sum_{n=0}^{\infty} -\frac{\left(qp^{-1-n/2} - pq^{-1/3-n/3}\right) z^n}{p^3 - q^2} + \sum_{n=0}^{\infty} -\frac{\left(p^{3/2} - q\right) p^{-n-3/2} z^{2\,n+1}}{p^3 - q^2} + \sum_{n=0}^{\infty} -\frac{q^{-1-n} p \left(q^{2/3} - p\right) z^{3\,n}}{p^3 - q^2} + \sum_{n=0}^{\infty} \frac{\left(q^{2/3} - p\right) q^{-n-2/3} z^{3\,n+1}}{p^3 - q^2}$$

```
> FPS[FPS](arctan(sqrt(z))+arcsinh(z^(1/3)),z,n)
```

$$\sum_{n=0}^{\infty} \frac{(-1)^n (2n)! \, 4^{-n} z^{2/3\,n+1/3}}{(2n+1)(n!)^2} + \sum_{n=0}^{\infty} \frac{(-1)^n z^{n+1/2}}{2n+1}$$

As one can observe, our implementation linearizes the computation of hypergeometric type power series. None of the above examples can directly be computed using the built-in Maple `convert/FormalPowerSeries`.

4 Conclusion

We have presented an algorithm and its Maple implementation to compute hypergeometric type power series as defined in Definition 1. We have shown that this extends Maple's capabilities to consider more expressions for which power series could not be directly computed before. We believe that this is an important advancement that should be integrated into computer algebra systems.

Besides other important steps like computing Puiseux numbers, the Laurent polynomial part, or finding a linear combination of hypergeometric type power series, the main ingredient of the algorithm is **mfoldHyper** which computes m-fold hypergeometric term solutions of holonomic recurrence equations. The latter was the main deficiency of the algorithm in [5] on which Maple's original implementation is based.

There are further types of series considered in [8]. The question is, what to do when it turns out that a given expression is not of hypergeometric type, or is even non-holonomic. We mention for example the case where the given expression is not of hypergeometric type but holonomic. That means a recurrence equation can be computed. Some trivial cases like $\exp(z + z^2) \cdot \cos(z)$ are well handled as we could find a recursive representation that could be used for fast computations of larger order Taylor expansions.

Example 22.

```
> f:=exp(z+z^2)*cos(z):
> FPS[FPS](f,z,n,fpstype=Holonomic)
```

$$\sum_{n=0}^{\infty} A(n)\, z^n,$$

$$A(n+4) = -\frac{4A(n) + 4A(n+1) + (-4n-8)A(n+2) + (-2n-6)A(n+3)}{(n+3)(n+4)}$$
$$, 0 \le n, \; [A(0) = 1, A(1) = 1, A(2) = 1, A(3) = 2/3] \quad (21)$$

```
> FPS[Taylor](f,z,0,6)
```

$$1 + z + z^2 + 2/3\, z^3 + 1/3\, z^4 + 2/15\, z^5$$

```
> T:=Time():FPS[Taylor](f,z,0,1000):Time()-T
```

250 ms

```
> T:=Time():series(f,z=0,1000):Time()-T
```

9332 ms

On the other hand, we also look for techniques to consider more special functions. We mention Mathieu functions (see this Maple link[7]) for which we are able to recover differential equations that they satisfy using our code `FPS[LinearDE]` which does not necessarily look for polynomial coefficients inside the differential equation sought.

Example 23.

```
> FPS[LinearDE](MathieuC(a, q, x),F(x))
```

$$(-2\,q\cos(2\,x)+a)\,F(x)+\frac{\mathrm{d}^2}{\mathrm{d}x^2}F(x)=0$$

```
> FPS[LinearDE](MathieuFloquet(a, q, x^2),F(x))
```

$$-4\,x^3\left(2\,q\cos\left(2\,x^2\right)-a\right)F(x)-\frac{\mathrm{d}}{\mathrm{d}x}F(x)+x\frac{\mathrm{d}^2}{\mathrm{d}x^2}F(x)=0$$

Further steps to consider rely on change of variables transformations (when possible) of such types of equations to holonomic equations. We are grateful to have gotten a question regarding Mathieu functions during our presentation at the 2020 Maple conference.

References

1. Gruntz, D., Koepf, W.: Maple package on formal power series. Maple Tech. Newsl. **2**(2), 22–28 (1995)
2. Hendricks, P.A., Singer, M.F.: Solving difference equations in finite terms. J. Symb. Comput. **27**(3), 239–259 (1999)
3. Horn, P., Koepf, W., Sprenger, T.: m-fold hypergeometric solutions of linear recurrence equations revisited. Math. Comput. Sci. **6**(1), 61–77 (2012). https://doi.org/10.1007/s11786-012-0107-8
4. Kauers, M., Paule, P.: The Concrete Tetrahedron. Symbolic Sums, Recurrence Equations, Generating Functions, Asymptotic Estimates. Springer, Vienna (2011). https://doi.org/10.1007/978-3-7091-0445-3
5. Koepf, W.: Power series in computer algebra. J. Symb. Comput. **13**(6), 581–603 (1992)
6. Petkovšek, M., Salvy, B.: Finding all hypergeometric solutions of linear differential equations. In: Bronstein, M. (ed.) ISSAC, pp. 27–33. Association for Computing Machinery, New York (1993)
7. Ryabenko, A.: Special formal series solutions of linear ordinary differential equations. In: Krob, D., Mikhalev, A.A., Mikhalev, A.V. (eds.) Formal Power Series and Algebraic Combinatorics, pp. 356–366. Springer, Heidelberg (2000). https://doi.org/10.1007/978-3-662-04166-6_32
8. Teguia Tabuguia, B.: Power series representations of hypergeometric types and non-holonomic functions in computer algebra. Ph.D. thesis, University of Kassel (2020). https://kobra.uni-kassel.de/handle/123456789/11598

[7] https://de.maplesoft.com/support/help/Maple/view.aspx?path=examples/Mathieu.

9. Teguia Tabuguia, B.: A variant of van Hoeij's algorithm to compute hypergeometric term solutions of holonomic recurrence equations. arXiv:2012.11513 [cs.SC] preprint (2020)
10. Teguia Tabuguia, B., Koepf, W.: Symbolic computation of hypergeometric type and non-holonomic power series. arXiv:2102.04157 [cs.SC] preprint (2021)
11. Van Hoeij, M.: Finite singularities and hypergeometric solutions of linear recurrence equations. J. Pure Appl. Algebra **139**(1–3), 109–131 (1999)

Rational Cone of Norm-Invariant Vectors Under a Matrix Action

Juan Tolosa(✉)

Stockton University, Galloway, USA
juan.tolosa@stockton.edu

Abstract. Starting from one example suggested by Dr. Robert Lopez in a *Maple* webinar, we study directions along which the norms of vectors are preserved under a matrix, regarded as a linear map. In particular, we find families of 2×2 matrices for which these directions are determined by integer vectors. We also explore a few examples of 3×3 matrices.

We used *Maple* both to explore this topic and to generate the included figures.

1 Introduction

In the 2015 Webinar talk *Eigenpairs in Maple* [3], Dr. Robert Lopez discussed how to use *Maple* to find eigenvalues and eigenvectors (eigenpairs) of a matrix A. An eigenvector of A is a (nonzero) vector whose direction is preserved under multiplication by A. As an aside, by the end of the talk, Dr. Lopez, posed the question: what about preserving the *magnitude* of the vector, rather than its direction? In other words, what about (nonzero) vectors $\mathbf{v}$ such that

$$\|\mathbf{v}\| = \|A\mathbf{v}\|, \tag{1}$$

where $\|\mathbf{v}\|$ is the Euclidean norm?

In search of an example, Dr. Lopez used a "for loop" in *Maple*, looking for integer matrices A which had vectors $\mathbf{v}$ with integer coordinates, satisfying the equation $\|\mathbf{v}\|^2 = \|A\mathbf{v}\|^2$. He came up with the following example:

$$A = \begin{pmatrix} 4 & 3 \\ -2 & -3 \end{pmatrix}.$$

Regarded as a map from $\mathbb{R}^2$ to itself, this matrix preserves the norms, but not the directions, of the vectors with integer coordinates $\mathbf{v}_1 = \langle 1, -1\rangle$ and $\mathbf{v}_2 = \langle 17, -19\rangle$.

Exploring this intriguing idea, I obtained several families of such "nice" 2×2 matrices, then considered a few 3×3 examples, and discussed a couple of related quadratic Diophantine equations. A detailed account is in [4].

R. M. Corless et al. (Eds.): MC 2020, CCIS 1414, pp. 394–409, 2021.
https://doi.org/10.1007/978-3-030-81698-8_26

First of all, here are some general comments.

- Since $\|\lambda\mathbf{v}\| = |\lambda| \cdot \|\mathbf{v}\|$, the entire line generated by any nonzero solution of (1) will consist of vectors whose norm is preserved by A; let us call these lines the *norm-preserving lines.*
- If A has an eigenvalue of 1, or -1, then the corresponding eigenspace will consist entirely of solutions of (1) as well.
- If A is orthogonal, then it acts as an isometry and norm is always preserved. Therefore, *every* line through the origin is norm-preserving. In the 2×2 case these isometries are typically rotations.

The interesting case, though, is when there are nonzero solutions of (1) that are *not* eigenvectors (and also when A is not orthogonal). This may happen *even* if A has an eigenvalue ± 1. For example,

$$A = \begin{pmatrix} 1 & -8 \\ 0 & 3 \end{pmatrix}$$

has eigenvalue 1 with eigenline determined by $\mathbf{v} = \langle 1, 0 \rangle$, but also has another norm-preserving line, determined by $\mathbf{w} = \langle 9, 2 \rangle$. Along this line, A acts like a rotation.

2 The 2×2 Case

We are interested in nonzero solutions of equation (1) for a general real-valued 2×2-matrix

$$A = \begin{pmatrix} a & b \\ c & d \end{pmatrix}.$$

Equation (1) is equivalent to $\|\mathbf{v}\|^2 = \|A\mathbf{v}\|^2$, or $(\mathbf{v}, \mathbf{v}) = (A\mathbf{v}, A\mathbf{v})$, where $(\mathbf{v}, \mathbf{w})$ is Euclidean inner product. The right-hand side becomes

$$\|A\mathbf{v}\|^2 = (A\mathbf{v}, A\mathbf{v}) = (\mathbf{v}, A^t A\mathbf{v}) = (\mathbf{v}, B\mathbf{v}),$$

where A^t is the transpose of A, and $B = A^t A$. Thus, equation (1) is equivalent to

$$(\mathbf{v}, (B - I)\mathbf{v}) = 0, \tag{2}$$

where I is the identity matrix. Denoting $\mathbf{v} = \langle x, y \rangle$ the coordinates of $\mathbf{v}$, the left-hand side of (2) is the quadratic form

$$\Phi(x, y) = (a^2 + c^2 - 1)x^2 + 2(ab + cd)xy + (b^2 + d^2 - 1)y^2; \tag{3}$$

with this notation, (1), or (2), is equivalent to $\Phi(x, y) = 0$.

Does every matrix A have norm-preserving lines at all? It turns out, a condition is needed:

Theorem 1. *Norm-preserving lines exist if and only if*

$$a^2 + b^2 + c^2 + d^2 \geq 1 + \det(A)^2. \tag{4}$$

The details of the proof are in [4]. Geometrically, this condition reflects the fact that for the existence of such lines, the ellipse (or degenerate ellipse) $(\mathbf{v}, B\mathbf{v}) = 1$ must intersect the unit circle. In terms of the eigenvalues of $B = A^t A$ (which are real and non-negative), $\lambda_1 \leq \lambda_2$, this happens if, and only if, $\lambda_1 \leq 1 \leq \lambda_2$. Indeed, by the extreme properties of eigenvalues (see, for example, [1]), we have

$$\begin{aligned}\lambda_1 &= \min_{\|\mathbf{v}\|=1} \|A\mathbf{v}\|^2 = \min_{\|\mathbf{v}\|=1} (\mathbf{v}, B\mathbf{v}) \\ &\leq \max_{\|\mathbf{v}\|=1} \|A\mathbf{v}\|^2 = \max_{\|\mathbf{v}\|=1} (\mathbf{v}, B\mathbf{v}) = \lambda_2.\end{aligned}$$

Figure 1 illustrates the intersection of this ellipse with the unit circle, for R. Lopez's matrix.

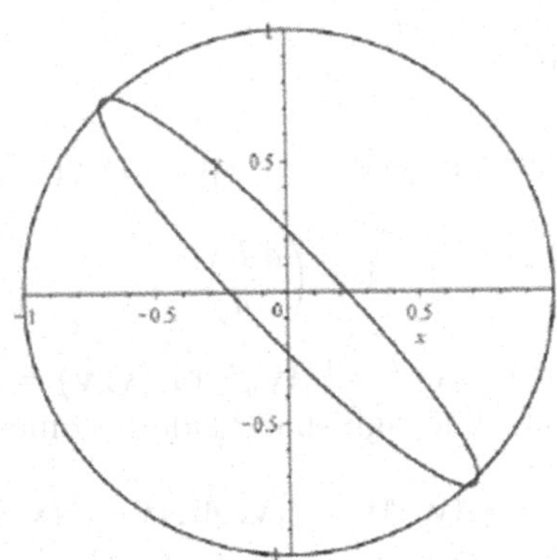

Fig. 1. Illustration of R. Lopez's example.

The norm-preserving lines will then pass through the intersections of the ellipse with the unit circle (see Fig. 2 below).

Families of matrices with integer solutions.

We want to find matrices

$$A = \begin{pmatrix} a & b \\ c & d \end{pmatrix}$$

with integer (or occasionally rational) entries, for which the norm-preserving lines can be determined by nonzero vectors $\mathbf{v} = \langle x, y \rangle$ with integer coordinates

x, y; we will call these *integer solution lines.* For such matrix, the Diophantine equation $\Phi(x, y) = 0$, with $\Phi(x, y)$ given by (3), must have nontrivial solutions.

If either $a^2 + c^2 = 1$, or $b^2 + d^2 = 1$, it is not hard to get families of matrices with rational entries and integer solution lines. For example, for the two-parameter family

$$A = \begin{pmatrix} \frac{3}{5} & b \\ \frac{4}{5} & d \end{pmatrix},$$

we get norm-preserving lines determined by $\mathbf{v}_1 = \langle 1, 0 \rangle$ and

$$\mathbf{v}_2 = \langle 5(1 - b^2 - d^2), 2(3b + 4d) \rangle.$$

The interesting case, however, is when the columns of A are not unit vectors. Assuming, for example, that $b^2 + d^2 \neq 1$, and working with the corresponding Diophantine equation, we obtained in [4] the following four two-parameter families of matrices:

$$\begin{pmatrix} a & a \pm 1 \\ c & c \pm 1 \end{pmatrix}.$$

Incidentally, their transposes,

$$\begin{pmatrix} a & c \\ a \pm 1 & c \pm 1 \end{pmatrix},$$

also have integer solution lines.

For each family, we get in [4] concrete formulas for the integer solution vectors. The example of Robert Lopez corresponds to the family

$$A = \begin{pmatrix} a & a - 1 \\ c & c - 1 \end{pmatrix}, \tag{5}$$

for $a = 4$ and $c = -2$. For this particular family, the norm-preserving lines are determined by the vectors

$$\mathbf{v}_1 = \langle 1, -1 \rangle$$

and

$$\mathbf{v}_2 = \langle (a - 1)^2 + (c - 1)^2 - 1, 1 - a^2 - c^2 \rangle.$$

We get similar formulas for the remaining cases.

Figure 2 illustrates the case $a = 2$, $c = -3$; the two length-preserving lines are shown. The direction vectors are $\mathbf{v}_1 = \langle 1, -1 \rangle$ and $\mathbf{v}_2 = \langle 4, -3 \rangle$. As expected, the solution lines pass through the intersections of the ellipse $\|A\mathbf{v}\| = 1$ with the unit circle. The solution lines in Fig. 1 are not depicted, since they are rather close to each other. If the ellipse is tangent to the unit circle, as for example when $a = 3$, $c = -2$, we get only one integer solution line.

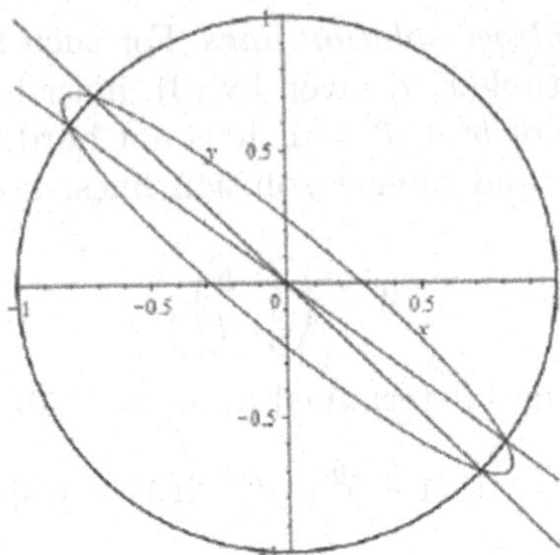

Fig. 2. Norm-preserving directions shown.

3 The 3 × 3 Case

The 3×3 case is considerably more complicated, as well as more interesting.

If A is a 3×3 real-valued matrix, also regarded as a linear map from $\mathbb{R}^3$ to itself, then in general $\|A\mathbf{v}\|^2 = 1$ is an ellipsoid, in terms of the coordinates of $\mathbf{v} = \langle x, y, z \rangle$. As in the 2×2 case, we have

$$\|A\mathbf{v}\|^2 = (A\mathbf{v}, A\mathbf{v}) = (\mathbf{v}, B\mathbf{v}),$$

where $B = A^t A$ is a symmetric matrix with real nonnegative eigenvalues. The equation for norm-preserving vectors, $\|A\mathbf{v}\| = \|\mathbf{v}\|$, or $(\mathbf{v}, B\mathbf{v}) = (\mathbf{v}, \mathbf{v})$, is equivalent to the equation of the cone

$$(\mathbf{v}, (B - I)\mathbf{v}) = 0, \tag{6}$$

where I is the identity matrix.

Like in the 2×2 case, if we denote the eigenvalues of B by $0 \leq \lambda_1 \leq \lambda_2 \leq \lambda_3$, then there is a solution of (6) if and only if

$$\lambda_1 \leq 1 \leq \lambda_3, \tag{7}$$

which guarantees a nonempty intersection of the ellipsoid (or degenerate ellipsoid) $\|B\mathbf{v}\|^2 = 1$ with the unit sphere $\|\mathbf{v}\|^2 = 1$.

If condition (7) is satisfied, then the cone determined by (6) will pass through this intersection of the ellipsoid and the unit sphere. As an illustration, for the matrix in Example 1 below, Fig. 3 (a) shows the ellipsoid $\|A\mathbf{v}\|^2 = 1$ and the unit sphere, and Fig. 3 (b) has the added solution cone $\|A\mathbf{v}\| = \|\mathbf{v}\|$; compare with Fig. 2 for the 2×2 case.

Unlike the 2×2 case, in the 3×3 case one cannot hope that in general *all* the lines in the cone (6) for a given matrix A with integer or rational coefficients will turn out to be integer solution lines. Examples 2 and 3 show that, however, we can still get infinitely many such lines.

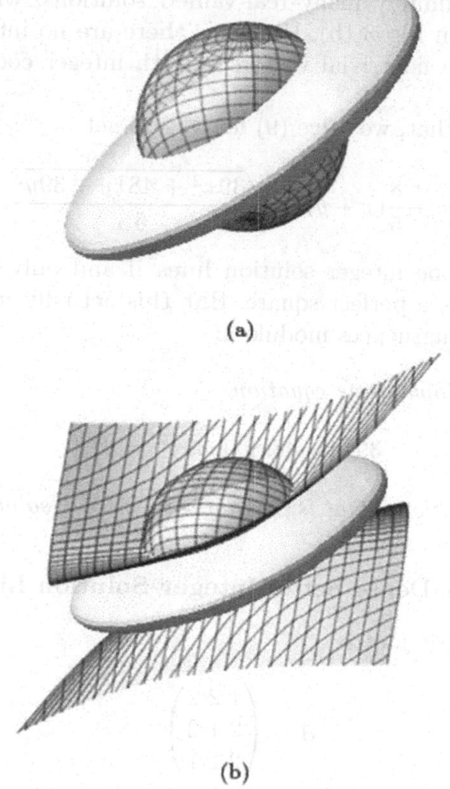

(a)

(b)

Fig. 3. Illustration for Example 1: (a) ellipsoid and unit sphere; (b) same, plus solution cone.

3.1 Example 1: No Integer Solution Lines

Let us consider the symmetric matrix

$$A = \begin{pmatrix} 1 & 1 & \frac{1}{2} \\ 1 & \frac{1}{2} & 1 \\ \frac{1}{2} & 1 & 1 \end{pmatrix} \tag{8}$$

The form $\|A\mathbf{v}\|^2 = (A\mathbf{v}, A\mathbf{v}) = (\mathbf{v}, B\mathbf{v})$ will in this case have matrix

$$B = A^t A = A^2 = \begin{pmatrix} \frac{9}{4} & 2 & 2 \\ 2 & \frac{9}{4} & 2 \\ 2 & 2 & \frac{9}{4} \end{pmatrix}$$

so that Eq. (6) will be

$$\frac{5}{4}x^2 + 4xy + 4xz + \frac{5}{4}y^2 + 4yz + \frac{5}{4}z^2 = 0. \tag{9}$$

This equation has infinitely many real-valued solutions, which constitute the solution cone shown in Fig. 3 (b). However, there are no integer solution lines, that is, there exist no nontrivial vectors $\mathbf{v}$ with integer coordinates such that $\|A\mathbf{v}\| = \|\mathbf{v}\|$.

In order to prove that, we solve (9) for z, and get

$$z = -\frac{8}{5}(x+y) \pm \frac{\sqrt{39x^2 + 48xy + 39y^2}}{5}.$$

Therefore, there will be integer solution lines if and only if the discriminant $39x^2 + 48xy + 39y^2$ is a perfect square. But this actually never happens, as is proved in [4] using congruences modulo 4:

Theorem 2. *The Diophantine equation*

$$39x^2 + 48xy + 39y^2 = u^2 \tag{10}$$

has no nontrivial solutions, that is, no nonzero integer solutions.

3.2 Example 2: A Dense Set of Integer Solution Lines

Consider the symmetric matrix

$$A = \begin{pmatrix} 1 & 2 & 2 \\ 2 & 1 & 2 \\ 2 & 2 & 1 \end{pmatrix}. \tag{11}$$

Here

$$B = A^t A = A^2 = \begin{pmatrix} 9 & 8 & 8 \\ 8 & 9 & 8 \\ 8 & 8 & 9 \end{pmatrix}$$

so that Eq. (6) will be

$$8x^2 + 16xy + 16xz + 8y^2 + 16yz + 8z^2 = 0,$$

or (after dividing by 8)

$$(x+y+z)^2 = 0. \tag{12}$$

In our case, the cone degenerates into the plane $x + y + z = 0$. Picking, for example, $\mathbf{v}_1 = \langle 1, 0, -1\rangle$ and $\mathbf{v}_2 = \langle 0, 1, -1\rangle$, we conclude that every integer norm-preserving line is generated by $\alpha\mathbf{v}_1 + \beta\mathbf{v}_2$, with integer coefficients α, β such that $\alpha^2 + \beta^2 > 0$. This constitutes a dense set of integer solution lines, in the plane $x + y + z = 0$. The ellipsoid $\|A\mathbf{v}\|^2 = 1$ lies inside the unit sphere, and is tangent to it along the intersection of the sphere with the solution plane; Fig. 4 depicts the situation.

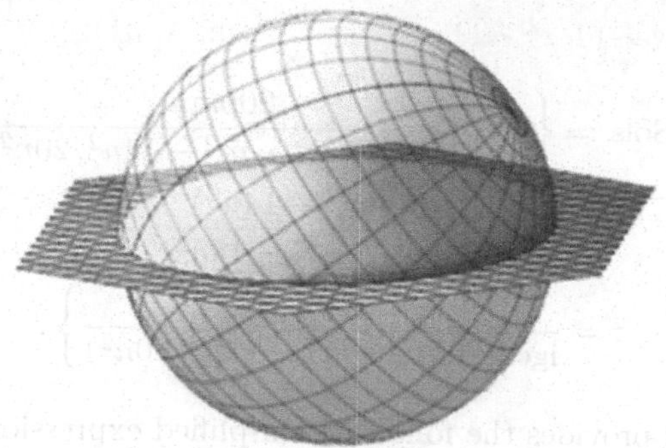

Fig. 4. Solution cone is degenerate.

3.3 Example 3: Infinitely Many Integer Solution Lines

Finally, let us consider an example when there are still infinitely many integer solution lines, yet we cannot guarantee they are dense in the cone of all solution lines. Consider the matrix

$$A = \begin{pmatrix} 1 & 2 & 3 \\ 2 & 1 & 1 \\ 1 & 1 & 1 \end{pmatrix}. \tag{13}$$

One eigenvalue of A is -1, with eigenvector $=\langle -1, 1, 0\rangle$, which therefore provides one integer solution line. Are there any other such lines? The matrix $B = A^t A$ is

$$B = \begin{pmatrix} 6 & 5 & 6 \\ 5 & 6 & 8 \\ 6 & 8 & 11 \end{pmatrix}$$

and consequently Eq. (6) becomes

$$5x^2 + 10xy + 12xz + 5y^2 + 16yz + 10z^2 = 0.$$

Solving for x yields

$$x = -y - \frac{6}{5}z \pm \frac{\sqrt{-20yz - 14z^2}}{5}. \tag{14}$$

We conclude that there will be integer solutions lines if and only if the discriminant $-20yz - 14z^2$ is a perfect square. This leads to the Diophantine equation

$$-20yz - 14z^2 = u^2. \tag{15}$$

An excellent tool for exploration of Diophantine equations is the *Maple* `isolve` command. In this case, we obtain

> intSols:=isolve(-14*z^2 - 20*$y*z = u^2$, {m, n, p});

$$\text{intSols} := \left\{ u = \frac{20pmn}{\text{igcd}(20mn, -m^2 - 14n^2, 20n^2)}, \right.$$
$$y = \frac{p(-m^2 - 14n^2)}{\text{igcd}(20mn, -m^2 - 14n^2, 20n^2)},$$
$$\left. z = \frac{20pn^2}{\text{igcd}(20mn, -m^2 - 14n^2, 20n^2)} \right\}.$$

Setting $p = 1$, this provides the following simplified expressions for y, z, u:

$$y = -m^2 - 14n^2; \qquad z = 20n^2, \qquad u = 20mn,$$

where m, n are arbitrary integers. And using (14), we find two corresponding values for x:

$$x_1 = m^2 + 4mn - 10n^2 \qquad \text{and} \qquad x_2 = m^2 - 4mn - 10n^2.$$

These formulas generate infinitely many solutions. For example, choosing $m = 1$ and $n = 2$, we obtain the integer solution vectors $\langle -31, -57, 80\rangle$ and $\langle -47, -57, 80\rangle$.

It remains to check that indeed all solutions can be generated in this way.

A detailed solution of the Diophantine equation (15) carried up in [4] leads to the following expression, which does contain all solutions:

$$\begin{cases} x = \frac{1}{10}(r^2 - 10v^2 \pm 4vr) \\ y = -\frac{1}{10}(14v^2 + r^2) \\ z = 2v^2, \end{cases}$$

where v and r are arbitrary integers.

It is not hard to reconcile both solutions (and, therefore, show that *Maple* indeed provides all solutions): it suffices to set

$$v^2 = 10n^2 \qquad \text{and} \qquad r^2 = 10\,m^2.$$

Here are a few more sample solutions:

Using *Maple*'s formulas (after dividing by the gcd):

- For $m = 1$ and $n = 1$, we get $\langle -1, -3, 4\rangle$ and $\langle -13, -15, 20\rangle$.
- For $m = 4$ and $n = 1$: $\langle -1, -3, 2\rangle$ and $\langle 11, -15, 10\rangle$.

Using the second set of formulas:

- For $(v, r) = (1, 1)$ we obtain

$$\left\langle -\frac{1}{2}, -\frac{3}{2}, 2 \right\rangle \sim \langle 1, 3, -4\rangle \qquad \text{and} \qquad \left\langle -\frac{13}{10}, -\frac{3}{2}, 2 \right\rangle \sim \langle 13, 15, -20\rangle.$$

- For $(v, r) = (1, 4)$ we get the vectors

$$\left\langle \frac{11}{5}, -3, 2 \right\rangle \sim \langle 11, -15, 10\rangle \qquad \text{and} \qquad \langle 1, 3, -2\rangle;$$

3.4 A General Method

There is a method for obtaining infinitely many integer solution lines, which can be applied to a general 3×3 matrix with integer or rational coefficients; the drawback is that one must know (at least) one nontrivial solution. This is an idea by T. Piezas [5]. Namely, if we know one particular integer solution $(y, z, u) = (m, n, p)$ of the Diophantine equation

$$ay^2 + byz + cz^2 = du^2,$$

then a two-parameter family of solutions is given by

$$\begin{aligned} y &= (am + bn)s^2 + 2cnst - cmt^2, \\ z &= -ans^2 + 2amst + (bm + cn)t^2, \\ u &= p(as^2 + bst + ct^2), \end{aligned}$$

where s and t are arbitrary integers.

For example, for the matrix

$$A = \begin{pmatrix} 1\,2\,3 \\ 3\,4\,5 \\ 2\,3\,4 \end{pmatrix},$$

equation (6) for $\mathbf{v} = \langle x, y, z \rangle$ is

$$13x^2 + 40xy + 52xz + 28y^2 + 76yz + 49z^2 = 0,$$

which solved with respect to x produces

$$x = \frac{-20y}{13} - 2z \pm \frac{\sqrt{36y^2 + 52yz + 39z^2}}{13}, \tag{16}$$

so to get integer solutions we need to solve the Diophantine equation

$$36y^2 + 52yz + 39z^2 = u^2. \tag{17}$$

Setting $z = 0$, it is not hard to guess the particular solution $(y, z, u) = (m, n, p) = (1, 0, 6)$. The corresponding two-parameter solution family looks like

$$\begin{aligned} y &= 36s^2 - 39t^2 \\ z &= 72st + 52t^2 \\ u &= 216s^2 + 312st + 234t^2. \end{aligned} \tag{18}$$

Each choice of (s, t) produces two integer solution lines $\mathbf{v} = \langle x, y, z \rangle$, using the two x-values provided by (16). For example, for $(s, t) = (1, 1)$ we get

$$\left\langle -\frac{2402}{13}, -3, 124 \right\rangle \sim \langle -2402, -39, 1612 \rangle, \qquad \text{and} \qquad \langle -302, -3, 124 \rangle;$$

and $(s, t) = (1, 2)$ yields

$$\left\langle -\frac{4976}{13}, -120, 352 \right\rangle \sim \langle -4976, -1560, 4576 \rangle, \qquad \text{and} \qquad \langle -656, -120, 352 \rangle.$$

4 Questions for Further Study, Applications

- Unlike the case of eigenpairs, the solution lines to equation (1) are very much dependent on the chosen norm in $\mathbb{R}^n$. It would be interesting to discuss similar solutions for other norms in Euclidean space.
- We have only grazed the case of a 3×3 matrix A. Can we find solvability conditions of (1) in terms of the coefficients of A, as we did in the 2×2 case? Can one find nontrivial families of integer matrices A for which (1) has integer solutions?
- **Application to toral automorphisms.** Many of the 2×2 integer matrices we studied, for example, the subfamily

$$A = \begin{pmatrix} q+1 & q \\ q & q-1 \end{pmatrix}, \tag{19}$$

with q an integer, are symmetric and have determinant -1; therefore, they can be regarded also as linear automorphisms of the (flat) 2-torus, which possess very interesting dynamical properties; see, for example [2], p. 42. Nontrivial such automorphisms have eigenvectors with irrational slopes; on the other hand, the integer solution lines of (19) bisect the eigendirections. We can therefore use integer arithmetic to compute iterates of vectors in the stable and in the unstable manifolds of such automorphisms. For example, the matrix

$$A = \begin{pmatrix} 3 & 2 \\ 2 & 1 \end{pmatrix} \tag{20}$$

has integer solution lines generated by $\mathbf{v}_1 = \langle 1, -1 \rangle$ and $\mathbf{v}_2 = \langle -1, 3 \rangle$, and irrational eigenvalues

$$\lambda_1 = 2 + \sqrt{5} \qquad \text{and} \qquad \lambda_2 = 2 - \sqrt{5}.$$

If we choose the equal-norm vectors

$$\mathbf{v}_1 = \langle 1, -1 \rangle \qquad \text{and} \qquad \mathbf{v}_3 = \frac{1}{\sqrt{5}}\mathbf{v}_2 = \frac{1}{\sqrt{5}}\langle -1, 3 \rangle,$$

then $\mathbf{u} = \mathbf{v}_1 + \mathbf{v}_3$ will be along the unstable direction of A, and $\mathbf{w} = \mathbf{v}_1 - \mathbf{v}_3$ will be along the stable direction. Therefore, on the one hand

$$A^n\mathbf{u} = \lambda_1^n\mathbf{u} = (1 + \sqrt{5})^n\mathbf{u},$$

and on the other hand

$$A^n\mathbf{u} = A^n\mathbf{v}_1 + \frac{1}{\sqrt{5}}A^n\mathbf{v}_2.$$

For example,

$$A^{10}\mathbf{u} = \begin{pmatrix} 1346269 & 832040 \\ 832040 & 514229 \end{pmatrix}\mathbf{u} =$$

$$= \left\langle 514229 + \frac{1}{\sqrt{5}}1149851,\ 317811 + \frac{710647}{\sqrt{5}} \right\rangle,$$

provides a way to compute $(2+\sqrt{5})^{10}\mathbf{u}$ using only integer arithmetic. A similar calculation can be used for iterates of vectors in the stable direction.

5 Sample Maple Code Used

In Sect. 3.3 we showed how one can use *Maple* to successfully solve a Diophantine equation. Here I want to showcase other *Maple* commands used in my research of this topic. Most of them are suitable for motivated Linear Algebra students. With the hope this part might be used for that purpose, you will find some redundancy here, and even comments of additional interesting results one gets during the process.

Calculation of the cone of norm-invariant directions:

```
> restart; with(LinearAlgebra):
>  A:=<a, b; c, d>; v:= <x, y>;
```

Cone $\|A\mathbf{v}\|^2 - \|\mathbf{v}\|^2 = 0$:

```
> one:=Norm(A.v, 2, conjugate=false)^2 - Norm(v,2,conjugate=false)^2;
> cone:=collect(one, {x^2, y^2});
```

A second way to generate the cone, via quadratic forms:

```
> B:=Transpose(A).A- IdentityMatrix(2);
  expand(DotProduct(v, B.v, conjugate=false)):
  QuadForm:=collect(%, {x^2, y^2});
```

Calculations for one of our two-parameter families of matrices.

```
> restart; with(LinearAlgebra):
  v:=<x,y>;
  A:=<a,a-1; c,c-1>;
  B:=Transpose(A).A;
> quadFormProto:=expand(Transpose(v).B.v):
  quadForm:= collect(quadFormProto, {x^2, y^2});
  quadFormZero:=collect(quadForm - x^2 - y^2, {x^2, y^2});
```

Rewriting denominator (and checking):

```
> den:=(a^2+c^2-2*a-2*c+1); denToo:=(a-1)^2+(c-1)^2-1;
differenceIs:=expand(den-denToo);
```

Find length-preserving lines:
```
> solve(quadFormZero=0,y);
```
Corresponding slopes:
```
>m1:=-1; m2:= -(a^2+c^2-1)/((a-1)^2+(c-1)^2-1);
```
Particular case: matrix in R. Lopez's paper:
```
> Alopez:= eval(A, {a=4, c=-2});
```
The corresponding slopes are `m1 = -1` and
```
>m20:=eval(m2, {a=4, c=-2});
```
Corresponding symmetric matrix $B = A^t A$, and its eigenvalues; as expected, 1 lies between them:

```
> B0:= Transpose(Alopez).Alopez;
  Eigenvalues(B0); evalf(%);
```

Corresponding quadratic form:
```
quadForm0:=eval(quadForm, {a=4, c=-2});
```
Here is the plot for this particular case, which corresponds to Fig. 1 in the paper. First I show the ellipse $\|A\mathbf{v}\|^2 = 1$, the unit circle, and the solution lines. For Fig. 1 in the paper, I removed the solution lines (second plot below).

```
> with(plots): with(plottools):
  Window:= x = -1..1, y= -1..1;
> p1:=plot(m1*x, Window,color=blue):
  p2:=plot(m20*x, Window,color=blue):
  p3:=implicitplot(quadForm0=1, Window,
  color= blue, grid = [100, 100], gridrefine=2):
  p4:=plot([cos(t), sin(t), t=0..2*Pi], color=black):
  display(p1, p2, p3, p4, scaling=constrained);
  display(p3, p4);
```

Another particular case is the matrix

$$\begin{pmatrix} 3 & 2 \\ -2 & -3 \end{pmatrix},$$

which has only one solution line. This answers a question from one reviewer: *In the 2D case, is it true that, if there is such a vector v, there are always two such?* This example shows the answer is negative. As the picture generated below shows, the ellipse is tangent to the unit circle from the inside, so there is only one norm-preserving direction. On the other hand, if there are two solution lines, and one of them is integer, then the second one must necessarily be integer as well, if the matrix A has integer or rational coefficients.[1]

[1] The code is purposefully redundant: `p1` and `p2` are the same, since both solutions lines coincide.

```
> partic:= {a=3, c=-2};
  A0:=eval(A, partic);
  m20:=eval(m2, partic);
> B0:= Transpose(A0).A0; Eigenvectors(B0);
> quadForm0:=eval(quadForm, partic);
> p1:=plot(m1*x, Window,color=blue):
  p2:=plot(m20*x, Window,color=blue):
  p3:=implicitplot(quadForm0=1, Window,
 color= blue, grid = [100, 100], gridrefine=2):
 p4:=plot([cos(t), sin(t), t=0..2*Pi], color=black):
display(p1, p2, p3, p4, scaling=constrained);
```

The third particular case correspond to Fig. 2 in the paper, which is also generated below:

```
> partic:= {a=2, c=-3};
  A0:=eval(A, partic);
  m20:=eval(m2, partic);
  B0:= Transpose(A0).A0; Eigenvalues(B0); evalf(%);
> p1:=plot(m1*x, Window,color=blue):
  p2:=plot(m20*x, Window,color=blue):
  p3:=implicitplot(quadForm0=1, Window,
  color= blue, grid = [100, 100], gridrefine=2):
  p4:=plot([cos(t), sin(t), t=0..2*Pi], color=black):
  display(p1, p2, p3, p4, scaling=constrained);
```

Here is an example of using Maple to experiment with the 3×3-case; the corresponding solution cone produces a lengthy expression.

```
> restart; with(Student[LinearAlgebra]): with(plots):
  vv:=<x,y,z>; uno:=Norm(vv,2)^2;
> A:=<q,q+1, q+2; s,s+1, s+2; r, r+1, r+2>;
  determinantOfAIs:=Determinant(A);
> uu:=A.vv; normImage:=Norm(uu,2)^2;
  sols:=solve(normImage=uno,x);
```

The particular choice of the parameters below leads to a degenerate matrix, for which we manage to find integer solution lines. Two of the eigenvectors of A are messy.

```
> A0:= eval(A, {q=1, r=2, s=3});
> eigsA0:= Eigenvalues(A0);
   approxEigsA0:=evalf(eigsA0);
    Eigenvectors(A0);
> solPart:=eval(sols, {q=1, r=2, s=3});
```

The set of these solutions is an entire cone in 3-space. Let us pick particular lines in this cone (four sets); the first pair of solutions actually has rational coordinates; we use the rational slopes obtained, to create the two integer solution vectors:

```
> solOne:=eval(solPart, {y=1, z=0});
  solTwo:=eval(solPart, {y=0, z=1});
> rationalLine1:=simplify(solOne[1]);
   rationalLine12:=simplify(solOne[2]);
> v1:=<-20+6, 13, 0>; v2:= <-20 -6, 13, 0>;
```

A direct check that the norm of $\mathbf{v}_1 = \langle -14, 13, 0\rangle$ is preserved, leads to the interesting identity $10^2 + 11^2 + 12^2 = 13^2 + 14^2 = 365$; the direct check for $\mathbf{v}_2$ is less interesting:

```
> v1Is:=v1;a1:=norm(v1, 2); w1:=A0.v1; b1:= Norm(w1, 2);
   v2Is:=v2;
   a2:=norm(v2, 2); w2:=A0.v2; b2:= Norm(w2, 2);
```

Here is Example 1 in the paper, and the corresponding pictures used to generate Fig. 3 (a) and (b). As shown in [4], the corresponding discriminant is not a perfect square, so we have no solution lines. To plot, we try both explicit and implicit plots. The second plot generated below, suitably rotated, is Fig. 3 (a) in the paper.

```
> restart; with(Student[LinearAlgebra]): with(plots):
  vv:=<x,y,z>; uno:=Norm(vv,2)^2;
  a:=1; b:=1/2;
  A:=<a,a, b; a, b,a; b,a,a>;
> eigsOfA:=Eigenvalues(A);  Eigenvectors(A);
> ImageIs:=A.vv;
  normImage:=Norm(ImageIs, 2)^2:
  quadForm:=expand(normImage);
  quadFormZero:=expand(normImage-uno);
> sols:=solve(quadFormZero=0, z)
> maxVal:=2.0:
   p1imp:=implicitplot3d(quadForm = 1, x=-maxVal..maxVal, y=-maxVal..maxVal,
   z=-maxVal..maxVal, grid=[50,50,50], style=surface):
   p1:=plot3d({explicit[1], explicit[2]}, x=-2..2, y=-2..2,
   view=-2..2, grid=[220,220]):
  sphere:=<sin(u)*sin(v), sin(u)*cos(v), cos(u)>:
  p2:=plot3d([sphere], u=0..Pi, v=0..2*Pi,
  axes=boxed, scaling=constrained, transparency = 0.4):
  display(p1,p2);
  display(p1imp, p2, axes=none);
```

And the second plot below, suitably rotated, produced Fig. 3(b):

```
> zFunction:=solve(quadFormZero=0, z);
  p3:= plot3d({zFunction[1], zFunction[2]}, x=-2..2, y=-2..2, view= -2..2,
  axes=normal, scaling=constrained):
  display(p1,p2,p3);
  display(p1imp, p2, p3, axes=none);
```

Here is Example 2; the solution cone in this case degenerates into a plane. As expected, when solving explicitly, both solutions coincide. The last plot, suitably rotated, produces Fig. 4 in the paper.

```
> restart; with(Student[LinearAlgebra]): with(plots):
  vv:=<x,y,z>; uno:=Norm(vv,2)^2;
  A:= <1,2,2;2,1,2;2,2,1>;
> Eigenvectors(A);
  B:=Transpose(A).A;
> ImageIs:=A.vv;
  normImage:=Norm(ImageIs, 2)^2;
  quadForm:=expand(normImage);
  quadFormZero:=expand(normImage-uno);
  factor(quadFormZero);
> sols:=solve(quadFormZero=0,z);
> zFunction:=solve(quadFormZero=0, z);
> p1:= implicitplot3d(quadForm = 1, x=-1.1..1.1, y=-1.1..1.1,
   z=-1.1..1.1, grid=[50,50,50], style=surface):
  smaller:=x=-1.1..1.1, y=-1.1..1.1, view=-1.1..1.1:
  sphere:=<sin(u)*sin(v), sin(u)*cos(v), cos(u)>:
  p2:=plot3d([sphere], u=0..Pi, v=0..2*Pi, axes=boxed,
  scaling=constrained, transparency = 0.5):
  p3:= plot3d(zFunction[1], smaller, axes=normal, scaling=constrained):
  display(p1, p2, p3, axes=none);
```

Acknowledgments. I want to thank the organizers of the *Maple 2020* conference, for providing a forum for fruitful discussions, and a platform for learning interesting results, as well as recent advances in *Maple*. I am also grateful to Dr. Robert Lopez for showcasing the intriguing idea of vectors whose norms are preserved by a linear map. Finally, I want to thank the reviewers for valuable comments and suggestions that helped improve this paper.

References

1. Gel'fand, I.M.: Lectures on Linear Algebra. Dover Publications, New York (1989)
2. Katok, A., Hasselblatt, B.: Introduction to the Modern Theory of Dynamical Systems. Cambridge University Press, New York (1995)
3. Lopez, R.: Eigenpairs in Maple (2015). https://www.maplesoft.com/webinars/recorded/featured.aspx?id=1181
4. Tolosa, J.: Length-preserving directions and some diophantine equations. Am. Math. Monthly **128**(2), 125–139 (2021). https://doi.org/10.1080/00029890.2021.1851564
5. Weisstein, E.W.: Diophantine equation–2nd powers. MathWorld–A Wolfram Web Resource (2019). mathworld.wolfram.com/DiophantineEquation2ndPowers.html

Branching Out into Structural Identifiability Analysis with Maple: Interactive Exploration of Uncontrolled Linear Time-Invariant Structures

Jason M. Whyte(✉)

Australian Research Council Centre of Excellence for Mathematical and Statistical Frontiers (ACEMS), School of Mathematics and Statistics, and Centre of Excellence for Biosecurity Risk Analysis (CEBRA), School of BioSciences, University of Melbourne, Parkville Victoria, Australia
jason.whyte@unimelb.edu.au

Abstract. Suppose we wish to predict a physical system's behaviour. We represent the system by model structure S (a set of related mathematical models defined by parametric relationships between variables), and parameter set Θ. Each parameter vector in Θ corresponds to a completely specified model in S. We use S with system data in estimating the "true" (unknown) parameter vector. Inconveniently, S may approximate our data equally well for multiple parameter vectors. If we cannot distinguish between alternatives, we may be unable to use S in decision making. If so, our efforts in data collection and modelling are fruitless.

This outcome occurs when S is not structurally global identifiable (SGI). Fortunately, we can test various structure classes for SGI prior to data collection. A non-SGI result may inform a remedy to the problem.

We aim to assist SGI testing with suitable Maple 2020 procedures. We consider a class of "state-space" structure where a state-variable vector $\mathbf{x}$ is described by constant-coefficient, ordinary differential equations, and outputs depend linearly on $\mathbf{x}$. The "transfer function" approach is suitable here, and also for the "compartmental" subclass (mass is conserved).

Our use of Maple's "Explore" permits an interactive consideration of a parent structure, and variants of this produced by user choices. Results of the SGI test may differ for different variants. Our approach may inform the interactive analysis of structures from other classes.

Keywords: Experimental design · Input-output relationships · Inverse problems · Laplace transform · Structural property · Symbolic algebra

1 Introduction

Suppose we wish to predict the behaviour of some physical system so that (for example) we can investigate the system's response to novel situations. Should we wish to utilise our system knowledge, we would formulate a mathematical model structure ("structure" for brevity), say S, to represent the system. Broadly

R. M. Corless et al. (Eds.): MC 2020, CCIS 1414, pp. 410–428, 2021.
https://doi.org/10.1007/978-3-030-81698-8_27

speaking, a structure has two main parts. The first is a collection of parametric relationships (e.g. differential equations) relating system features (state variables, $\mathbf{x}$, which may not be observable), any inputs (or controls, $\mathbf{u}$), and observable quantities (outputs, $\mathbf{y}$). The second is a parameter space Θ. Prior to predicting system behaviour with S, we must estimate the true parameter vector $\boldsymbol{\theta}^* \in \Theta$ from system observations.

Parameter estimation may return multiple (even infinitely-many) equally valid estimates of $\boldsymbol{\theta}^*$. Inconveniently, distinct estimates may lead S to produce very different predictions, either for state variables, or for outputs beyond the range of our data. In such a case, an inability to distinguish between alternative estimates renders us unable to confidently use S for prediction. Consequently, if we cannot address the question which motivated our study, our efforts in data collection and modelling are unproductive.

The problem of non-unique parameter estimates may follow inexorably from the combination of a study design (including planned inputs), and S. (To explain further, features of S, such as outputs and initial conditions, may follow from the study design. We illustrate this effect for an "open-loop" system where outputs do not influence state variables or inputs in Fig. 1.) If so, we can anticipate this problem by testing S subject to its planned inputs for the property of structural global identifiability (SGI). We emphasise that such a test does not require data. Instead, we assume that "data" is provided by S under idealised conditions. These conditions depend on the class of structure under consideration. However, typical assumptions include: an infinite, error-free data record is available; and, our structure correctly represents the system. When S is an uncontrolled structure, we also assume that the initial state is not an equilibrium state. Solving algebraic equations derived from S will show whether it is possible (but not certain) for us to obtain a unique estimate of $\boldsymbol{\theta}^*$ under our idealised conditions. We do not expect a better result for real (noisy, limited) data.

There are other potential rewards for testing S for SGI. Test results may guide the reparameterisation of S into some alternative S', which may enable parameter estimation to produce a more favourable result than that achievable for S. Similarly, when a structure is not SGI under a given experimental design, one can iteratively examine the potential for alternative designs—which may produce a modified form of S—to produce more useful results.

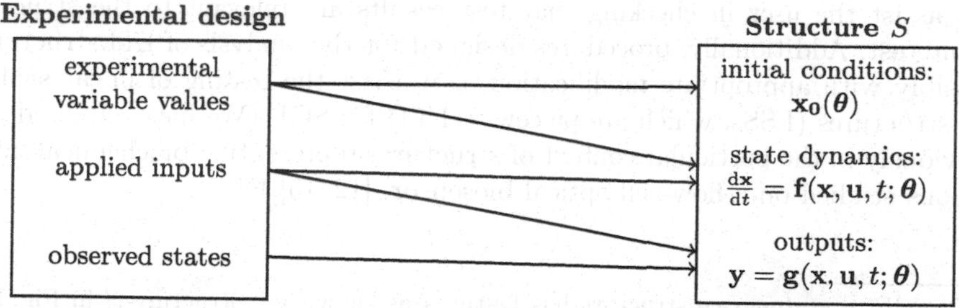

Fig. 1. An illustration of how experimental design in the study of an open-loop system can determine features of a model structure S aiming to represent the system.

Despite these benefits, the testing of structures for SGI remains uncommon in various domains. This may reflect the specialised nature of identifiability analysis, which requires skills unlike those employed in mathematical modelling. Based on experience, we expect that modellers seeking to scrutinise their model structures will appreciate easy-to-use software tools. We may characterise such tools as those which do not require a somewhat esoteric knowledge of mathematics, or extensive experience with a symbolic algebra package.

We shall use procedures written in the Maple 2020 programming language [8] to illuminate the testing of structures for SGI. We demonstrate key concepts through a consideration of continuous-time, uncontrolled, linear time-invariant state-space (henceforth, for brevity, ULTI) structures.[1] More particularly, we consider the "compartmental" (that is, subject to conservation of mass conditions) subclass of ULTI structures, which arise in various modelling applications. Some standard test methods may not be appropriate for compartmental structures, which guides our choice of test method here. From an educational standpoint, testing LTI structures for SGI motivates the study of various topics, including: systems theory; the Laplace transform; and algebraic equations.

To further extend the value of our procedures, we incorporate these into a routine which automates the testing of a "parent" structure for SGI, requiring the user only to define the structure. Further, when used with Maple's `Explore`, this routine permits an interactive assessment of the SGI test results obtained for variants of the parent structure (where these variants may be determined by alternative experimental designs). Experimentation only requires the user to specify (via input boxes) the initial conditions of state variables, and which of these are observed, producing a modified structure and a new application of the SGI test. We are unaware of any other software designed for this purpose.

We also intend to assist those conversant with identifiability analysis. We note recent concerns around reproducibility in computational biology (see, for example, Laubenbacher and Hastings [6]). Reproducibility is impeded when symbolic algebra packages behave inconsistently (as noted for Maple's `assume` command by Armando and Ballarin [1]). We intend that our routines will facilitate the checking of SGI test results obtained from either an alternative testing method, or from code written in another language. We also seek to aid reproducibility with procedures designed to eliminate a source of potential error in structure specification, or to aid the user in recognising other specification errors. This can assist the user in checking that test results are relevant to the structure of interest. Additionally, procedures designed for the analysis of LTI structures, possibly with appropriate modification, can assist the testing of linear switching structures (LSSs, which are piecewise LTI) for SGI. (We have explored this previously in the particular context of structures representing biochemical interactions studied on a flow-cell optical biosensor: [12–15].)

---

[1] Broadly, a state-space structure has features as shown for Structure S in Fig. 1. A ULTI structure includes a collection of linear, constant-coefficient ordinary differential equations that describe the time evolution of state variables.

The remainder of this chapter is organised as follows. We present essential definitions pertinent to LTI state-space structures, and an outline of concepts useful in testing a (general state-space) structure for SGI in Sect. 2. We shall focus on the "transfer function" (TF) approach—one of the earliest methods, yet found in relatively recent textbooks (e.g. [4]), and one which suits our interest in compartmental structures. Section 3 summarises our implementation of the TF approach in Maple 2020 by outlining our procedures and presenting code listings. We demonstrate the use of our code and its output by application to a test-case structure in Sect. 4. Section 5 offers concluding remarks. In Appendix 1 and Appendix 2 we provide the Maple code used to draw a compartmental diagram, and launch the interactive SGI test, respectively.

We conclude this section by introducing notation.

1.1 Notation

We denote the field of real numbers by $\mathbb{R}$, and its subset containing only positive (non-negative) values by $\mathbb{R}_+$ ($\bar{\mathbb{R}}_+$). The natural numbers $\{1, 2, 3, \ldots\}$ are denoted by $\mathbb{N}$. The field of complex numbers is denoted by $\mathbb{C}$. Given field $\mathbb{F}$ and some indeterminate w, $\mathbb{F}(w)$ denotes the field of rational functions in w over $\mathbb{F}$. Given $r, c \in \mathbb{N}$ and $\mathbb{F}$, we use $\mathbb{F}^{r\times c}$ to denote the set of matrices of r rows and c columns having elements in $\mathbb{F}$.

We use a bold lower-case (upper-case) symbol such as $\mathbf{a}$ ($\mathbf{A}$) to denote a vector (matrix), and a superscript $\top$ associated with any such object indicates its transpose. Given vector $\mathbf{x}$, $\dot{\mathbf{x}}$ denotes its derivative with respect to time. To specify the (i, j)-th element of a matrix, say $\mathbf{A}$, we may use a lower-case symbol such as $a_{i,j}$, or $(\mathbf{A})_{i,j}$ when this is easier to interpret. For $n \in \mathbb{N}$, we use $\mathbf{I}_n$ to represent the $n \times n$ identity matrix.

2 Preliminaries

In this section we present selected concepts necessary for the development to follow. We begin in Sect. 2.1 by introducing features of ULTI structures. In Sect. 2.2 we provide general definitions for structural global identifiability, and outline a process for testing a general state-space structure for this property. We provide details of how to adapt this for ULTI structures in Sect. 2.3. These details inform the Maple code we shall present subsequently.

2.1 Linear Time-Invariant State-Space Structures

LTI state-space structures are appropriate for modelling aspects of various physical applications. These include quantifying the interconversion of forms of matter in the pyrolysis of oil-bearing rock (e.g. [16]), or predicting the time evolution of drug concentrations in distinct compartments (say, tissues) of a living subject (e.g. Godfrey [5]). A key assumption is that the system's state variables (say concentrations) change (e.g. due to metabolic processes, including elimination from the system) according to first-order kinetics (for examples, see Rescigno [9]).

Definition 1. *An* **uncontrolled linear time-invariant (ULTI) state-space structure** M *with indices* $n, k \in \mathbb{N}$ *and parameter set* $\Theta \subset \mathbb{R}^p$ $(p \in \mathbb{N})$ *has mappings*

$$\mathbf{A} : \Theta \to \mathbb{R}^{n \times n}, \quad \mathbf{C} : \Theta \to \mathbb{R}^{k \times n}, \quad \mathbf{x_0} : \Theta \to \mathbb{R}^n .$$

The state variables and outputs at any time belong to the "state space" $X = \mathbb{R}^n$ *and "output space"* $Y = \mathbb{R}^k$*, respectively. Then, given some unspecified* $\boldsymbol{\theta} \in \Theta$*,* M *has "representative system"* $M(\boldsymbol{\theta})$ *given by*

$$\begin{aligned} \dot{\mathbf{x}}(t; \boldsymbol{\theta}) = \mathbf{A}(\boldsymbol{\theta})\mathbf{x}(t; \boldsymbol{\theta}) , \quad \mathbf{x}(0; \boldsymbol{\theta}) = \mathbf{x_0}(\boldsymbol{\theta}) , \\ \mathbf{y}(t; \boldsymbol{\theta}) = \mathbf{C}(\boldsymbol{\theta})\mathbf{x}(t; \boldsymbol{\theta}) . \end{aligned} \tag{1}$$

An **uncontrolled positive LTI state-space structure** *with indices* $n, k \in \mathbb{N}$ *is a ULTI state-space structure having representative system of the form given in* (1)*, where states and outputs are restricted to non-negative values. That is, the structure has* $X = \bar{\mathbb{R}}^n_+$ *and* $Y = \bar{\mathbb{R}}^k_+$.

An **uncontrolled compartmental LTI state-space structure** *with indices* $n, k \in \mathbb{N}$ *is an uncontrolled positive LTI state-space structure composed of systems having system matrices subject to "conservation of mass" conditions:*

- *all elements of* $\mathbf{C}$ *are non-negative, and*
- *for* $\mathbf{A} = (a_{i,j})_{i,j=1,\ldots,n}$,

$$\begin{aligned} a_{ij} &\geq 0 , \qquad\qquad i, j \in \{1, \ldots, n\} , \quad i \neq j , \\ a_{ii} &\leq -\sum_{\substack{j=1 \\ j \neq i}}^{n} a_{ji} , \quad i \in \{1, \ldots, n\} . \end{aligned} \tag{2}$$

2.2 Structural Identifiability of Uncontrolled Structures

In their consideration of LTI state-space structures, Bellman and Åström [2] outlined what we may consider as the 'classical' approach to testing structures for SGI. Essentially, this involves solving a set of test equations informed by the structure's output, and using the solution set to judge the structure as SGI or otherwise. We pursue this approach following the treatment of ULTI structures in [15], which was influenced by Denis-Vidal and Joly-Blanchard [3].

Definition 2 (From Whyte [18, Definition 7]). *Suppose we have a structure of uncontrolled state-space systems* M*, having parameter set* Θ *(an open subset of* $\mathbb{R}^p$*,* $p \in \mathbb{N}$*), and time set* $T \subseteq [0, \infty)$*. For some unspecified* $\boldsymbol{\theta} \in \Theta$*,* M *has representative system* $M(\boldsymbol{\theta})$*, which has state function* $\mathbf{x}(\cdot; \boldsymbol{\theta}) \in \mathbb{R}^n$ *and output* $\mathbf{y}(\cdot; \boldsymbol{\theta}) \in \mathbb{R}^k$*. Adapting the notation of Fig. 1 for this uncontrolled case, suppose that the state-variable dynamics and output of system* $M(\boldsymbol{\theta})$ *are determined by functions* $\mathbf{f}(\mathbf{x}, \cdot; \boldsymbol{\theta})$ *and* $\mathbf{g}(\mathbf{x}, \cdot; \boldsymbol{\theta})$*, respectively. Suppose that* M *satisfies conditions:*

1. $\mathbf{f}(\mathbf{x},\cdot;\boldsymbol{\theta})$ *and* $\mathbf{g}(\mathbf{x},\cdot;\boldsymbol{\theta})$ *are real and analytic for every* $\boldsymbol{\theta} \in \Theta$ *on* $\mathcal{S}$ *(a connected open subset of* $\mathbb{R}^n$ *such that* $\mathbf{x}(t;\boldsymbol{\theta}) \in \mathcal{S}$ *for every* $t \in [0,\tau]$, $\tau > 0$*).*
2. $\mathbf{f}(\mathbf{x}_0(\boldsymbol{\theta}),0;\boldsymbol{\theta}) \neq \mathbf{0}$ *for almost all* $\boldsymbol{\theta} \in \Theta$.

Then, for some finite time $\tau > 0$, *we consider the set*

$$\mathcal{I}(M) \triangleq \left\{\boldsymbol{\theta}' \in \Theta : \mathbf{y}(t;\boldsymbol{\theta}') = \mathbf{y}(t;\boldsymbol{\theta}) \quad \forall t \in [0,\tau]\right\}. \tag{3}$$

If, for almost all $\boldsymbol{\theta} \in \Theta$*:*

$\mathcal{I}(M) = \{\boldsymbol{\theta}\}$, *M is structurally globally identifiable (SGI);*
$\mathcal{I}(M)$ *is a countable set, M is structurally locally identifiable (SLI);*
$\mathcal{I}(M)$ *is not a countable set, M is structurally unidentifiable (SU).*

In testing structures from various classes (including the LTI class) for SGI we employ a variant of Definition 2 that is easier to apply. We take advantage of the fact that certain "invariants", $\boldsymbol{\phi}(\boldsymbol{\theta})$, (see Vajda, [10]), completely determine our output function. As such, we may replace (the functional equation) Eq. (3) with a system of algebraic equations in these invariants.

Definition 3 (Whyte [18, Definition 8]**).** *Suppose that structure M satisfies Conditions 1 and 2 of Definition 2. Then, for some arbitrary* $\boldsymbol{\theta} \in \Theta$, *we define*

$$\mathcal{I}(M,\boldsymbol{\phi}) \triangleq \left\{\boldsymbol{\theta}' \in \Theta : \boldsymbol{\phi}(\boldsymbol{\theta}') = \boldsymbol{\phi}(\boldsymbol{\theta})\right\} \equiv \mathcal{I}(M), \tag{4}$$

and determination of this allows classification of M according to Definition 2.

Remark 1. In the analysis of (say, uncontrolled) LSS structures, there are some subtleties to Definition 3. It is appropriate to consider the response on independent time intervals between switching events as the same parameter vector does not apply across all such intervals. It is appropriate to re-conceptualise invariants as a collection of features across the time domain; each interval between switching events contributes features which define the structure's output on that interval ([12,13]).

When Definition 3 is appropriate for the class of structure at hand, we may employ this at the end of a well-defined process, which we summarise below.

Proposition 1 (A general algorithm for testing a structure for SGI, from Whyte [18, Proposition 1]**).**

Given some model structure M with parameter set Θ, *having representative system* $M(\boldsymbol{\theta})$ *for unspecified* $\boldsymbol{\theta} \in \Theta$*:*

Step 1 *Obtain invariants* $\boldsymbol{\phi}(\boldsymbol{\theta})$*: there are various approaches, some having conditions (e.g. that M is structurally minimal, see Remark 2) that may be difficult to check.*

Step 2 *Form alternative invariants* $\boldsymbol{\phi}(\boldsymbol{\theta}')$ *by substituting* $\boldsymbol{\theta}'$ *for* $\boldsymbol{\theta}$ *in* $\boldsymbol{\phi}(\boldsymbol{\theta})$.

Step 3 *Form equations* $\boldsymbol{\phi}(\boldsymbol{\theta}') = \boldsymbol{\phi}(\boldsymbol{\theta})$.

Step 4 *Solve these equations to obtain $\boldsymbol{\theta}' \in \Theta$ in terms of $\boldsymbol{\theta}$ to determine $\mathcal{I}(M, \boldsymbol{\phi})$.*
Step 5 *Scrutinise $\mathcal{I}(M, \boldsymbol{\phi})$ so as to judge M according to Definition 3.*

The particularities of Proposition 1 depend on both the class of the structure under investigation, and the testing method we will employ. In the next subsection we provide an overview of the TF method, which is appropriate for the compartmental LTI structures of interest to us here.

2.3 The Transfer Function Method of Testing Uncontrolled LTI State-Space Structures for SGI

The TF method makes use of the Laplace transform of a structure's output function (causing an alternative name, e.g. [5]). As such, it is appropriate to recall the Laplace transform of a real-valued function.

Definition 4. *Suppose some real-valued function f is defined for all non-negative time. (That is, $f : \bar{\mathbb{R}}_+ \mapsto \mathbb{R}$, $t \mapsto f(t)$.) We represent the (unilateral) Laplace transform of f with respect to the transform variable $s \in \mathbb{C}$ by*

$$\mathcal{L}\{f\}(s) \triangleq \int_0^\infty f(t) \cdot e^{-st} \mathrm{d}t ,$$

if this exists on some domain of convergence $\mathcal{D} \subset \mathbb{C}$.

When applying the TF to the output of a controlled LTI structure, we must check to ensure that $\mathcal{D}$ exists. However, given an ULTI structure having finitely-valued parameters (a physically realistic assumption), each component of $\mathbf{x}$ or $\mathbf{y}$ is a sum of exponentials with finite exponents which depend linearly on t. As such, the Laplace transform does exist on some domain of convergence, the specific nature of which is unimportant for our purposes here. (We direct the reader interested in details to Sects. 2.3.1 and 3.1 of Whyte [18].)

Given ULTI structure S having representative system $S(\boldsymbol{\theta})$ informed by $\mathbf{A}(\boldsymbol{\theta}) \in \mathbb{R}^{n\times n}$ and $\mathbf{C}(\boldsymbol{\theta}) \in \mathbb{R}^{k\times n}$, we may write the Laplace transform of the output function of $S(\boldsymbol{\theta})$ as:

$$\mathcal{L}\{\mathbf{y}(\cdot;\boldsymbol{\theta})\}(s;\boldsymbol{\theta}) = \mathbf{H_2}(s;\boldsymbol{\theta}) , \tag{5}$$

where (5) exists on domain of convergence $\mathcal{C}_0$, and the "transfer matrix" is[2]

$$\mathbf{H_2}(s;\boldsymbol{\theta}) \triangleq \mathbf{C}(\boldsymbol{\theta})\Big(s\mathbf{I}_n - \mathbf{A}(\boldsymbol{\theta})\Big)^{-1} \mathbf{x_0}(\boldsymbol{\theta}) \in \mathbb{R}(s)^{k\times 1} . \tag{6}$$

The elements of $\mathbf{H_2}$ ("transfer functions") are rational functions in s. We refer to these functions as "unprocessed" if we have not attempted certain actions. We must undertake one or more of these in order to obtain invariants from $\mathbf{H_2}$ for testing S for SGI. We shall describe these steps and their result for the case of compartmental ULTI structures in the following definition.

[2] We have adapted the notation of [11, Chapter 2] to include $\mathbf{x_0}$, as otherwise initial-condition parameters do not appear in the SGI test.

Definition 5 (Canonical form of a transfer function (adapted from [18, Definition 9])). *Given compartmental ULTI structure S of $n \in \mathbb{N}$ states, suppose that associated with $S(\boldsymbol{\theta})$ is a transfer matrix $\mathbf{H_2}$ (as in (6)), composed of unprocessed transfer functions. (Recall that we know $\mathcal{L}\{\mathbf{y}\}$ exists on some domain $\mathcal{C}_0 \subset \mathbb{C}$, and hence that $\mathbf{H_2}$ is defined.) Given element $(\mathbf{H_2}(s;\boldsymbol{\theta}))_{i,j} \in \mathbb{C}(s)$, we must cancel any common factors between the numerator and denominator polynomials ("pole-zero cancellation"). Following this, we may choose to obtain the associated transfer function in "canonical form" by rewriting it to ensure that the denominator is monic. The result is an expression of the form:*

$$(\mathbf{H_2}(s;\boldsymbol{\theta}))_{i,j} = \frac{\omega_{i,j,r+p}(\boldsymbol{\theta})s^p + \cdots + \omega_{i,j,r}(\boldsymbol{\theta})}{s^r + \omega_{i,j,r-1}(\boldsymbol{\theta})s^{r-1} + \cdots + \omega_{i,j,0}(\boldsymbol{\theta})}, \quad \forall s \in \mathcal{C}_0\,, \qquad r \in \{1,\ldots,n\}, \quad p \in \{0,\ldots,r-1\}. \tag{7}$$

The coefficients $\omega_{i,j,0},\ldots,\omega_{i,j,r+p}$ in (7) *contribute invariants towards $\phi(\boldsymbol{\theta})$.*

We may prefer to retain a non-monic denominator if this is desirable, such as when coefficients of s are polynomial in $\boldsymbol{\theta}$, but would not be if the denominator was rewritten to become monic. Given a non-monic denominator, we obtain all coefficients of the transfer function to use as invariants. In our procedures we give the user choice on whether or not to require that transfer functions have monic denominators.

Remark 2. Various approaches to testing an LTI structure S for SGI (e.g. the similarity transform method) are only applicable to a "structurally minimal" S. We see that S is not structurally minimal if we can reduce it to a structure $\tilde{S}$ of $n_1 < n$ state variables (and, say, parameter set $\tilde{\Theta}$) where, for almost all $\boldsymbol{\theta} \in \Theta$, there is some $\tilde{\boldsymbol{\theta}} \in \tilde{\Theta}$ such that the outputs of $S(\boldsymbol{\theta})$ and $\tilde{S}(\tilde{\boldsymbol{\theta}})$ are identical. The TF method has the advantage of not requiring structural minimality. Instead, undertaking any possible pole-zero cancellation in transfer functions (as required by Definition 5) allows the test to access the parameter information available in a structurally minimal form of S.

In the testing of an uncontrolled LSS structure for SGI using procedures presented here, checking for pole-zero cancellation in the constituent LTI structures in effect after the first switching event is typically not trivial. This has led to indirect ([14]) and direct ([15]) approaches involving far greater algebraic complexity.

In the next section we present the Maple procedures we shall use in testing a ULTI structure for SGI. The source code is available for download from [17].

3 An Implementation of the Transfer Function Method for Uncontrolled LTI State-Space Structures

In Sect. 3.1 we show our procedures for an implementation of the TF method in order of use (according to a general scheme such as Proposition 1), and explain certain key features in our specific context. In Sect. 3.2 we combine these component procedures into a complete SGI test procedure. We validated our procedures by applying them to structures scrutinised in the literature (e.g. two variants of a three-compartment LTI structure in DiStefano [4, Example 10.3]), and confirming that our results were equivalent.

3.1 Component Procedures

Procedures `process_matrix` (Listing 1.1), `collect_invariants` (Listing 1.2) and `identifiability_eqn_list` (Listing 1.4) were adapted from Maple 2015 ([7]) routines presented in Whyte [15, Appendix B]. Here we have updated those original routines for Maple 2020 [8]. We have also taken steps to make the original routines more efficient and concise, such as by replacing some loops with `map` commands, or using more appropriate types of data structures. Further, we have improved upon `process_matrix`; previously the routine merely flagged a non-monic denominator in a transfer function. The revised procedure uses the logical parameter `canonical_form`, which specifies whether or not transfer function denominators should be made monic. This choice may influence the procedure's output: a processed transfer function matrix. As this matrix is passed to `collect_invariants`, we have adapted this procedure accordingly.

Remark 3. Aside from its role in the 'classical' approach to testing structures for SGI, historically there was another reason to ensure that each transfer function had a monic denominator. This step enabled the comparison of elements of transfer matrices drawn from two different structures. If each pair of rational functions have the same coefficients, then the two structures produce the same output. (Finding the parameter vectors which cause this equality relates to whether or not two structures have the property of "structural indistinguishability", a generalisation of structural identifiability; we test for the former using methods similar to those used for the latter.) However, given the symbolic algebra packages available now, monic denominators are not essential. Our code permits the user to allow non-monic denominators in transfer functions by setting `canonical_form:=false`.

Procedure `process_matrix` (Listing 1.1, the start of Step 1 of Proposition 1 in this setting) prepares the transfer matrix associated with a structure S (`transfer_matrix`) for the extraction of invariants. (Recall the discussion in Sect. 2.3.) The `sort_order` list parameter directs `sort` in how to order parameters and the complex variable (say s) which appear in the processed transfer functions. For each of these, the procedure returns the numerator and denominator polynomials, stored in a matrix.

Listing 1.1. Procedure `process_matrix` for processing a matrix of transfer functions obtained from a LTI structure to enable the subsequent extraction of invariants.

```
process_matrix:=proc(sort_order::list(symbol), transfer_matrix::Matrix,
    canonical_form::truefalse, s::symbol:="s", $)
local i:=0, j:=0, colMAX, rowMAX, current_element, leading_denom_coeff,
    processed_matrix, new_numer, new_denom;
```

```
description "Prepare matrices of transfer functions for extraction of
    invariants. This involves pole-zero cancellation, and conversion of
    transfer functions into their canonical form (by ensuring denominators
    are monic) if ''canonical_form'' is set to true.";
rowMAX, colMAX:=LinearAlgebra[Dimensions](transfer_matrix);
processed_matrix:=Matrix(rowMAX, colMAX);
for i to rowMAX do;
for j to colMAX do;
current_element:=normal(transfer_matrix[i, j]);
leading_denom_coeff:=lcoeff(denom(current_element), s);
if (canonical_form=true and leading_denom_coeff<>1) then new_numer:=eval
    (numer(current_element)/leading_denom_coeff);
new_denom:=eval(denom(current_element)/leading_denom_coeff);
else new_numer:=numer(current_element);
new_denom:=denom(current_element);
fi;
new_numer:=sort(collect(new_numer, s), sort_order, plex);
new_denom:=sort(collect(new_denom, s), sort_order, plex);
processed_matrix[i, j]:=[new_numer, new_denom];
od;
od;
return processed_matrix;
end proc;
```

Procedure `collect_invariants` (Listing 1.2, the conclusion of Step 1 of Proposition 1) extracts the coefficients from a processed transfer matrix, including the invariants. (Later in Listing 1.6 we process the returned object to disregard any numeric coefficients.)

Listing 1.2. Procedure `collect_invariants` which extracts the invariants from a processed transfer matrix.

```
collect_invariants:=proc(processed_matrix::Matrix, s::symbol, $)
local i:=0, j:=0, colMAX, rowMAX, latest, coeff_set:={},
    element_numer_coeffs, element_denom_coeffs;
description "Extract the invariants from a matrix of transfer
    functions placed in the canonical form.";
rowMAX, colMAX:=LinearAlgebra[Dimensions](processed_matrix);
for i from 1 to rowMAX do;
for j from 1 to colMAX do;
latest:={};
element_numer_coeffs:={coeffs(processed_matrix[i,j][1],s)};
element_denom_coeffs:={coeffs(processed_matrix[i,j][2],s)};
latest:=element_numer_coeffs union element_denom_coeffs;
coeff_set:=coeff_set union latest;
od;
od;
return map(primpart, coeff_set);
end proc;
```

The procedure `theta_prime_creation` (Listing 1.3, the start of Step 2 of Proposition 1) is new. This routine intends to remove a point in SGI analysis at which human error could cause a mismatch between the ordering of parameters in $\boldsymbol{\theta}$ and $\boldsymbol{\theta}'$, potentially causing an inaccurate test result. The list of the structure's parameters `theta` is modified to return the alternative parameter list `theta_prime`. This process ensures that there is a clear relationship between corresponding elements of $\boldsymbol{\theta}$ and $\boldsymbol{\theta}'$ (to aid interpretation of (4)), and the correspondences are correct. When `theta_mod_type` equals "underscore", an element of `theta_prime` is defined by adding an underscore suffix to the corresponding `theta` element (line 6). Alternatively, when `theta_mod_type` equals "Caps" `theta_prime` is populated by capitalised versions of `theta` (line 7). This option is appropriate when `theta` only contains entries which begin with a lower-case alphabetic character.

Listing 1.3. Procedure **`theta_prime_creation`** creates a recognisable alternative parameter from each element of the original parameter vector $\boldsymbol{\theta}$.

```
theta_prime_creation:=proc(theta::list(symbol), theta_mod_type::identical("
      underscore","Caps"):="Caps", $)::list(symbol);
local i, theta_prime, common_params;
description "A list of symbols theta is modified to create list theta_prime
      such that the connection between original and alternative parameters is
       apparent. For theta_mod_type=''Caps'' (the default), modify to upper
      case. For theta_mod_type=''underscore'', append an underscore (_) to
      each symbol. To specify a subscripted parameter in theta, use two
      underscores, e.g. k__1, not k[1].";
theta_prime:=Array(theta);
for i from 1 to numelems(theta) do;
if (theta_mod_type="underscore") then theta_prime[i]:=convert(StringTools[
      Insert](theta[i],length(theta[i]),"_"),symbol);
elif (theta_mod_type="Caps") then theta_prime[i]:=convert(StringTools[
      UpperCase](theta[i]), symbol); fi;
od;
# Check that the use of theta_mod_type has created elements of theta_prime
      which differ from all elements of theta. (For example, using
      theta_mod_type="Caps", if theta contained upper-case symbols, then,
      inappropriately, theta_prime would also contain these.)
common_params:=convert(theta,set) intersect convert(theta_prime,set);
if (numelems(common_params) >0) then error "Inappropriate theta parameter(s)
      for nominated theta_mod_type:", common_params fi;
return convert(theta_prime,list);
end proc;
```

Procedure `identifiability_eqn_list` (Listing 1.4, concluding Step 2 and Step 3 of Proposition 1) uses the structure's invariants $\phi(\boldsymbol{\theta})$, and parameter vectors $\boldsymbol{\theta}$ and $\boldsymbol{\theta}'$, and returns the necessary SGI test equations $\phi(\boldsymbol{\theta}) = \phi(\boldsymbol{\theta}')$.

Listing 1.4. Procedure **`identifiability_eqn_list`** forms the SGI test equations.

```
identifiability_eqn_list:=proc(invariants::list, theta::list(symbol),
      theta_prime::list(symbol), $)
description "Use a structure's invariants in forming equations
      necessary for the structural global identifiability test.";
return invariants =~ subs(theta =~ theta_prime, invariants);
end proc;
```

Procedure `classify_solutions` (Listing 1.5) is also new. It addresses Step 5 of Proposition 1 by scrutinising the solutions of the SGI test equations.

Listing 1.5. Procedure **`classify_solutions`** classifies the structure as SGI, SLI, SU, or "unknown" if classification is not possible.

```
classify_solutions:= proc(solset::list, theta::list(symbol), theta_prime::
      list(symbol), $)
local num_soln_families:=numelems(solset), i, lhsides, rhsides,
      free_param_by_soln:=Vector(num_soln_families), classification:=["
      Unknown: inspect solutions",magenta], difference, RootOf_check,
      RootOf_count;
description "Use the solution set of the SGI test equations in classifying
      the structure under investigation.";
# Initially the structure is unclassifed. It is SGI if there is exactly one
      solution family, and it is theta_prime equals theta.
if (num_soln_families =1 and verify(solset[1],theta_prime=~theta)=true) then
      classification:=["SGI",green]; return (classification); fi;
```

```
# The structure is SU if any of the solution families contain free
    parameters. If so, the difference of the left and right-hand-sides of
    an equation in solset is zero. Look for free parameters over each
    solution family in solset in turn. As soon as we find a free parameter,
    classify the structure as SU and exit the routine.
for i from 1 to num_soln_families do;
lhsides := map(lhs, solset[i]);
rhsides := map(rhs, solset[i]);
difference := simplify(lhsides - rhsides);
free_param_by_soln[i] := select(x -> x = 0, difference);
if (numelems(free_param_by_soln[i])>0) then classification:=["SU",red];
    return (classification); fi;
od;
# If the flow has continued to this point, any solution family in solset
    does not have free parameters (not SU) and does not have a unique
    solution (not SGI). If there are multiple families, or a family
    contains roots, the structure is SLI.
if (num_soln_families > 1) then classification:= ["SLI (multiple solutions)"
    ,yellow]; return (classification); fi;
# If we reach this point, we have a single solution family. Does it contain
    roots?
rhsides := map(rhs, solset);
RootOf_check:= map(type,rhsides,RootOf);
RootOf_count:= select(x -> x = true, RootOf_check);
if (RootOf_count >0) then classification:=["SLI (solution contains 'RootOf')
    ",yellow]; return (classification); fi;
# If we reach this point, we have not classified the structure.
return (classification);
end proc;
```

Remark 4. The procedures `classify_solutions`, `identifiability_eqn_list`, and `theta_prime_creation` are not restricted to use in testing LTI structures for SGI. Also, each of the procedures in this section may be used in testing a controlled LTI state-space structure for SGI.

In the next subsection we combine our component procedures into a complete procedure for testing a ULTI state-space structure for SGI. Subsequent use of this with `Explore` allows us to interactively test a parent structure and its variants.

3.2 A Complete SGI Test Procedure for ULTI State-Space Structures

Given some defined structure, Listing 1.6 forms the transfer matrix $\mathbf{H_2}(s;\boldsymbol{\theta})$, then draws on Listings 1.1 to 1.5 in applying steps of the SGI test. We call our procedure `Uncontrolled_Lin_Comp_Fig` (Appendix 1) to draw a (modified) compartmental diagram associated with the structure as part of the output, which also shows $\boldsymbol{\theta}$, $\boldsymbol{\theta}'$, the solution set of the SGI test Eqs. (4), and a classification of the structure. Use of the procedure via Listing 1.8 produces text-input boxes which permit the user to modify values of observation gains (elements of **C**) and initial conditions. Also, drop-down menus permit the user to select the value of `theta_mod_type` used in creating $\boldsymbol{\theta}'$, or the select the diagram layout style from the options provided by `DrawGraph`.

Listing 1.6. `Explore_SGI_test` combines routines from Section 3.1 resulting in a procedure suitable for testing an ULTI structure for SGI.

```
Explore_SGI_test:=proc(A::Matrix, obs_gains::list, ICs::list,
    outflow_params::list, layout_style, theta_mod_type::identical("
    underscore","Caps"), s::symbol, canonical_form::truefalse,
    tracing::truefalse, $)
local C, n, x0, x_colour, y_colour, outgraph, G, prelim1, H2, H2_proc,
    theta, sort_order, coeff_collection, phi_list, i, eqn_list,
    phi1vec_list, new_list, theta_prime, solset, classification,
    solset_Matrix, textmatrix, textplot1, textplot2, textplot3;
description "A procedure (for use with Explore) to allow interactive
    testing of a parent (continuous time, uncontrolled, linear, time-
    invariant state-space structure and variants derived from it. The
    user creates variants by setting the structure's initial
    conditions (ICs) or observation gains (obs_gains) to constants,
    including zero. The output includes the original parameters theta
    and alternative parameters theta_prime, the solution set of the
    test equations, and a type of compartmental diagram showing the
    dependencies between outputs and state variables.";
interface(rtablesize = 15);
C:=LinearAlgebra[DiagonalMatrix](obs_gains);
n:=LinearAlgebra[RowDimension](A);
x0:=Vector[column](ICs);
x_colour:="LightBlue";
y_colour:="LightGreen";
outgraph:=Uncontrolled_Lin_Comp_Fig(A, C, x0, x_colour, y_colour,
    outflow_params, tracing);
G:=GraphTheory[DrawGraph](outgraph, layout=layout_style, stylesheet=[
    vertexhighlightborder=false, vertexborder=false]);
prelim1:=LinearAlgebra[MatrixAdd](LinearAlgebra[IdentityMatrix](n), A,
    s, -1);
H2:=LinearAlgebra[Multiply](C, LinearAlgebra[Multiply](LinearAlgebra[
    MatrixInverse](prelim1), x0));
# Ensure that we have a transfer matrix, not a scalar (as could happen
    when C is a row vector) or a vector
if (type(H2,`+`)=true) then H2:=convert([H2], Matrix); else
H2:=ArrayTools[Reshape](H2, [n, 1]); fi;
theta:=convert(`union`(indets(A), indets(x0), indets(C)), list);
sort_order:=[s, op(theta)];
H2_proc:=process_matrix(sort_order,H2,canonical_form,s);
coeff_collection:=convert(collect_invariants(H2_proc,s),list);
# Here retain only the coefficients that depend on system parameters:
    exclude elements that are type numeric.
phi_list:=remove(type, coeff_collection,numeric);
theta_prime:=theta_prime_creation(theta,theta_mod_type);
eqn_list:=identifiability_eqn_list(phi_list,theta,theta_prime);
solset:=solve(eqn_list,theta_prime);
classification := classify_solutions(solset, theta, theta_prime);
solset_Matrix:=convert(solset,Matrix);
# Display the outputs as an array of objects.
textmatrix:=Matrix([["theta",op(theta)],["theta_prime",op(theta_prime)
    ]]):
textplot1:=plots[textplot]([0,0,textmatrix],axes=none):
textplot2:=plots[textplot]([0,0,solset_Matrix],axes=none,title="
    Solutions for theta_prime in terms of theta"):
textplot3:=plots[textplot]([0,0,classification[1],'font'=["times","
    roman",20]],axes=none,title="Structure classification:",
    background=classification[2]):
plots[display](Array(1..4,1..1, [[textplot1],[textplot2],[G],[
    textplot3]]));
end proc;
```

4 Towards Interactive Inspection of the Effect of Changing Experimental Designs on the SGI Test

We consider a parent compartmental ULTI state-space structure (as in Definition 1) of three compartments, as we may find in pharmacological applications.

We assume that we can observe each state variable. We may obtain simpler variants of the structure (reflecting changes to the experimental design, but not the nature of the physical system) by setting any parameter in $\mathbf{x_0}$ or $\mathbf{C}$ to a non-negative constant. We employ notation for parameters in $\mathbf{A}$ (rate constants) common to pharmacological applications: k_{ij}, $(i \neq j,\, j \neq 0)$ relates to the flow of mass from x_j to x_i, and k_{0j} relates to the outflow of mass from x_j to the environment (see Godfrey [5, Chapter 1].)

We specify the structure by:

$$\mathbf{x}(\cdot;\boldsymbol{\theta}) = \begin{bmatrix} x_1(\cdot;\boldsymbol{\theta}) \\ x_2(\cdot;\boldsymbol{\theta}) \\ x_3(\cdot;\boldsymbol{\theta}) \end{bmatrix}, \qquad \mathbf{x_0}(\boldsymbol{\theta}) = \begin{bmatrix} x_{0_1} \\ x_{0_2} \\ x_{0_3} \end{bmatrix}, \qquad \mathbf{y}(\cdot;\boldsymbol{\theta}) = \begin{bmatrix} y_1(\cdot;\boldsymbol{\theta}) \\ y_2(\cdot;\boldsymbol{\theta}) \\ y_3(\cdot;\boldsymbol{\theta}) \end{bmatrix},$$
$$\mathbf{A}(\boldsymbol{\theta}) = \begin{bmatrix} -(k_{21}+k_{01}) & k_{12} & 0 \\ k_{21} & -(k_{12}+k_{32}) & k_{23} \\ 0 & k_{32} & -k_{23} \end{bmatrix}, \qquad \mathbf{C}(\boldsymbol{\theta}) = \begin{bmatrix} c_1 & 0 & 0 \\ 0 & c_2 & 0 \\ 0 & 0 & c_3 \end{bmatrix}, \tag{8}$$

where the parameter vector is

$$\boldsymbol{\theta} = \left(k_{01}\; k_{12}\; k_{21}\; k_{23}\; k_{32}\; x_{0_1}\; x_{0_2}\; x_{0_3}\; c_1\; c_2\; c_3\right)^{\top} \in \bar{\mathbb{R}}_{+}^{11}\,.$$

For simplicity, we have chosen to consider a parent structure that has a diagonal $\mathbf{C}$. By setting any $c_i = 0$ $(i = 1, 2, 3)$, we readily produce an alternative structure (associated with an alternative experimental design) which models observations that are independent of x_i. For drawing the compartmental diagram associated with the parent structure or its variants (using procedure `Uncontrolled_Lin_Comp_Fig`, Appendix 1), $\mathbf{A}$ directs us to record the parameters associated with flows out of the system with `outflow_params` $\triangleq [k_{01}, 0, 0]$.

Figure 2 shows the SGI test results (the result sections of the `Explore` window) for the parent structure illustrated by (8). The top panel shows $\boldsymbol{\theta}$ and $\boldsymbol{\theta}'$ for ease of comparison. The third panel presents a compartmental diagram of the structure under consideration. The bottom panel shows the structure's classification.

The second panel shows the solution set of the test equations. Here we see that some parameters are uniquely identifiable (e.g. $K_{01} = k_{01}$). Other parameters are free (e.g. $X_{20} = X_{20}$), leading to the structure's classification as SU. The solution also provides other insights. We note that we may rearrange the expression for C_1 to yield $C_1 X_{20} = c_1 x_{20}$. That is, whilst we cannot uniquely estimate c_1 and x_{20} individually, we may be able to obtain a unique estimate of their product. This insight may guide the reparameterisation of the parent structure so as to remove one contributor to the structure's SU status.

We can readily consider variants of the parent structure. Using the appropriate input box on the Explore dashboard, setting $c_1 = 1$ results in an SGI structure. Alternatively, modifying the parent structure by setting $c_1 = c_2 = 1$ and $c_3 = 0$ yields an SLI structure.

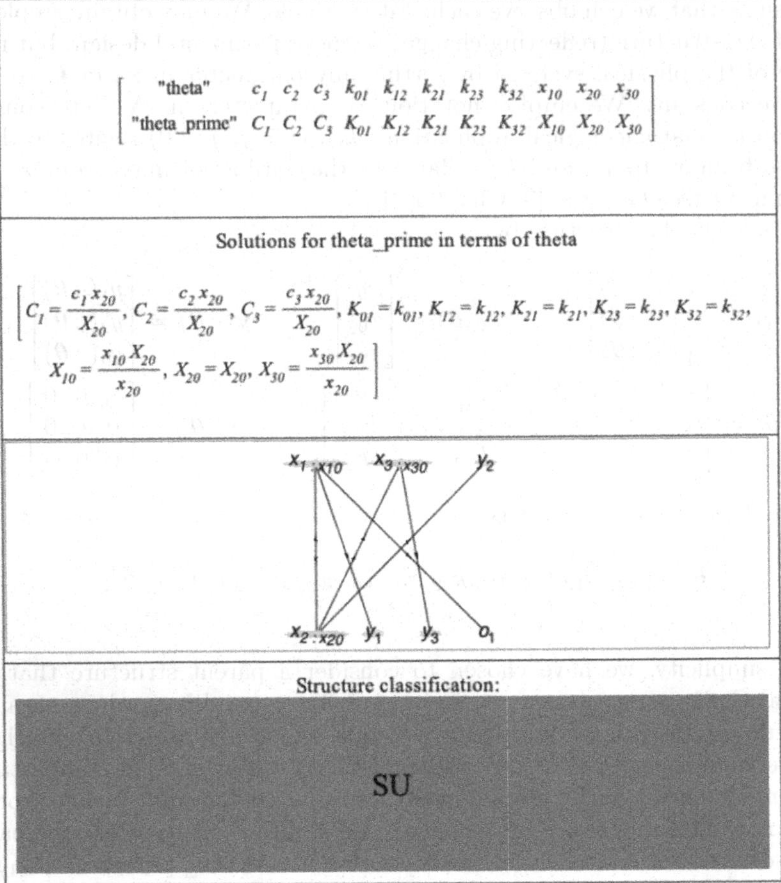

Fig. 2. Key features of the output window produced by application of Maple's `Explore` to `Explore_SGI_test` (Listing 1.6) in the study of our parent structure having representative system (8).

Remark 5. Our procedures were designed for ULTI structures, however, we can also accommodate the experimental case where the initial condition of any state variable is set by an impulsive input, and there are no other applied inputs.

5 Concluding Remarks

We have presented Maple 2020 code to allow the interactive testing of a parent ULTI structure and its variants for SGI. Whilst we believe this to be a novel contribution, there are still opportunities to improve upon the presentation here.

- We used the workaround of an `Array` so that `Explore` could display multiple objects (not merely test results) in our interactive panel. This choice limited our control over object layout. Our presentation may be improved by designing an interactive application which uses "embedded components".
- A diagram produced by `Uncontrolled_Lin_Comp_Fig` will be more informative if it could show each edge labelled with the appropriate parameter. At present, `DrawGraph` is limited to showing numerical weights on edges. Hence, it will be useful to produce a new procedure (based on `DrawGraph`) that does not have this restriction.

We also see opportunities to further the contributions of this chapter. An extension of `Uncontrolled_Lin_Comp_Fig` to suit controlled LTI structures will require modifications to include the influence of inputs on states. Certain complexities in the testing of controlled structures (see [18, Section 4]) will necessitate substantial changes to how our interactive application processes arguments. For example, it may be desirable to consider an SGI test where output is available for (the often realistic case of) a limited number of inputs that do not permit us to obtain the structure's invariants. The testing of structures of non-linear systems for SGI will require new methods for extracting invariants, and for displaying any edges which depend on state variables in a non-linear manner.

Acknowledgements. The author thanks the organisers of "Identifiability problems in systems biology" at the American Institute of Mathematics (San Jose, California, August 19–23, 2019) for the invitation to attend, and participants for useful discussions. This chapter's presentation benefited from the language definition for Maple code (for LaTeX's listings package) by Maplesoft's Erik Postma. Appreciation also goes to an anonymous reviewer for helpful comments which informed or inspired various improvements to the original Maple code.

Appendix 1 Maple Code for Drawing a Modified Compartmental Diagram

We use Listing 1.7 in drawing a modified compartmental diagram of the model structure currently under investigation. When the `Explore` window associated with Listing 1.8 is launched, the diagram displayed is updated in response to user selections from the drop-down "layout" menu or changes to the input boxes which set parameter values.

Listing 1.7. Maple code which uses the definition of a model structure and some user-specified parameters drawn from this in drawing a modified compartmental diagram

```
Uncontrolled_Lin_Comp_Fig:=proc(A::Matrix, C::Matrix, x0::Vector, x_colour::
    string, y_colour::string, outflow_params::list, tracing::truefalse, $)
local AugmentedA, Cmod, num_xy, zero_row_padding, AdjacencyA, AugA_rows,
    outflow_mat, outflow_detect, outflow_col, outflow_det_cols, AugA_cols,
    i, j, m, n, x0_list, xseq, xvertex_colouring, yvertex_colouring,
    vertex_colouring, num_outflows, outflow_labels, num_outrows, xlabels,
    ylabels, outflow_tag_indices, labels_list, G, outflow_indices,
    outflow_colour;
description "Combine matrices A and C and initial state vector x0 of a
    defined linear time-invariant continuous-time structure to produce a
    diagram showing interconnections of state variables (x), initial
    conditions, outflows, and dependence of outputs (y) on x.";
```

```
n:=LinearAlgebra[RowDimension](A);
if (type(C,Vector[row])=true) then Cmod:=convert(C,Matrix);
else Cmod:=C; fi;
m:=LinearAlgebra[RowDimension](Cmod);
# As A is specified using the compartmental modelling convention, construct an adjacency matrix from transpose(A). This allows us to use the conventions of graph theory, and hence Maple's GraphTheory without modification.
AugmentedA:=LinearAlgebra[Transpose](A);
# Add to AugmentedA new columns showing dependence of y on x.
AugmentedA:=<AugmentedA | LinearAlgebra[Transpose](Cmod)>;
num_xy:=LinearAlgebra[ColumnDimension](AugmentedA);
zero_row_padding:=LinearAlgebra[ZeroMatrix](m, num_xy);
AugmentedA:=<AugmentedA, zero_row_padding>;
if (tracing=true) then print("AugmentedA with outputs", AugmentedA); fi;
# Note outflows by the presence of outflow_params in A's main diagonal.
if (outflow_params<>[]) then
AugA_rows:=LinearAlgebra[RowDimension](AugmentedA);
outflow_mat:=Matrix(AugA_rows,0);
for i from 1 to n do;
outflow_detect:={-outflow_params[i]} intersect {op(A[i,i])};
if (outflow_detect<>{}) then outflow_col:=Matrix(AugA_rows,1); outflow_col[i,1]:=op(outflow_detect); outflow_mat:=<outflow_mat | outflow_col>; fi; od;
if (tracing=true) then print("outflow_mat",outflow_mat); fi;
# Append this to AugmentedA to add columns for the outflows; each gets a node, made invisible later.
AugmentedA:=<AugmentedA|outflow_mat>;
if (tracing=true) then print("AugA with outflows", AugmentedA); fi;
# Now add some more zero padding rows
outflow_det_cols:=LinearAlgebra[ColumnDimension](outflow_mat);
AugA_cols:=LinearAlgebra[ColumnDimension](AugmentedA);
zero_row_padding:=LinearAlgebra[ZeroMatrix](outflow_det_cols, AugA_cols);
AugmentedA:=<AugmentedA, zero_row_padding>;
if (tracing=true) then print("AugA fully padded", AugmentedA); fi;
fi; # end of block for processing outflows
AdjacencyA:=AugmentedA;
# Create an adjacency matrix for the graph, setting diagonal elements to zero to avoid self-loops
for i to LinearAlgebra[RowDimension](AdjacencyA) do;
    for j to LinearAlgebra[ColumnDimension](AdjacencyA) do;
if (AdjacencyA[i, j]<>0) then if (i=j) then AdjacencyA[i, j]:=0;
else AdjacencyA[i, j]:=1; fi; fi;
od; od;
# Define the colouring of vertices by type
xvertex_colouring:=seq(x_colour, 1..n);
# We need to establish which outputs are active as a result of user inputs.
yvertex_colouring:=seq(y_colour, 1..m);
vertex_colouring:=[xvertex_colouring, yvertex_colouring];
# Include any outflow with an "invisible" node.
num_outflows:=LinearAlgebra[ColumnDimension](outflow_mat);
if (num_outflows>0) then outflow_colour:="white";
vertex_colouring:=[op(vertex_colouring),seq(outflow_colour,1..num_outflows)];
outflow_labels:=outflow; fi;
G:=GraphTheory[Graph](AdjacencyA, vertexcolor=vertex_colouring);
# Customise the vertex labels
xseq:=[seq(x[i], i=1..n)]; xseq:=map(convert,xseq,symbol);
x0_list:=convert(x0,list); x0_list:=map(convert,x0_list,symbol);
xlabels:=convert(Vector[row](n),list);
for i from 1 to n do; xlabels[i]:=convert(StringTools[Join]([xseq[i],x0_list[i]]," : "),symbol); od;
if (tracing=true) then print("xlabels",xlabels); fi;
ylabels:=seq(y[i], i=1..m);
labels_list:=[op(xlabels), ylabels];
if (tracing=true) then print("labels_list",labels_list); fi;
if (num_outflows>0) then outflow_indices:=[`$`(num_xy+1..num_xy+num_outflows)];
```

```
outflow_tag_indices:=[seq(`if`(outflow_params[i] <> 0, i, NULL), i = 1..
    numelems(outflow_params))];
if (tracing=true) then print("outflow_tag_indices", outflow_tag_indices); fi
    ;
labels_list:=[op(labels_list),seq(o[i],i=outflow_tag_indices)];
if (tracing=true) then print("final labels_list", labels_list); fi;
fi;
G:=GraphTheory[RelabelVertices](G, labels_list);
return G;
end proc
```

Appendix 2 Maple Code to Launch an Explore Window

Listing 1.8 presents the `Explore` command which launches our interactive SGI test dashboard by invoking Listing 1.6. Here we consider the case of three state variables and three outputs; the user can readily change these details. To explain the parameters: `A` is the structure's $\mathbf{A}(\boldsymbol{\theta})$, `p1`, `p2`, `p3` are the observation gain parameters on the leading diagonal of $\mathbf{C}(\boldsymbol{\theta})$, and `p4`, `p5`, `p6` are the initial state parameters in $\mathbf{x_0}(\boldsymbol{\theta})$. Initially, each of `p1`,...,`p6` are assigned a parameter symbol appropriate for their relationship to $\boldsymbol{\theta}$. Each of these six parameters may be changed through a text-input box. Parameter `p7` supplies a graph output style understood by `DrawGraph`, initially (the widely applicable) "default". Output from other options (such as "spring") may be easier to interpret, but return an error when any of `p1`, `p2`, or `p3` are set to zero, causing the removal of a link between a state variable and its corresponding output. Parameter `p8` takes one of the two pre-defined values for `theta_mod_type`, which dictates the method employed in creating `theta_prime` from `theta` (used by `theta_prime_creation`). The user changes `p7` and `p8` values by selecting an option from the relevant drop-down menu. If logical-type parameter `tracing:=true`, Maple will show the output of steps used in constructing the structure's compartmental diagram.

Listing 1.8. Maple code using Maple's `Explore` with `Explore_SGI_routine` (Listing 1.6) to produce an interactive panel.

```
Explore(Explore_SGI_test(A, [p1, p2, p3], [p4, p5, p6], outflow_params, p7,
    p8, s, canonical_form, tracing), parameters = [[p1, controller =
    textarea, label = "x1 observation gain",placement=left], [p2,
    controller = textarea, label = "x2 observation gain",placement=left], [
    p3, controller = textarea, label = "x3 observation gain",placement=left
    ], [p4, controller = textarea, label = "x1 initial condition",placement
    =left], [p5, controller = textarea, label = "x2 initial condition",
    placement=left], [p6, controller = textarea, label = "x3 initial
    condition",placement=left], [p7=[default, bipartite, circle, planar,
    spectral, spring, tree], label=Layout, placement=left], [p8=["Caps","
    underscore"],label="theta_mod_type",placement=left]], initialvalues = [
    p1 = c__1, p2 = c__2, p3 = c__3, p4 = x__10, p5 = x__20, p6 = x__30, p7
    = default], size = [650, 750], echoexpression = false, insert = true,
    title = "A use of 'Explore' to interactively test variants of a parent
    uncontrolled, linear time-invariant structure for structural global
    identifiability", overview = "Explore is used to generate the solutions
     for the test of a structure for SGI. Starting from a parent structure,
     by specifying values (or parameters in theta) we may change which
    states are observed, and which have non-zero initial conditions.");
```

References

1. Armando, A., Ballarin, C.: A reconstruction and extension of Maple's assume facility via constraint contextual rewriting. J. Symb. Comput. **39**(5), 503–521 (2005). https://doi.org/10.1016/j.jsc.2004.12.010
2. Bellman, R., Åström, K.J.: On structural identifiability. Math. Biosci. **7**(3–4), 329–339 (1970). https://doi.org/10.1016/0025-5564(70)90132-X
3. Denis-Vidal, L., Joly-Blanchard, G.: Equivalence and identifiability analysis of uncontrolled nonlinear dynamical systems. Automatica **40**(2), 287–292 (2004). https://doi.org/10.1016/j.automatica.2003.09.013
4. DiStefano III, J.: Dynamic systems biology modeling and simulation, 1st edn. Elsevier, Academic Press, Amsterdam (2013)
5. Godfrey, K.: Compartmental Models and Their Application. Academic Press Inc. (1983). https://doi.org/10.1002/bdd.2510060312
6. Laubenbacher, R., Hastings, A.: Editorial. Bull. Math. Biol. **80**(12), 3069–3070 (2018). https://doi.org/10.1007/s11538-018-0501-8
7. Maplesoft, a division of Waterloo Maple Inc., Waterloo, Ontario, Maple 2015.2, 20 December 2015, X86 64 LINUX, Build ID 1097895
8. Maplesoft, a division of Waterloo Maple Inc., Waterloo, Ontario: Maple 2020.1, 30 June 2020, APPLE UNIVERSAL OSX, Build ID 1482634
9. Rescigno, A.: Compartmental analysis revisited. Pharmacol. Res. **39**(6), 471–478 (1999). https://doi.org/10.1006/phrs.1999.0467
10. Vajda, S.: Structural equivalence of linear systems and compartmental models. Math. Biosci. **55**(1–2), 39–64 (1981). https://doi.org/10.1016/0025-5564(81)90012-2
11. Walter, É., Pronzato, L.: Identification of Parametric Models from Experimental Data. Communication and Control Engineering. Springer, San Diego (1997)
12. Whyte, J.M.: On deterministic identifiability of uncontrolled linear switching systems. WSEAS Trans. Syst. **6**(5), 1028–1036 (2007)
13. Whyte, J.M.: A preliminary approach to deterministic identifiability of uncontrolled linear switching systems. In: Pham, T. (ed.) 3rd WSEAS International Conference on Mathematical Biology and Ecology (MABE 2007). Proceedings of the WSEAS International Conferences, Gold Coast, Queensland, Australia, January 2007
14. Whyte, J.M.: Inferring global a priori identifiability of optical biosensor experiment models. In: Li, G.Z., Hu, X., Kim, S., Ressom, H., Hughes, M., Liu, B., McLachlan, G., Liebman, M., Sun, H. (eds.) IEEE International Conference on Bioinformatics and Biomedicine, IEEE BIBM 2013, Shanghai, China, pp. 17–22, December 2013. https://doi.org/10.1109/BIBM.2013.6732453
15. Whyte, J.M.: Global a priori identifiability of models of flow-cell optical biosensor experiments. Ph.D. thesis, School of Mathematics and Statistics, University of Melbourne, Victoria, Australia, December 2016
16. Whyte, J.M., Metcalfe, A.V., Sugden, M.A., Abbott, G.D., Pearce, C.E.M.: Estimation of parameters in pyrolysis kinetics. In: Gulati, C., Lin, Y.-X., Mishra, S., Rayner, J. (eds.) Advances in Statistics, Combinatorics and Related Areas, pp. 361–373. World Scientific (2002). https://doi.org/10.1142/9789812776372_0032
17. Whyte, J.M.: Maple 2020 procedures and a dashboard for interactive testing of uncontrolled linear time-invariant structures for structural global identifiability, March 2021. https://maple.cloud/app/5312540069855232
18. Whyte, J.M.: Model structures and structural identifiability: What? Why? How? In: Wood, D.R., de Gier, J., Praeger, C.E., Tao, T. (eds.) 2019-20 MATRIX Annals. MXBS, vol. 4, pp. 185–213. Springer, Cham (2021). https://doi.org/10.1007/978-3-030-62497-2_10

A Maple Exploration of Problem 6 of the IMO 88

Zhenbing Zeng[1(✉)], Xiang Sun[1(✉)], Yong Huang[2,3], Yaochen Xu[1], Xiaoru Chen[2], and Lu Yang[4]

[1] Department of Mathematics, Shanghai University, Shanghai 200444, China
zbzeng@shu.edu.cn, yaochen@picb.ac.cn

[2] South China Institute of Software Engineering, Guangzhou University, Guangzhou 510990, China
cxr@sise.com.cn

[3] Institute of Computing Sciences and Technology, Guangzhou University, Guangzhou 510006, China

[4] Chengdu Institute of Computer Applications, Chinese Academy of Sciences, Chengdu 610041, China
luyang@casit.ac.cn

Abstract. In this paper, we show that by using Maple software, some direct searching computation could derive a solution to Problem 6 of the 1988 International Mathematics Olympiad, which asks to prove that if a and b are integers such that $ab+1$ divides a^2+b^2, then $(a^2+b^2)/(ab+1)$ is the square of an integer.

Keywords: International Mathematics Olympiad · Maple software · Integer · Square

1 Problem 6 of the IMO 1988

The 29th International Mathematical Olympiad was held on July 9–21, 1988 at Canberra, Australia. 268 contestants from 49 countries participated in the Olympiad, and 17 people of them got the golden prize. Problem 6 of IMO 1988 was called "The Legend Problem 6 of IMO" (see [1]) since only 11 among 268 participants answered it correctly (that means that they obtained 7 points, the highest score, for this question), which is significantly lower than the correct ratios of other problems of this Olympiad and also the problems of other years. To see this, recall that in this year, the numbers of people (in percentages) who got 7 points on Problem 1 to Problem 6 are:

$$36.57\%,\quad 26.87\%,\quad 11.57\%,\quad 24.63\%,\quad 32.09\%,\quad 4.10\%,$$

and the averages for 1981 to 2000 are as follows:

$$30.70\%,\quad 31.02\%,\quad 13.97\%,\quad 25.47\%,\quad 25.98\%,\quad 11.90\%,$$

Supported by the National Natural Science Foundation of China Project No. 11471209.

R. M. Corless et al. (Eds.): MC 2020, CCIS 1414, pp. 429–437, 2021.
https://doi.org/10.1007/978-3-030-81698-8_28

Problem 6 of the IMO 1988 reads as follows:

Problem. *Let a and b be positive integers such that $ab+1$ divides a^2+b^2, show that $(a^2+b^2)/(ab+1)$ is the square of an integer.*

There are already many discussions on solving this problem on the internet, for example, see [2–4]. Most discussions concentrate on training the techniques for solving this hard IMO problem. In this paper, we show that by using Maple software as a computation tool for searching solutions of the derived Diophantine equation

$$a^2 + b^2 = k \cdot (ab + 1), \tag{1}$$

an intuitive approach to the IMO problem can be constructed from the Maple computation data.

The paper is organized as follows: In Sect. 2 we present a Maple program to generate solutions to the IMO problem and describe some properties we observed from the data, in Sect. 3 we give a simple proof for the IMO problem based on the Maple experiment, in Sect. 4 we construct a recursion formula for the Diophantine equation related to Problem 6 of IMO 1988.

2 Maple Experiment

The maple computation is starting from the following short program:

```
Sab:=[]
   for a to 1000 do
      for b to 1000 do
         if a^2+b^2 mod a*b+1 = 0 then
            Sab:=[op(Sab),[a,b,(a^2+b^2)/(a*b+1)]]
         end if
      end do
   end do;
Sab
```

This procedure will yield the following result Sab in 1 s:

```
[[1, 1, 1], [2, 8, 4], [3, 27, 9], [4, 64, 16], [5, 125, 25],
 [6, 216, 36], [7, 343, 49], [8, 2, 4], [8, 30, 4], [8, 512, 64],
 [9, 729, 81], [10, 1000, 100], [27, 3, 9], [27, 240, 9], [30, 8, 4],
 [30, 112, 4], [64, 4, 16], [112, 30, 4], [112, 418, 4], [125, 5, 25],
 [216, 6, 36], [240, 27, 9], [343, 7, 49], [418, 112, 4], [512, 8, 64],
 [729, 9, 81], [1000, 10, 100]
]
```

The list Sab contains 27 integer triples (a, b, k) that satisfies the Diophantine equation $a^2 + b^2 = k(ab + 1)$. It is apparent that if (a, b, k) is a solution of this Diophantine equation, then (b, a, k) is also a solution of the equation. Try the above program for a from 1 to 10000 and for b from a to 10000, the computation outputs the following 31 triples in about 93 s:

```
[[1, 1, 1], [2, 8, 4], [3, 27, 9], [4, 64, 16], [5, 125, 25], [6, 216, 36],
 [7, 343, 49], [8, 30, 4], [8, 512, 64], [9, 729, 81], [10, 1000, 100],
 [11, 1331, 121], [12, 1728, 144], [13, 2197, 169], [14, 2744, 196],
 [15, 3375, 225], [16, 4096, 256], [17, 4913, 289], [18, 5832, 324],
 [19, 6859, 361], [20, 8000, 400], [21, 9261, 441], [27, 240, 9],
 [30, 112, 4], [64, 1020, 16], [112, 418, 4], [125, 3120, 25],
 [216, 7770, 36], [240, 2133, 9], [418, 1560, 4], [1560, 5822, 4]
]
```

Do further experiment for a from 1 to 100000 and for b from a to 100000. Then, the computation took 16990 s (on a desktop computer with Intel ® 2 Duo CPU T7500, 2.0 GB RAM, Maple 15). The results contains 65 solutions (a, b, k) with $a \le b$. The result can be rearranged into 46 chains $G_1, G_2, \ldots, G_{46}$ as follows:

G_1: `[1, 1, 1];`

G_2: `[2, 8, 4]`$\lhd$`[8, 30, 4]`$\lhd$`[30, 112, 4]`$\lhd$`[112, 418, 4]`$\lhd$`[418, 1560, 4]`
$\lhd$`[1560, 5822, 4]`$\lhd$`[5822, 21728, 4]`$\lhd$`[21728, 81090, 4];`

G_3: `[3, 27, 9]`$\lhd$`[27, 240, 9]`$\lhd$`[240, 2133, 9]`$\lhd$`[2133, 18957, 9];`

G_4: `[4, 64, 16]`$\lhd$`[64, 1020, 16]`$\lhd$`[1020, 16256, 16];`

G_5: `[5, 125, 25]`$\lhd$`[125, 3120, 25]`$\lhd$`[3120, 77875, 25];`

G_6: `[6, 216, 36]`$\lhd$`[216, 7770, 36];`

G_7: `[7, 343, 49]`$\lhd$`[343, 16800, 49];`

G_8: `[8, 512, 64]`$\lhd$`[512, 32760, 64];`

G_9: `[9, 729, 81]`$\lhd$`[729, 59040, 81];`

G_{10}: `[10, 1000, 100]`$\lhd$`[1000, 99990, 100];`

G_k **for** $k = 11, 12, \ldots, 46$:

```
[11, 1331, 121];   [12, 1728, 144];   [13, 2197, 169];   [14, 2744, 196];
[15, 3375, 225];   [16, 4096, 256];   [17, 4913, 289];   [18, 5832, 324];
[19, 6859, 361];   [20, 8000, 400];   [21, 9261, 441];   [22,10648, 484];
[27,19683, 729];   [28,21952, 784];   [29,24389, 841];   [30,27000, 900];
[31,29791, 961];   [32,32768,1024];   [33,35937,1089];   [34,39304,1156];
[35,42875,1225];   [36,46656,1296];   [37,50653,1369];   [38,54872,1444];
[39,59319,1521];   [40,64000,1600];   [41,68921,1681];   [42,74088,1764];
[43,79507,1849];   [44,85184,1936];   [45,91125,2025];   [46,97336,2116];
```

Note here $(a, b, k) \in G_p$ if and only if $k = p^2$, the initial element of the chain G_p is (p, p^3, p^2), and $(a, b, k) \lhd (a', b', k')$ if $b = a'$ and $k = k'$ (we may call (a', b', k') the successor of (a, b, k)). From the above data, we can observe the following facts:

1. For any positive integer p, the following three integers

$$a = p, \quad b = p^3, \quad k = p^2 \tag{2}$$

form a solution of the Diophantine equation $a^2 + b^2 = k(ab + 1)$. Indeed, this property can be verified very easily.
2. It seems that, for any positive integer k, if (a, b, k) is a solution of the Diophantine equation, then there exists an integer c so that $c > b$ and (b, c, k) is also a solution of the Diophantine equation.
3. Further, if $a < b < c$ and $(a, b, k), (b, c, k)$ are solutions of the Diophantine equation, then we have

$$a^2 + b^2 = k(ab + 1), \quad b^2 + c^2 = k(bc + 1), \tag{3}$$

and therefore,

$$c^2 - a^2 = kb(c - a), \quad \text{and} \quad c = -a + kb. \tag{4}$$

4. Assume that $a < b$ and $k > 1$. If (a, b, k) is a solution of $a^2 + b^2 = k(ab + 1)$, then

$$-a + kb \geq -a + 2b = 2b - a > a. \tag{5}$$

Let $c = -a+kb$. Then $c > b$ and (b, c, k) is indeed a solution of the Diophantine equation, since

$$\begin{aligned} b^2 + c^2 - k(bc + 1) &= b^2 + (-a + kb)^2 - k\,(b(-a + kb) + 1) \\ &= -abk + a^2 + b^2 - k = -k(ab + 1) + a^2 + b^2 = 0. \end{aligned} \tag{6}$$

5. For any integer k with $k = p^2$ where p is an integer with $p > 1$, a monotonely increasing sequence of integers:

$$s_1(k) < s_2(k) < s_3(k) < \cdots < s_n(k) < \cdots,$$

can be constructed recursively, from the two initial values

$$s_1(k) = p, \quad s_2(k) = p^3, \tag{7}$$

and the following recursive formular

$$s_{n+1} = ks_n - s_{n-1}, \tag{8}$$

so that all following triples

$$(s_1(k), s_2(k), k), \quad (s_2(k), s_3(k), k), \quad \cdots, \quad (s_n(k), s_{n+1}(k), k), \quad \cdots$$

are solutions of the Diophantine equation $a^2 + b^2 = k(ab + 1)$.

3 A Solution to the Problem 6

Based on the observation from the Maple experiment described in the previous section, now we can give proof to Problem 6 of IMO 1988. Note that from a solution $(a, b, k)\,(a < b, k > 1)$ we constructed another solution (b, c, k) so that $b < c$, and therefore, an infinitely many "increasing" solutions can be constructed. "Decreasing" solutions (but finitely many in this direction) can also be constructed in a similar way. Namely, we can prove a proposition as follows:

Proposition 1. *If $a < b, k > 1$, and (a, b, k) is a solution of the Diophantine equation $a^2 + b^2 = k(ab + 1)$, then $(ka - b, a, k)$ and $(b, -a + kb, k)$ are also solutions of this equation.*

The proof is very easy, so we omit it here. Noticed here we do not need to assume that $k = p^2$ for some integer p. Let

$$S(k) := \{(a, b, k) | a^2 + b^2 = k(ab + 1)\}.$$

Then Proposition 1 implies that

$$S(k) \neq \emptyset \Longrightarrow \#S(k) = \infty. \tag{9}$$

A Solution of the 1988 IMO Problem 6. Define a partial order $\preccurlyeq$ on $S(k)$ as follows: $(a_1, b_1, k), (a_2, b_2, k) \in S(k)$, then

$$(a_1, b_1, k) \preccurlyeq (a_2, b_2, k) \Leftrightarrow a_1 \leq a_2.$$

Therefore, for $a < b$ and $k > 1$, we have

$$(ka - b, a, k) \preccurlyeq (a, b, k) \preccurlyeq (b, -a + kb, k). \tag{10}$$

It is clear that $\preccurlyeq$ is also a total order on $S(k)$, since for any given triples $(a_1, b_1, k), (a_2, b_2, k) \in S(k)$, we have either $(a_1, b_1, k) \preccurlyeq (a_2, b_2, k)$ or $(a_2, b_2, k) \preccurlyeq (a_1, b_1, k)$. Apparently, the infinite set $S(k)$ has a minimal element in order $\preccurlyeq$, and (a, b, k) is a minimal solution if and only if $ka - b = 0$. Thus,

$$k = \frac{a^2 + b^2}{ab + 1} = \frac{a^2 + (ka)^2}{ka^2 + 1}, \tag{11}$$

which immediately implies that

$$k(ka^2 + 1) = a^2 + k^2a^2, \tag{12}$$

and hence, $k = a^2$, i.e., k is the square of an integer. □

4 Recursion Formula for the Diophantine Equation

In the previous section, we have seen that

$$\#S(k) = \begin{cases} 0, & \text{if } k \neq p^2, \\ \infty, & \text{if } k = p^2, \end{cases} \tag{13}$$

and, for $k = p^2$ the set $S(k) = S(p^2)$ is formed by $(s_n, s_{n+1}, p^2), n = 1, 2, \ldots$ with initial values

$$s_1 = p, \quad s_2 = p^3, \tag{14}$$

and the following recursion formula

$$s_{n+1} = p^2 s_n - s_{n-1}. \tag{15}$$

A direct way to reconstruct s_n is a generating function (cf. [5]). For this, we may assume that

$$s_n = Ax^n + By^n, \quad n = 1, 2, \ldots, \tag{16}$$

where x, y are two zeros of the characteristics equation

$$z^2 - p^2 z + 1 = 0, \tag{17}$$

that is,

$$x = \frac{1}{2}\left(p^2 + \sqrt{p^4 - 4}\right), \quad y = \frac{1}{2}\left(p^2 - \sqrt{p^4 - 4}\right), \tag{18}$$

and A, B are numbers determined by

$$A \cdot \frac{p^2 + \sqrt{p^4 - 4}}{2} + B \cdot \frac{p^2 - \sqrt{p^4 - 4}}{2} = p^2, \tag{19}$$

$$A \cdot \left(\frac{p^2 + \sqrt{p^4 - 4}}{2}\right)^2 + B \cdot \left(\frac{p^2 - \sqrt{p^4 - 4}}{2}\right)^2 = p^3. \tag{20}$$

It is easy to learn that

$$A = -\frac{p^2\left(p^2 - 2p - \sqrt{p^4 - 4}\right)}{\sqrt{p^4 - 4}\left(p^2 + \sqrt{p^4 - 4}\right)}, \quad B = \frac{p^2\left(p^2 - 2p + \sqrt{p^4 - 4}\right)}{\sqrt{p^4 - 4}\left(p^2 - \sqrt{p^4 - 4}\right)}. \tag{21}$$

Therefore, we have

$$\begin{aligned} s_n(p^2) = &-\frac{p^2\left(p^2 - 2p - \sqrt{p^4 - 4}\right)}{2^n\sqrt{p^4 - 4}} \cdot (p^2 + \sqrt{p^4 - 4})^{n-1} \\ &+ \frac{p^2\left(p^2 - 2p + \sqrt{p^4 - 4}\right)}{2^n\sqrt{p^4 - 4}} \cdot (p^2 - \sqrt{p^4 - 4})^{n-1}. \end{aligned} \tag{22}$$

The above formula can also be produced by using Maple recurrence equation solver `rsolve` directly.

We can also use Maple to construct polynomial form of $s_n = s_n(p^2)$. For this we can compute s_n for $n = 3, 4, \ldots$ recursively. The result for $n = 3, 4, \ldots, 15$ is as follows:

$$
\begin{aligned}
s_3 &= p^5 - p, \\
s_4 &= p^7 - 2p^3, \\
s_5 &= p^9 - 3p^5 + p, \\
s_6 &= p^{11} - 4p^7 + 3p^3, \\
s_7 &= p^{13} - 5p^9 + 6p^5 - p, \\
s_8 &= p^{15} - 6p^{11} + 10p^7 - 4p^3, \\
s_9 &= p^{17} - 7p^{13} + 15p^9 - 10p^5 + p \\
s_{10} &= p^{19} - 8p^{15} + 21p^{11} - 20p^7 + 5p^3, \\
s_{11} &= p^{21} - 9p^{17} + 28p^{13} - 25p^9 + 15p^5 - p, \\
s_{12} &= p^{23} - 10p^{19} + 36p^{15} - 56p^{11} + 35p^7 - 6p^3, \\
s_{13} &= p^{25} - 11p^{21} + 45p^{17} - 84p^{13} + 70p^9 - 21p^5 + p, \\
s_{14} &= p^{27} - 12p^{23} + 55p^{19} - 120p^{15} + 126p^{11} - 56p^7 + 7p^3, \\
s_{15} &= p^{29} - 13p^{25} + 66p^{21} - 165p^{17} + 210p^{13}120 - p^9 + 28p^5 - p,
\end{aligned}
$$

We can find that for a fixed n, s_n is a polynomial of p of degree $2n-1$:

$$s_n(p^2) = p^{2n-1} - (n-2)p^{2n-5} + \cdots = \sum_{0 \le j < (2n-1)/4} c_{n,j} p^{2n-1-4j}, \tag{23}$$

and

$$c_{n,1} = -(n-2), \quad c_{n,2} = \frac{1}{2}(n-3)(n-4), \tag{24}$$

and the absolute values of $c_{n,j}$ for other n, j also appear in the Pascal's Triangle (also known as the Yanghui triangle in China). To see this, we display the Pascal's Triangle in the following form:

```
  1
  1    1
  1    2    1
  1    3    3   [1]
  1    4   [6]   4    1
  1   [5]  10   10    5   (1)
 [1]   6   15   20  (15)   6    1
  1    7   21  (35)  35   21    7    1
  1    8  (28)  56   70   56   28    8    1
  1   (9)  36   84  126  126   84   36    9    1
 (1)  10   45  120  210  252  210  120   45   10    1
```

This form of Pascal's Triangle can be seen in [6]. Some numbers are marked by brackets `[ ]` or `( )`, we will explain immediately. Looking at the Pascal's Triangle in this form as an infinite matrix, we may find that for fixed n, the absolute values of s_n are distributed in a line passing the element 1 at position $(n, 1)$ and the element $n-2$ at position $(n-1, 2)$. For example, for

$$s_7(p^2) = p^{13} - 5p^9 + 6p^5 - p, \tag{25}$$

the absolute values of its coefficients $1, 5, 6, 1$, marked by brackets `[ ]` in the Pascal's Triangle, are appearing precisely on a line passing through bracketed numbers `[1]`, `[5]`, `[6]`, and `[1]`, and the absolute values of the coefficients

$$s_{11}(p^2) = p^{21} - 9p^{17} + 28p^{13} - 25p^9 + 15p^5 - p, \tag{26}$$

marked by brackets `( )`, are appearing on the line through bracketed numbers `(1)`, `(9)`, `(28)`, `(35)`, `(15)` and `(1)` in the Pascal's Triangle. The observation can be summarized as the following proposition:

Proposition 2. *Let $p > 1$ be an integer $s_n = s_n(p^2)$ the sequence constructed by*

$$s_1 = p, \quad s_2 = p^3, \quad \text{and} \quad s_{n+1} = -s_{n-1} + p^2 \cdot s_n.$$

Then

$$s_n(p^2) = \sum_{0 \le j < (2n-1)/4} (-1)^j \binom{n-1-j}{j} \cdot p^{2n-1-4j}. \tag{27}$$

This proposition can be proved by mathematical induction according to $s_1 = p, s_2 = p^3$ and $s_{n+1} = p^2 s_n - s_{n-1}$. Since our main purpose in this paper is to give a solution to Problem 6 of the IMO 1988, we leave the proof of Proposition 2 to readers. Note also that from the recursion formula (15) and (14) it is easy to prove that the sum of the absolute values of coefficients of $s_n(p^2)$ equals to the n-th Fibonacci number F_n, i.e.,

$$\begin{aligned}
s_1 &= p \longrightarrow 1 = 1 = F_1,\\
s_2 &= p^3 \longrightarrow 1 = 1 = F_2,\\
s_3 &= p^5 - p \longrightarrow 1 + 1 = 2 = F_3,\\
s_4 &= p^7 - 2p^3 \longrightarrow 1 + 2 = 3 = F_4,\\
s_5 &= p^9 - 3p^5 + p \longrightarrow 1 + 3 + 1 = 5 = F_5,\\
s_6 &= p^{11} - 4p^7 + 3p^3 \longrightarrow 1 + 4 + 3 = 8 = F_6,\\
s_7 &= p^{13} - 5p^9 + 6p^5 - p \longrightarrow 1 + 5 + 6 + 1 = 13 = F_7,\\
s_8 &= p^{15} - 6p^{11} + 10p^7 - 4p^3 \longrightarrow 1 + 6 + 10 + 4 = 21 = F_8,\\
s_9 &= p^{17} - 7p^{13} + 15p^9 - 10p^5 \longrightarrow 1 + 7 + 15 + 10 = 34 = F_9,\\
s_{10} &= p^{19} - 8p^{15} + 21p^{11} - 20p^7 + 5p^3 \longrightarrow 1 + 8 + 21 + 20 + 5 = 55 = F_{10}.
\end{aligned}$$

A proof of this fact can be derived from the following equality:

$$\sum_{j=0}^{\lfloor n/2 \rfloor} \binom{n-j}{j} = F_{n+1} \tag{28}$$

A proof of (28) can be found from Maplesoft's Application webpage [7].

5 Conclusion

In this paper, we presented a simple Maple program for searching the solution of the Diophantine equation derived from the IMO 1988 Problem 6, and via observation of the data, we gave proof to the IMO problem. We also demonstrated a connection between the general solution of the IMO problem and the Pascal's Triangle. The Maple experiment shows that sufficient numeric solutions of a very difficult Diophantine equation, like Problem 6 of IMO 1988, is very crucial for making a correct answer to the question.

Acknowledgment. The authors would like to express their appreciation to the anonymous reviewers for their valuable comments and suggestions.

References

1. Crew, B.: The legend of question six: one of the hardest maths problems ever. https://www.sciencealert.com/the-legend-of-question-six-one-of-the-hardest-maths-problems-ever. Accessed 14 Jan 2021
2. AoPSonline: 1988 IMO Problems/Problem 6 - Art of Problem Solving. https://artofproblemsolving.com/wiki/index.php?title=1988_IMO_Problems/Problem_6. Accessed 14 Jan 2021
3. Mathematics Stack Exchange: Simple solution to Question 6 from the 1988 Math Olympiad, https://artofproblemsolving.com/wiki/index.php?title=1988_IMO_Problems/Problem_6. Accessed 14 Jan 2021
4. Maths and Musings: 1988 IMO Question Six: Solving the Hardest Problem on the Hardest Test. https://medium.com/cantors-paradise/1988-imo-question-six-2ef095cd23c6. Accessed 14 Jan 2021
5. Seaborn, J.B.: Generating functions and recursion formulas. In: Hypergeometric Functions and Their Applications. Texts in Applied Mathematics, vol. 8. Springer, New York. https://doi.org/10.1007/978-1-4757-5443-8_11. Accessed 14 Jan 2021
6. Stover, C., Weisstein, E.W.: Pascal's Triangle. From MathWorld-A Wolfram Web Resource. https://mathworld.wolfram.com/PascalsTriangle.html. Accessed 27 Mar 2021
7. Waterloo Maple Inc.: Pascal's Triangle and it's Relationship to the Fibonacci Sequence. https://www.maplesoft.com/applications/view.aspx?SID=3617&view=html. Accessed 27 Mar 2021

An Isometric Embedding of the Impossible Triangle into the Euclidean Space of Lowest Dimension

Zhenbing Zeng[1(✉)], Yaochen Xu[1(✉)], Zhengfeng Yang[2], and Zhi-bin Li[3]

[1] Department of Mathematics, Shanghai University, 99 Shangda Rd, 200444 Shanghai, China
zbzeng@shu.edu.cn, yaochen@picb.ac.cn

[2] School of Software Engineering, East China Normal University, 3663 North Zhongshan Rd, 200062 Shanghai, China
zfyang@sei.ecnu.edu.cn

[3] School of Computer Science and Technology, East China Normal University, 3663 North Zhongshan Rd, 200062 Shanghai, China
zbli@cs.ecnu.edu.cn

Abstract. The impossible triangle, invented independently by Oscar Reutersvärd and Roger Penrose in 1934 and 1957, is a famous geometry configuration that cannot be realized in our living space. Many people admitted that this object could be constructed in the four-dimensional Euclidean space without rigorous proof. In this paper, we prove that the isometric embedding problem can be decided by finite points on the configuration, then applying Menger and Blumenthal's classical method of Euclidean embedding of finite metric space we determined the lowest Euclidean dimension, and finally using MAPLE obtained the coordinates of the isometric embedding. Our investigation shows that the impossible triangle is impossible to be isometrically embedded in the dimension four Euclidean space, but there is an isometric embedding to the dimension five space.

Keywords: Isometric embedding · Impossible triangle · Euclidean space · Simplex

1 Introduction

The impossible triangle was firstly painted in 1934 by the Swedish painter Oscar Reutersvärd who was born in 1915 in Stockholm and was trained in arts by a Russian immigrant professor of the Academy of Arts in St.Petersburg at that time. Oscar Reutersvärd drew his version of triangle as a set of cubes in parallel projection, as shown in Fig. 1(left). Actually, he started this figure by placing a perfect six-pointed star shape in the middle, and around the star, he added nine

Financial support from the Chinese National Science Foundation Project No. 61772203.

R. M. Corless et al. (Eds.): MC 2020, CCIS 1414, pp. 438–457, 2021.
https://doi.org/10.1007/978-3-030-81698-8_29

cubes, filling the empty spaces between the stars for creating the 3D illusion. He soon realized that what he'd drawn was paradoxical: something that couldn't be built in the real world. (See [1]).

Reutersvärd was diagnosed with dyslexia at a young age, which prevented him from accurately estimating the size and distance of objects, but he was determined to follow in the footsteps of his artistic family. He continued to design thousands of impossible figures throughout his life. Reutersvärd's achievements were honored in 1982 by a series of Swedish postage stamps.

A different version of this impossible triangle was independently created by the English physicist and mathematician Roger Penrose in 1954. Unlike Reutersvärd's figure, he painted triangle as three bars connected with right angles (later known as the Penrose tribar or Penrose triangle), as shown in Fig. 1(right).

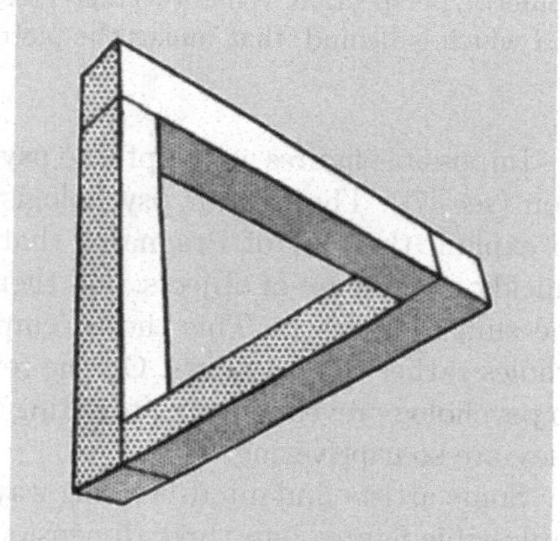

Fig. 1. (*Left*): The impossible figure drawn by Oscar Reutersvärd. (*Right*). The impossible structure in L. S. Penrose and R. Penrose's article published in 1958 in the British Journal of Psychology [2].

L. Penrose and R. Penrose sent a copy of the article to M.C. Escher. Note, neither Penrose nor Escher had known about artworks by Reutersvärd at that time (cf. [3]). Escher created his famous lithographs "Ascending and Descending" in 1960 and "Waterfall" in 1961. M.C. Escher provided many popular examples of impossible figures in his drawings and woodcuts. Perhaps the weirdest structure to mathematicians is the impossible cube in "Belvedere" (1958) as shown in Fig. 2(left). To understand impossible figures, we need first to understand two-dimensional representations of three-dimensional objects. A simple line drawing, such as the *Necker cube* illustrated in Fig. 2(right), could be interpreted in two ambiguous ways. The Belvedere' toy cube can be regarded as a version of the Necker cube where the edges cross in inconsistent ways.

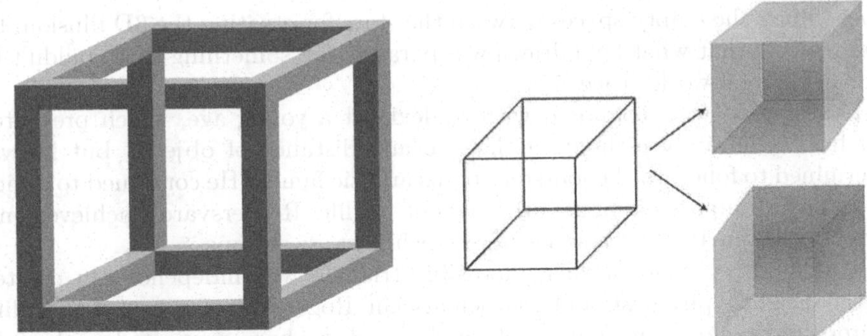

Fig. 2. (*Left*): The toy cube held in hands of the boy in M.C. Escher's lithograph "Belvedere" (1958). (*Right*): The Necker Cube is first published as a rhomboid in 1832 by Swiss crystallographer Louis Albert Necker. It is a wire-frame drawing of a cube in isometric perspective. When two lines cross, the picture does not show which is in front and which is behind, that makes the picture can be interpreted two different ways.

Impossible figures are helpful to psychology research on human visual perception (see [7]). The Gestalt psychologists used the impossible triangle and cube to explain the Law of Pragnanz, that the human mind loves to simplify, and quickly make sense of objects, and therefore, human see the whole image, before the sum of its parts. This theory emphasizes that human perceives objects as wholes rather than as parts. On the reverse, they say that the Gestalt approach to psychology reveals some interesting insights about impossible figures and why they are so captivating.

Some artists find mind-bending ways to bring the Penrose Triangle and other impossible figures into three-dimensional reality. They create a clever design on certain objects so that they look like the proper impossible figure when viewed from the correct angle. An example is the Impossible Triangle sculpture in Perth, Australia, shown in Fig. 3, created in November 1999 by artist Brian McKay in collaboration with architect Ahmad Abas. Another impossible triangle is located in the center of belgian village Ophoven, built by dutch artist and mathematician Mathieu Heamekers in 1995.

As we have seen, for the Impossible Triangle sculpture, there are only two appropriate positions from where people could see the proper Penrose triangle. It is curious to ask that if there are other ways to install a certain structure, say, in higher dimensional space, so that people (in the usual three-dimensional space) can see the impossible figure from a larger viewing angle? The question can be rephrased more precisely as below: if anybody can build a geometric configuration in four or higher-dimensional Euclidean space so that people could see an image of the impossible triangle in the real world?

Some people argued (cf. [4]) that since each local part of the Penrose triangle is 3-dimensional, so lies in some 3-dimensional subspace, and that the edges are straight lines, every piece lies on the same 3-dimensional subspace, so if we don't

Fig. 3. An Impossible Triangle sculpture was designed by artist Brian McKay and architect Ahmad Abas, which was built in Claisebrook Square in East Perth, Australia. It is 13.5 m high, and has remained an East Perth landmark for 20 years.

allow the edges to bend, the figure is also not possible in higher dimensions when it is not possible in the 3-dimensions.

An opposing viewpoint is that figures like the Penrose triangle which seems impossible in our three-dimensional space might be possible in the fourth dimension. For example, Blue Sam [5] indicated as the surface of the Penrose triangle is (up to taking a smooth approximation at the edges) a smooth 2-manifold, so by the Whitney Embedding Theorem, it must be embeddable in 5-dimensional space. Sam also claimed that if we are not bothered about keeping the edges straight, the embedding can be done in three dimensions, and therefore he believed in the four dimension space it is possible to construct an embedding with straight edges. Vlad Alexeev [6] claimed that the bars of a four-dimensional impossible triangle can be connected at right angles and it will not be distorted from any point of view as distinct from the three-dimensional impossible triangle. However, to the best of our knowledge, we have not found a rigorous proof of the embedding neither to 5 nor 4 dimensional space in literature.

In this paper, we will construct an explicit embedding of the Penrose triangle in the Euclidean space. We will prove that the minimal n for constructing an isometric embedding of the Penrose triangle in $\mathbb{R}^n$ is $n = 5$. The paper is organized as follows: In Sect. 2 we show that the embedding problem can be reduced to a set of finite points, and the finite points can be isometrically embedded into the Euclidean space $\mathbb{R}^n$ with minimal $n = 5$, in Sect. 3 we solve the system of equations derived from the reduced embedding problem, so to give an explicit embedding of the Penrose triangle into $\mathbb{R}^n$ for $n = 5$, in Sect. 4 we present an intuitive explanation to the five-dimensional configuration of the Penrose triangle, and show an analog isometric embedding of the Möbius band to $\mathbb{R}^4$.

2 The Minimal Isometric Embedding

In view of mathematics, making an imagined object in higher-dimensional Euclidean space $\mathbb{R}^n$ means to construct points set $\mathfrak{S}$ that is isometric to the configuration in imagination, or topologically looks like the imagined object, and to see a point set of higher dimensional space from the usual $\mathbb{R}^3$ just means to compute the image of certain function $f : \mathbb{R}^n \to \mathbb{R}^3$ that acts like a camera we used every day for taking $2D$ images of $3D$ objects. As we often take several photos of a three-dimensional object from several different view angles to obtain more information on the $3D$ shape, it is also necessary to take several $3D$ images of one higher dimensional object to percept its whole structure.

The idea for lifting the impossible figures in higher dimensional space is very natural. As we all know, any $2D$ animal living in a plane world $\mathbb{R}^2$ is not able to build a Möbius trip from a paper band, since any movement of the paper in the plane world just cannot twist the paper band, and lift it as depicted in Fig. 4. Though the task to twist and lift a paper band is an impossible mission for any $2D$ animals, human in $\mathbb{R}^3$ can do this job very easily, and mathematicians even can write down the coordinates of points on the Möbius band as in [8].

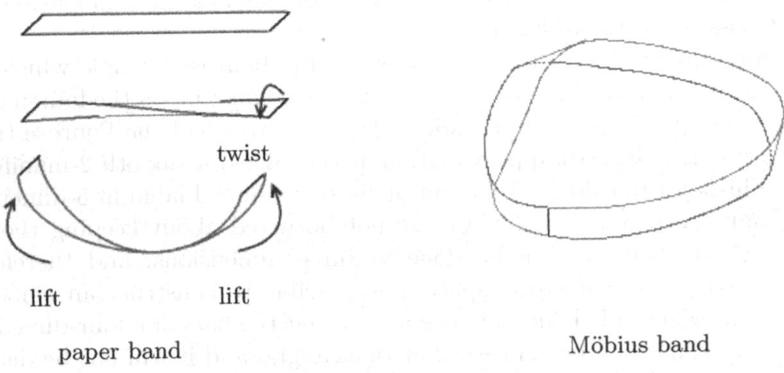

Fig. 4. When making a Möbius band from a paper band, the twist and lift operations must be done in the three dimensional space.

For constructing the Penrose triangle, we may start from three equal copies of an L-shape object L, formed by two perpendicular cylinders in the three-dimensional space, as depicted in Fig. 5. Just like that in the plane world $2D$ animals cannot twist or lift up a paper band, in our living $3D$ space we can connect L_1, L_2 at point A without problem, but we cannot connect L_1, L_3 at C and L_2, L_3 at B simultaneously. So the real difficulty for making the Penrose triangle is that we are not able to *move* the three objects *out* of the $3D$ space where we are living.

Now imagine that some of us (say, Yog Sothoth) happened to know the gate to a higher dimensional world, then he could build the Penrose triangle from the $3D$ components L_1, L_2, L_3 in some higher dimensional space which is invisible to us. To show that he had done the craft correctly, Yog could also cut his product again into several pieces $3D$ figures, possibly new ones, as shown in Fig. 6, and brought them back to the three-dimensional world where we are living, as evidence of his work in the higher dimensional space.

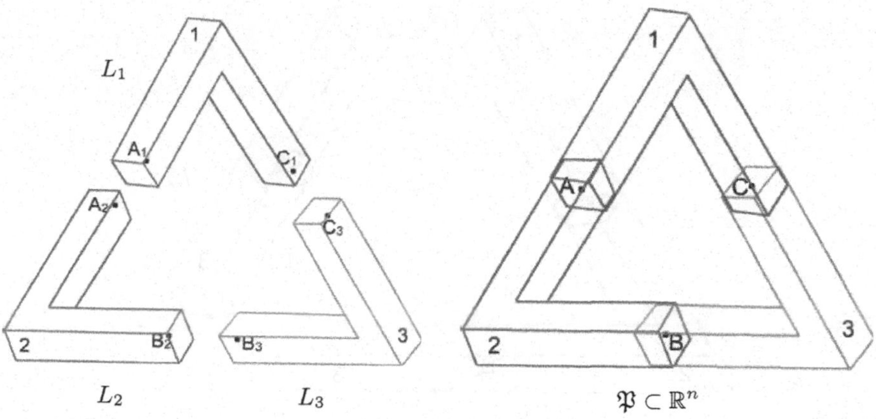

Fig. 5. L_1, L_2, L_3 in the left are three equal copies of $3D$ materials for making the Penrose triangle. If anybody knows the gate to the higher dimension, he could take L_1, L_2, L_3 out of the three dimensional space and construct a right Penrose triangle in some higher dimensional space with the $3D$ components.

In mathematics, the above actions can be understood as decomposing an object that is impossible into finitely many $3D$ components, and constructing the impossible configuration (an isometric embedding) in $\mathbb{R}^n$ for some $n > 3$ with the components, and then partitioning the point set into some disjoint parts that can be displayed in the three-dimensional space (a piecewise isometric immersion of the impossible configuration in $\mathbb{R}^3$).

For doing the isometric embedding and piecewise immersion, we may decompose the Penrose triangle into the union of three regular cubes and three right cylinders as depicted in Fig. 7. Whereas altogether the cubes and cylinders have 24 vertices, we use the 24 lower case Greek letters to denote them. The cubes and cylinders are constructed as follows. Notice that the Penrose is a non-convex polytope, and its convex hull $\mathfrak{P}$ (i.e., the smallest convex set that contains the Penrose triangle) has six extremals. Here a point P is called an extremal point of a convex set $K \subset \mathbb{R}^n$, if there exists no $P_1, P_2 \in K$ such that $P_1 \neq P_2 \in K$ and $P = c \cdot P_1 + (1-c)P_2$ for some $c \in (0,1)$. Let $\alpha, \beta, \lambda, o, \phi, \omega$ denote the six extremals of $\mathfrak{P}$ Let C_1, C_2, C_3 be the maximal cylinders contained in the Penrose triangle. It is clear that

$$A_1 = C_2 \cap C_3, \quad A_2 = C_3 \cap C_1, \quad A_3 = C_1 \cap C_2,$$

are three disjoint regular cubes contained in the Penrose triangle,

$$B_i = C_i \setminus (A_1 \cup A_2 \cup A_3), i = 1, 2, 3,$$

are three right cylinders contained in the Penrose triangle, and the three cubes and three cylinders form a disjoint partition of the Penrose triangle. We may write

$$A_1 = \alpha\beta\gamma\delta\varepsilon\zeta\eta\theta, A_2 = \iota\kappa\lambda\mu\nu\xi o\pi, A_3 = \rho\sigma\varsigma\tau\upsilon\phi\chi\psi\omega, \tag{1}$$

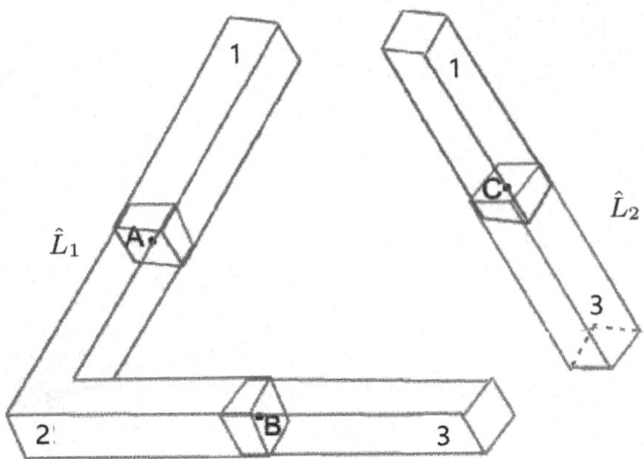

Fig. 6. If any body had built a proper Penrose triangle in higher dimensional space $\mathbb{R}^n$ from three equal $3D$ objects L_1, L_2, L_3, they could cut their product into two $3D$ objects $\hat{L}_1, \hat{L}_2$ and brought back to the real world.

and

$$B_2 = \gamma\eta\theta\delta\text{-}\nu\iota\kappa\xi, B_1 = \nu\iota\mu\pi\text{-}\chi\upsilon\tau\psi, C_3 = \rho\upsilon\tau\sigma\text{-}\varepsilon\zeta\eta\theta. \tag{2}$$

Without loss of generality, we may assume that the cubes are isometric to $[0,1]\times[0,1]\times[0,1]$, and the cylinders are isometric to $[0,1]\times[0,1]\times(0,a)$, for appropriate $a \geq 3$. Note that The last requirement $a \geq 3$ is pre-assumed according to the most of drawings of Reutersvärd or Penrose's impossible figures. For convenience, we shall denote the Penrose triangle with $\alpha\beta = 1, \gamma\kappa = a$ by notation $\Delta(a)$.

Note also that the Penrose triangle $\Delta(a)$ is contained in the polytype shell formed by removing the convex hull (we shall denote it by $\mathfrak{Q}$) of the six points $\{\eta, \theta, \iota, \nu, \tau, \upsilon\}$ from $\mathfrak{P}$, the convex hull of $\{\alpha, \beta, \lambda, o, \phi, \omega\}$. That is,

$$\Delta(a) \subset \mathfrak{P} \setminus \mathfrak{Q} \quad (a \geq 3). \tag{3}$$

For convenience, we shall call the extremals of $\mathfrak{P}$ and $\mathfrak{Q}$ the extremal point of $\Delta(a)$. As we have seen from Fig. 7, the decomposition

$$\Delta(a) = (A_1 \cup A_2 \cup A_3) \cup (\overline{B}_1 \cup \overline{B}_2 \cup \overline{B}_3), \tag{4}$$

where $\overline{B}_i$ $(i = 1, 2, 3)$ are the closure of B_i, shows that $\Delta(a)$ is a polyhedral-complex (cf. [9]). Therefore, if we can isometrically embed the vertices of the cubes A_1, A_2, A_3 and cylinders B_1, B_2, B_3, that is, the 24 points $\alpha, \beta, \gamma, \cdots, \omega \in \Delta(a)$, into any Euclidean space $\mathbb{R}^n$, then we can construct a Penrose triangle configuration in that space, too. Indeed, we can prove that the extremals of $\Delta(a)$ are essential for constructing such isometric embedding. Namely, we have the following result.

Theorem 1. *Assume that $\Delta(a)$ is the Penrose triangle as in Fig. 7, so that $\alpha\beta = 1, \gamma\kappa = a \geq 3$, and*

$$F : \{\alpha, \beta, \eta, \theta; \iota, \lambda, o, \nu; \tau, \upsilon, \phi, \omega\} \to \mathbb{R}^n$$

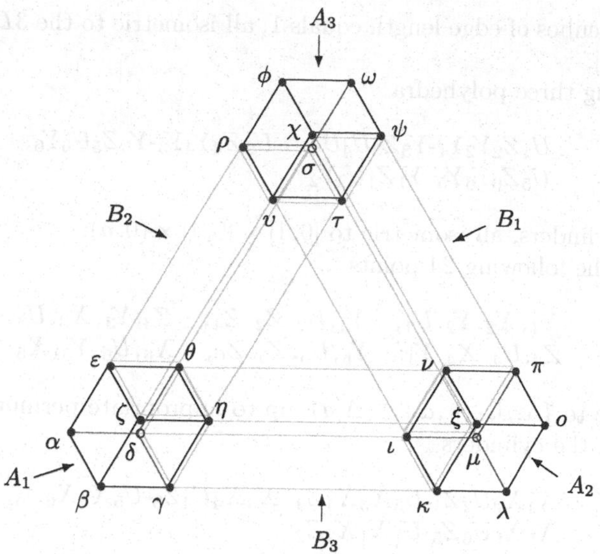

Fig. 7. The Penrose triangle $\Delta(a)$ is decomposed into the disjoint union of three regular cubes A_1, A_2, A_3 and three right (open) cylinders B_1, B_2, B_3.

is any isometric embedding, then F can be extended to an isometric mapping:

$$\hat{F} : \Delta(a) \to \mathbb{R}^n.$$

Proof(outline) Without loss of generality we may assume that

$$\begin{aligned} X_1 &= F(\alpha), X_2 = F(\beta), Y_1 = F(\theta), Y_2 = F(\eta);\\ X_3 &= F(\lambda), X_4 = F(o), Y_3 = F(\iota), Y_4 = F(\nu);\\ X_5 &= F(\omega), X_6 = F(\phi), Y_5 = F(\tau), Y_6 = F(\upsilon). \end{aligned} \tag{5}$$

Let $G, H : \mathbb{R} \times \mathbb{R}^n \to \mathbb{R}^n$ be functions defined by

$$G(V, W) = \frac{a+1}{a+2}V + \frac{1}{a+2}W, H(V, W) = \frac{a}{a+1}V + \frac{1}{a+1}W.$$

Then we can construct points $Y_i, Z_i, U_i (i = 1, 2, \cdots, 6)$ in the space $\mathbb{R}^n$ as follows:

$$\begin{aligned} Z_1 &= G(X_1, X_6), Z_2 = G(X_2, X_3), Z_3 = G(X_3, X_2),\\ Z_4 &= G(X_4, X_5), Z_5 = G(X_5, X_4), Z_6 = G(X_6, X_1), \end{aligned} \tag{6}$$

$$\begin{aligned} U_2 &= H(X_2, Y_6), U_4 = H(X_4, Y_2), U_6 = H(X_6, Y_4),\\ U_1 &= H(X_1, Y_3), U_3 = H(X_3, Y_5), U_5 = H(X_5, Y_1). \end{aligned} \tag{7}$$

Figure 8 shows the generated points. We can verify the following facts:

1. the following three polyhedra

$$X_1X_2Z_2U_1Z_1U_2Y_2Y_1,\ Y_3Z_3X_3U_3Y_4U_4X_4Z_4,$$
$$Z_6U_5Y_5Y_6X_6U_6Z_5X_5$$

are regular cubes of edge length equals 1, all isometric to the $3D$ cube $[0,1] \times [0,1] \times [0,1]$;

2. the following three polyhedra

$$U_1Z_2Y_2Y_1\text{-}Y_3Z_3U_4U_4, \quad U_3Z_4Y_4Y_3\text{-}Y_5Z_5U_6Y_6,$$
$$U_5Z_6Y_6Y_5\text{-}Y_1Z_1U_2Y_2$$

are right cylinders, all isometric to $[0,1] \times [0,1] \times (0,a)$;

3. the set of the following 24 points

$$X_1, X_2, Y_2, U_1, \quad Y_1, U_2, Z_2, Z_1; \quad Z_3, Y_3, X_3, U_3,$$
$$Z_4, U_4, X_4, Y_4; \quad Y_6, U_5, Z_5, Z_6, \quad X_6, U_6, Y_5, X_5$$

is isometric to $\{\alpha, \beta, \cdots, \omega\} \subset \Delta(a)$, up to appropriate permutation;

4. and finally, the cylinders

$$X_1X_2U_2Z_1\text{-}U_3X_3X_4Z_4, \; X_3X_4U_4Z_3\text{-}U_5X_5X_6Z_6,$$
$$X_5X_6U_6Z_5\text{-}U_1X_1X_2Z_2$$

are mutually perpendicular, all isometric to $[0,1] \times [0,1] \times [0, a+2]$. There union $\hat{F}(\Delta(a))$ forms a Penrose triangle (actually, tribar) in $\mathbb{R}^n$. □

In the rest of this section, we prove that the 12 extremal points of the Penrose triangle $\Delta(a)$ can be isometrically embedded in $\mathbb{R}^n$ for $n = 5$, and $n = 5$ is the least dimension for embedding the Penrose triangle into Euclidean space. For this, we need to consider the metric on the extremals of $\Delta(a)$. Let

$$X_{12} := \{\alpha, \beta, \theta, \eta; \lambda, o, \iota, \nu; \omega, \phi, \tau, \upsilon\}$$

be the set of the 12 extremals of $\Delta(a)$. As $\Delta(a)$ can be isometrically immersed (projected) in the three dimensional Euclidean space $\mathbb{R}^3$, in a piecewise way, as depicted in Fig. 6 and Fig. 7, we can define the distance $d(x,y)$ between any two extremal points $x, y \in \Delta(a)$ by

$$d(x,y) := \max_{I \in \Pi} D(I(x), I(y)), \tag{8}$$

here Π is the set of all piecewise isometric immersion of $\Delta(a)$ in $\mathbb{R}^3$, and $D(X,Y)$ is the usual distance between two points X, Y in the three dimensional Euclidean space. It is clear that (X_{12}, d) is a metric space, i.e.,

$$\begin{aligned} &d(x,y) \geq 0, \text{ and } d(x,y) = 0 \text{ if and only if } x = y, \\ &d(x,y) = d(y,x), \\ &d(x,y) + d(y,z) \geq d(x,z), \end{aligned} \tag{9}$$

hold for all $x, y, z \in X_{12}$. Since Π contains all projection from all possible $3D$ components of $\Delta(a)$ to $\mathbb{R}^3$, we have the following inequality:

$$d(x,y) \geq \max_{P \in \Pi(x,y)} D(P(x), P(y)) \tag{10}$$

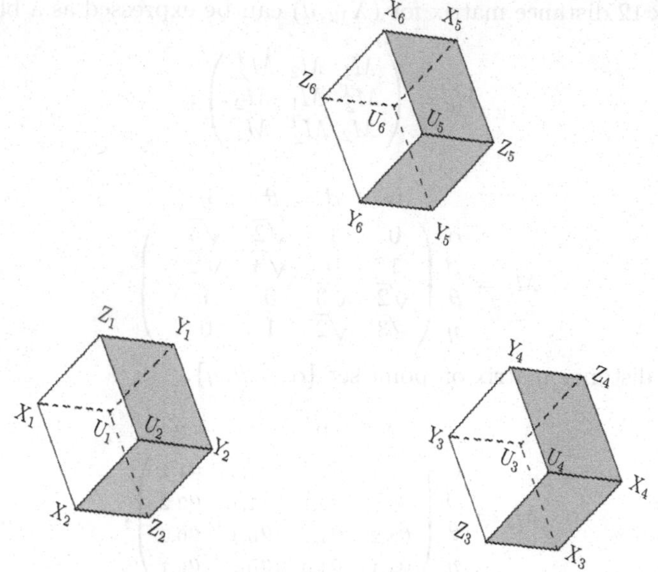

Fig. 8. $X_1, X_2, \cdots, X_6; Y_1, Y_2, \cdots, Y_6$ are the images of F, an isometric map from the 12 extremals of the Penrose triangle $\Delta(a)$ to $\mathbb{R}^n$, and three cubes are generated from $X_i, Y_i (1 \leq i \leq 6)$ and functions G, H.

here $\Pi(x, y)$ is the set of all mapping that projects a $3D$ polyhedral component of $\Delta(a)$ which contains x, y to $\mathbb{R}^3$.

Taking an example for $x = \alpha, y = \beta$, it is easy to see that any projection $P : \Delta(a) \to \mathbb{R}^3$ we have $D(P(\alpha), P(\beta)) \leq \alpha\beta = 1$, and there is also a projection that project the component $A_1 \subset \Delta(a)$ to

$$P(\alpha) = (0, 0, 0), P(\beta) = (1, 0, 0),$$

thus $d(\alpha, \beta) \geq D(P(\alpha), P(\beta)) = 1$. For points $x = \alpha, y = o$, we see that $\alpha, o \in C_1 = A_1 \cup B_3 \cup A_2$, and a point initially at position $\alpha \in \Delta(a)$ can move to position $o \in \Delta(a)$ as follows: first along $\boldsymbol{u} = \alpha\beta$ to the position $\beta \in \Delta(a)$, then turn 90° on the place $\alpha\beta\gamma$, continue to move along the line $\boldsymbol{v} = \beta\gamma\kappa\lambda$ to the position $\lambda \in \Delta(a)$, then turn 90° on the plane $\kappa\lambda o$, and move to position $o \in \Delta(a)$ finally. Apply the following Pythagoras Theorem we can compute the distance between α and o is $\sqrt{a^2 + 4a + 6}$.

Theorem 2. *Assume that a point X started to move along the direction $\boldsymbol{u}$ for a straight distance a, then move along the direction $\boldsymbol{v}$ for a straight distance b, and so on, and move along the direction $\boldsymbol{w}$ for a distance c, and finally arrived the point Y. Assume that $\boldsymbol{u}, \boldsymbol{v}, \cdots, \boldsymbol{w}$ are pair-wisely perpendicular to each other, Then the straight length between points X and Y is $\sqrt{a^2 + b^2 + \cdots + c^2}$.*

The 12×12 distance matrix for (X_{12}, d) can be expressed as a block matrix as follows:

$$M_d = \begin{pmatrix} M_1 & M_2 & M_2^T \\ M_2^T & M_1 & M_2 \\ M_2 & M_2^T & M_1 \end{pmatrix}, \tag{11}$$

where

$$M_1 = \begin{matrix} \\ \alpha \\ \beta \\ \theta \\ \eta \end{matrix} \begin{matrix} \begin{matrix} \alpha & \beta & \theta & \eta \end{matrix} \\ \begin{pmatrix} 0 & 1 & \sqrt{2} & \sqrt{3} \\ 1 & 0 & \sqrt{3} & \sqrt{2} \\ \sqrt{2} & \sqrt{3} & 0 & 1 \\ \sqrt{3} & \sqrt{2} & 1 & 0 \end{pmatrix} \end{matrix}, \tag{12}$$

is the 4×4 distance matrix on point set $\{\alpha, \beta, \theta, \eta\}$,

$$M_2 = \begin{matrix} \\ \alpha \\ \beta \\ \theta \\ \eta \end{matrix} \begin{matrix} \begin{matrix} \lambda & o & \iota & \nu \end{matrix} \\ \begin{pmatrix} a_{2,1} & a_{2,2} & a_{1,0} & a_{1,1} \\ a_{2,0} & a_{2,1} & a_{1,1} & a_{2,2} \\ a_{1,2} & a_{1,1} & a_{0,1} & a_{0,0} \\ a_{1,1} & a_{1,0} & a_{0,2} & a_{0,1} \end{pmatrix} \end{matrix}, \tag{13}$$

and

$$a_{i,j} := \sqrt{(a+i)^2 + j} \ (i, j = 0, 1, 2)$$

for shorter.

Isometric embeddability in the Euclidean space has been well understood since the classical works of Menger, von Neumann, Schoenberg, and others (see, e.g., [10–13]). Given a set of finite points $X = \{p_0, p_1, \cdots, p_N\}$, and a metric $d : X \times X \to \mathbb{R}_{\geq 0}$, the problem of isometric embedding (X, d) in the Euclidean space $\mathbb{R}^n$ can be characterized by the Cayley-Menger determinant of X (and its subset). For (X, d), let $d_{i,j} = d(p_i, p_j)$ for $i, j = 0, 1, \cdots, N$. The *Cayley-Menger determinant* is defined by

$$D(X) := \det \begin{pmatrix} 0 & 1 & 1 & 1 & \cdots & 1 \\ 1 & 0 & d_{0,1}^2 & d_{0,2}^2 & \cdots & d_{0,N}^2 \\ 1 & d_{1,0}^2 & 0 & d_{1,2}^2 & \cdots & d_{1,N}^2 \\ 1 & d_{2,0}^2 & d_{2,1}^2 & 0 & \cdots & d_{2,N}^2 \\ \vdots & \vdots & \vdots & \vdots & \ddots & \vdots \\ 1 & d_{N,0}^2 & d_{N,1}^2 & d_{N,2}^2 & \cdots & 0 \end{pmatrix},$$

Following result can be found in [11].

Theorem 3. *(Blumenthal, [11]) A finite metric space (X, d) is Euclidean with dimension n if and only if there are $p_0, p_1, \cdots, p_n \in X$ such that*

$$\begin{aligned} &\text{(i) } (-1)^{j+1} D(p_0, \ldots, p_j) > 0 \ \text{ for } 1 \leq j \leq n, \text{ and} \\ &\text{(ii) } D(p_0, \cdots, p_n, x) = D(p_0, \cdots, p_n, y) \\ &\quad = D(p_0, \cdots, p_n, x, y) = 0 \ \text{ for all } x, y \in X. \end{aligned} \tag{14}$$

Applying the above theorem, we can search maximal subset $p_0, p_1, \cdots, p_n$ of X_{12} that satisfies the conditions (i) and (ii) of the Theorem 3. As $\#X_{12} = 12$ and X_{12} has only $2^{12} = 4096$ subsets, we wrote a MAPLE program to find all subsets of X_{12} which satisfy (i) and (ii) simultaneously. The first step is to generate all 4096 subsets of X_{12} by using $\texttt{powerset}(X_{12})$ in MAPLE package `combinat`, and the second step is a loop to check if a subset $S \in \texttt{powerset}(X_{12})$ satisfies (i) and (ii). It is clear that if a subset S_1 failed in checking, and $S_1 \subset S_2$, then S_2 is void to check. Since the computation is very straightforward, we omit to print the MAPLE program here for saving space. The searching result is that there are altogether 64 different subsets satisfying the required conditions, and each of them contains 6 points. Actually, we have used `Maple` software running on a notebook computer with Intel(R) Core(TM) i7 CPU and 8 GB and found that the 64 subsets of $p_0, p_1, \cdots, p_5$ that satisfy the conditions (i) and (ii) of Theorem 3 can be represented as members of the Cartesian product

$$\{\alpha, \theta\} \times \{\beta, \eta\} \times \{\lambda, \iota\} \times \{o, \nu\} \times \{\omega, \tau\} \times \{\phi, \upsilon\}.$$

Applying MAPLE computation, we can prove the isometric embeddability of X_{12} in the five-dimensional Euclidean space $\mathbb{R}^5$ as follows.

Theorem 4. *The point set X_{12} with distance matrix M_d given by* (11) *can be isometrically embedded into $\mathbb{R}^n$ for $n \geq 5$, and $n = 5$ is the least dimension for the isometric embedding.*

Proof. Here we prove this theorem by a constructive method. Consider points $\alpha, \beta, \lambda, o, \omega, \phi \in X_{12}$. Then for $D(\alpha, \beta)$, $-D(\alpha, \beta, \lambda)$ and $D(\alpha, \beta, \lambda, o)$ we have:

$$+\det \begin{array}{c} \begin{array}{ccc} & \alpha & \beta \end{array} \\ \begin{pmatrix} 0 & 1 & 1 \\ 1 & 0 & 1 \\ 1 & 1 & 0 \end{pmatrix} \end{array} \begin{array}{l} \\ \\ \alpha \\ \beta \end{array} = 2 > 0, \tag{15}$$

$$-\det \begin{array}{c} \begin{array}{cccc} & \alpha & \beta & \lambda \end{array} \\ \begin{pmatrix} 0 & 1 & 1 & 1 \\ 1 & 0 & 1 & (a+2)^2+1 \\ 1 & 1 & 0 & (a+2)^2 \\ 1 & (a+2)^2+1 & (a+2)^2 & 0 \end{pmatrix} \end{array} \begin{array}{l} \\ \\ \alpha \\ \beta \\ \lambda \end{array} = 4(a+2)^2 > 0, \tag{16}$$

$$+\det \begin{pmatrix} 0 & 1 & 1 & 1 & 1 \\ 1 & 0 & 1 & (a+2)^2+1 & (a+2)^2+2 \\ 1 & 1 & 0 & (a+2)^2 & (a+2)^2+1 \\ 1 & (a+2)^2+1 & (a+2)^2 & 0 & 1 \\ 1 & (a+2)^2+2 & (a+2)^2+1 & 1 & 0 \end{pmatrix} = 8(a+2)^2 > 0, \tag{17}$$

For $-D(\alpha,\beta,\lambda,o,\omega)$ and $D(\alpha,\beta,\lambda,o,\omega,\phi)$, using notation

$$a_{2,0}=a+2, a_{2,1}=\sqrt{(a+2)^2+1}, a_{2,2}=\sqrt{(a+2)^2+2}.$$

for better print quality. Then applying the built-in MAPLE commands `det` for computing determinants and `factor` for factorizing polynomials, we have

$$-\det\begin{pmatrix} 0 & 1 & 1 & 1 & 1 & 1 \\ 1 & 0 & 1 & a_{2,1}^2 & a_{2,2}^2 & a_{2,1}^2 \\ 1 & 1 & 0 & a_{2,0}^2 & a_{2,1}^2 & a_{2,2}^2 \\ 1 & a_{2,1}^2 & a_{2,0}^2 & 0 & 1 & a_{2,1}^2 \\ 1 & a_{2,2}^2 & a_{2,1}^2 & 1 & 0 & a_{2,0}^2 \\ 1 & a_{2,1}^2 & a_{2,2}^2 & a_{2,1}^2 & a_{2,0}^2 & 0 \end{pmatrix} = 4\,(a+3)\,(a+1)\left(3\,a^2+12\,a+13\right) > 0, \tag{18}$$

and

$$+\det\begin{pmatrix} 0 & 1 & 1 & 1 & 1 & 1 & 1 \\ 1 & 0 & 1 & a_{2,1}^2 & a_{2,2}^2 & a_{2,1}^2 & a_{2,0}^2 \\ 1 & 1 & 0 & a_{2,0}^2 & a_{2,1}^2 & a_{2,2}^2 & a_{2,1}^2 \\ 1 & a_{2,1}^2 & a_{2,0}^2 & 0 & 1 & a_{2,1}^2 & a_{2,2}^2 \\ 1 & a_{2,2}^2 & a_{2,1}^2 & 1 & 0 & a_{2,0}^2 & a_{2,1}^2 \\ 1 & a_{2,1}^2 & a_{2,2}^2 & a_{2,1}^2 & a_{2,0}^2 & 0 & 1 \\ 1 & a_{2,0}^2 & a_{2,1}^2 & a_{2,2}^2 & a_{2,1}^2 & 1 & 0 \end{pmatrix} = 24\,(a+3)^2\,(a+1)^2 > 0. \tag{19}$$

Do this computation with MAPLE further, it is easily verified that

$$D(\alpha,\beta,\lambda,o,\omega,\phi,x)=0, \quad D(\alpha,\beta,\lambda,o,\omega,\phi,x,y)=0, \tag{20}$$

hold for all $x,y\in\{\eta,\theta,\iota,\nu,\tau,\upsilon\}$. Indeed, according to symmetry, we need only to check this for $x\in\{\eta,\theta\}, y\in\{\iota,\nu\}$. Combine (15) to (20) and Theorem 3, we proved that X_{12} can be isometrically embedded into $\mathbb{R}^5$. This implies also that the Penrose triangle cannot be isometrically embedded into $\mathbb{R}^4$ as people generally believed. □

3 Solving Embedding Equations Using Symbolic Computation

Theorem 4 confirms the existence of an isometric embedding of (the 12 extremal points on) the Penrose triangle in space $\mathbb{R}^5$. One may use the general method given by Blumenthal [11] or Lu Yang and Jingzhong Zhang [14] to create an explicit representation of the embedding, i.e., the concrete coordinates $(x_i, y_i, z_i, u_i, v_i) \in \mathbb{R}^5$ $(i = 1, 2, \cdots, 12)$ as the images of

$$X_{12} = \{\alpha, \beta, \eta, \theta, \iota, \lambda, \nu, o, \tau, \upsilon, \varphi, \omega\}$$

in $\mathbb{R}^5$ under the isometric embedding. Due to the symmetry of the Penrose triangle, we can also find a solution of the embedding equations using symbolic computation. We will show this in this section. Suppose that the Penrose triangle has been realized in the five-dimensional Euclidean space, then the projection of the configuration in the $\mathbb{R}^n$ along any direction (an oriented straight line) into the three-dimensional space would be the three cubes (that is a rigid movement of $[0,1] \times [0,1] \times [0,1] \subset \mathbb{R}^4$) connected by three bars (that is isometric to the cylinder $[0,1] \times [0,1] \times [0,a]$. As depicted in Fig. 7, we may assume the cubes are

$$\Delta := \alpha\beta\gamma\delta\epsilon\varepsilon\zeta\eta, \quad \Pi := \iota\kappa\lambda\mu\nu\xi o\pi, \quad \Sigma := \rho\upsilon\tau\sigma\phi\chi\psi\omega.$$

With a unitary orthogonal transform, we may change that the vertex δ on the first cube Δ is lying on the origin $(0,0,0,0,0)$, and the coordinates of other vertices are

$$\begin{aligned} &\alpha = (1,0,0,0,0), \quad \beta = (1,1,0,0,0), \quad \gamma = (0,1,0,0,0), \\ &\varepsilon = (1,0,1,0,0), \quad \zeta = (1,1,1,0,0), \\ &\eta = (0,1,1,0,0), \quad \theta = (0,0,1,0,0). \end{aligned} \tag{21}$$

As shown in Fig. 9, the original design of Reutersvärd, the cube Π can move along a straight line ℓ_s (in the appropriate space, here, we assume it is a line in the $\mathbb{R}^5$) to Δ, so that Π coincides Δ with

$$\begin{aligned} &\rho \to \alpha, \quad \upsilon \to \beta, \quad \tau \to \gamma, \quad \sigma \to \delta, \\ &\phi \to \varepsilon, \quad \chi \to \zeta, \quad \psi \to \eta, \quad \omega \to \theta, \end{aligned} \tag{22}$$

after movement, and the other cube Π can be moved to Δ along a line ℓ_p so that

$$\begin{aligned} &\iota \to \alpha, \quad \kappa \to \beta, \quad \lambda \to \gamma, \quad \mu \to \delta, \\ &\nu \to \varepsilon, \quad \xi \to \zeta, \quad o \to \eta, \quad \pi \to \theta. \end{aligned} \tag{23}$$

Without loss of generality, we may assume that line ℓ_s and line ℓ_p are lying in the plane

$$\{(x, y, z, u, v) | x = 0, y = 0, z = 0\} \subset \mathbb{R}^5,$$

and the coordinates of vertices $\delta \in \Delta, \mu \in \Sigma, \sigma \in \Pi$ are

$$\delta = (0,0,0,u_1,v_1), \mu = (0,0,0,u_2,v_2), \sigma = (0,0,0,u_3,v_3), \tag{24}$$

respectively. We can take $u_1 = 0, v_1 = 0$ as in (21), here we use this form just for symmetry. Therefore, the coordinates of all other 21 vertices of Δ, Π, Σ can be determined by (21) and (24). In particular, we have

$$\begin{aligned} &\alpha = (1,0,0,u_1,v_1), \beta = (1,1,0,u_1,v_1), \lambda = (0,1,0,u_2,v_2), \\ &o = (0,1,1,u_2,v_2), \omega = (0,0,1,u_3,v_3), \phi = (1,0,1,u_3,v_3). \end{aligned} \tag{25}$$

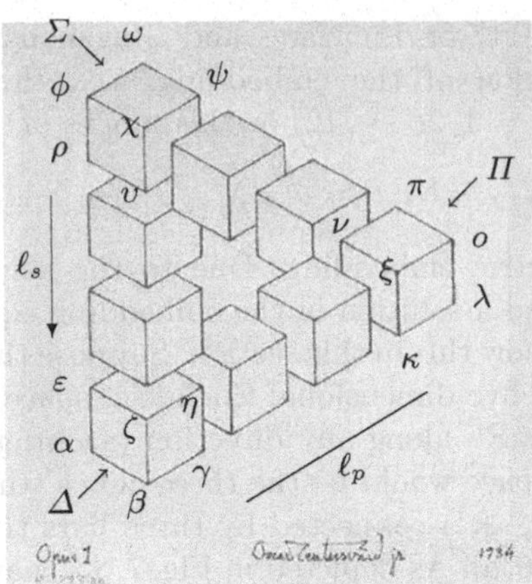

Fig. 9. ℓ_s, ℓ_p are two lines in $\mathbb{R}^5$ so that the cube Σ can be moved to coincide Δ in parallel to ℓ_s, and the cube can be moved to Δ in parallel to ℓ_p. Some vertices of the cubes Δ, Π, Σ are not marked in the picture. When viewing Δ, Π, Σ as necker cubes, take ζ, ξ, χ to the front-most positions. We may assume that ℓ_s is the line determined by points $(0,0,0,0,0)$ and $(0,0,0,u_3,v_3)$, and ℓ_p is the line determined by $(0,0,0,0,0)$ and $(0,0,0,u_2,v_2)$, here u_2, u_2, u_3, v_3 are numbers in (27).

Using the coordinates to compute the distance $\alpha\beta, \alpha\lambda, \alpha o, \alpha\omega, \alpha\phi$, we establish the following system of equations:

$$\begin{aligned} &1+(u_1-u_2)^2+(v_1-v_2)^2-(a+2)^2=0,\\ &1+(u_1-u_3)^2+(v_1-v_3)^2-(a+2)^2=0,\\ &1+(u_2-u_3)^2+(v_2-v_3)^2-(a+2)^2=0,\\ &u_1=0, v_1=0. \end{aligned} \tag{26}$$

Using MAPLE software it is easy to solve this system of equations. It is clear that (26) has infinitely many solutions, from each solution we can construct an equilateral triangle ABC of edge $\sqrt{a^2+4a+3}$ in the plane by taking $A = (0,0), B = (u_2, v_2), C = (u_3, v_3)$. For our purpose we need only one real solution, so we take the following symmetric one:

$$u_2 = u_3 = \frac{\sqrt{3}}{2}\sqrt{a^2+4a+3}, v_2 = -v_3 = \frac{1}{2}\sqrt{a^2+4a+3}, \tag{27}$$

and therefore, we can write the coordinates of the 24 vertices of the cube Δ, Π, Σ as follows:

$$\begin{aligned} &\alpha=(1,0,0,0,0), && \iota=(1,0,0,\sqrt{3}b/2,b/2), && \rho=(1,0,0,\sqrt{3}b/2,-b/2),\\ &\beta=(1,1,0,0,0), && \kappa=(1,1,0,\sqrt{3}b/2,b/2), && \upsilon=(1,1,0,\sqrt{3}b/2,-b/2),\\ &\gamma=(0,1,0,0,0), && \lambda=(0,1,0,\sqrt{3}b/2,b/2), && \tau=(0,1,0,\sqrt{3}b/2,-b/2),\\ &\delta=(0,0,0,0,0), && \mu=(0,0,0,\sqrt{3}b/2,b/2), && \sigma=(0,0,0,\sqrt{3}b/2,-b/2), \end{aligned}$$

$$\begin{aligned}
&\varepsilon = (1,0,1,0,0), \quad &&\nu = (1,0,1,\sqrt{3}b/2,b/2), \quad &&\phi = (1,0,1,\sqrt{3}b/2,-b/2),\\
&\zeta = (1,1,1,0,0), \quad &&\xi = (1,1,1,\sqrt{3}b/2,b/2), \quad &&\chi = (1,1,1,\sqrt{3}b/2,-b/2),\\
&\eta = (0,1,1,0,0), \quad &&o = (0,1,1,\sqrt{3}b/2,b/2), \quad &&\psi = (0,1,1,\sqrt{3}b/2,-b/2),\\
&\theta = (0,0,1,0,0), \quad &&\pi = (0,0,1,\sqrt{3}b/2,b/2), \quad &&\omega = (0,0,1,\sqrt{3}b/2,-b/2).
\end{aligned} \tag{28}$$

here $b = \sqrt{a^2+4a+3}$.
The distance matrix of the above 24 points is as follows:

$$DM_{24} = \begin{pmatrix} D & M & M \\ M^T & D & M \\ M^T & M^T & D \end{pmatrix}, \tag{29}$$

where

$$D = \begin{pmatrix}
0 & 1 & \sqrt{2} & 1 & 1 & \sqrt{2} & \sqrt{3} & \sqrt{2} \\
1 & 0 & 1 & \sqrt{2} & \sqrt{2} & 1 & \sqrt{2} & \sqrt{3} \\
\sqrt{2} & 1 & 0 & 1 & \sqrt{3} & \sqrt{2} & 1 & \sqrt{2} \\
1 & \sqrt{2} & 1 & 0 & \sqrt{2} & \sqrt{3} & \sqrt{2} & 1 \\
1 & \sqrt{2} & \sqrt{3} & \sqrt{2} & 0 & 1 & \sqrt{2} & 1 \\
\sqrt{2} & 1 & \sqrt{2} & \sqrt{3} & 1 & 0 & 1 & \sqrt{2} \\
\sqrt{3} & \sqrt{2} & 1 & \sqrt{2} & \sqrt{2} & 1 & 0 & 1 \\
\sqrt{2} & \sqrt{3} & \sqrt{2} & 1 & 1 & \sqrt{2} & 1 & 0
\end{pmatrix},$$

and

$$M = \begin{pmatrix}
b & \sqrt{b^2+1} & \sqrt{b^2+2} & \sqrt{b^2+1} & \sqrt{b^2+1} & \sqrt{b^2+2} & \sqrt{b^2+3} & \sqrt{b^2+2} \\
\sqrt{b^2+1} & b & \sqrt{b^2+1} & \sqrt{b^2+2} & \sqrt{b^2+2} & \sqrt{b^2+1} & \sqrt{b^2+2} & \sqrt{b^2+3} \\
\sqrt{b^2+2} & \sqrt{b^2+1} & b & \sqrt{b^2+1} & \sqrt{b^2+3} & \sqrt{b^2+2} & \sqrt{b^2+1} & \sqrt{b^2+2} \\
\sqrt{b^2+1} & \sqrt{b^2+2} & \sqrt{b^2+1} & b & \sqrt{b^2+2} & \sqrt{b^2+3} & \sqrt{b^2+2} & \sqrt{b^2+1} \\
\sqrt{b^2+1} & \sqrt{b^2+2} & \sqrt{b^2+3} & \sqrt{b^2+2} & b & \sqrt{b^2+1} & \sqrt{b^2+2} & \sqrt{b^2+1} \\
\sqrt{b^2+2} & \sqrt{b^2+1} & \sqrt{b^2+2} & \sqrt{b^2+3} & \sqrt{b^2+1} & b & \sqrt{b^2+1} & \sqrt{b^2+2} \\
\sqrt{b^2+3} & \sqrt{b^2+2} & \sqrt{b^2+1} & \sqrt{b^2+2} & \sqrt{b^2+2} & \sqrt{b^2+1} & b & \sqrt{b^2+1} \\
\sqrt{b^2+2} & \sqrt{b^2+3} & \sqrt{b^2+2} & \sqrt{b^2+1} & \sqrt{b^2+1} & \sqrt{b^2+2} & \sqrt{b^2+1} & b
\end{pmatrix}.$$

We can use MAPLE to verify the following inequalities

$$D(\alpha,\beta) = -1 < 0,$$
$$D(\alpha,\beta,\lambda) = 2\left(b^2+2\right)\left(b^2+1\right) > 0,$$

$$D(\alpha,\beta,\lambda,o) = -4\,b^4 - 16\,b^2 - 12 < 0,$$
$$D(\alpha,\beta,\lambda,o,\omega) = 8\,b^6 + 28\,b^4 + 24\,b^2 > 0,$$
$$D(\alpha,\beta,\lambda,o,\omega,\phi) = -16\,b^6 - 36\,b^4 < 0, \tag{30}$$

and that the equality

$$D(\alpha,\beta,\lambda,o,\omega,\phi,x) = D(\alpha,\beta,\lambda,o,\omega,\phi,x) = D(\alpha,\beta,\lambda,o,\omega,\phi,x,y) = 0,$$

holds for any x, y from all other 18 points given in (28). Note that the inequality (30) implies that the 24 points can not be embedded into $\mathbb{R}^n$ for $n < 5$.

Remark 1. Let *proj* : $\mathbb{R}^5 \to \mathbb{R}^3$ be the projection defined by the following matrix:

$$proj := \begin{pmatrix} 1 & 0 & 0 \\ 0 & 1 & 0 \\ 0 & 0 & 1 \\ 0 & \sqrt{3}/3 & \sqrt{3}/3 \\ 0 & 1 & -1 \end{pmatrix}. \tag{31}$$

Then, the eight points $\alpha,\beta,\gamma,\delta,\varepsilon,\zeta,\eta,\theta$, the vertices of $\Delta \subset \mathbb{R}^5$, are mapped to the following points in $\mathbb{R}^3$:

$$[1,0,0],\ [1,1,0],\ [0,1,0],\ [0,0,0],\ [1,0,1],\ [1,1,1],\ [0,1,1],\ [0,0,1],$$

respectively, clearly they are vertices of a cube (say, Δ') in $\mathbb{R}^3$; the eight points $\iota,\kappa,\lambda,\mu,\nu,\xi,o,\pi$ (i.e., the vertices of $\Pi \subset \mathbb{R}^5$) are mapped to the vertices of a three-dimensional cube (say, Π'):

$$[1,b,0],\ [1,1{+}b,0],\ [0,1{+}b,0],\ [0,b,0],\ [1,b,1],\ [1,1{+}b,1],\ [0,1{+}b,1],\ [0,b,1],$$

respectively, and the eight points $\rho,\upsilon,\tau,\sigma,\phi,\chi,\psi,\omega$ (the vertices of $\Sigma \subset \mathbb{R}^5$) are mapped to the vertices of a three-dimensional cube (say, Σ'):

$$[1,0,b],\ [1,1,b],\ [0,1,b], [0,0,b],\ [1,0,1{+}b],\ [1,1,1{+}b],\ [0,1,1{+}b],\ [0,0,1{+}b],$$

respectively. It is also clear that in the space $\mathbb{R}^3$, we can move Δ' to Π' by $L_p : (x,y,z) \longrightarrow (x, y+b, z)$, and move the cube Δ' to Σ' by $L_s := (x,y,z) \longrightarrow (x,y,z+b)$.

Therefore, what we have seen from the impossible in the three-dimensional space, can be understood as an shadow of a five-dimensional geometric object in the three-dimensional world, under the projection defined by (31).

4 Epilogue

We have proved, the Penrose triangle has an isometric embedding in lowest dimension Euclidean space $\mathbb{R}^5$, as a subset of $\mathfrak{P} \setminus \mathfrak{Q}$, the difference set of two simplexes, where $\mathfrak{P}$ is formed by $\alpha, \beta, \lambda, o, \omega, \phi$, and $\mathfrak{Q}$ formed by $\theta, \eta, \iota, \nu, \tau, \upsilon$. Figure 10 gives an intuitive explanation of this fact.

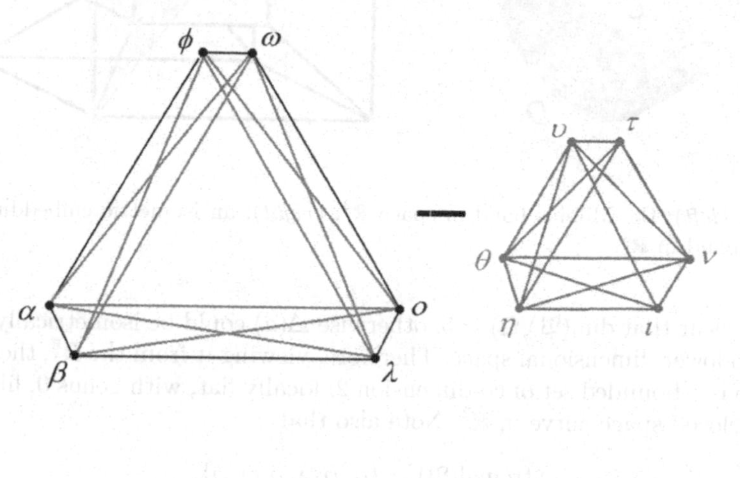

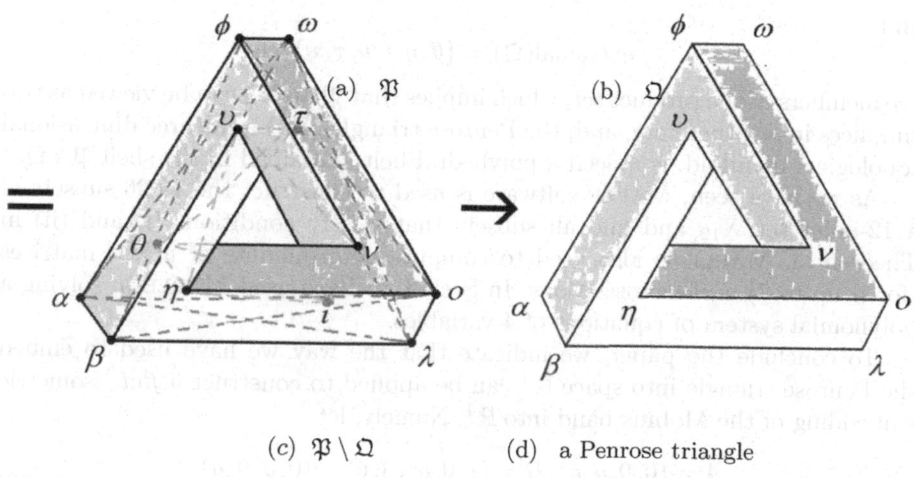

Fig. 10. (a): $\mathfrak{P}$: the convex hull of the Penrose triangle, also a simplex formed by $\alpha, \beta, \lambda, o, \omega, \phi$ in the space $\mathbb{R}^5$; (b): $\mathfrak{Q}$, the simplex formed by points $\theta, \eta, \iota, \nu, \tau, \upsilon$ in $\mathbb{R}^5$; (c): $\Delta(a) \subset \mathfrak{P} \setminus \mathfrak{Q}$; (d): the piecewise isometric immersion of the Penrose triangle in the three dimensional space as we see.

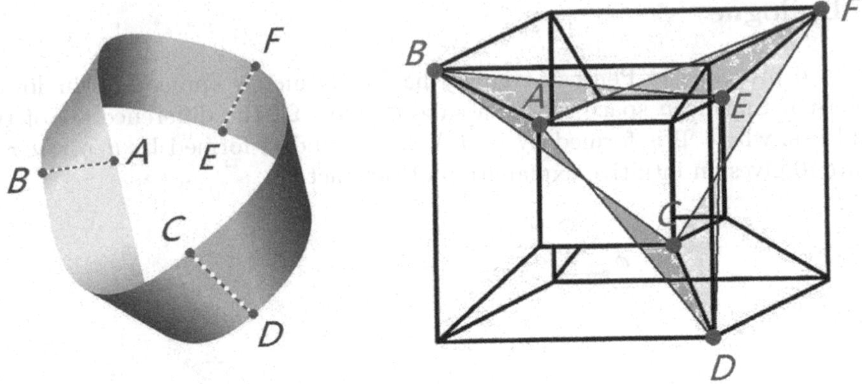

Fig. 11. (*left*): the Möbius band in space $\mathbb{R}^3$; (*right*): an isometric embedding of the Möbius band in $\mathbb{R}^4$.

It is clear that $\dim(\mathfrak{P}\setminus\mathfrak{Q}) = 5$, otherwise $\Delta(a)$ could be isometrically embedded into lower dimensional space. Therefore, viewing it from the $\mathbb{R}^5$, the Penrose triangle is a bounded set of co-dimension 2, locally flat, with genus 0, like a non-planar closed space curve in $\mathbb{R}^3$. Note also that

$$\text{extremal}(\mathfrak{P}) = \{\alpha, \beta, \lambda, o, \omega, \phi\},$$

and

$$\text{extremal}(\mathfrak{Q}) = \{\theta, \eta, \iota, \nu, \tau, \upsilon\}$$

are members of the product set, which implies that $\mathfrak{P}$ and $\mathfrak{Q}$ can be viewed as two simplices in $\mathbb{R}^5$, therefore, and, the Penrose triangle $\Delta(a)$ as a three dimensional topological manifold, is indeed a polyhedral belt contained in the shell $\mathfrak{P}\setminus\mathfrak{Q}$.

As we have seen, MAPLE software is used to construct the 4,096 subsets of a 12-point set X_{12} and find all subsets that satisfy conditions (i) and (ii) in Theorem 3. MAPLE is also used to compute determinants of $n \times n$ matrices (for n up to 7) and factorizations. In Sect. 3 we have used MAPLE in solving a polynomial system of equations of 4 variables.

To conclude the paper, we indicate that the way we have used to embed the Penrose triangle into space $\mathbb{R}^n$ can be applied to construct a *flat* isometric embedding of the Möbius band into $\mathbb{R}^4$. Namely, let

$$\begin{aligned} A = (0,0,a,a), B = (1,0,a,a), C = (0,a,0,a),\\ D = (1,a,0,a), E = (0,a,a,0), F = (1,a,a,0), \end{aligned} \tag{32}$$

then, the rectangles $ABCD, CDEF, EFAB$ in the 4-dim space form a Möbius band in $\mathbb{R}^4$ so that every interior point of the rectangles has a flat neighborhood, as shown in Fig. 11. See [15] for more works on isometric embeddings and immersions of Möbius bands. We wonder if a similar method can be applied to construct an isometric embedding of the impossible cube (Fig. 2(left)) into $\mathbb{R}^n$ so that the embedded cube produces a weird view from three-dimensional space.

Acknowledgment. We are grateful to the anonymous reviewers for their insightful comments and valuable improvements to our paper.

References

1. Wikipedia: Oscar Reutersvärd. https://en.wikipedia.org/wiki/Oscar_Reutersv%C3%A4rd. Accessed 11 Jan 2020
2. Penrose, L.S., Penrose, R.: Impossible objects: a special type of visual illusion. Br. J. Psychol. **49**(1), 1–88 (1958)
3. Penrose, R.: The Art of the Impossible: M.C. Escher and Me. Escher in het Paleis. https://www.escherinhetpaleis.nl/escher-today/the-roger-penrose-puzzle/?lang=en. Accessed 11 Jan 2020
4. Anonymous: Could a penrose triangle be constructed in higher dimensions? https://www.reddit.com/r/math/comments/8une0j/topology_could_a_penrose_triangle_be_constructed/. Accessed 29 Dec 2019
5. Blue Sam: Would this be mathematically possible if we had a fourth spacial dimension. https://www.reddit.com/r/math/comments/98nege/. Accessed 29 Dec 2019
6. Alexeev, V.: Hypercube. https://im-possible.info/english/articles/hypercube/. Accessed 29 Dec 2019
7. Danesi, M.: The Puzzle Instinct: The Meaning of Puzzles in Human Life, pp. 82–83. Indiana University Press, Bloomington (2004)
8. Weisstein, E.W.: Möbius Strip. From MathWorld-A Wolfram Web Resource. http://mathworld.wolfram.com/MoebiusStrip.html. Accessed 11 Jan 2020
9. Aleksandrov, P.S.: Combinatorial Topology. Graylock, Rochester (1956). (Translated from Russian)
10. Schoenberg, I.J.: On certain metric spaces arising from Euclidean space by a change of metric and their embedding in Hilbert space. Ann. Math. **38**(2), 787–793 (1937)
11. Blumenthal, L.M.: Theory and Applications of Distance Geometry. Clarendon Press, Oxford (1953)
12. Maehara, H.: Euclidean embeddings of finite metric spaces. Discrete Math. **313**(23), 2848–2856 (2013)
13. Bowers, J.C., Bowers, P.L., Bowers, J.C., Bowers, P.L.: A menger redux: embedding metric spaces isometrically in Euclidean space. Am. Math. Mon. **124**(7), 621–636 (2017)
14. Yang, L., Zhang, J.Z.: The rank concept of abstract metric space. J. Chin. Univ. Sci. Technol. **10**(4), 52–65 (1980)
15. Sabitov, I.K.: Isometric immersions and embeddings of a flat Möbius strip in Euclidean spaces. Izv. Math. **71**(5), 1049 (2007)

Errata for Maple Conference 2019 Proceedings

"A Poly-algorithmic Quantifier Elimination Package in Maple" Erratum

Zak Tonks(✉)

University of Bath, Bath, England
z.p.tonks@bath.ac.uk

In "A Poly-algorithmic Quantifier Elimination Package in Maple" in the proceedings of Maple Conference 2019, it is claimed that one can distribute quantifiers from $Q_{n-m+1}x_{n-m+1}, \ldots, Q_i x_i$ for some $n-m+1 \leq i \leq n$ into a boolean disjunction or conjunction B in order to propagate Virtual Term Substitution (VTS). However, it would be an error to distribute an existential quantifier into a conjunction, or similarly distribute a universal quantifier into a disjunction in order to achieve quantifier elimination (QE). One notes that $\forall x\,(A(x) \vee B(x)) \Rightarrow (\forall x\,A(x) \vee \forall x\,B(x))$, but the two expressions are not equivalent, and in particular the latter does not imply the former, which is what would be required for the distributivity. Similarly $\exists x\,(A(x) \wedge B(x)) \not\Leftarrow (\exists x\,A(x) \wedge \exists x\,B(x))$. Hence the tree structure for VTS suggested in the paper is valid within any one block of quantifiers, i.e. those quantifiers that share the same quantifier symbol, but the quantifier free equivalent of tree must be collapsed to one QE problem for a next subsequent block. This is as elimination of an existential quantifier in VTS canonically forms a disjunction of the results of virtual substitution, where further distributivity is allowable (and similar for a universal quantifier). Therefore what is Fig. 2 from that work should be replaced by a "layered tree"—demonstrated in Fig. 1. The poly-algorithm discussed in the work is valid within any one block of quantifiers, so Fig. 2 from that work is perfectly canonical if the quantifiers are homogeneous, i.e. share the same quantifier symbol $Q \in \{\forall, \exists\}$. In fact, the package now uses the poly-algorithm within the last block of quantifiers having used solely VTS for elimination beforehand, so the sentiment of the work remains for the package `QuantifierElimination`, under these slightly more restrictive circumstances.

R. M. Corless et al. (Eds.): MC 2020, CCIS 1414, pp. 461–462, 2021.
https://doi.org/10.1007/978-3-030-81698-8

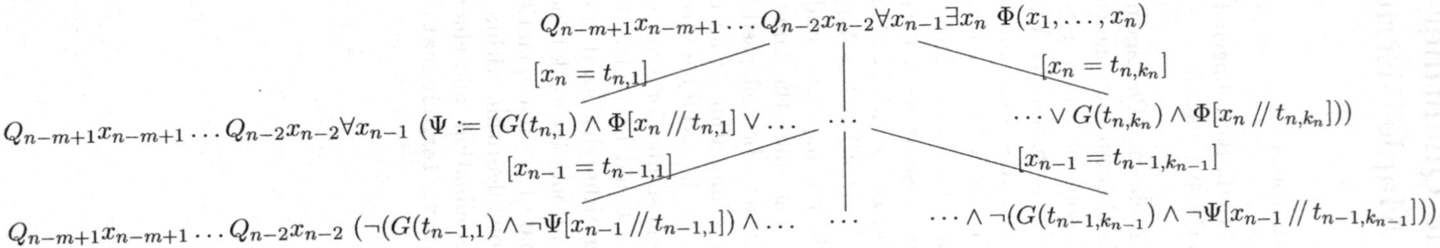

$$Q_{n-m+1}x_{n-m+1}\cdots Q_{n-2}x_{n-2}\forall x_{n-1}\exists x_n\ \Phi(x_1,\ldots,x_n)$$

$[x_n = t_{n,1}]$ $\quad\cdots\quad$ $[x_n = t_{n,k_n}]$

$$Q_{n-m+1}x_{n-m+1}\cdots Q_{n-2}x_{n-2}\forall x_{n-1}\ (\Psi := (G(t_{n,1}) \wedge \Phi[x_n\,/\!/\,t_{n,1}] \vee \ldots \quad \cdots \quad \ldots \vee G(t_{n,k_n}) \wedge \Phi[x_n\,/\!/\,t_{n,k_n}]))$$

$[x_{n-1} = t_{n-1,1}]$ $\quad\cdots\quad$ $[x_{n-1} = t_{n-1,k_{n-1}}]$

$$Q_{n-m+1}x_{n-m+1}\cdots Q_{n-2}x_{n-2}\ (\neg(G(t_{n-1,1}) \wedge \neg\Psi[x_{n-1}\,/\!/\,t_{n-1,1}]) \wedge \ldots \quad \cdots \quad \ldots \wedge \neg(G(t_{n-1,k_{n-1}}) \wedge \neg\Psi[x_{n-1}\,/\!/\,t_{n-1,k_{n-1}}]))$$

Fig. 1. The generic layered VTS tree formed by QE on $Q_{n-m+1}x_{n-m+1}\cdots Q_{n-2}x_{n-2}\forall x_{n-1}\exists x_n\ \Phi(x_1,\ldots,x_n)$, where the last two quantifiers are forced as $\forall$ and $\exists$ to demonstrate the consolidation of the disjunction formed by elimination of $\exists x_n$ into one implicitly universally quantified formula, Ψ for VTS to traverse next.

Author Index

Printed in the United States
by Baker & Taylor Publisher Services